Advances in Experimental Medicine and Biology

Cellular Neuroscience, Neural Circuits and Systems Neuroscience

Volume 1473

This book subseries publishes reviews and original articles covering all aspects of cellular and systems neuroscience, particularly neural circuits. It is also open to contributions on neurological disorders and their molecular and biochemical basis. A subseries of the highly successful Advances in Experimental Medicine and Biology, it provides a publication vehicle for articles focusing on the latest research, developments and methods, as well as opinions and principles. Contributors are experienced neuroscientists and researchers with extensive knowledge of the embryology, histology, anatomy, physiology and pharmacology of central and peripheral neurons. It endeavors to highlight the importance of the correct choice of animal models and the implementation of the 3Rs principles to translate molecular and cellular findings of basic research into clinical applications. This series is a valuable source of state-of-art knowledge and inspiration for future research ideas for a wide range of researchers and animal care professionals working in the field of neuroscience.

C. Fernando Valenzuela
David N. Linsenbardt • Jeffrey L. Weiner
Editors

Effects of Alcohol on the Brain across the Lifespan

Recent Advances from Preclinical and Clinical Studies

Springer

Editors
C. Fernando Valenzuela
Department of Neurosciences
and New Mexico Alcohol Research
Center, School of Medicine
University of New Mexico Health
Sciences Center
Albuquerque, NM, USA

David N. Linsenbardt
Department of Neurosciences
and New Mexico Alcohol Research
Center, School of Medicine
University of New Mexico Health
Sciences Center
Albuquerque, NM, USA

Jeffrey L. Weiner
Department of Physiology and
Pharmacology
Wake Forest School of Medicine
Winston-Salem, NC, USA

ISSN 0065-2598 ISSN 2214-8019 (electronic)
Advances in Experimental Medicine and Biology
ISSN 2524-6577 ISSN 2524-6585 (electronic)
Cellular Neuroscience, Neural Circuits and Systems Neuroscience
ISBN 978-3-031-81907-0 ISBN 978-3-031-81908-7 (eBook)
https://doi.org/10.1007/978-3-031-81908-7

This Springer imprint is published by the registered company Springer Nature Switzerland AG
The registered company address is: Gewerbestrasse 11, 6330 Cham, Switzerland

Preface

Alcohol's influence on the human brain is profound, multifaceted, and enduring, affecting neural development, function, and behavior across the lifespan. From prenatal exposure to the challenges faced in adulthood, the journey of alcohol's impact on the brain is marked by a complex interplay of genetic, molecular, and environmental factors. This book, a comprehensive exploration of the effects of alcohol on the brain, provides a wealth of knowledge, emphasizing critical developmental stages and employing insights from both clinical and preclinical studies.

The first seven chapters focus on ethanol exposure's short- and long-term effects during early development. The chapter of Granato focuses on the fundamental role of excitatory pyramidal neurons and GABAergic interneurons in the neocortex, which are responsible for higher cognitive functions. It reviews experimental studies with animal models of fetal alcohol spectrum disorders (FASD), highlighting how prenatal alcohol exposure (PAE) disrupts these crucial cortical circuits, leading to intellectual disabilities. The chapter emphasizes the complexity of alcohol's impact on brain development, including the timing of exposure and species differences, and discusses both primary alterations and compensatory mechanisms in neocortical neurons.

Vella et al. thoroughly examine how PAE affects neuroimmune function throughout life. This chapter traces the ontogeny of neuroimmune changes, from early life through adolescence to adulthood, using rodent models to provide detailed insights. It explores the potential mechanisms behind these changes, including the role of the gut microbiota, and underscores the significance of understanding these processes for developing targeted interventions.

Clark et al. focus on the neurobiological underpinnings of learning and memory impairments in FASD, particularly emphasizing the hippocampus. The chapter reviews rodent model research and highlights how developmental alcohol exposure affects hippocampal circuitry and neural representations of learning and memory. The discussion extends to less-studied areas such as visual-spatial and object learning, proposing future research directions that promise to deepen our understanding of memory deficits in FASD.

Valenzuela and collaborators contribute to an in-depth analysis of the impact of PAE on the thalamus. Integrating findings from animal models and human studies, the chapter details how alcohol exposure across various developmental stages specifically alters thalamic structure and function. It emphasizes the significance of these alterations for cognitive functions and calls for advanced imaging techniques and further research to elucidate the thalamus's role in FASD pathophysiology fully.

Cunningham et al. explore the concept of adult hippocampal neurogenesis as a therapeutic target for FASD. This chapter reviews the mechanisms of neurogenesis under normal conditions and its disruption by developmental alcohol exposure. It discusses potential therapeutic approaches, such as pharmacological interventions and neurophysiological techniques, highlighting the need for preclinical and clinical studies to validate these strategies for FASD treatment.

The chapter of Licheri and Brigman addresses the often-overlooked issue of sleep disturbances in individuals with FASD. It reviews clinical findings on sleep issues, including short sleep duration, sleep fragmentation, and parasomnias, and their correlation with cognitive impairments. It also examines the limited preclinical research on sleep physiology in FASD, advocating for more studies to uncover the molecular and physiological mechanisms underlying these disturbances.

The contribution of Pritha and collaborators shifts focus to the molecular level, discussing the role of non-coding RNAs (ncRNAs) in the neurodevelopmental deficits associated with FASD. The chapter reviews current knowledge from preclinical studies on ncRNA-mediated mechanisms and their potential as biomarkers and therapeutic targets. It highlights the need for standardized protocols and further research to harness the full potential of ncRNAs in understanding and treating FASD-related CNS dysfunction.

The next two chapters address the neurobiological effects of alcohol during adolescence, a critical developmental period associated with heightened vulnerability to alcohol exposure. Crowley and her colleagues provide a detailed synopsis of our current understanding of how adolescent alcohol exposure impacts the development of the prefrontal cortex (PFC). This brain region plays a critical role in the etiology of AUD and also undergoes major changes during adolescence. Their review focuses primarily on preclinical studies characterizing the impact of adolescent alcohol on developing GABAergic and peptidergic PFC circuitry and how these early life adaptations contribute to negative outcomes later in life, with an emphasis on the sexually dimorphic nature of some of these changes.

The chapter by Vetreno and collaborators examines the impact of adolescent alcohol exposure on the developing neuroimmune system. They review numerous studies noting that binge drinking is common in adolescence and that these risky drinking patterns can have profound effects on the maturation of glial cells and innate immune signaling cascades. They also highlight neural circuitry impacted by these maladaptive alterations, with a major focus on the basal forebrain cholinergic system and the hippocampal neurogenic niche, and present evidence that these changes may increase the risk of AUD and neurodegenerative diseases later in life. Finally, they draw attention to recent findings suggesting that some of the deleterious effects of alcohol on the adolescent neuroimmune system may be reversible and that anti-inflammatory therapeutics may represent a promising treatment option to mitigate these adverse effects of alcohol.

Finally, the last four chapters review the actions of ethanol during adulthood/aging. Maphis and her colleagues explore the impact of alcohol use on risk for Alzheimer's Disease, with a focus on neurobiological mechanisms. Matthews and his colleagues explore the broad range of impacts aging has on neural function and behavior in the context of alcohol use in rodents. Deak and his colleagues detail the role of neuroinflammation in aging and alcohol use, discuss the consequences of alcohol-mediated thiamine deficiency, and address emerging evidence on the intersection of alcohol and dementia. Aguayo and his colleagues review our current understanding of the impact of age on synaptic transmission and neuronal excitability, focusing on alterations to ion channels and their interactions with alcohol and emphasizing the need for additional studies evaluating the impact of alcohol in aged brains.

The chapters in this book collectively underscore the intricate and far-reaching effects of alcohol on the brain across the lifespan. By integrating insights from diverse approaches—including electrophysiological, biochemical, molecular biological, histological, neuroimaging, and behavioral studies—this book aims to summarize a large body of experimental evidence on alcohol's impact on various brain cellular components. This comprehensive exploration not only advances our knowledge but also sets the stage for developing more effective interventions and support strategies for individuals affected by alcohol exposure. As we continue to unravel the complexities of alcohol's influence on the brain, interdisciplinary collaboration and innovative research will be crucial in mitigating its detrimental effects and improving outcomes for affected individuals.

Albuquerque, NM, USA — C. Fernando Valenzuela
Albuquerque, NM, USA — David N. Linsenbardt
Winston-Salem, NC, USA — Jeffrey L. Weiner

Contents

About the Editors

C. Fernando Valenzuela is a Regents' Professor of Neurosciences at the University of New Mexico Health Sciences Center and is the Director of the New Mexico Alcohol Research Center. He earned his M.D. with honors from the Colombian School of Medicine (Universidad El Bosque) and later obtained his Ph.D. in Biomedical Sciences at the University of California, Riverside, under the mentorship of Dr. David A. Johnson. Following this, he completed postdoctoral training in Dr. Adron Harris's laboratory within the Department of Pharmacology at the University of Colorado Health Sciences Center in Denver. Dr. Valenzuela's research primarily explores the effects of alcohol on synaptic transmission and plasticity during brain development, with a particular focus on how alcohol influences the maturation of neuronal circuits.

David N. Linsenbardt is an Assistant Professor of Neurosciences at the University of New Mexico Health Sciences Center. He earned his Ph.D. from Indiana University–Purdue University Indianapolis in Addiction Neuroscience under the mentorship of Dr. Stephen Boehm. He then completed postdoctoral training in the laboratory of Dr. Christopher Lapish in the same Department. Dr. Linsenbardt's research primarily explores the effects of alcohol on neurobehavioral function, with an emphasis on the motivating attributes of alcohol that lead to binge and high-intensity alcohol drinking, as well as the consequences of excessive alcohol drinking throughout the lifespan.

Jeffrey L. Weiner is a Professor in the Department of Translational Neuroscience at the Wake Forest University School of Medicine. He is also the Director of the NIAAA-supported Wake Forest Translational Alcohol Research Center and the inaugural institutional Faculty Ombuds. He pursued his academic training at the University of Toronto, completing a B.S. in Neuroscience, an M.S. in Pharmacology, and a Ph.D. in Pharmacology working with Dr. Peter L. Carlen. He then completed a postdoctoral fellowship with Dr. Thomas V. Dunwiddie in the Department of Pharmacology at the University of Colorado Health Sciences Center in Denver. Dr. Weiner's laboratory uses rodent models and an integration of behavioral and neurobiological approaches to elucidate the circuit mechanisms underlying vulnerability and resilience to alcohol use disorder.

Part I

Effects of Alcohol on the Fetal Brain

1 Defects of Cortical Microcircuits Following Early Exposure to Alcohol

Alberto Granato

Abstract

The interplay between excitatory pyramidal neurons and GABAergic interneurons is the basic building block of neocortical microcircuits and plays a critical role in carrying out higher cognitive functions. Cortical circuits are deeply and permanently disrupted by exposure to alcohol during brain development, the main non-genetic cause of intellectual disability. Here, I review experimental studies of fetal alcohol spectrum disorders, dealing with permanent cellular and molecular alterations of neocortical neurons and their connections.

Keywords

Pyramidal neurons · Neocortex · Interneurons · Disinhibition · Canonical circuit · Dendrites · Fetal alcohol · Intellectual disability

Exposure to ethanol during development is a well-known cause of intellectual disability (ID) and represents a serious social and economic issue (Popova et al. 2023). The multi-layered neocortical mantle of mammals is the key structure devoted to higher cognitive functions (Phillips 2023). Therefore, it is not surprising that all the clinical syndromes characterized by ID, including fetal alcohol spectrum disorders (FASD), feature widespread disruption of neocortical neural circuits (Granato and Merighi 2022; Granato et al. 2024). The present chapter is aimed at providing a summary of the main permanent alterations of cortical neurons and their connections, as demonstrated in studies based on experimental models of FASD.

A. Granato (✉)
Department of Veterinary Sciences, University of Turin, Grugliasco, TO, Italy
e-mail: alberto.granato@unito.it

1.1 Cortical Microcircuit

1.1.1 Pyramidal Neurons

Most neocortical cells are excitatory pyramidal neurons (PNs), located in layers (L) 2/3, 5, and 6. They display an extensive dendritic arbor with distinct basal and apical domains. The apical dendrite of most PNs ascends through more superficial cortical layers and, before reaching the pial surface, branches extensively to form an apical tuft in L1. Excitatory spiny stellate and star pyramidal neurons populate the granular layer (L4) and represent the main target of ascending sensory afferents from the thalamus. The long axons of PNs are the anatomical substrate of associative and callosal cortico-cortical connections, as well as of cortico-subcortical connections. The so-called "canonical cortical circuit" (reviewed in Lübke and Feldmeyer 2007) sus-

C. F. Valenzuela et al. (eds.), *Effects of Alcohol on the Brain across the Lifespan*, Advances in Experimental Medicine and Biology 1473, https://doi.org/10.1007/978-3-031-81908-7_1

tains local inter- and intralaminar connections and represents the stereotyped mainstream of information processing in cortical modules. Thalamo-cortical specific afferents synapse onto L4 neurons and directly on the basal dendrites of L5 PNs (Constantinople and Bruno 2013), while those from higher-order thalamic nuclei, such as the posterior complex, feed into L1 (Wimmer et al. 2010). Sensory information is then relayed to L2/3 and back to the infragranular layers (L5/6), which provide the main output from the neocortex.

Besides higher-order thalamic afferents, L1 also receives axons from higher-order cortical areas (feedback or top-down cortico-cortical connections; Cauller 1995). Due to its low cellular density, the role of L1 in cortical microcircuits has been long undervalued. Nevertheless, more recently, the simultaneous presence of apical tufts of PNs and top-down projections from other cortical areas revived the interest for L1 (Larkum 2013a; Schuman et al. 2021). Top-down projections to the apical dendritic domain in L1 are thought to convey information related to prediction, attention, context, and past experience (Larkum 2013b). Thus, PNs are ideally suited to integrate bottom-up sensory information (by means of their basal dendritic arbor) and contextual cues (by means of the apical dendrite). Interestingly, some of the behavioral anomalies observed in FASD might involve these "apically modulated" functions, such as attention (Louth et al. 2016) and figure-ground segregation (Castillo Castejón et al. 2019). Other anomalies might imply disorders of the bottom-up (basal) side of sensory processing (e.g., Carr et al. 2010).

The rich repertoire of receptors and ion channels of PNs, as well as the consequent electrophysiological properties, reflect the complexity of the dendritic tree. Ion channels responsible for active conductances are distributed throughout the dendritic tree of PNs (reviewed in Migliore and Shepherd 2002). Voltage-gated sodium channels are involved in the backpropagation of action potentials (Stuart and Sakmann 1994), while potassium channels are thought to modulate the intrinsic excitability of PNs (e.g., Guan et al. 2007; Benhassine and Berger 2009). Hyperpolarization-activated (HCN) channels mediating the I_h current display higher density at distal locations along the apical dendrite and are crucial for the communication between the apical and basal compartment of PNs (reviewed in Phillips et al. 2016). HCN channels are developmentally regulated, and their density increases dramatically during postnatal development (Atkinson and Williams 2009).

Both the basal dendrites and the apical tuft are endowed with NMDA receptors, while voltage-gated calcium channels (VGCC), responsible for the apical electrogenesis, are concentrated just beneath the tuft (Larkum et al. 2009). NMDA receptors and VGCC play a critical role in different types of dendritic synaptic plasticity (e.g., Verhoog et al. 2013; Cichon and Gan 2015). Similarly to what is described above for HCN channels, other receptors and ion channels, such as glutamatergic AMPA receptors and potassium channels, are developmentally regulated and may represent a target of teratogenic agents (Kumar et al. 2002; Guan et al. 2011).

Finally, it is worth noting that the PNs occupying L2/3, L5, and L6 are not homogeneous, as they feature several distinctive electrophysiological and anatomical properties (see Larkum et al. 2007; Spruston 2008; Ledergerber and Larkum 2010).

1.1.2 Interneurons

The interaction between excitatory PNs and inhibitory GABAergic interneurons (INs) plays a pivotal role for the cortical machinery. Although representing only about 10–15% of the cortical cells, INs display a great diversity of morphofunctional features and synaptic targets. For the purposes of the present review, they can be conveniently subdivided into three non-overlapping populations (Tremblay et al. 2016). Fast-spiking interneurons expressing parvalbumin (PV) synapse onto the soma or axon initial segment of PNs. Dendrite-targeting INs express somatostatin (SOM) and often the calcium-binding protein calbindin (CB; Wang et al. 2004; Ma et al. 2006). Finally, INs expressing vasoactive intestinal peptide (VIP), which often coexpress calretinin (CR; Cauli et al. 2014), target other inhibitory INs. Therefore, their action results in a disinhibition of PNs. Such a disinhibitory mechanism has been postulated to be particularly relevant for the modulation of contextual information and of the apical dendritic function operant in L1 (Pfeffer et al. 2013; Karnani et al. 2014; Pfeffer 2014; see also Fig. 1.1). Furthermore, INs and IN-mediated disinhibition play a central role in cortical development and plasticity (Hensch 2005; Williams and Holtmaat 2019; Goff and Goldberg 2021; Liguz-Lecznar et al. 2022).

Unlike PNs, which are generated in the ventricular zone and migrate towards the cortical plate following a radial route, IN precursors are located in the ganglionic eminence of the ventral telencephalon and, prior to their final radial migration, follow a tangential path (Parnavelas 2000). Furthermore, while PV and SOM INs originate from the medial ganglionic eminence (MGE), most VIP/CR cells are generated later than other IN classes in the caudal ganglionic eminence (CGE; Xu et al. 2004; Butt et al. 2005). Interestingly, despite their different sites of origin and migration paths, both PNs and INs share the inside-out neurogenetic gradient. In fact, late-born PNs of supragranular layers migrate past early-born PNs residing in the deep layers. The same inside-out gradient also applies to MGE-

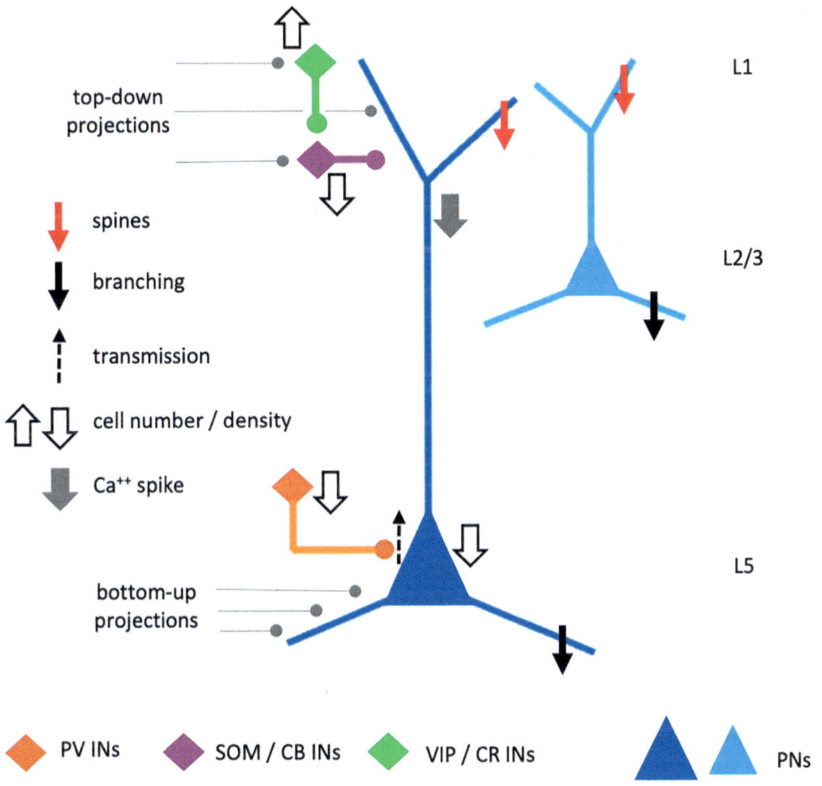

Fig. 1.1 Summary diagram showing some features of cortical microcircuits and their anomalies after early postnatal exposure to alcohol in rodents. Basal dendrites of pyramidal neurons (PNs) receive bottom-up input from the thalamus, while the apical tuft is the main recipient of feedback, top-down projections from higher cortical areas. After alcohol exposure, the density of infragranular pyramidal neurons (L5 PNs), parvalbumin interneurons (PV INs), and somatostatin/calbindin interneurons (SOM / CB INs) is reduced. Conversely, VIP / calretinin interneurons (VIP / CR INs), which disinhibit PNs, are increased. GABAergic transmission of PV INs is enhanced. The branching of PN basal dendrites and the spine density of the apical dendrites are reduced. Apical and basal compartments of PNs communicate through dendritic calcium spikes, which are decreased after alcohol exposure

derived INs (Miyoshi and Fishell 2011). Moreover, CGE-derived INs, which are generated later than MGE ones, are found chiefly in superficial layers. It is worth mentioning that the final layering of cortical INs is completed only during early postnatal life in rodents (Miyoshi and Fishell 2011).

1.2 Early Alcohol Exposure and Cortical Microcircuit

Any discussion about the effects of early alcohol exposure on cortical circuitry should take into account that ethanol triggers massive apoptosis in the developing cortical neurons. This phenomenon is thought to be the consequence of the opposite effects of ethanol on GABA and NMDA receptors (Olney 2014). According to Olney's hypothesis, the positive modulation of GABA and the antagonistic effect of alcohol on NMDA receptors trigger apoptosis through the inhibition of neuronal activity (see also Bird et al. 2020). The reduced network activity is responsible for a pro-apoptotic increase of the BAX/BCL-2 ratio. The anti-apoptotic effect of the sustained electrical activity seems to be not directly related to a reduced caspase 3 level (Schroer et al. 2023). Other mechanisms, such as the activation of astrocytes and/or microglia with release of pro-inflammatory cytokines, can contribute to the neurodegeneration observed after early exposure to alcohol (e.g., Topper et al. 2015).

The apoptotic neurodegeneration seems to be relatively more pronounced for neurons of the infragranular layers (Olney et al. 2002; Toesca et al. 2003; Granato et al. 2003). Besides neurons, glial cells can also undergo apoptotic death after developmental exposure to ethanol (Creeley et al. 2013).

Neurons that survive the apoptotic death undergo deep remodeling of their connections and functional properties, as some molecules involved in apoptosis, such as the p75 low-affinity neurotrophin receptor and caspase 3, can also mediate plastic, possibly maladaptive changes (reviewed in Granato and Dering 2018). The main alterations of cortical microcircuits following early postnatal exposure to alcohol in rodent models of FASD are described below and summarized in Fig. 1.1.

1.2.1 Altered Lamination and Afferent-Efferent Organization

The migration of neurons from the site of origin to their final destination is one of the key phenomena of cortical development. As discussed in Sect. 1.1.2, excitatory PNs and GABAergic INs are generated in different forebrain regions and follow distinct migratory routes (Parnavelas 2000). Prenatal exposure to alcohol interferes with the migratory process and results in misplacement of cortical neurons. In utero exposure to ethanol delays the radial migration of both early- and late-generated PNs (Miller 1986, 1993; Delatour et al. 2019). Neurons destined to layer 2/3 can terminate their migration in infragranular layers (Miller 1993). Overall, the disrupted and desynchronized migratory pattern may represent one of the leading mechanisms affecting the correct establishment of cortical circuitries.

The tangential migration of interneurons appears to be also affected by prenatal exposure to alcohol. Using a transgenic mouse line harboring tdTomato fluorescence in MGE-derived interneurons, Skorput et al. (2015) demonstrated an increased density of these neurons in the prefrontal cortex of alcohol-treated animals at E16. The anomaly of MGE-derived interneurons persisted into the adult life, as the same authors also observed an increased number of PV interneurons in layer 5 of adult mice exposed to ethanol in utero (Skorput et al. 2015).

Multiple mechanisms might account for the alcohol-induced alteration of neuronal migration. Ethanol antagonizes NMDA receptors, whose role in the stimulation of cortical neuron migration is well established (Behar et al. 1999; Reiprich et al. 2005). Another proposed mechanism is the interference with the transforming growth factor (Siegenthaler and Miller 2004).

The L4 zone recipient of thalamo-cortical terminals is much narrower in rats prenatally exposed to alcohol than in controls (Granato et al. 1995). The number of neurons giving rise to cortico-cortical associative projections is strongly reduced in adult rats exposed to ethanol during the first week of postnatal life, corresponding to the third trimester of gestation in humans (Granato et al. 2003). Most associative neurons of alcohol-exposed animals are located in supragranular layers, an observation that can be due either to selective apoptosis of infragranular layers (see above) or to defective connection of surviving L5/6 neurons (Granato et al. 2003). Conversely, owing to a partial lack of naturally occurring developmental pruning, other projecting systems remain permanently exuberant following early exposure to ethanol. This is the case for neurons of origin of the cortico-spinal tract, which display increased density and a wider tangential/laminar distribution in adult rats prenatally exposed to alcohol (Miller 1987). Altogether, these observations point to alcohol exposure disrupting the general laminar and input-output organization of the neocortex.

1.2.2 Alterations of PNs

Dendritic alterations represent the main anatomical counterpart of ID (Kaufmann and Moser 2000; Granato and Merighi 2022). In fact, anomalies of the extensive dendritic tree of PNs represent one of the main landmarks of FASD. Moreover, dendritic alterations observed in experimental FASD parallel the basal versus apical dualism described in physiological conditions. Rats exposed to ethanol during early postnatal life show a reduction of size and branching of basal dendrites, both in the somatosensory and in the prefrontal cortex, while the spine density is not affected (Granato et al. 2003, 2012; Granato and Van Pelt 2003; Hamilton et al. 2010; De Giorgio and Granato 2015). A specular alteration is observed, following the same experimental protocol, in the apical dendrites, where normal metric and branching properties are associated to a reduced spine density (Whitcher and Klintsova 2008; Granato et al. 2012; De Giorgio and Granato 2015). Finally, using an experimental paradigm of combined pre- and early postnatal exposure to alcohol, Louth et al. (2018) observed layer-specific alterations of the dendritic tree of prefrontal PNs: supragranular and layer 6 PNs showed increased and decreased dendritic size, respectively, while layer 5 neurons were unaffected.

The electrophysiological properties of PNs are also modified by early exposure to ethanol. As the structure of the dendritic tree is a key determinant of firing properties (Mainen and Sejnowski 1996), the electrophysiological changes can represent the consequence of the alterations of the dendritic arbor. Furthermore, the altered trafficking and distribution of dendritic ion channels can account for the anomalous functional properties of PNs (e.g., Kim et al. 2007).

The Hermes Yeh's group used a protocol based on prenatal exposure to demonstrate specular modifications of the balance between excitation and inhibition in layer 5/6 PNs of different cortical areas. While the inhibitory transmission on PNs of the medial prefrontal cortex is favored (Skorput et al. 2015), the excitatory input on the same neurons of the somatosensory cortex is enhanced, as compared to the inhibitory input (Delatour et al. 2020). These interareal differences can be partly explained by the aberrant tangential migration of INs, resulting in the increased density of L5 PV cells in the prefrontal cortex (Skorput et al. 2015; see above, Sect 1.2.1). In L5 PNs of adult rats exposed to ethanol during the third-trimester equivalent, we have demonstrated a shutdown of calcium electrogenesis at the level of the apical dendrite caused by an impairment of voltage-gated calcium channels (Granato et al. 2012). Apical calcium spikes are crucial for the communication between apical and basal compartments of PNs (Larkum et al. 2009). As pointed out above, the I_h current sustained by HCN channels is also involved in the modulation of the interplay between basal and apical domains (Phillips et al. 2016). However, to our knowledge, there are no experimental FASD studies dealing with the distribution and function of HCN channels in the apical dendrite of PNs.

However, Yao et al. (2023) observed an increased expression of HCN1 channels in synaptosomes from the prefrontal cortex of mice prenatally exposed to alcohol. Therefore, the effects of early exposure to alcohol on the I_h current in PNs deserve further investigation.

As described above, both the basal dendrites of PNs and the apical tuft feature NMDA glutamate receptors. The effects of early exposure to ethanol on NMDA ligand-gated channels are reviewed in Costa et al. (2000). Licheri et al. (2021) demonstrated that prenatal exposure in mice leads to a sex-specific dysregulation of NMDA-mediated excitatory postsynaptic currents. Glutamate receptors of the AMPA type are also affected. Layer 6 neurons of the medial prefrontal cortex display an increased response to AMPA receptor stimulation after developmental ethanol exposure (Louth et al. 2016).

The alterations of PN dendrites and ion channels affect the communication between apical and basal compartments of the principal neocortical cells, with deep functional and behavioral consequences. The presence of segregated basal and apical dendrites that can interact and integrate bottom-up sensory information with top-down contextual cues offers a clear computational advantage for learning and memory (e.g., Guerguiev et al. 2017; Capone et al. 2023). Therefore, the dysregulated crosstalk between dendritic domains of PNs can explain the behavioral defects observed in developmental learning disabilities (see, for review, Granato et al. 2024).

1.2.3 FASD and Cortical INs

Reports on the density and distribution of PV INs in experimental FASD are contrasting. They are not affected in the sensori-motor cortex of rats after early postnatal exposure (from postnatal day 2 - P2 - through P6; Granato 2006). The number of PV INs is strongly reduced throughout the rostro-caudal extent of the mouse cortex after alcohol exposure at P7 (Smiley et al. 2015). Their density is increased in the prefrontal cortex after in utero exposure (Skorput et al. 2015), while it is decreased in the same cortical areas after neonatal exposure (Hamilton et al. 2017). To explain the contradictory findings of the last two studies, we can assume that alcohol interferes with different mechanisms during different developmental stages (notably, leading to aberrant migration during prenatal life and to apoptosis during early postnatal life). Reduction of the number of PV INs has been observed in the posterior parietal cortex of mice exposed to alcohol in utero, even though compensatory mechanisms can be operant to reach normal levels of inhibitory input to PNs (Licheri et al. 2023). An enhancement of GABAergic neurotransmission by PV INs has been demonstrated in the retrosplenial cortex of mice exposed during postnatal life (Bird et al. 2021). This effect can be partly due to a postsynaptic mechanism mediated by the increased expression of GABA(A) receptor subunits observed in adult guinea pigs after prenatal exposure (Bailey et al. 2001), whereas the fast-spiking properties of PV cells might be impaired due to the reduction of Kv3.4 potassium channels (Tavian et al. 2011). PV INs were shown to be reduced also in the cingulate cortex of adult rats after prenatal exposure (Moore et al. 1998).

The density of INs expressing SOM and/or CB, supposed to target the dendrites of PNs, is reduced after P2-P6 exposure in rats (Granato 2006) and after P7 exposure in mice as well (Smiley et al. 2019).

Conflicting results are reported for INs expressing CR and/or VIP, most of which target other INs, thus mediating PN disinhibition. The density of CR INs is increased following P2-P6 exposure (Granato 2006) or moderate prenatal exposure (Kenton et al. 2020). Interestingly, examining a human fetus exposed to alcohol, Léger et al. (2020) observed the same aberrant radial distribution of CR neurons found in the study by Granato (2006). It is also worth mentioning that an increase of VIP INs has been described in experimental models of Down syndrome (e.g., Hernández et al. 2012). Conversely, a decrease of CR IN density has been observed after P7 alcohol exposure (Smiley et al. 2015). VIP INs are also decreased in the fetal forebrain of sheep prenatally exposed to ethanol (Anderson et al. 2008).

Inputs synapsing onto different dendritic compartments of PNs are modulated by different classes of GABAergic INs (see Sect. 1.1.2 and Tremblay et al. 2016, for review). The cooperation between PNs and INs is critical for the functions of the neocortex. Both VIP disinhibiting INs and SOM dendrite-targeting INs are involved in the induction of synaptic plasticity (Cichon and Gan 2015; Williams and Holtmaat 2019). Moreover, an elegant work carried out in vivo demonstrated that VIP INs are recruited by reinforcement signals during an auditory discrimination task, increasing the gain of PNs through disinhibition (Pi et al. 2013). Similarly, Lee et al. (2013) have shown that VIP INs are activated during whisking by feedback projections from the motor cortex to the somatosensory cortex (Lee et al. 2013). Therefore, the disruption of cortical circuits mediated by INs can strongly impair goal-oriented behavior. The deletion of the tyrosine kinase receptor ERBB4 specifically from VIP INs disrupts their function and results in altered sensory learning (Batista-Brito et al. 2017). On the other hand, the deletion of the *Tsc1* gene in a CGE-derived IN group leads to an increased density of VIP INs (Hu et al. 2024). *Tsc1* mutations are responsible for tuberous sclerosis, a clinical condition in which ID affects about 50% of individuals.

1.3 Concluding Remarks

Experimental models of precocious exposure to alcohol represent a convenient way to study the mechanisms leading to ID, as FASD and other neurodevelopmental disorders share several cortical anomalies (Granato and Merighi 2022). Moreover, experimental models of FASD may prove useful to study what amount of ethanol is harmful to the developing brain, although a safe drinking level has not been established (Valenzuela et al. 2012; Charness et al. 2016). However, the detrimental effects of early experimental exposure to alcohol on cortical circuits are a function of several variables, including time of exposure (prenatal vs. early postnatal), considered area, species under investigation, modes of administration, etc. A striking example is provided by the apparently conflicting results discussed in Sect. 1.2.3 regarding the number of PV INs. This is not necessarily a drawback, as the morpho-functional effects of exposure at different developmental stages can be related to different neurobehavioral consequences (e.g., Sadrian et al. 2014). However, owing to the complexity of the experimental protocols, that require long survival times, there are only a few systematic investigations dealing with permanent alterations following different paradigms of alcohol exposure. Another critical point for the interpretation of experimental results is that any mechanistic explanation might be confounded by the lack of a clear distinction between primary alterations and those that result from compensatory homeostatic mechanisms. For instance, the increase of disinhibiting CR INs (Granato 2006) can represent an attempt to counterbalance the reduced excitability of PN apical dendrites (Granato et al. 2012) or vice versa. However, there is now compelling evidence that PNs can no longer be considered integrate and fire point neurons, but their computational properties and cognitive functions are based on the communication between apical and basal dendritic compartments and on their modulation by (dis)inhibiting interneurons (see, for review, Granato et al. 2024). Therefore, whatever the causative link, the disruption of cortical microcircuit can explain the behavioral sequelae observed in FASD. Furthermore, despite their limitations, animal models of developmental alcohol exposure made it possible to achieve notable results and to provide valuable therapeutic hints (see, for instance, Bird and Valenzuela 2023, for the effects of positive allosteric modulation of NMDA receptors).

References

Anderson DJ, Mondares RL, Born DE, Gleason CA (2008) The effect of binge fetal alcohol exposure on the number of vasoactive intestinal peptide-producing neurons in fetal sheep brain. Dev Neurosci 30:276–284. https://doi.org/10.1159/000110349

Atkinson SE, Williams SR (2009) Postnatal development of dendritic synaptic integration in rat neocorti-

cal pyramidal neurons. J Neurophysiol 102:735–751. https://doi.org/10.1152/jn.00083.2009
Bailey CD, Brien JF, Reynolds JN (2001) Chronic prenatal ethanol exposure increases GABA(a) receptor subunit protein expression in the adult Guinea pig cerebral cortex. J Neurosci 21:4381–4389. https://doi.org/10.1523/JNEUROSCI.21-12-04381.2001
Batista-Brito R, Vinck M, Ferguson KA, Chang JT, Laubender D, Lur G, Mossner JM, Hernandez VG, Ramakrishnan C, Deisseroth K, Higley MJ, Cardin JA (2017) Developmental dysfunction of VIP interneurons impairs cortical circuits. Neuron 95:884–895.e9. https://doi.org/10.1016/j.neuron.2017.07.034
Behar TN, Scott CA, Greene CL, Wen X, Smith SV, Maric D, Liu QY, Colton CA, Barker JL (1999) Glutamate acting at NMDA receptors stimulates embryonic cortical neuronal migration. J Neurosci 19:4449–4461. https://doi.org/10.1523/JNEUROSCI.19-11-04449.1999
Benhassine N, Berger T (2009) Large-conductance calcium-dependent potassium channels prevent dendritic excitability in neocortical pyramidal neurons. Pflugers Arch 457:1133–1145. https://doi.org/10.1007/s00424-008-0569-3
Bird CW, Valenzuela CF (2023) The positive allosteric modulator of NMDA receptors, GNE-9278, blocks the ethanol-induced decrease of excitability in developing retrosplenial cortex neurons from mice. Neuropsychopharmacol Rep 43:77–84. https://doi.org/10.1002/npr2.12306
Bird CW, Barber MJ, Post HR, Jacquez B, Chavez GJ, Faturos NG, Valenzuela CF (2020) Neonatal ethanol exposure triggers apoptosis in the murine retrosplenial cortex: role of inhibition of NMDA receptor-driven action potential firing. Neuropharmacology 162:107837. https://doi.org/10.1016/j.neuropharm.2019.107837
Bird CW, Chavez GJ, Barber MJ, Valenzuela CF (2021) Enhancement of parvalbumin interneuron-mediated neurotransmission in the retrosplenial cortex of adolescent mice following third trimester-equivalent ethanol exposure. Sci Rep 11:1716. https://doi.org/10.1038/s41598-021-81173-z
Butt SJB, Fuccillo M, Nery S, Noctor S, Kriegstein A, Corbin JG, Fishell G (2005) The temporal and spatial origins of cortical interneurons predict their physiological subtype. Neuron 48:591–604. https://doi.org/10.1016/j.neuron.2005.09.034
Capone C, Lupo C, Muratore P, Paolucci PS (2023) Beyond spiking networks: the computational advantages of dendritic amplification and input segregation. Proc Natl Acad Sci USA 120:e2220743120. https://doi.org/10.1073/pnas.2220743120
Carr JL, Agnihotri S, Keightley M (2010) Sensory processing and adaptive behavior deficits of children across the fetal alcohol spectrum disorder continuum. Alcohol Clin Exp Res 34:1022–1032. https://doi.org/10.1111/j.1530-0277.2010.01177.x
Castillo Castejón O, González I, Prieto E, Pérez T, Pablo LE, Pueyo V (2019) Visual cognitive impairments in children at risk of prenatal alcohol exposure. Acta Paediatr 108:2222–2228. https://doi.org/10.1111/apa.14904
Cauli B, Zhou X, Tricoire L, Toussay X, Staiger JF (2014) Revisiting enigmatic cortical calretinin-expressing interneurons. Front Neuroanat 8:52. https://doi.org/10.3389/fnana.2014.00052
Cauller L (1995) Layer I of primary sensory neocortex: where top-down converges upon bottom-up. Behav Brain Res 71:163–170. https://doi.org/10.1016/0166-4328(95)00032-1
Charness ME, Riley EP, Sowell ER (2016) Drinking during pregnancy and the developing brain: is any amount safe? Trends Cogn Sci 20:80–82. https://doi.org/10.1016/j.tics.2015.09.011
Cichon J, Gan WB (2015) Branch-specific dendritic ca(2+) spikes cause persistent synaptic plasticity. Nature 520:180–185. https://doi.org/10.1038/nature14251
Constantinople CM, Bruno RM (2013) Deep cortical layers are activated directly by thalamus. Science 340:1591–1594. https://doi.org/10.1126/science.1236425
Costa ET, Savage DD, Valenzuela CF (2000) A review of the effects of prenatal or early postnatal ethanol exposure on brain ligand-gated ion channels. Alcohol Clin Exp Res 24:706–715. https://doi.org/10.1111/j.1530-0277.2000.tb02043.x
Creeley CE, Dikranian KT, Johnson SA, Farber NB, Olney JW (2013) Alcohol-induced apoptosis of oligodendrocytes in the fetal macaque brain. Acta Neuropathol Commun 1:23. https://doi.org/10.1186/2051-5960-1-23
De Giorgio A, Granato A (2015) Reduced density of dendritic spines in pyramidal neurons of rats exposed to alcohol during early postnatal life. Int J Dev Neurosci 41:74–79. https://doi.org/10.1016/j.ijdevneu.2015.01.005
Delatour LC, Yeh PW, Yeh HH (2019) Ethanol exposure in utero disrupts radial migration and pyramidal cell development in the somatosensory cortex. Cereb Cortex 29:2125–2139. https://doi.org/10.1093/cercor/bhy094
Delatour LC, Yeh PWL, Yeh HH (2020) Prenatal exposure to ethanol alters synaptic activity in layer V/VI pyramidal neurons of the somatosensory cortex. Cereb Cortex 30:1735–1751. https://doi.org/10.1093/cercor/bhz199
Goff KM, Goldberg EM (2021) A role for vasoactive intestinal peptide interneurons in neurodevelopmental disorders. Dev Neurosci 43:168–180. https://doi.org/10.1159/000515264
Granato A (2006) Altered organization of cortical interneurons in rats exposed to ethanol during neonatal life. Brain Res 1069:23–30. https://doi.org/10.1016/j.brainres.2005.11.024
Granato A, Van Pelt J (2003) Effects of early ethanol exposure on dendrite growth of cortical pyramidal neurons: inferences from a computational model. Brain Res Dev Brain Res 142:223–227. https://doi.org/10.1016/s0165-3806(03)00094-4

Granato A, Dering B (2018) Alcohol and the developing brain: why neurons die and how survivors change. Int J Mol Sci 19:2992. https://doi.org/10.3390/ijms19102992

Granato A, Merighi A (2022) Dendrites of neocortical pyramidal neurons: the key to understand intellectual disability. Cell Mol Neurobiol 42:147–153. https://doi.org/10.1007/s10571-021-01123-1

Granato A, Santarelli M, Sbriccoli A, Minciacchi D (1995) Multifaceted alterations of the thalamo-cortico-thalamic loop in adult rats prenatally exposed to ethanol. Anat Embryol (Berl) 191:11–23. https://doi.org/10.1007/BF00215293

Granato A, Di Rocco F, Zumbo A, Toesca A, Giannetti S (2003) Organization of cortico-cortical associative projections in rats exposed to ethanol during early postnatal life. Brain Res Bull 60:339–344. https://doi.org/10.1016/s0361-9230(03)00052-2

Granato A, Palmer LM, De Giorgio A, Tavian D, Larkum ME (2012) Early exposure to alcohol leads to permanent impairment of dendritic excitability in neocortical pyramidal neurons. J Neurosci 32(4):1377–1382. https://doi.org/10.1523/JNEUROSCI.5520-11.2012

Granato A, Phillips WA, Schulz JM, Suzuki M, Larkum ME (2024) Dysfunctions of cellular context-sensitivity in neurodevelopmental learning disabilities. Neurosci Biobehav Rev 161:105688. https://doi.org/10.1016/j.neubiorev.2024.105688

Guan D, Lee JCF, Higgs MH, Spain WJ, Foehring RC (2007) Functional roles of Kv1 channels in neocortical pyramidal neurons. J Neurophysiol 97:1931–1940. https://doi.org/10.1152/jn.00933.2006

Guan D, Horton LR, Armstrong WE, Foehring RC (2011) Postnatal development of A-type and Kv1- and Kv2-mediated potassium channel currents in neocortical pyramidal neurons. J Neurophysiol 105:2976–2988. https://doi.org/10.1152/jn.00758.2010

Guerguiev J, Lillicrap TP, Richards BA (2017) Towards deep learning with segregated dendrites. elife 6:e22901. https://doi.org/10.7554/eLife.22901

Hamilton GF, Whitcher LT, Klintsova AY (2010) Postnatal binge-like alcohol exposure decreases dendritic complexity while increasing the density of mature spines in mPFC layer II/III pyramidal neurons. Synapse 64:127–135. https://doi.org/10.1002/syn.20711

Hamilton GF, Hernandez IJ, Krebs CP, Bucko PJ, Rhodes JS (2017) Neonatal alcohol exposure reduces number of parvalbumin-positive interneurons in the medial prefrontal cortex and impairs passive avoidance acquisition in mice deficits not rescued from exercise. Neuroscience 352:52–63. https://doi.org/10.1016/j.neuroscience.2017.03.058

Hensch TK (2005) Critical period plasticity in local cortical circuits. Nat Rev Neurosci 6:877–888. https://doi.org/10.1038/nrn1787

Hernández S, Gilabert-Juan J, Blasco-Ibáñez JM, Crespo C, Nácher J, Varea E (2012) Altered expression of neuropeptides in the primary somatosensory cortex of the down syndrome model Ts65Dn. Neuropeptides 46:29–37. https://doi.org/10.1016/j.npep.2011.10.002

Hu JS, Malik R, Sohal VS, Rubenstein JL, Vogt D (2024) Tsc1 loss in VIP-lineage cortical interneurons results in more VIP+ interneurons and enhanced excitability. Cells 13:52. https://doi.org/10.3390/cells13010052

Karnani MM, Agetsuma M, Yuste R (2014) A blanket of inhibition: functional inferences from dense inhibitory connectivity. Curr Opin Neurobiol 26:96–102. https://doi.org/10.1016/j.conb.2013.12.015

Kaufmann WE, Moser HW (2000) Dendritic anomalies in disorders associated with mental retardation. Cereb Cortex 10:981–991. https://doi.org/10.1093/cercor/10.10.981

Kenton JA, Ontiveros T, Bird CW, Valenzuela CF, Brigman JL (2020) Moderate prenatal alcohol exposure alters the number and function of GABAergic interneurons in the murine orbitofrontal cortex. Alcohol 88:33–41. https://doi.org/10.1016/j.alcohol.2020.06.001

Kim J, Jung SC, Clemens AM, Petralia RS, Hoffman DA (2007) Regulation of dendritic excitability by activity-dependent trafficking of the A-type K+ channel subunit Kv4.2 in hippocampal neurons. Neuron 54:933–947. https://doi.org/10.1016/j.neuron.2007.05.026

Kumar SS, Bacci A, Kharazia V, Huguenard JR (2002) A developmental switch of AMPA receptor subunits in neocortical pyramidal neurons. J Neurosci 22:3005–3015. https://doi.org/10.1523/JNEUROSCI.22-08-03005.2002

Larkum ME (2013a) The yin and yang of cortical layer 1. Nat Neurosci 16:114–115. https://doi.org/10.1038/nn.3317

Larkum ME (2013b) A cellular mechanism for cortical associations: an organizing principle for the cerebral cortex. Trends Neurosci 36:141–151. https://doi.org/10.1016/j.tins.2012.11.006

Larkum ME, Waters J, Sakmann B, Helmchen F (2007) Dendritic spikes in apical dendrites of neocortical layer 2/3 pyramidal neurons. J Neurosci 27:8999–9008. https://doi.org/10.1523/JNEUROSCI.1717-07.2007

Larkum ME, Nevian T, Sandler M, Polsky A, Schiller J (2009) Synaptic integration in tuft dendrites of layer 5 pyramidal neurons: a new unifying principle. Science 325:756–760. https://doi.org/10.1126/science.1171958

Ledergerber D, Larkum ME (2010) Properties of layer 6 pyramidal neuron apical dendrites. J Neurosci 30:13031–13044. https://doi.org/10.1523/JNEUROSCI.2254-10.2010

Lee S, Kruglikov I, Huang ZJ, Fishell G, Rudy B (2013) A disinhibitory circuit mediates motor integration in the somatosensory cortex. Nat Neurosci 16:1662–1670. https://doi.org/10.1038/nn.3544

Léger C, Dupré N, Laquerrière A et al (2020) In utero alcohol exposure exacerbates endothelial protease activity from pial microvessels and impairs GABA interneuron positioning. Neurobiol Dis 145:105074. https://doi.org/10.1016/j.nbd.2020.105074

Licheri V, Chandrasekaran J, Bird CW, Valenzuela CF, Brigman JL (2021) Sex-specific effect of prenatal alcohol exposure on N-methyl-D-aspartate receptor function in orbitofrontal cortex pyramidal neurons of

mice. Alcohol Clin Exp Res 45:1994–2005. https://doi.org/10.1111/acer.14697
Licheri V, Jacquez BJ, Castillo VK, Sainz DB, Valenzuela CF, Brigman JL (2023) Long-term effects of low prenatal alcohol exposure on GABAergic interneurons of the murine posterior parietal cortex. Alcohol Clin Exp Res (Hoboken) 47:2248–2261. https://doi.org/10.1111/acer.15210
Liguz-Lecznar M, Dobrzanski G, Kossut M (2022) Somatostatin and somatostatin-containing interneurons-from plasticity to pathology. Biomol Ther 12:312. https://doi.org/10.3390/biom12020312
Louth EL, Bignell W, Taylor CL, Bailey CDC (2016) Developmental ethanol exposure leads to long-term deficits in attention and its underlying prefrontal circuitry. eNeuro 3:ENEURO.0267-16.2016. https://doi.org/10.1523/ENEURO.0267-16.2016
Louth EL, Luctkar HD, Heney KA, Bailey CDC (2018) Developmental ethanol exposure alters the morphology of mouse prefrontal neurons in a layer-specific manner. Brain Res 1678:94–105. https://doi.org/10.1016/j.brainres.2017.10.005
Lübke J, Feldmeyer D (2007) Excitatory signal flow and connectivity in a cortical column: focus on barrel cortex. Brain Struct Funct 212:3–17. https://doi.org/10.1007/s00429-007-0144-2
Ma Y, Hu H, Berrebi AS, Mathers PH, Agmon A (2006) Distinct subtypes of somatostatin-containing neocortical interneurons revealed in transgenic mice. J Neurosci 26:5069–5082. https://doi.org/10.1523/JNEUROSCI.0661-06.2006
Mainen ZF, Sejnowski TJ (1996) Influence of dendritic structure on firing pattern in model neocortical neurons. Nature 382:363–366. https://doi.org/10.1038/382363a0
Migliore M, Shepherd GM (2002) Emerging rules for the distributions of active dendritic conductance. Nat Rev Neurosci 3:362–370. https://doi.org/10.1038/nrn810
Miller MW (1986) Effects of alcohol on the generation and migration of cerebral cortical neurons. Science 233:1308–1311. https://doi.org/10.1126/science.3749878
Miller MW (1987) Effect of prenatal exposure to alcohol on the distribution and time of origin of corticospinal neurons in the rat. J Comp Neurol 257:372–382. https://doi.org/10.1002/cne.902570306
Miller MW (1993) Migration of cortical neurons is altered by gestational exposure to ethanol. Alcohol Clin Exp Res 17:304–314. https://doi.org/10.1111/j.1530-0277.1993.tb00768.x
Miyoshi G, Fishell G (2011) GABAergic interneuron lineages selectively sort into specific cortical layers during early postnatal development. Cereb Cortex 21:845–852. https://doi.org/10.1093/cercor/bhq155
Moore DB, Quintero MA, Ruygrok AC, Walker DW, Heaton MB (1998) Prenatal ethanol exposure reduces parvalbumin-immunoreactive GABAergic neuronal number in the adult rat cingulate cortex. Neurosci Lett 249:25–28. https://doi.org/10.1016/s0304-3940(98)00378-4
Olney JW (2014) Focus on apoptosis to decipher how alcohol and many other drugs disrupt brain development. Front Pediatr 2:81. https://doi.org/10.3389/fped.2014.00081
Olney JW, Tenkova T, Dikranian K, Qin Y, Labruyere J, Ikonomidou C (2002) Ethanol-induced apoptotic neurodegeneration in the developing C57BL/6 mouse brain. Dev Brain Res 133:115–126. https://doi.org/10.1016/s0165-3806(02)00279-1
Parnavelas JG (2000) The origin and migration of cortical neurones: new vistas. Trends Neurosci 23:126–131. https://doi.org/10.1016/s0166-2236(00)01553-8
Pfeffer CK (2014) Inhibitory neurons: vip cells hit the brake on inhibition. Curr Biol 24:R18–R20. https://doi.org/10.1016/j.cub.2013.11.001
Pfeffer CK, Xue M, He M, Huang ZJ, Scanziani M (2013) Inhibition of inhibition in visual cortex: the logic of connections between molecularly distinct interneurons. Nat Neurosci 16:1068–1076. https://doi.org/10.1038/nn.3446
Phillips WA (2023) The cooperative neuron: cellular foundations of mental life. Oxford University Press, Oxford
Phillips WA, Larkum ME, Harley CW, Silverstein SM (2016) The effects of arousal on apical amplification and conscious state. Neurosci Conscious 2016:niw015. https://doi.org/10.1093/nc/niw015
Pi HJ, Hangya B, Kvitsiani D, Sanders JI, Huang ZJ, Kepecs A (2013) Cortical interneurons that specialize in disinhibitory control. Nature 503:521–524. https://doi.org/10.1038/nature12676
Popova S, Charness ME, Burd L, Crawford A, Hoyme HE, Mukherjee RAS, Riley EP, Elliott EJ (2023) Fetal alcohol spectrum disorders. Nat Rev Dis Primers 9:11. https://doi.org/10.1038/s41572-023-00420-x
Reiprich P, Kilb W, Luhmann HJ (2005) Neonatal NMDA receptor blockade disturbs neuronal migration in rat somatosensory cortex in vivo. Cereb Cortex 15:349–358. https://doi.org/10.1093/cercor/bhh137
Sadrian B, Lopez-Guzman M, Wilson DA, Saito M (2014) Distinct neurobehavioral dysfunction based on the timing of developmental binge-like alcohol exposure. Neuroscience 280:204–219. https://doi.org/10.1016/j.neuroscience.2014.09.008
Schuman B, Dellal S, Prönneke A, Machold R, Rudy B (2021) Neocortical layer 1: an elegant solution to top-down and bottom-up integration. Annu Rev Neurosci 44:221–252. https://doi.org/10.1146/annurev-neuro-100520-012117
Schroer J, Warm D, De Rosa F, Luhmann HJ, Sinning A (2023) Activity-dependent regulation of the BAX/BCL-2 pathway protects cortical neurons from apoptotic death during early development. Cell Mol Life Sci 80:175. https://doi.org/10.1007/s00018-023-04824-6
Siegenthaler JA, Miller MW (2004) Transforming growth factor beta1 modulates cell migration in rat cortex: effects of ethanol. Cereb Cortex 14:791–802. https://doi.org/10.1093/cercor/bhh039
Skorput AG, Gupta VP, Yeh PW, Yeh HH (2015) Persistent Interneuronopathy in the prefrontal cortex

of young adult offspring exposed to ethanol in utero. J Neurosci 35:10977–10988. https://doi.org/10.1523/JNEUROSCI.1462-15.2015

Smiley JF, Saito M, Bleiwas C, Masiello K, Ardekani B, Guilfoyle DN, Gerum S, Wilson DA, Vadasz C (2015) Selective reduction of cerebral cortex GABA neurons in a late gestation model of fetal alcohol spectrum disorder. Alcohol 49:571–580. https://doi.org/10.1016/j.alcohol.2015.04.008

Smiley JF, Bleiwas C, Masiello K, Petkova E, Betz J, Hui M, Wilson DA, Saito M (2019) Effects of neonatal ethanol on cerebral cortex development through adolescence. Brain Struct Funct 224:1871–1884. https://doi.org/10.1007/s00429-019-01881-1

Spruston N (2008) Pyramidal neurons: dendritic structure and synaptic integration. Nat Rev Neurosci 9:206–221. https://doi.org/10.1038/nrn2286

Stuart GJ, Sakmann B (1994) Active propagation of somatic action potentials into neocortical pyramidal cell dendrites. Nature 367:69–72. https://doi.org/10.1038/367069a0

Tavian D, De Giorgio A, Granato A (2011) Selective underexpression of Kv3.2 and Kv3.4 channels in the cortex of rats exposed to ethanol during early postnatal life. Neurol Sci 32:571–577. https://doi.org/10.1007/s10072-010-0446-7

Toesca A, Giannetti S, Granato A (2003) Overexpression of the p75 neurotrophin receptor in the sensori-motor cortex of rats exposed to ethanol during early postnatal life. Neurosci Lett 342:89–92. https://doi.org/10.1016/s0304-3940(03)00258-1

Topper LA, Baculis BC, Valenzuela CF (2015) Exposure of neonatal rats to alcohol has differential effects on neuroinflammation and neuronal survival in the cerebellum and hippocampus. J Neuroinflammation 12:160. https://doi.org/10.1186/s12974-015-0382-9

Tremblay R, Lee S, Rudy B (2016) GABAergic interneurons in the neocortex: from cellular properties to circuits. Neuron 91:260–292. https://doi.org/10.1016/j.neuron.2016.06.033

Valenzuela CF, Morton RA, Diaz MR, Topper L (2012) Does moderate drinking harm the fetal brain? Insights from animal models. Trends Neurosci 35:284–292. https://doi.org/10.1016/j.tins.2012.01.006

Verhoog MB, Goriounova NA, Obermayer J, Stroeder J, Hjorth JJJ, Testa-Silva G, Baayen JC, de Kock CPJ, Meredith RM, Mansvelder HD (2013) Mechanisms underlying the rules for associative plasticity at adult human neocortical synapses. J Neurosci 33:17197–17208. https://doi.org/10.1523/JNEUROSCI.3158-13.2013

Wang Y, Toledo-Rodriguez M, Gupta A, Wu C, Silberberg G, Luo J, Markram H (2004) Anatomical, physiological and molecular properties of Martinotti cells in the somatosensory cortex of the juvenile rat. J Physiol 561(Pt 1):65–90. https://doi.org/10.1113/jphysiol.2004.073353

Whitcher LT, Klintsova AY (2008) Postnatal binge-like alcohol exposure reduces spine density without affecting dendritic morphology in rat mPFC. Synapse 62:566–573. https://doi.org/10.1002/syn.20532

Williams LE, Holtmaat A (2019) Higher-order Thalamocortical inputs gate synaptic long-term potentiation via disinhibition. Neuron 101(1):91–102.e4. https://doi.org/10.1016/j.neuron.2018.10.049

Wimmer VC, Bruno RM, de Kock CPJ, Kuner T, Sakmann B (2010) Dimensions of a projection column and architecture of VPM and POm axons in rat vibrissal cortex. Cereb Cortex 20:2265–2276. https://doi.org/10.1093/cercor/bhq068

Xu Q, Cobos I, De La Cruz E, Rubenstein JL, Anderson SA (2004) Origins of cortical interneuron subtypes. J Neurosci 24:2612–2622. https://doi.org/10.1523/JNEUROSCI.5667-03.2004

Yao H, Wang C, Xia Z (2023) Prenatal alcohol exposure enhanced alcohol preference and susceptibility to PTSD in a sex-dependent manner through the synaptic HCN1 channel. J Affect Disord 324:143–152. https://doi.org/10.1016/j.jad.2022.12.069

Ontogenetic Neuroimmune Changes Following Prenatal Alcohol Exposure: Implications for Neurobehavioral Function

2

Victoria R. Vella, Parker J. Holman, Tamara S. Bodnar, and Charlis Raineki

Abstract

This chapter reviews the enduring effects of prenatal alcohol exposure (PAE) on neuroimmune function across the lifespan, including discussion of associated neurobehavioral alterations. Alcohol has potent teratogenic effects, with a large body of work linking PAE to perturbations in neuroimmune function. These PAE-related neuroimmune disturbances may have downstream effects on neurobehavioral function given the critical role of the neuroimmune system in central nervous system development. The neuroimmune system matures over time, playing distinct roles depending on the developmental processes occurring within that maturational stage. This chapter thus takes an ontogenetic approach to understanding how PAE induces unique neuroimmune changes across the lifespan, beginning with a review of changes in early life before moving into adolescence and ending in adulthood. The focus will be on work utilizing rodent models, which allow for more tightly controlled conditions than are possible in human research. The chapter concludes with a discussion of possible mechanisms underlying the developmental changes in neuroimmune function following PAE, with a specific focus on the role of the gut microbiota.

Keywords

Prenatal alcohol exposure · Developmental alcohol exposure · Neuroimmune · Microglia · Cytokines · Preclinical models · Microbiota

2.1 Introduction

2.1.1 Overview of Prenatal Alcohol Exposure

Prenatal alcohol exposure (PAE) has been linked to widespread and long-term structural and functional alterations to developing offspring (Connor et al. 2000; Jones and Smith 1973; Lebel et al. 2011; Marquardt and Brigman 2016; Mattson et al. 2019; Norman et al. 2009). Fetal blood

V. R. Vella · P. J. Holman
Department of Psychology, Brock University, St. Catharines, ON, Canada

T. S. Bodnar
Department of Biological Sciences, University of Calgary, Calgary, AB, Canada

Alberta Children's Hospital Research Institute, Calgary, AB, Canada

Hotchkiss Brain Institute, Calgary, AB, Canada

C. Raineki (✉)
Department of Psychology, Brock University, St. Catharines, ON, Canada

Centre for Neuroscience, Brock University, St. Catharines, ON, Canada
e-mail: craineki@brocku.ca

C. F. Valenzuela et al. (eds.), *Effects of Alcohol on the Brain across the Lifespan*, Advances in Experimental Medicine and Biology 1473, https://doi.org/10.1007/978-3-031-81908-7_2

alcohol levels rise to those of the pregnant individual within two hours post-alcohol consumption; however, reduced metabolic capacity, accumulation of ethanol within the amniotic fluid, and reabsorption/reuptake of this ethanol-laden amniotic fluid leave the fetus exposed to alcohol for a prolonged period (Burd et al. 2012). This exposure to alcohol can critically alter the developmental trajectory of the offspring, a phenomenon first described in the late 1960s and early 1970s by two independent groups, one in France (Lemoine et al. 2003—English translation of the original 1968 French paper) and another in the United States (Jones et al. 1973). While PAE can lead to wide-ranging perturbations in the development of every bodily system within the fetus (Akison et al. 2019; Himmelreich et al. 2020; Reid et al. 2019; Young et al. 2022), the effects on the central nervous system (CNS) are especially pervasive (Caputo et al. 2016; Valenzuela et al. 2012). Past work has noted reductions in overall brain volume and agenesis of the corpus callosum among other structural abnormalities to regions including the cerebellum, cerebral cortex, and various subcortical structures following PAE (Lebel et al. 2011; Mattson et al. 2019; Norman et al. 2009; Sowell et al. 2008; Wozniak et al. 2017). These structural alterations may give rise to co-occurring functional changes, including deficits in cognitive, sensory, motor, social, adaptive, and affective domains (Hellemans et al. 2010; Hoyme et al. 2016; Kully-Martens et al. 2012; Mattson et al. 2019; Rasmussen 2005). Given the particular vulnerability of the CNS to the teratogenic effects of PAE, it is crucial to better understand the mechanisms mediating these disturbances. The neuroimmune system may be a mechanistic link between PAE and associated neurobehavioral impairments, with a body of preclinical work demonstrating PAE-related alterations in neuroimmune function (Drew and Kane 2014). Indeed, early-life immune perturbations have been linked to later neurological dysfunction, due in part to the important roles the immune system plays in both the development and continued homeostatic functioning of the CNS (See Fig. 2.1) (Bilbo and Schwarz 2009).

Importantly, various anti-inflammatory treatments such as minocycline, pioglitazone, and ibuprofen have shown promise in attenuating the physiological and behavioral consequences of PAE-induced neuroimmune dysfunction (Chastain et al. 2019; Drew et al. 2015; Goodfellow et al. 2018; Kane et al. 2011; Ren et al. 2019). Not only do these findings lend weight to the hypothesis that early neuroimmune disturbances may underlie many PAE-induced effects, but they also provide a promising avenue for future work to develop more effective and targeted treatment strategies to ameliorate PAE-induced neurobehavioral deficits.

2.1.2 The Role of the Neuroimmune System in Typical and Atypical Development

The immune system is a complex conglomerate of diverse cells, proteins, and organs working in synchrony to defend the body against both endogenous and exogenous threats. However, the role of the immune system extends beyond that of surveillance and protection in times of threat, with bidirectional communication between the brain and immune system occurring throughout life to maintain a course of typical development and functioning (Bilbo and Schwarz 2009, 2012; Tian et al. 2012). The neuroimmune system is comprised of a dedicated network of cells—such as glial cells—and centrally synthesized immune factors which function independently of classic immune activation to guide and support the ongoing maturation of the CNS. Microglia, noted for their function as the resident immune cells of the CNS, are less appreciated for their role in CNS development (Bilimoria and Stevens 2015; Lenz and Nelson 2018). However, in part through their phagocytic capacity, microglia play important roles in synaptic pruning and refinement, as well as in clearing apoptotic neurons during developmental cell death (Bilimoria and Stevens 2015; Lenz and Nelson 2018; Tay et al. 2017). Microglia are also crucial for other neurodevelopmental processes, including neurogenesis and neuronal survival, synapse formation, myelination, and

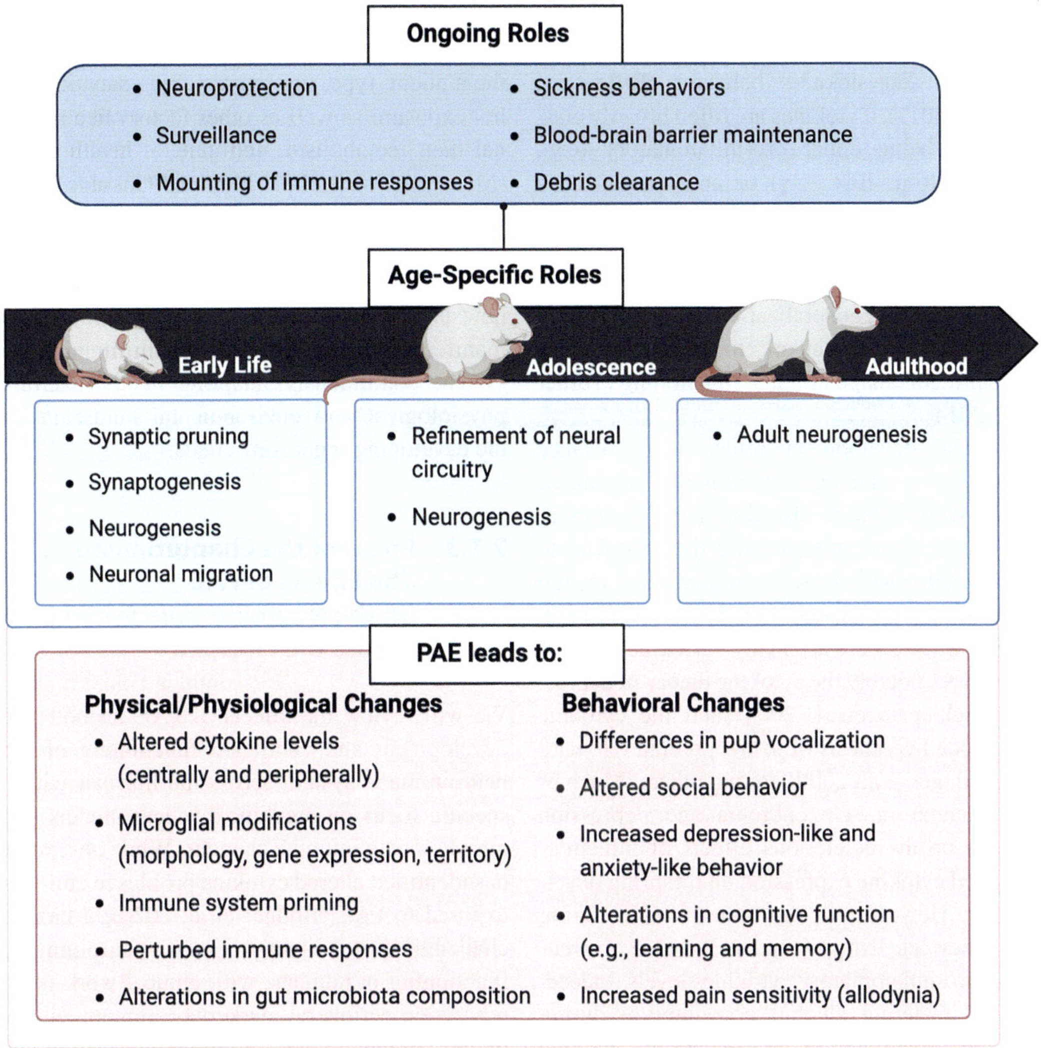

Fig. 2.1 Roles of the neuroimmune system and impact of PAE The neuroimmune system plays both ongoing and age-specific roles, which extend far beyond that of protection and surveillance, working under homeostatic conditions to support the development and functioning of the CNS. PAE can disrupt many of these functions, leading to enduring physical/physiological and behavioral changes across the lifespan. Created with BioRender

oligodendrogenesis, accomplished in part via the production of signaling proteins called cytokines (Bilimoria and Stevens 2015; Lenz and Nelson 2018; Tay et al. 2017; Turner et al. 2014).

Cytokines are pivotal to the coordinated mounting of immune responses, helping to regulate immune cell activity both centrally and in the periphery (Deverman and Patterson 2009; Turner et al. 2014). As such, cytokines both act on and are released by various cell types, including, but not limited to, neurons, microglia, and astrocytes (See Szelényi 2001 and Turner et al. 2014 for reviews of the sources, targets, and functions of specific cytokines). Cytokines are extensively involved in typical neurodevelopment, with roles in neural induction, neurogenesis, neuronal migration, gliogenesis, and synaptogenesis among numerous other crucial roles (Reviewed in Deverman and Patterson 2009). Beyond being critical for early CNS development, cytokines are also essential for typical neural function throughout life, playing important roles in key cognitive processes such as learning and memory, as well as for emotion regulation and sickness behavior (Bilbo and Schwarz 2012). Cytokines are often broadly classified as being either pro-inflammatory (e.g., IL-1β, TNF-α, IFN − ɣ) or anti-inflammatory (e.g., IL- 10) (Turner et al. 2014). While this pro/anti-inflammatory binary classification can oversimplify the often situation-dependent roles cytokines play, this generalization does provide a useful starting framework to understand how these proteins shape immune functioning (Turner et al. 2014).

Typical development relies on a balance between pro- and anti-inflammatory signaling, with a skew in either direction away from cytokine homeostasis predisposing the offspring to later health challenges, including poor mental health outcomes (Bauer et al. 2007; Meyer et al. 2008). In fact, there are many 'cytokine theories of disease', notably the cytokine theory of depression (Schiepers et al. 2004) and the cytokine imbalance hypothesis of schizophrenia (Howard 2013; Meyer et al. 2009) among others. Much of the research in schizophrenia and depression focuses on the deleterious effects of infection-triggered cytokine expression on offspring development. However, alcohol consumption during pregnancy can also promote an increase in circulating pro-inflammatory cytokines levels. Indeed, chronic maternal alcohol consumption during pregnancy has been shown to elevate cytokine levels, including IL-1β, TNF-α, and IL-6 (Ahluwalia et al. 2000). The combination of prenatal exposure, not only to alcohol itself but also to heightened inflammatory signaling, can perturb later neuroimmune function in the offspring (See Fig. 2.1) (Han et al. 2021; Shimizu et al. 2023). This may render the offspring uniquely vulnerable to later insults given the critical role of the neuroimmune system in development, with consequences not only for the neuroimmune function itself but for the unfolding of subsequent developmental processes in which this system plays a role (Bilbo and Schwarz 2012; Bodnar et al. 2022a; Noor et al. 2017; Raineki et al. 2017; Zhang et al. 2012). However, the manifestation of these disturbances is likely to vary as a function of multiple factors, including, but not limited to, the amount, type, timing, and frequency of alcohol exposure as well as other factors like maternal diet, metabolism, and general health status (Maier and West 2001; May and Gossage 2011; O'Leary et al. 2010). While many pre- and postnatal factors play a role in the complex neurobehavioral effects of PAE, this chapter will focus on how PAE impacts on the neuroimmune system manifest as a function of age, with the immune system continuously adapting to the shifting physiological and environmental landscape of the developing organism.

2.1.3 Focus of the Chapter: The Effects of PAE on the Neuroimmune System Across the Lifespan

We will review the effects of PAE on both the development and continued functioning of the neuroimmune system across the lifespan with a specific focus on preclinical rodent models and mainly innate immune changes. While past work has identified altered cytokine profiles in children exposed to PAE (Bodnar et al. 2020), it can be challenging to directly assess neuroimmune functioning in humans, with clinical work often relying on peripheral markers of immune alterations. Animal models represent powerful tools for investigating the effects of PAE under consistent environmental conditions and controlling for confounding variables as well as allowing for direct access to brain tissue. While the time course of development differs between rodents and humans, rodents progress through similar developmental stages in terms of brain and immune maturation (for a comparison of brain and immune development across species, see Semple et al. 2013). Thus, rodents are appropriate models for studying the effects of PAE on these systems, so long as the relative stage of development is considered. There is a multitude of animal models of PAE, with model choice relying largely on the desired amount of exposure (low, moderate, high, or binge levels) and the desired duration of exposure (e.g., gestation or third-trimester model only). While there is comparably more work using binge-like PAE models, work using various rodent models has found that PAE can still lead to significant alterations throughout the body even at low to moderate doses (reviewed in Comasco et al. 2018; Noor and Milligan 2018), including changes in both basal immune function and response to immune challenge (Barbaccia et al. 2007; Bodnar et al. 2016; Pascual et al. 2017; Sanchez et al. 2017; Terasaki and Schwarz 2016; Terasaki and Schwarz 2017). We will focus on preclinical data which demonstrate the effects of PAE on neuroimmune function across the lifespan, beginning in early life, before moving into adolescence, and then into adulthood. When evaluating the effects of PAE on neuroimmune function, it is important to consider the developmental context of the individual. The immune system matures over time, with distinctions in both the makeup and function relative to the developmental stage of the individual identified (Adkins et al. 2004; Dorshkind and Crooks 2023; Georgountzou and Papadopoulos 2017). For example, microglia are morphologically distinct in early life compared to adulthood, with differences in gene expression relative to developmental stage (Bennett et al. 2016; Lenz and Nelson 2018; Matcovitch-Natan et al. 2016). Given the shifting landscape of the neuroimmune system over time, it follows that the effects of PAE may present differently depending on the stage of maturation and development. Indeed, the effects of PAE on immune function change across development, with differences in cytokine levels detected in PAE animals at specific timepoints and not others. For example, it has been shown that female PAE rats had a pro-inflammatory bias at postnatal day (P) 8, which was no longer detected at P22 (Bodnar et al. 2016). Finally, we will review possible mechanisms underlying the PAE-induced alterations to immune system function, focusing specifically on alterations in gut microbiota composition and short-chain fatty acid production.

2.2 PAE-Induced Alterations in Neuroimmune Function in Early Life

Early development, including that which occurs in utero, is critical in establishing a trajectory of growth that follows a characteristic pattern toward maturity in typically developing individuals. A disruption in the unfolding of development processes can have downstream effects on the attainment of successive milestones, with early life development (~P0-P20 in rodents) providing the foundation for both neural and immune maturation. The brain and immune system develop over time, beginning in utero and extending well into the adolescent period (Semple et al. 2013). The rodent is born immunologically immature, as various postnatal experiences, including microbial colonization, help to educate and shape ongoing immune function (Adkins et al. 2004; Belkaid and Hand 2014; Georgountzou and Papadopoulos 2017). In turn, the neuroimmune system is critical in shaping neural development, particularly during the early postnatal period (Bilbo and Schwarz 2009; Deverman and Patterson 2009; Morimoto and Nakajima 2019). Rodent brain development in the first 2 weeks of life is akin to that seen in the third trimester of human development and includes the brain growth spurt (Dobbing and Sands 1979; Zeiss 2021). This is a period of rapid growth, with a multitude of relevant developmental processes, including synaptogenesis, synaptic pruning, myelination, programed cell death, and neurogenesis (Zeiss 2021) unfolding via the action of coordinated neuroimmune activity (Bilbo and Schwarz 2009). As mentioned previously, microglia play a key role in all of these neural processes, in part, through their phagocytic activity as well as via the production of cytokines (Bilbo and Schwarz 2009; Bilimoria and Stevens 2015; Lenz and Nelson 2018). For example, microglia phagocytize synapses, dead or dying neurons, and other debris (Bilimoria and Stevens 2015; Lenz and Nelson 2018).

To model alcohol exposure during the human third trimester equivalent, many rodent models utilize a developmental alcohol exposure paradigm whereby pups are exposed to alcohol during the first 1–2 weeks of life. Developmental alcohol exposure can lead to region-specific alterations in microglial morphology (Drew et al. 2015; Kane et al. 2011; Topper et al. 2015), even when only a single dose of ethanol is administered (Lowery et al. 2021). For example, developmental ethanol exposure (P4–9) resulted in altered microglial morphology in the hippocampal CA1 region, lobule V of the cerebellar cortex, and the parietal cerebral cortex, with somal hypertrophy and short, branched processes compared to microglia from control animals at P10 (Drew et al. 2015). There was also a co-occurring rise in the expression of pro-inflammatory cytokines, including IL-1β, TNF-α, and the chemokine CCL2 in the hippocampus and cerebellum and elevated IL-1β and TNF-α in the cerebral cortex of ethanol-treated rats compared to vehicle-treated controls (Drew et al. 2015). This rise in cytokine/chemokine expression in combination with the noted changes in microglial morphology indicates a neuroinflammatory state; specifically, CCL2 acts as a chemoattractant to facilitate the recruitment of microglia, as well as peripheral immune cells, to sites of inflammation (Selenica et al. 2013; Semple et al. 2010), while both IL-1β and TNF-α are potent inflammatory mediators able to influence microglial activity as well as the synthesis and release of other inflammatory factors (Hewett et al. 2012; Mendiola and Cardona 2018; Neniskyte et al. 2014; Raffaele et al. 2020). However, IL-1β and TNF-α also have important regulatory roles under physiological conditions, specifically for synaptic plasticity and hippocampal-associated processes such as learning and memory (Albensi and Mattson 2000; Beattie et al. 2002; Bourgognon and Cavanagh 2020; Prieto et al. 2019; Schneider et al. 1998). This makes the finding of increased PAE-related hippocampal IL-1β and TNF-α particularly important, as imbalanced expression of these key immune mediators could potentially impact critical cognitive processes, including learning and memory, both of which are commonly impacted in individuals exposed to PAE (Brady et al. 2011; Manji et al. 2009; Schambra et al. 2017). Interestingly, postnatal administration of pioglitazone (PPAR-γ agonist with anti-inflammatory effects) was able to prevent this rise in pro-inflammatory cytokines and normalize the morphological alterations in the microglia of developmental ethanol-exposed rats, as compared to their no-drug counterparts (Drew et al. 2015).

Beyond these morphological and cytokine-based microglial changes, other studies have found additional evidence for alcohol-induced microglial alterations, such as decreases in microglia territory (one indicator of an activated state) in the hippocampus of P10 rats following developmental alcohol exposure (Boschen et al. 2016) and in the hippocampus, cerebellum, and parietal cortex of P10 mice (Drew et al. 2015). Given the role of microglia in the ongoing development of the CNS, it follows that alterations to the functioning of these cells, particularly in early life, could have lasting neurobehavioral consequences (Bilimoria and Stevens 2015; Lenz and Nelson 2018). This could be through PAE-related perturbations in the ability of microglia to maintain their typical physiological functions, such as synaptic pruning and clearance of cellular debris, or via exaggerated inflammatory responses to later challenges following this early life priming (Bilimoria and Stevens 2015; Lenz and Nelson 2018). Indeed, microglial dysfunction has been linked to neurodevelopmental disorders such as autism spectrum disorder, as well as to psychiatric disorders commonly seen in individuals with PAE, including anxiety and depression (Bilimoria and Stevens 2015; O'Connor and Paley 2009). Early priming of microglia has also been implicated in the pathogenesis of a number of neurodegenerative disorders such as Alzheimer's disease and multiple sclerosis (Distéfano-Gagné et al. 2023; Leng and Edison 2021; Perry and Holmes 2014). However, further work is needed to uncover whether microglia are the initiating source of neuroinflammation or reacting to, and then perpetuating, pre-existing inflammation.

While we have thus far focused on studies using developmental alcohol exposure paradigms that model exposure during the human third trimester, prenatal exposure to alcohol in rodents during the first two-trimester human equivalent has also demonstrated significant neuroimmune alterations in early life, including alterations to basal postnatal cytokine expression in various brain regions. Bodnar et al. (2016) identified an early-life pro-inflammatory bias in female rats following moderate PAE, with increased cytokine levels in the prefrontal cortex, hippocampus, and serum at P8 compared to controls. The only change observed at P0 was a decrease in IFN-γ in the whole brain of PAE pups, though by P8, the PAE pups had higher serum TNF-α, higher hippocampal IFN-γ, IL-4, IL-5, TNF-α, and IL-1β, and higher prefrontal cortex IL-5 and IL-6 (Bodnar et al. 2016). They also found generally decreased cytokine expression in the hypothalamus and spleens of PAE animals at P8, with lower IL-2, TNF-α, and IL-1β in the hypothalamus and lower IL-13, IL-4, IL-5, IL-10, and IL-6 in the spleen (Bodnar et al. 2016). However, by weaning (P22), cytokine levels in the serum, prefrontal cortex, hippocampus, hypothalamus, and spleen had normalized in PAE animals (Bodnar et al. 2016). Data from our lab using the same model of PAE found similar results in males, with male PAE rats exhibiting widespread increases in basal cytokine levels in both the amygdala (IFN-γ, IL-10, IL-13, IL-1β, and IL-4) and hypothalamus (IFN-γ, IL-10, and IL-1β) at P8 compared to control males, indicative of an early life neuroinflammatory bias (Vella et al. 2023). Likewise, we also found that by P22 there was a reduction in the breadth of PAE-induced neuroimmune alterations, with the only difference being an increase in IL-10 in the hypothalamus of PAE compared to control animals, potentially as a remnant of the earlier inflammation within this region given the anti-inflammatory function of IL-10 (Moore et al. 2001; Ouyang et al. 2011). While the early-life PAE-induced pro-inflammatory bias appears to diminish with age in this rodent model, these results reflect basal neuroimmune activity, and it remains to be seen if a later-life challenge would induce further alterations to cytokine levels in the brain.

Alterations to basal cytokine levels and immune cell functioning in early life could program the developing immune system, such that later functioning, including responses to later challenges, becomes perturbed and either inappropriately enhances or suppresses immune function. Though the PAE literature tends to focus on immune challenges in adulthood, there is some evidence that even early-life responses to immune challenges may be altered by PAE. Following low-level developmental ethanol exposure (P2–16 via vapor inhalation) and subsequent low-dose lipopolysaccharide (LPS) injection on P17 to mimic bacterial challenge, ethanol-exposed female rats failed to show the same increase in IL-1β mRNA in the frontal cortex as female controls (Topper and Valenzuela 2014). This suggests a blunting of the neuroimmune response to LPS, as central levels of IL-1β would typically rise in response to peripheral LPS administration (Tonelli and Postolache 2005). However, no other cytokine levels were significantly different between groups, and this effect was not seen in males or other brain regions assessed (cerebellar vermis, dentate gyrus), potentially due to the low levels of ethanol exposure (~80 mg/dL blood alcohol concentration) (Topper and Valenzuela 2014). Further work is needed to assess how exposure to various levels of alcohol at different points during gestation and early life could impact subsequent responses to immune challenges.

Non-immune challenges, such as those induced by early life stress or adversity, could also trigger a neuroimmune response, particularly in the context of a primed immune system. Moderate gestation-long PAE decreased basal levels of the chemokine KC/GRO and increased the anti-inflammatory cytokine IL-10 in the amygdala of PAE rats at P12 (Raineki et al. 2017). Following early life adversity using a low-bedding model from P7–12, PAE pups had an altered immune response, failing to show the same reduction in peripheral cytokine expression (KC/GRO, TNF-α, IL-4, and IL-10) as control pups exposed to the same early life adversity (Raineki et al. 2017). These neuroimmune changes may mediate some of the behavioral differences observed, as typically reared PAE pups vocalized less than controls and did not show the same increased vocalization following early life adversity (Raineki et al. 2017). Likewise, even a single bout of maternal separation (4 h on P10) can lead to sex-specific changes in the mRNA expression of a number of immune factors in the hypothalamus, hippocampus, and amygdala of PAE mice at P10 (Ruffaner-Hanson et al. 2023). Notably, a greater number of changes in mRNA expression were seen in PAE females—particularly in the amygdala—whereby PAE females exposed to early life stress had elevated mRNA levels of the microglial marker TMEM119, IL-1β, and the innate immune receptor, toll-like receptor 4 (TLR4)-related factors, including high mobility group box 1(HMGB1), TLR4, NF- B inhibitor alpha (NFKBIA), and NOD-like receptor family pyrin domain-containing protein 3 (NLRP3) (Ruffaner-Hanson et al. 2023). Interestingly, none of these changes were seen in the amygdala of the male PAE mice. Meanwhile, in the absence of early life stress, PAE alone generally had opposing effects, blunting the expression of transmembrane protein 119 (TMEM119), HMGB1, IL-1R1, NFKBIA, NLRP3, and IL-10 in the hypothalamus of PAE female mice but did not result in any changes in male mice (Ruffaner-Hanson et al. 2023). However, the PAE males experiencing early life stress did show decreased IL-1β and increased IL-10 expression in the hippocampus and a decrease in CCL2 mRNA expression in the hypothalamus following PAE alone (Ruffaner-Hanson et al. 2023). This suggests that—while not as robustly affected by PAE or early life stress as the females—males were still experiencing significant alterations, albeit in a different pattern than the females. There was some overlap among the sexes, with both male and female PAE mice having blunted expression of TLR4 mRNA levels in the hypothalamus (Ruffaner-Hanson et al. 2023). Taken together, these results not only highlight the potent effects of PAE, particularly when combined with a later challenge, but also demonstrate the importance of assessing both sexes given the heterogeneity of outcomes following PAE. While the above studies demonstrate that the effects of PAE on neuroimmune function closely follow the initial insult, the effects of early-life immune perturbations can be pervasive and long-lasting, and, in some cases, these alterations only begin to manifest later in life, as will be discussed in the following sections.

2.3 PAE-Induced Alterations in Neuroimmune Function in Adolescence

Adolescence marks a transitional phase of development during which pivotal changes in both physical maturation and behavior take place (McCormick and Mathews 2007; M. Schneider 2013). Though difficult to clearly delineate in terms of age, adolescence is typically defined as occurring roughly from P21-P59 in rodents, depending on the defining criteria used (McCormick and Mathews 2007). The combination of a shifting social landscape, hormonal changes, sexual maturation, and ongoing neural development brings about unique behavioral patterns during this period (Schneider 2013; Spear 2000), including increased social behavior, risk-taking, impulsivity, and reward-seeking (Chambers et al. 2003; Laviola et al. 2003; Sachser et al. 2020; Schneider 2013; Spear 2000). PAE has been linked to altered social behavior, with descriptions in the literature including impaired social recognition memory (Holman et al. 2018, 2021), altered social play behavior (Holman et al. 2018; Lawrence et al. 2008), and differences in social investigation (Mooney and Varlinskaya 2011) in adolescent rats exposed to PAE or developmental alcohol exposure. Whether this impairment in social behavior in PAE rodents is due to alterations in neuroimmune function remains to be determined. However, a large body of research into the pathophysiology of autism spectrum disorder has linked immune dysfunction with social behavior deficits (Hughes et al. 2023; Meltzer and Van De Water 2017; Onore et al. 2012), making it plausible that similar pathways could be playing a role in the PAE-induced alterations in social behavior.

Adolescents have a unique immune profile compared to that of early life or adulthood, as the close interplay between the immune and endocrine systems leads to a sex-specific shift in immune function during puberty (Brenhouse and Schwarz 2016). While males tend to have increased innate inflammatory responses in early life, a switch occurs after puberty, whereby females tend to mount more intensive immune responses compared to males (Klein and Flanagan 2016). Indeed, androgens have been found to be immunosuppressive, while the effects of estradiol vary depending on the immune function under study (Foo et al. 2017; Klein and Flanagan 2016). Though these changes are better characterized peripherally rather than centrally, there is evidence of sex-specific differences in neuroimmune function as well (Osborne et al. 2018; Schwarz et al. 2012). For example, there are sex-specific differences in the number, morphology, and cytokine expression of microglia (Osborne et al. 2018; Schwarz et al. 2012). Likewise, PAE can disturb neuroendocrine–neuroimmune interactions in a sexually dimorphic manner (reviewed in Bodnar and Weinberg 2013), making the inclusion of both sexes in research of paramount importance.

The neuroimmune system continues to play a key role in ongoing neural development during adolescence, with microglia important for the continued maturation of the prefrontal cortex (Schalbetter et al. 2022), helping to regulate synaptic pruning and refine cortical circuits. Continued synaptic pruning, myelination, and refining of neural circuitry not only necessitates a distinct immune milieu but also renders the adolescent uniquely vulnerable to insults, as perturbations in these key processes can adversely affect later health outcomes (Brenhouse and Schwarz 2016). Indeed, many mental health disorders begin to emerge during human adolescence, and there is increasing evidence for a role of the immune system in the pathogenesis of psychiatric disorders, particularly in depression (Kohler et al. 2016; Miller and Raison 2016; Tay et al. 2018). Humans exposed to PAE have disproportionately high rates of mental health disorders, with up to 90% experiencing some form of mental health problem during their lifetime (Pei et al. 2011). The preclinical literature also finds disrupted behavior, with PAE rodents showing increased anxiety-like and depressive-like behaviors in a variety of behavioral tasks (Brolese et al. 2014; Raineki et al. 2016; Marquardt and Brigman 2016).

These noted behavioral alterations make adolescence a key time point of interest for understanding how and why PAE can lead to poor mental health outcomes and whether alterations to the neuroimmune system could be playing a role. However, the breadth of research assessing PAE-induced alterations to neuroimmune function during the adolescent period is fairly limited in comparison to those focused either in early life or adulthood. This is unfortunate, given the evidence of an age-dependent impact of PAE on immune system function, with unique effects in adolescents versus adults. One study assessed the effects of PAE (Gestational day (G)11–20) on offspring immune gene expression in the olfactory bulb of Long Evans rats during adolescence (P40) and adulthood (P90) (Gano et al. 2020). Notably, PAE induced far more gene changes in adulthood compared to adolescence, and for genes that showed changes in both age groups, PAE adolescents showed opposite patterns of activation compared to PAE adults (Gano et al. 2020). For example, there was downregulation of Chi3l3 (M2 microglial marker), Egr1 (involved in neuronal plasticity), and the Nra4 family (involved in the acute stress response and cytokine signaling) in adolescence and upregulation of these same genes in adulthood (Gano et al. 2020). Similar results were found following developmental ethanol exposure (P4–9) in Sprague Dawley rats, as while adolescents (P35) had higher hippocampal cytokine levels compared to adults (P60) regardless of developmental ethanol exposure, most of the ethanol-induced changes in cytokine levels were seen in the adult rats, such as increases in TNF-α, KC/GRO, and IFN-γ compared to sham-treated controls (Baker et al. 2023). An exception was a rise in the cytokine IL-6 in the hippocampus of ethanol-exposed rats in both adolescence and adulthood (Baker et al. 2023). Interestingly, choline supplementation from P10–30 was able to mitigate the rise in hippocampal IFN-γ in adult rats with developmental ethanol exposure (Baker et al. 2023).

Additionally, when looking at the ratio of pro- and anti-inflammatory cytokines, adult, but not adolescent, alcohol-exposed rats had an increased TNF-α:IL-10 ratio—indicative of a pro-inflammatory bias—however, choline supplementation normalized this ratio to that of sham-treated controls (Baker et al. 2023). Choline is an essential nutrient found in a variety of foods, such as eggs, and plays a key role in typical brain development (Zeisel 2004). There is evidence that choline functions through anti-inflammatory mechanisms, with these results suggesting that postnatal choline supplementation may be a viable treatment option to mitigate ethanol-induced neuroinflammation. Indeed, there is an emerging body of both preclinical and clinical work demonstrating the beneficial effects of pre-and postnatal choline supplementation on neurodevelopment, particularly for the cognitive deficits associated with PAE (Reviewed in Ernst et al. 2022).

While the influence of early-life immune priming on central nervous system development persists through disruptions in the successive maturation of neural circuitry, immune activation during adolescence itself may yield distinct effects on brain development and function, particularly within the context of a previously primed immune system. Moderate gestation-long PAE increased basal TNF-α and IL-1β in the hippocampus of adolescent (P31) Sprague-Dawley rats (Wang et al. 2019). This PAE-induced rise in hippocampal TNF-α and IL-1β was also seen in work at earlier pre-adolescent timepoints (Drew et al. 2015). As mentioned previously, both TNF-α and IL-1β are important for the regulation of synaptic plasticity, including for long-term potentiation in the hippocampus (Bourgognon and Cavanagh 2020). Alterations to these cytokines at basal levels could impair these important processes, with downstream effects on learning and memory (Bourgognon and Cavanagh 2020). Indeed, work in both humans and rodents has linked PAE to learning and memory impairments during the adolescent period (Olson et al. 1997; Wagner et al. 2014; Willford et al. 2004; Wozniak et al. 2004). This rise in basal cytokine levels may also impair later immune responses, as following injection with LPS at P30, PAE rats failed to show the same rise in TNF-α and IL-1β as controls, with downregulation of TLR4—the receptor involved in LPS recognition—in PAE rats relative to controls (Wang et al. 2019).

However, there have been conflicting results within the literature regarding the effects of PAE on neuroimmune function in adolescence, with some studies finding little or no impact of PAE either at basal levels or in response to subsequent challenge. For example, PAE from G11–20 reduced basal fractalkine receptor (CX_3CL-1R) levels in the hippocampus of adolescent (P35) Long Evans rats; however, there were no further differences in cytokine response to a subsequent ethanol exposure in adolescence between PAE and control rats (Doremus-Fitzwater et al. 2020). Ethanol exposure (P4–9) did not affect the microglia of adolescent mice in terms of density, distribution, morphology, or motility in the visual cortex (Wong et al. 2018) or motility in the somatosensory cortex (Wong et al. 2021). Similarly, binge-level developmental ethanol exposure (P4–9) did not affect microglial motility, surveillance, or morphology in the cerebellum of transgenic adolescent (P28) L7cre/Ai9+/−/Cx3cr1G/+ mice (which allows for in vivo fluorescent imaging of microglia) (Cealie et al. 2023).

Adolescents appear to be less sensitive to alcohol challenge overall, with work on the effects of alcohol consumption in typically developing adolescents and adults finding that adolescent C57BL/6 mice (P35–46) did not show the same increase in central cytokine expression as adults (P84–120) following 10 days of alcohol consumption (Kane et al. 2014). While adults had increased cytokine and chemokine expression in the hippocampus (CCL2), cerebellum (CCL2 and IL-6), and cerebral cortex (CCL2), adolescents did not show any changes in these brain regions, suggesting that the adolescent cytokine response may be less sensitive to alcohol challenge (Kane et al. 2014). This blunted response observed in typically developing animals could in part explain the lack of PAE effects during adolescence (Cealie et al. 2023; Doremus-Fitzwater et al. 2020; Wong et al. 2018, 2021), with most PAE-induced alterations in neuroimmune functioning occurring in early life or adulthood. However, the limited number of studies, in combination with differing methodologies, could also explain the lack of significant neuroimmune alterations in adolescent PAE rodents. More research is needed to elucidate whether PAE at different doses, developmental periods, or administration patterns would yield differences in the neuroimmune response of adolescents.

2.4 PAE-Induced Alterations in Neuroimmune Function in Adulthood

The neuroimmune system plays an ongoing role in the continued health and functioning of the adult CNS. The immune system continues to work in a defensive capacity to respond to potential threats, such as viral or bacterial invasion, with bidirectional communication between the CNS and immune system being important for the coordination of immune responses as well as sickness behaviors (see Fig. 2.1) (Dantzer 2018). Microglia support CNS function in a non-immunological role as well by clearing cellular debris, refining synaptic connections, and supporting adult neurogenesis through the removal of excess apoptotic neurons via phagocytosis (Gemma and Bachstetter 2013; Sato 2015; Sierra et al. 2010; Tay et al. 2017). As mentioned above, adverse prenatal environments, such as those induced by PAE, could prime the immune system such that later life responses to immune challenge become perturbed. These alterations in immune functioning have been demonstrated through 'second hit' studies, where PAE animals are exposed to either an immune challenge or to various stress paradigms later in life. For example, PAE altered basal cytokine expression in the brain of fetal rats (G17), which preceded an excessive immune response to LPS challenge in adulthood (P90), potentially owing to priming of the neuroimmune system from this early life insult (Terasaki and Schwarz 2016). Specifically, PAE fetuses of both sexes had elevated CCL6, CCR6, IL-21, IL-10ra, and TNF-α in the hippocampus and cortex compared to controls (Terasaki and Schwarz 2016). In a second experiment, rats were given a low dose of LPS (25 μg/kg) or saline in adulthood to see how PAE affects long-term immune functioning (Terasaki and Schwarz 2016). The PAE rats had exaggerated cytokine production both centrally (hippocampus and medial prefrontal cortex) and peripherally (spleen) following LPS injection, suggesting that their early-life exposure to PAE induced long-lasting alterations in how the immune system responds to later challenge (Terasaki and Schwarz 2016). To assess whether this exaggerated response to the LPS immune challenge would affect cognition, rats were tested in the novel object recognition task, a measure of recognition memory (Terasaki and Schwarz 2016). Neither male nor female PAE rats injected with LPS could discriminate between a novel object and one they had seen previously, suggesting that the exaggerated LPS-induced inflammation (owing to a priming of the immune system in early life) resulted in impaired recognition memory in the PAE animals in adulthood (Terasaki and Schwarz 2016). A later study using a second hit of alcohol in place of LPS found that female PAE rats had higher IL-6 expression in the prefrontal cortex compared to control females after adult (P90+) exposure to alcohol (4.5 g/kg) (Terasaki and Schwarz 2017). IL-6 is a pleiotropic cytokine, with the potential to demonstrate both neuroprotective and neurodegenerative effects (Kummer et al. 2021; Rothaug et al. 2016). IL-6 can act as a neurotrophic factor, with its increased expression in the prefrontal cortex of PAE rats potentially exerting protective effects following adult alcohol consumption (Kummer et al. 2021; Rothaug et al. 2016). However, central rises in IL-6 have also been linked to significant neurological disorders, including Alzheimer's disease and Parkinson's disease (Kummer et al. 2021; Rothaug et al. 2016). Additionally, this exposure to alcohol in adulthood impaired performance in the novel object recognition task, but it did so for all rats exposed to alcohol in adulthood regardless of whether they had experienced PAE (Terasaki and Schwarz 2017).

Given the consequences of neuroimmune alterations for cognitive functioning, a critical next step is the identification of potential interventions that can mitigate this rise in central cytokine expression. As discussed above, choline is one emerging intervention option demonstrating promising results. Rats who received choline supplementation for twenty days (P10–30) following developmental alcohol exposure (P4–9) did not show the same increase in basal hippocampal cytokines (IFN-γ and TNF-α) in adulthood (P60) as rats who experienced developmental alcohol exposure alone (Baker et al. 2022). Furthermore, alcohol-exposed rats differed in their response to LPS depending on whether or not they also received choline supplementation: those without choline supplementation had a generally blunted response to LPS, while those with choline supplementation had similar responses to that of the control groups not exposed to alcohol (Baker et al. 2022).

The effect of early-life immune priming following PAE is a robust phenomenon that appears in response to a variety of challenges, including in response to physical trauma or injury. For example, male PAE rats experience enhanced allodynia following varying degrees of damage to the sciatic nerve in adulthood (Noor et al. 2017; Sanchez et al. 2017). A mild single suture constriction injury led to allodynic responses in PAE males, accompanied by increases in astrocyte activation compared to controls; however, there were no differences in microglial activation between groups (Sanchez et al. 2017). Astrocytes have been implicated in chronic pain states, as they produce pro-inflammatory cytokines along with other metabolites, which influence neuronal activity and contribute to the onset of pain (Tang et al. 2021). Thus, increased astrocyte activation may be contributing to the increased allodynia experienced by the PAE animals following damage to the sciatic nerve (Sanchez et al. 2017; Tang et al. 2021). Meanwhile, a standard 4 suture constriction injury led to persistent allodynia, increased spinal glia activation (both astrocytes and microglia), elevated pro-inflammatory cytokines/chemokines (IL-1β, IL-6, TNF-α, CXCL1) and decreased levels of the anti-inflammatory cytokine IL-10 in PAE males compared to controls (Noor et al. 2017). Moreover, control nerve-damaged rats had increased IL-10, likely in an attempt to mitigate excess inflammation (Noor et al. 2017). An imbalance in pro- and anti-inflammatory signaling could impair the recovery process in PAE animals, as IL-10 helps to mitigate host damage following inflammatory responses (Moore et al. 2001; Ouyang et al. 2011). While overexpression of pro-inflammatory cytokines can undoubtedly be harmful, a blunting of the immune response can be of equal detriment, as a tailored immune response is key to successful recovery. Following a CNS wound in adulthood, male PAE rats had decreased TNF-α at the wound site, a response that was further blunted in PAE rats who received a subsequent exposure to alcohol before and after wound infliction (DeVito and Stone 2001). This decreased TNF-α has been noted in PAE rodents following other types of challenges as well: after in vivo priming with LPS, peripheral blood mononuclear cells from male but not female PAE rats showed a blunted release of TNF-α in response to LPS in vitro (Chiappelli et al. 1997). Typically, TNF-α levels rise rapidly in response to tissue damage, playing a role in mounting inflammatory responses which, when appropriately tailored to the insult, can assist in wound healing and repair, notwithstanding that an overexpression of TNF-α has been linked to poor health outcomes (Probert 2015; Ritsu et al. 2017; Shohami 1999). A blunting of this response in the PAE animals could impair their ability to mount an appropriate inflammatory response following an immune challenge such as wound infliction or LPS. Inappropriate inflammatory responses following PAE are exhibited peripherally as well, with female PAE rats experiencing a more severe and prolonged course of inflammation following an arthritis challenge in adulthood (Bodnar et al. 2022b; Zhang et al. 2012). Moreover, PAE rats show basal differences in myeloid cell proportions in peripheral lymphoid organs (spleen and lymph nodes), with in vitro stimulation with LPS leading to increased production of pro-inflammatory cytokines from PAE-derived leukocytes (Sanchez et al. 2017).

While second-hit studies clearly demonstrate the long-lasting effects of PAE on neuroimmune function in adulthood, there is evidence of PAE-induced alterations even in the absence of any challenge. PAE and prenatal and lactational alcohol exposure (PLAE) led to activation of several inflammatory pathways in adult C57BL/6 mice, with PAE increasing NLRP3, caspase-1, NF-κB/p65, IL-1β, and TLR4 expression in the prefrontal cortex and PLAE increasing TLR4 expression in the prefrontal cortex and hippocampus, NLRP3, caspase-1 in the prefrontal cortex and IL-1β in the hippocampus (Cantacorps et al. 2017). Additionally, PLAE mice scored lower on a spatial working memory task, the Y-maze spontaneous alternation test, in adulthood (P60) compared to controls (Cantacorps et al. 2017). The increase in TLR4 in the PAE and PLAE mice is of particular interest as there is evidence for the role of this receptor in the mounting of neuroinflammatory responses following adult alcohol consumption (Alfonso-Loeches et al. 2010; Fernandez-Lizarbe et al. 2009; Montesinos et al. 2015). Activation of TLR4 signaling pathways following alcohol consumption in adulthood can trigger microglial activation and pro-inflammatory cytokine release, with consequences for cognitive functioning, including learning and memory (Fernandez-Lizarbe et al. 2009; Montesinos et al. 2015). In a study assessing the effects of PAE across the lifespan (G15, P0, P20, P66), TLR4 knockout mice exposed to PAE did not experience any of the deleterious effects on neuroimmune function experienced by wild type (WT) C57BL/6 PAE mice (Pascual et al. 2017). The WT PAE mice had increased cortical IL-1β at all timepoints measured (G15, P0, P20, P66) (Pascual et al. 2017). Several cytokines and chemokines, including IL-17, MIP-1α (macrophage inflammatory proteins-1α), and fractalkine, were upregulated in the cortices of G15 PAE pups as well as in P0 and P20 female PAE pups (Pascual et al. 2017). These cytokine alterations co-occurred with an increase in markers indicative of microglial activation in the cerebral cortices of WT PAE mice, including CD11b on P0, P20, and P66, MHC-II at P0 and 20, and Iba-1 immunoreactivity at P20 (Pascual et al. 2017). To assess whether these neuroimmune alterations would have downstream effects on behavior and cognitive functioning, mice were tested in the elevated plus maze (to measure anxiety-like behavior) and the passive avoidance test (to measure short- and long-term memory) in adulthood (P66) (Pascual et al. 2017). WT PAE mice showed increased anxiety-like behavior in the elevated plus maze, along with learning and memory impairments in the passive avoidance test, effects not seen in the TLR4 knockout PAE mice (Pascual et al. 2017). The decrease in the scope of basal alterations with age reflects what has been discussed earlier in the chapter, whereby basal neuroimmune markers appear to be more impacted close to the initial insult. However, PAE still induced several alterations in the adult WT PAE mice, such as increased IL-6 and CD11b, with long-term cognitive and behavioral impairments noted in the WT PAE mice in adulthood (Pascual et al. 2017). Given the findings of altered immune responses in PAE animals both at basal levels and following later-life challenges, a crucial next step is identifying underlying mechanisms behind these alterations and how changes to other systems may contribute.

2.5 Possible Mechanisms Underlying PAE-Induced Alterations in Immune Function

While previous sections have focused on the effects of PAE on neuroimmune functioning and the consequences for neurobehavioral development, it is important to consider the mechanisms through which PAE may be exerting these effects. PAE-induced alterations to neuroimmune function are unlikely to occur in isolation and may be driven by changes to both the development and functioning of related systems. In addition to being essential for virtually all aspects of gastrointestinal function, the gut microbiome is also fundamental for overall host health, including in the development and function of the CNS and immune system (Belkaid and Hand 2014; Erny et al. 2015; Lynch et al. 2023; Margolis et al.

2021; Young and Schmidt 2008). Indeed, the gut microbiome is integral in the establishment, maturation, and ongoing function of the immune system and is being increasingly recognized for its impact on neurobehavioral function, particularly in the context of aberrant early life development (Belkaid and Hand 2014; Cruz-Pereira et al. 2020; Erny et al. 2015; Fung et al. 2017; Lynch et al. 2023; Margolis et al. 2021; Peirce and Alviña 2019; Purchiaroni et al. 2013; Sittipo et al. 2022). Through their significant metabolic activity, the gut microbiota produce key metabolites, such as short-chain fatty acids (SCFAs). SCFAs represent one mechanistic link through which the gut microbiome can affect host function, with these metabolites acting through various immune and endocrine pathways to affect not only local gut structure and function but also to act more distally in the CNS (See Fig. 2.2) (reviewed in Koh et al. 2016). These metabolites are able to enter host circulation and modulate immune function via activation of G-protein-coupled receptors on both immune and enteric neuronal cells to increase cytokine and chemokine production, recruit leukocytes to sites of inflammation, and regulate the functioning of T regulatory cells, key immune cells involved in adaptive immune functioning (Corrêa-Oliveira et al. 2016; Kim et al. 2013; Koh et al. 2016; Schirmer et al. 2016; Van Der Hee and Wells 2021; Vinolo et al. 2011). Additionally, SCFAs are important for the maintenance of blood-brain barrier integrity, with the administration of SCFAs helping to mitigate the severity of damage induced by insults such as stroke and traumatic brain injury (Fock and Parnova 2023; Li et al. 2016; Park and Sohrabji 2016).

Studies utilizing germ-free (GF) rodent models have been instrumental in understanding the importance of microbial colonization for neuroimmune development (Margolis et al. 2021). GF rodents exhibit a wide array of physical, physiological, and behavioral abnormalities, some of which can be rescued following selective recolonization during sensitive developmental periods (Luczynski et al. 2016; Margolis et al. 2021). For example, the microglia of adult GF rodents are morphologically distinct, with altered gene

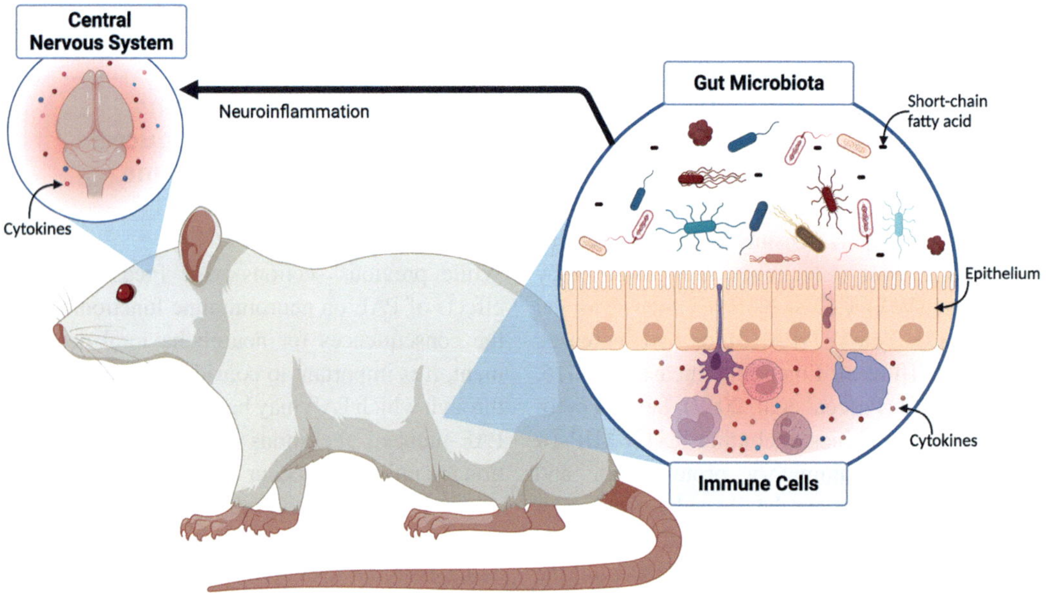

Fig. 2.2 Mechanisms underlying PAE-induced inflammation. PAE-induced alterations in gut microbiota composition and short-chain fatty acid production may lead to both peripheral and central immune responses, including increased neuroinflammation. This can be through alterations in gut barrier permeability, with translocation of luminal content driving a peripheral immune response, or via short-chain fatty acid-mediated communication between the gut and immune system. Created with BioRender

expression (Erny et al. 2015; Matcovitch-Natan et al. 2016). One study found that providing the SCFAs propionate, butyrate, and acetate in the drinking water of adult GF mice could restore their microglia, which differed in morphology at basal levels compared to specific pathogen-free mice and failed to take on an activated state following LPS injection (Erny et al. 2015). After 4 weeks of SCFA administration, the GF mice showed normalized microglia that no longer differed in terms of density, morphology, or maturity (Erny et al. 2015). GF animals have also shown alterations in gray matter volume (Lu et al. 2018), increased white matter volume and reduced oligodendrogenesis (Ahmed et al. 2021), increased blood-brain barrier permeability (Braniste et al. 2014), and sex-specific differences in adult hippocampal neurogenesis (Scott et al. 2020). GF mice show anxiolytic behavior in both the elevated plus maze and light dark box (Clarke et al. 2013; Neufeld et al. 2011a, b), however, behavioral alterations could be restored via microbial reconstitution—but only if it occurred early in life (e.g., 3 weeks; Clarke et al. 2013) as reconstitution at 10 weeks had no effect on anxiety-like behavior (Neufeld et al. 2011a, b). The wide-ranging spectrum of impairments seen in these GF animals highlights the importance of the microbiota during development, especially given the temporal overlap between microbial colonization and early-life brain and immune development.

While underlying microbial changes have been implicated in a range of physical (Carding et al. 2015), neurodevelopmental (Cao et al. 2021; Fattorusso et al. 2019), and mental (Jiang et al. 2015, 2018) health disorders, research into the effects of PAE on the microbiome is currently in its infancy. A landmark study by Bodnar et al. (2022a) was the first to use a rodent model to assess the effects of PAE on fecal microbiota composition in adulthood, finding that PAE resulted in long-lasting alterations. At P80, PAE rats had elevated bacterial richness compared to controls, along with distinct clustering in terms of community composition (Bodnar et al. 2022a). Analysis of differentially abundant taxa found that while both prenatal treatment groups were similar at the phylum level, there were differences at the genera level, with more abundant levels of *Bacteroides, Roseburia,* and *Proteus*—all important in the production of SCFAs—in PAE animals (Bodnar et al. 2022a). After stratifying by sex, there were significant sex- and prenatal treatment-specific differences in the relative abundance of several taxa at both the phylum and genus levels, highlighting the importance of including both sexes in research on the microbiota. Additionally, while control males and females showed distinct clustering with regard to community composition, this was not seen in PAE animals, with PAE males showing considerable overlap with PAE females (Bodnar et al. 2022a). These findings demonstrate the long-lasting impacts of PAE on microbiota composition.

Given that microbial composition shifts over time and is unique in early life compared to in adulthood (Flemer et al. 2017), work from our lab aimed to build on these adult-specific findings using the same rodent model of PAE as Bodnar et al. (2022a) to explore the effects of PAE on microbial composition at earlier timepoints in development (P8, P22, and P38) (Vella et al. 2023). We found that PAE led to age- and sex-specific changes to the gut microbiota, both in terms of compositional alterations as well as differences in SCFA production (Vella et al. 2023). Pups of both sexes had reduced bacterial diversity at P8 compared to controls, but only female PAE pups had taxa-specific alterations at this age. However, by P22 PAE animals of both sexes had alterations to various bacterial taxa. While alterations at lower taxonomic ranks varied by sex, PAE animals of both sexes had decreases in clades known to produce high levels of SCFA and butyric acid. Indeed, we found that butyric acid was decreased in PAE animals at P22 compared to controls. This finding is notable for several reasons, in part due to the critical role of butyric acid within the gut, as well as the timeframe in which this shift is occurring. Butyric acid mediates energy homeostasis by acting as a key energy source for colonocytes, inhibits the production of pro-inflammatory cytokines via inhibition of NF-κB, and supports gut barrier function through the modulation of tight junction proteins (Luhrs et al. 2002; Peng et al. 2009; Segain et al. 2000). One day prior to P22 is gut closure, a critical developmental process whereby the intestinal barrier decreases in permeability, preventing the passage of macromolecules from the gut lumen (Weström et al. 2020). Given the noted importance of butyric acid for gut barrier integrity, it is hypothesized that a reduction in butyric acid during this important milestone could have deleterious effects on the development and later permeability of the gut barrier. Alterations to gut barrier integrity have been linked to peripheral inflammation, as luminal content is able to leak into the bloodstream, potentially contributing to an inflammatory state (see Fig. 2.2) (Lopetuso et al. 2015). Indeed, there was also a rise in IL-6 and IL-13 in the serum of the PAE animals at P22, potentially in response to the alterations in gut microbiota composition and butyric acid levels given the role of these cytokines in inflammatory states within the gut (Al-Sadi et al. 2014; Guo et al. 2021; Mannon and Reinisch 2012; Suzuki et al. 2011). Peripheral inflammatory states, specifically an increase in circulating cytokine levels, can influence neuroimmune activity. For example, some cytokines (e.g., TNF-α, IL-1β, IL-6) are able to cross the blood-brain barrier, while others may alter barrier integrity (e.g., by modifying tight junctions located on brain endothelial cells) allowing for increased trafficking of immune cells across the blood-brain barrier and into the CNS (Banks et al. 2009; Huang et al. 2021).

Moving into adolescence (P38), PAE rats of both sexes continue to show differences in the relative abundance of various taxa as compared to controls (Vella et al. 2023). While there were numerous sex-specific differences, there was a decrease in the Prevotellaceae_NK3B31 family in PAE rats of both sexes compared to controls. Prevotellaceae appears to be sensitive to the effects of PAE, as Bodnar et al. (2022a) also found a decrease in Prevotellaceae_NK3B31 in PAE males in adulthood. Alterations to the Prevotellaceae clade have also been found in research on the microbiota of individuals with different neurodevelopmental disorders. For example, a reduction in unclassified Prevotellaceae, as well as *Prevotella* (a genus under Prevotellaceae) was reported in children with autism spectrum disorder compared to neurotypical controls (Kang et al. 2013). Similarly, children with ADHD had a significantly lower abundance of Prevotellaceae at the family level and *Prevotella* at the genus level compared to neurotypical control children (Prehn-Kristensen et al. 2018). These findings suggest that alterations to the Prevotellaceae clade may be linked to altered neurodevelopment. Together, these results suggest that PAE has enduring effects on the gut microbiota, which manifest as a function of both age and sex. Though there is no prototypical example of a healthy gut microbiome, shifts in the abundance of beneficial versus pathogenic bacteria, and subsequent alterations to key metabolites like SCFAs, could be detrimental to neuroimmune function and warrant further investigation within the context of PAE (Belkaid and Hand 2014; The Human Microbiome Project Consortium 2012). Given these promising findings, a key next step will be to assess the impacts of PAE on the gut microbiota using various models of PAE and at varying degrees of exposure. In doing so, it may become possible to identify unique microbial signatures of PAE, which could not only serve as a biomarker but also inform the development of more targeted treatment interventions.

2.6 Summary and Conclusions

This chapter focuses on the effects of PAE on neuroimmune function across the lifespan, with a specific emphasis on findings from the preclinical literature. While studies using human subjects must often rely on peripheral markers of immune activation, the use of rodent models affords the opportunity to directly assess the CNS. Exposure to alcohol during gestation or the third-trimester equivalent in rodents can lead to long-lasting alterations in neuroimmune functioning, with some differences seen at basal levels and others uncovered only after exposure to challenge. The effects of PAE are unique to the age of the individual and the current role of the immune system within that time frame, thus making it unsurprising that the effects of PAE manifest differently depending on age. Overall, the literature suggests that while PAE induces robust changes to neuroimmune function in early life as well as in adulthood, there were less changes seen in adolescence. Whether this is due to a lessened impact of PAE at this stage or simply due to a lack of research during this period remains to be answered. It is also important to consider whether alterations to functionally related systems, such as the gut microbiota, could impact how PAE alters neuroimmune function. The maturation and continued functioning of the immune system rely on input from the microbiota, making it plausible that alterations to gut microbiota composition could have downstream effects on immune system functioning, with further consequences for neural development and neurobehavioral functioning. Indeed, emerging evidence suggests that PAE uniquely alters the rodent gut microbiota across the lifespan. However, further work is needed to better understand the direction of this relationship (i.e., are changes to the gut microbiota leading to alterations in immune system functioning or is perturbed immune functioning leading to changes in microbial colonization, or both), as well as how different levels and durations of alcohol exposure could impact these outcomes. Increased understanding of the mechanisms behind the neuroimmune perturbations seen following PAE can help to inform the development of more targeted interventions while setting the stage for the identification of novel biomarkers of PAE.

Funding This work was supported by grants from the NIH/NIAAA (R01 AA022460) and The Azrieli Foundation to CR and TSB.

References

Adkins B, Leclerc C, Marshall-Clarke S (2004) Neonatal adaptive immunity comes of age. Nat Rev Immunol 4(7):553–564. https://doi.org/10.1038/nri1394

Ahluwalia B, Wesley B, Adeyiga O, Smith DM, Da-Silva A, Rajguru S (2000) Alcohol modulates cytokine secretion and synthesis in human fetus: an in vivo and in vitro study. Alcohol 21(3):207–213. https://doi.org/10.1016/s0741-8329(00)00076-8

Ahmed S, Travis SD, Díaz-Bahamonde FV, Porter DDL, Henry SN, Mykins J, Ravipati A, Booker A, Ju J, Ding H, Ramesh AK, Pickrell AM, Wang M, LaConte S, Howell BR, Yuan L, Morton PD (2021) Early influences of microbiota on white matter development in germ-free piglets. Front Cell Neurosci 15:807170. https://doi.org/10.3389/fncel.2021.807170

Akison LK, Reid N, Wyllie M, Moritz KM (2019) Adverse health outcomes in offspring associated with fetal alcohol exposure: a systematic review of clinical and preclinical studies with a focus on metabolic and body composition outcomes. Alcohol Clin Exp Res 43(7):1324–1343. https://doi.org/10.1111/acer.14078

Albensi BC, Mattson MP (2000) Evidence for the involvement of TNF and NF-kB in hippocampal synaptic plasticity. Synapse 35(2):151–159. https://doi.org/10.1002/(SICI)1098-2396(200002)35:2<151::AID-SYN8>3.0.CO;2-P

Alfonso-Loeches S, Pascual-Lucas M, Blanco AM, Sanchez-Vera I, Guerri C (2010) Pivotal role of TLR4 receptors in alcohol-induced neuroinflammation and brain damage. J Neurosci 30(24):8285–8295. https://doi.org/10.1523/JNEUROSCI.0976-10.2010

Al-Sadi R, Ye D, Boivin M, Guo S, Hashimi M, Ereifej L, Ma TY (2014) Interleukin-6 modulation of intestinal epithelial tight junction permeability is mediated by JNK pathway activation of claudin-2 gene. PLoS One 9(3):e85345. https://doi.org/10.1371/journal.pone.0085345

Baker JA, Bodnar TS, Breit KR, Weinberg J, Thomas JD (2023) Choline supplementation alters hippocampal cytokine levels in adolescence and adulthood in an animal model of fetal alcohol spectrum disorders. Cells 12(4):546. https://doi.org/10.3390/cells12040546

Baker JA, Breit KR, Bodnar TS, Weinberg J, Thomas JD (2022) Choline supplementation modifies the effects of developmental alcohol exposure on immune responses in adult rats. Nutrients 14(14):2868. https://doi.org/10.3390/nu14142868

Banks WA, Lynch JL, Price TO (2009) Cytokines and the blood–brain barrier. In: Siegel A, Zalcman SS (eds) The Neuroimmunological basis of behavior and mental disorders. Springer US, pp 3–17. https://doi.org/10.1007/978-0-387-84851-8_1

Barbaccia ML, Scaccianoce S, Del Bianco P, Campolongo P, Trezza V, Tattoli M, Cuomo V, Steardo L (2007) Cognitive impairment and increased brain neurosteroids in adult rats perinatally exposed to low millimolar blood alcohol concentrations. Psychoneuroendocrinology 32(8–10):931–942. https://doi.org/10.1016/j.psyneuen.2007.06.013

Bauer S, Kerr BJ, Patterson PH (2007) The neuropoietic cytokine family in development, plasticity, disease and injury. Nat Rev Neurosci 8(3):221–232. https://doi.org/10.1038/nrn2054

Beattie EC, Stellwagen D, Morishita W, Bresnahan JC, Ha BK, Von Zastrow M, Beattie MS, Malenka RC (2002) Control of synaptic strength by glial TNFα. Science 295(5563):2282–2285. https://doi.org/10.1126/science.1067859

Belkaid Y, Hand TW (2014) Role of the microbiota in immunity and inflammation. Cell 157(1):121–141. https://doi.org/10.1016/j.cell.2014.03.011

Bennett ML, Bennett FC, Liddelow SA, Ajami B, Zamanian JL, Fernhoff NB, Mulinyawe SB, Bohlen CJ, Adil A, Tucker A, Weissman IL, Chang EF, Li G, Grant GA, Hayden Gephart MG, Barres BA (2016) New tools for studying microglia in the mouse and human CNS. Proc Natl Acad Sci 113(12):E1738. https://doi.org/10.1073/pnas.1525528113

Bilbo SD, Schwarz JM (2009) Early-life programming of later-life brain and behavior: a critical role for the immune system. Front Behav Neurosci 3:14. https://doi.org/10.3389/neuro.08.014.2009

Bilbo SD, Schwarz JM (2012) The immune system and developmental programming of brain and behavior. Front Neuroendocrinol 33(3):267–286. https://doi.org/10.1016/j.yfrne.2012.08.006

Bilimoria PM, Stevens B (2015) Microglia function during brain development: new insights from animal models. Brain Res 1617:7–17. https://doi.org/10.1016/j.brainres.2014.11.032

Bodnar TS, Hill LA, Weinberg J (2016) Evidence for an immune signature of prenatal alcohol exposure in female rats. Brain Behav Immun 58:130–141. https://doi.org/10.1016/j.bbi.2016.05.022

Bodnar TS, Lee C, Wong A, Rubin I, Wegener Parfrey L, Weinberg J (2022a) Evidence for long-lasting alterations in the fecal microbiota following prenatal alcohol exposure. Alcohol Clin Exp Res 46(4):542–555. https://doi.org/10.1111/acer.14784

Bodnar TS, Mak DY, Hill LA, Ellis L, Yu W, Weinberg J (2022b) Modulatory role of prenatal alcohol exposure and adolescent stress on the response to arthritis challenge in adult female rats. EBioMedicine 77:103876. https://doi.org/10.1016/j.ebiom.2022.103876

Bodnar TS, Raineki C, Wertelecki W, Yevtushok L, Plotka L, Granovska I, Zymak-Zakutnya N, Pashtepa A, Wells A, Honerkamp-Smith G, Coles CD, Kable JA, Chambers CD, Weinberg J (2020) Immune network dysregulation associated with child neurodevelopmental delay: modulatory role of prenatal alcohol exposure. J Neuroinflammation 17(1):39. https://doi.org/10.1186/s12974-020-1717-8

Bodnar T, Weinberg J (2013) Prenatal alcohol exposure: impact on neuroendocrine–neuroimmune networks. In: Cui C, Grandison L, Noronha A (eds) Neural-immune interactions in brain function and alcohol related disorders. Springer, US, pp 307–357. https://doi.org/10.1007/978-1-4614-4729-0_10

Boschen KE, Ruggiero MJ, Klintsova AY (2016) Neonatal binge alcohol exposure increases microglial activation in the developing rat hippocampus. Neuroscience 324:355–366. https://doi.org/10.1016/j.neuroscience.2016.03.033

Bourgognon J-M, Cavanagh J (2020) The role of cytokines in modulating learning and memory and brain plasticity. Brain Neurosci Adv 4:239821282097980. https://doi.org/10.1177/2398212820979802

Brady ML, Allan AM, Caldwell KK (2011) A limited access mouse model of prenatal alcohol exposure that produces long-lasting deficits in hippocampal-dependent learning and memory. Alcohol Clin Exp Res 36(3):457–466. https://doi.org/10.1111/j.1530-0277.2011.01644.x

Braniste V, Al-Asmakh M, Kowal C, Anuar F, Abbaspour A, Tóth M, Korecka A, Bakocevic N, Ng LG, Kundu P, Gulyás B, Halldin C, Hultenby K, Nilsson H, Hebert H, Volpe BT, Diamond B, Pettersson S (2014) The gut microbiota influences blood-brain barrier permeability in mice. Sci Transl Med 6(263):263ra158. https://doi.org/10.1126/scitranslmed.3009759

Brenhouse HC, Schwarz JM (2016) Immunoadolescence: Neuroimmune development and adolescent behavior. Neurosci Biobehav Rev 70:288–299. https://doi.org/10.1016/j.neubiorev.2016.05.035

Brolese G, Lunardi P, Broetto N, Engelke DS, Lìrio R, Batassini C, Tramontina AC, Gonçalves C-A (2014) Moderate prenatal alcohol exposure laters behaviour and neuroglial parameters in adolescent rats. Behav Brain Res 269:175–184. https://doi.org/10.1016/j.bbr.2014.04.023

Burd L, Blair J, Dropps K (2012) Prenatal alcohol exposure, blood alcohol concentrations and alcohol elimination rates for the mother, fetus and newborn. J Perinatol 32:652–659. https://doi.org/10.1038/jp.2012.57

Cantacorps L, Alfonso-Loeches S, Moscoso-Castro M, Cuitavi J, Gracia-Rubio I, López-Arnau R, Escubedo E, Guerri C, Valverde O (2017) Maternal alcohol binge drinking induces persistent neuroinflammation associated with myelin damage and behavioural dysfunctions in offspring mice. Neuropharmacology 123:368–384. https://doi.org/10.1016/j.neuropharm.2017.05.034

Cao X, Liu K, Liu J, Liu Y-W, Xu L, Wang H, Zhu Y, Wang P, Li Z, Wen J, Shen C, Li M, Nie Z, Kong X-J (2021) Dysbiotic gut microbiota and dysregulation of cytokine profile in children and teens with autism Spectrum disorder. Front Neurosci 15. https://doi.org/10.3389/fnins.2021.635925

Caputo C, Wood E, Jabbour L (2016) Impact of fetal alcohol exposure on body systems: a systematic review. Birth Defects Res C Embryo Today 108(2):174–180. https://doi.org/10.1002/bdrc.21129

Carding S, Verbeke K, Vipond DT, Corfe BM, Owen LJ (2015) Dysbiosis of the gut microbiota in disease. Microbial Ecol Health Disease 26(0). https://doi.org/10.3402/mehd.v26.26191

Cealie MY, Douglas JC, Le LHD, Vonkaenel ED, McCall MN, Drew PD, Majewska AK (2023) Developmental ethanol exposure has minimal impact on cerebellar microglial dynamics, morphology, and interactions with Purkinje cells during adolescence. Front Neurosci 17:1176581. https://doi.org/10.3389/fnins.2023.1176581

Chambers RA, Taylor JR, Potenza MN (2003) Developmental neurocircuitry of motivation in adolescence: a critical period of addiction vulnerability. Am J Psychiatry 160(6):1041–1052. https://doi.org/10.1176/appi.ajp.160.6.1041

Chastain LG, Franklin T, Gangisetty O, Cabrera MA, Mukherjee S, Shrivastava P, Jabbar S, Sarkar DK (2019) Early life alcohol exposure primes hypothalamic microglia to later-life hypersensitivity to immune stress: possible epigenetic mechanism. Neuropsychopharmacology 44(9):1579–1588. https://doi.org/10.1038/s41386-019-0326-7

Chiappelli F, Kung MA, Tio DL, Tritt SH, Yirmiya R, Taylor AN (1997) Fetal alcohol exposure augments the blunting of tumor necrosis factor production in vitro resulting from in vivo priming with lipopolysaccharide in young adult male but not female rats. Alcohol Clin Exp Res 21(8):1542–1546. https://doi.org/10.1111/j.1530-0277.1997.tb04487.x

Clarke G, Grenham S, Scully P, Fitzgerald P, Moloney RD, Shanahan F, Dinan TG, Cryan JF (2013) The microbiome-gut-brain axis during early life regulates the hippocampal serotonergic system in a sex-dependent manner. Mol Psychiatry 18(6):666–673. https://doi.org/10.1038/mp.2012.77

Comasco E, Rangmar J, Eriksson UJ, Oreland L (2018) Neurological and neuropsychological effects of low and moderate prenatal alcohol exposure. Acta Physiol 222(1):e12892. https://doi.org/10.1111/apha.12892

Connor PD, Sampson PD, Bookstein FL, Barr HM, Streissguth AP (2000) Direct and indirect effects of prenatal alcohol damage on executive function. Dev Neuropsychol 18(3):331–354. https://doi.org/10.1207/S1532694204Connor

Corrêa-Oliveira R, Fachi JL, Vieira A, Sato FT, Vinolo MAR (2016) Regulation of immune cell function by short-chain fatty acids. Clinical Trans Immunol 5(4):e73. https://doi.org/10.1038/cti.2016.17

Cruz-Pereira JS, Rea K, Nolan YM, O'Leary OF, Dinan TG, Cryan JF (2020) Depression's unholy trinity: dysregulated stress, immunity, and the microbiome. Annu Rev Psychol 71(1):49–78. https://doi.org/10.1146/annurev-psych-122216-011613

Dantzer R (2018) Neuroimmune interactions: from the brain to the immune system and vice versa. Physiol Rev 98(1):477–504. https://doi.org/10.1152/physrev.00039.2016

Deverman BE, Patterson PH (2009) Cytokines and CNS development. Neuron 64(1):61–78. https://doi.org/10.1016/j.neuron.2009.09.002

DeVito WJ, Stone S (2001) Prenatal exposure to ethanol alters the neuroimmune response to a central nervous system wound in the adult rat. Alcohol 25(1):39–47. https://doi.org/10.1016/S0741-8329(01)00161-6

Distéfano-Gagné F, Bitarafan S, Lacroix S, Gosselin D (2023) Roles and regulation of microglia activity in multiple sclerosis: insights from animal models. Nat Rev Neurosci 24(7):397–415. https://doi.org/10.1038/s41583-023-00709-6

Dobbing J, Sands J (1979) Comparative aspects of the brain growth spurt. Early Hum Dev 3(1):79–83. https://doi.org/10.1016/0378-3782(79)90022-7

Doremus-Fitzwater TL, Youngentob SL, Youngentob L, Gano A, Vore AS, Deak T (2020) Lingering effects of prenatal alcohol exposure on basal and ethanol-evoked expression of inflammatory-related genes in the CNS of adolescent and adult rats. Front Behav Neurosci 14:82. https://doi.org/10.3389/fnbeh.2020.00082

Dorshkind K, Crooks G (2023) Layered immune system development in mice and humans. Immunol Rev 315(1):5–10. https://doi.org/10.1111/imr.13198

Drew PD, Johnson JW, Douglas JC, Phelan KD, Kane CJM (2015) Pioglitazone blocks ethanol induction of microglial activation and immune responses in the hippocampus, cerebellum, and cerebral cortex in a mouse model of fetal alcohol spectrum disorders. Alcohol Clin Exp Res 39(3):445–454. https://doi.org/10.1111/acer.12639

Drew PD, Kane CJM (2014) Fetal alcohol spectrum disorders and neuroimmune changes. Int Rev Neurobiol 118:41–80. https://doi.org/10.1016/B978-0-12-801284-0.00003-8

Ernst AM, Gimbel BA, De Water E, Eckerle JK, Radke JP, Georgieff MK, Wozniak JR (2022) Prenatal and postnatal choline supplementation in fetal alcohol spectrum disorder. Nutrients 14(3):688. https://doi.org/10.3390/nu14030688

Erny D, Hrabě De Angelis AL, Jaitin D, Wieghofer P, Staszewski O, David E, Keren-Shaul H, Mahlakoiv T, Jakobshagen K, Buch T, Schwierzeck V, Utermöhlen O, Chun E, Garrett WS, McCoy KD, Diefenbach A, Staeheli P, Stecher B, Amit I, Prinz M (2015) Host microbiota constantly control maturation and function of microglia in the CNS. Nat Neurosci 18(7):965–977. https://doi.org/10.1038/nn.4030

Fattorusso A, Di Genova L, Dell'Isola G, Mencaroni E, Esposito S (2019) Autism Spectrum disorders and the gut microbiota. Nutrients 11(3):521. https://doi.org/10.3390/nu11030521

Fernandez-Lizarbe S, Pascual M, Guerri C (2009) Critical role of TLR4 response in the activation of microglia induced by ethanol. J Immunol 183(7):4733–4744. https://doi.org/10.4049/jimmunol.0803590

Flemer B, Gaci N, Borrel G, Sanderson IR, Chaudhary PP, Tottey W, O'Toole PW, Brugère J-F (2017) Fecal microbiota variation across the lifespan of the healthy laboratory rat. Gut Microbes 8(5):428–439. https://doi.org/10.1080/19490976.2017.1334033

Fock E, Parnova R (2023) Mechanisms of blood–brain barrier protection by microbiota-derived short-chain fatty acids. Cells 12(4):657. https://doi.org/10.3390/cells12040657

Foo YZ, Nakagawa S, Rhodes G, Simmons LW (2017) The effects of sex hormones on immune function: a meta-analysis: sex hormones and immune function. Biol Rev 92(1):551–571. https://doi.org/10.1111/brv.12243

Fung TC, Olson CA, Hsiao EY (2017) Interactions between the microbiota, immune and nervous systems

in health and disease. Nat Neurosci 20(2):145–155. https://doi.org/10.1038/nn.4476
Gano A, Prestia L, Middleton FA, Youngentob SL, Ignacio C, Deak T (2020) Gene expression profiling reveals a lingering effect of prenatal alcohol exposure on inflammatory-related genes during adolescence and adulthood. Cytokine 133:155126. https://doi.org/10.1016/j.cyto.2020.155126
Gemma C, Bachstetter AD (2013) The role of microglia in adult hippocampal neurogenesis. Front Cell Neurosci 7. https://doi.org/10.3389/fncel.2013.00229
Georgountzou A, Papadopoulos NG (2017) Postnatal innate immune development: from birth to adulthood. Front Immunol 8:957. https://doi.org/10.3389/fimmu.2017.00957
Goodfellow MJ, Shin YJ, Lindquist DH (2018) Mitigation of postnatal ethanol-induced neuroinflammation ameliorates trace fear memory deficits in juvenile rats. Behav Brain Res 338:28–31. https://doi.org/10.1016/j.bbr.2017.09.047
Guo Y, Wang B, Wang T, Gao L, Yang Z, Wang F, Shang H, Hua R, Xu J (2021) Biological characteristics of IL-6 and related intestinal diseases. Int J Biol Sci 17(1):204–219. https://doi.org/10.7150/ijbs.51362
Han VX, Patel S, Jones HF, Dale RC (2021) Maternal immune activation and neuroinflammation in human neurodevelopmental disorders. Nat Rev Neurol 17(9):564–579. https://doi.org/10.1038/s41582-021-00530-8
Hellemans KGC, Sliwowska JH, Verma P, Weinberg J (2010) Prenatal alcohol exposure: fetal programming and later life vulnerability to stress, depression and anxiety disorders. Neurosci Biobehav Rev 34(6):791–807. https://doi.org/10.1016/j.neubiorev.2009.06.004
Hewett SJ, Jackman NA, Claycomb RJ (2012) Interleukin-1β in central nervous system injury and repair. European J Neurosci Dis 1(2):195–211
Himmelreich M, Luke CJ, Travis Hargrove E (2020) The lay of the land: fetal alcohol spectrum disorder (FASD) as a whole-body diagnosis. In: Begun AL, Murray MM (eds) The Routledge handbook of social work and addictive behaviors. Routledge, pp 191–216
Holman PJ, Ellis L, Morgan E, Weinberg J (2018) Prenatal alcohol exposure disrupts male adolescent social behavior and oxytocin receptor binding in rodents. Horm Behav 105:115–127. https://doi.org/10.1016/j.yhbeh.2018.08.004
Holman PJ, Raineki C, Chao A, Grewal R, Haghighat S, Fung C, Morgan E, Ellis L, Yu W, Weinberg J (2021) Altered social recognition memory and hypothalamic neuropeptide expression in adolescent male and female rats following prenatal alcohol exposure and/or early-life adversity. Psychoneuroendocrinology 126:105146. https://doi.org/10.1016/j.psyneuen.2021.105146
Howard J (2013) The cytokine hypothesis: a neurodevelopmental explanation for the emergence of schizophrenia later in life. Adv Biosci Biotechnol 04:81. https://doi.org/10.4236/abb.2013.48A2011
Hoyme HE, Kalberg WO, Elliott AJ, Blankenship J, Buckley D, Marais A-S, Manning MA, Robinson LK, Adam MP, Abdul-Rahman O, Jewett T, Coles CD, Chambers C, Jones KL, Adnams CM, Shah PE, Riley EP, Charness ME, Warren KR, May PA (2016) Updated clinical guidelines for diagnosing fetal alcohol spectrum disorders. Pediatrics 138(2):e20154256. https://doi.org/10.1542/peds.2015-4256
Huang X, Hussain B, Chang J (2021) Peripheral inflammation and blood–brain barrier disruption: effects and mechanisms. CNS Neurosci Ther 27(1):36–47. https://doi.org/10.1111/cns.13569
Hughes HK, Moreno RJ, Ashwood P (2023) Innate immune dysfunction and neuroinflammation in autism spectrum disorder (ASD). Brain Behav Immun 108:245–254. https://doi.org/10.1016/j.bbi.2022.12.001
Jiang H, Ling Z, Zhang Y, Mao H, Ma Z, Yin Y, Wang W, Tang W, Tan Z, Shi J, Li L, Ruan B (2015) Altered fecal microbiota composition in patients with major depressive disorder. Brain Behav Immun 48:186–194. https://doi.org/10.1016/j.bbi.2015.03.016
Jiang H, Zhang X, Yu Z, Zhang Z, Deng M, Zhao J, Ruan B (2018) Altered gut microbiota profile in patients with generalized anxiety disorder. J Psychiatr Res 104:130–136. https://doi.org/10.1016/j.jpsychires.2018.07.007
Jones KL, Smith DW (1973) Recognition of the fetal alcohol syndrome in early infancy. Lancet 302(7836):999–1001. https://doi.org/10.1016/S0140-6736(73)91092-1
Jones KL, Smith DW, Ulleland CN, Pytkowicz Streissguth A (1973) Pattern of malformation in offspring of chronic alcoholic mothers. Lancet 301(7815):1267–1271. https://doi.org/10.1016/s0140-6736(73)91291-9
Kane CJM, Phelan KD, Douglas JC, Wagoner G, Johnson JW, Xu J, Phelan PS, Drew PD (2014) Effects of ethanol on immune response in the brain: region-specific changes in adolescent versus adult mice. Alcohol Clin Exp Res 38(2):384–391. https://doi.org/10.1111/acer.12244
Kane CJM, Phelan KD, Han L, Smith RR, Xie J, Douglas JC, Drew PD (2011) Protection of neurons and microglia against ethanol in a mouse model of fetal alcohol spectrum disorders by peroxisome proliferator-activated receptor-γ agonists. Brain Behav Immun 25:S137–S145. https://doi.org/10.1016/j.bbi.2011.02.016
Kang D-W, Park JG, Ilhan ZE, Wallstrom G, LaBaer J, Adams JB, Krajmalnik-Brown R (2013) Reduced incidence of prevotella and other fermenters in intestinal microflora of autistic children. PLoS One 8(7):e68322. https://doi.org/10.1371/journal.pone.0068322
Kim MH, Kang SG, Park JH, Yanagisawa M, Kim CH (2013) Short-chain fatty acids activate GPR41 and GPR43 on intestinal epithelial cells to promote inflammatory responses in mice. Gastroenterology 145(2):396–406.e10. https://doi.org/10.1053/j.gastro.2013.04.056

Klein SL, Flanagan KL (2016) Sex differences in immune responses. Nat Rev Immunol 16(10):626–638. https://doi.org/10.1038/nri.2016.90

Koh A, De Vadder F, Kovatcheva-Datchary P, Bäckhed F (2016) From dietary fiber to host physiology: short-chain fatty acids as key bacterial metabolites. Cell 165(6):1332–1345. https://doi.org/10.1016/j.cell.2016.05.041

Kohler O, Krogh J, Mors O, Eriksen Benros M (2016) Inflammation in depression and the potential for anti-inflammatory treatment. Curr Neuropharmacol 14(7):732–742. https://doi.org/10.2174/1570159X14666151208113700

Kully-Martens K, Denys K, Treit S, Tamana S, Rasmussen C (2012) A review of social skills deficits in individuals with fetal alcohol spectrum disorders and prenatal alcohol exposure: profiles, mechanisms, and interventions. Alcohol Clin Exp Res 36(4):568–576. https://doi.org/10.1111/j.1530-0277.2011.01661.x

Kummer KK, Zeidler M, Kalpachidou T, Kress M (2021) Role of IL-6 in the regulation of neuronal development, survival and function. Cytokine 144:155582. https://doi.org/10.1016/j.cyto.2021.155582

Laviola G, Macrì S, Morley-Fletcher S, Adriani W (2003) Risk-taking behavior in adolescent mice: psychobiological determinants and early epigenetic influence. Neurosci Biobehav Rev 27(1–2):19–31. https://doi.org/10.1016/S0149-7634(03)00006-X

Lawrence RC, Cale Bonner H, Newsom RJ, Kelly SJ (2008) Effects of alcohol exposure during development on play behavior and c-Fos expression in response to play behavior. Behav Brain Res 188(1):209–218. https://doi.org/10.1016/j.bbr.2007.10.028

Lebel C, Roussotte F, Sowell ER (2011) Imaging the impact of prenatal alcohol exposure on the structure of the developing human brain. Neuropsychol Rev 21(2):102–118. https://doi.org/10.1007/s11065-011-9163-0

Lemoine P, Harousseau H, Borteyru JP, Menuet JC (2003) Children of alcoholic parents—observed anomalies: discussion of 127 cases. Ther Drug Monit 25(2):132–136. https://doi.org/10.1097/00007691-200304000-00002

Leng F, Edison P (2021) Neuroinflammation and microglial activation in Alzheimer disease: where do we go from here? Nat Rev Neurol 17(3):157–172. https://doi.org/10.1038/s41582-020-00435-y

Lenz KM, Nelson LH (2018) Microglia and beyond: innate immune cells as regulators of brain development and behavioral function. Front Immunol 9:698. https://doi.org/10.3389/fimmu.2018.00698

Li H, Sun J, Wang F, Ding G, Chen W, Fang R, Yao Y, Pang M, Lu Z-Q, Liu J (2016) Sodium butyrate exerts neuroprotective effects by restoring the blood-brain barrier in traumatic brain injury mice. Brain Res 1642:70–78. https://doi.org/10.1016/j.brainres.2016.03.031

Lopetuso LR, Scaldaferri F, Bruno G, Petito V, Franceschi F, Gasbarrini A (2015) The therapeutic management of gut barrier leaking: the emerging role for mucosal barrier protectors. Eur Rev Med Pharmacol Sci 19(6):1068–1076

Lowery RL, Cealie MY, Lamantia CE, Mendes MS, Drew PD, Majewska AK (2021) Microglia and astrocytes show limited, acute alterations in morphology and protein expression following a single developmental alcohol exposure. J Neurosci Res 99(8):2008–2025. https://doi.org/10.1002/jnr.24808

Luczynski P, McVey Neufeld K-A, Oriach CS, Clarke G, Dinan TG, Cryan JF (2016) Growing up in a bubble: using germ-free animals to assess the influence of the gut microbiota on brain and behavior. Int J Neuropsychopharmacol 19(8):pyw 020. https://doi.org/10.1093/ijnp/pyw020

Luhrs H, Gerke T, Muller JG, Melcher R, Schauber J, Boxberger F, Scheppach W, Menzel T (2002) Butyrate inhibits NF-μB activation in lamina propria macrophages of patients with ulcerative colitis. Scand J Gastroenterol 37:458. https://doi.org/10.1080/003655202317316105

Lu J, Synowiec S, Lu L, Yu Y, Bretherick T, Takada S, Yarnykh V, Caplan J, Caplan M, Claud EC, Drobyshevsky A (2018) Microbiota influence the development of the brain and behaviors in C57BL/6J mice. PLOS ONE 13(8):e0201829. https://doi.org/10.1371/journal.pone.0201829

Lynch CMK, Cowan CSM, Bastiaanssen TFS, Moloney GM, Theune N, van de Wouw M, Florensa Zanuy E, Ventura-Silva AP, Codagnone MG, Villalobos-Manríquez F, Segalla M, Koc F, Stanton C, Ross P, Dinan TG, Clarke G, Cryan JF (2023) Critical windows of early-life microbiota disruption on behaviour, neuroimmune function, and neurodevelopment. Brain Behav Immun 108:309–327. https://doi.org/10.1016/j.bbi.2022.12.008

Maier SE, West JR (2001) Drinking patterns and alcohol-related birth defects. Alcohol Res Health 25(3):168–174

Manji S, Pei J, Loomes C, Rasmussen. (2009) A review of the verbal and visual memory impairments in children with foetal alcohol spectrum disorder. Dev Neurorehabil 12(4):239–247. https://doi.org/10.1080/17518420902980118

Mannon P, Reinisch W (2012) Interleukin 13 and its role in gut defence and inflammation. Gut 61(12):1765–1773. https://doi.org/10.1136/gutjnl-2012-303461

Margolis KG, Cryan JF, Mayer EA (2021) The microbiota-gut-brain axis: from motility to mood. Gastroenterology 160(5):1486–1501. https://doi.org/10.1053/j.gastro.2020.10.066

Marquardt K, Brigman JL (2016) The impact of prenatal alcohol exposure on social, cognitive and affective behavioral domains: insights from rodent models. Alcohol 51:1–15. https://doi.org/10.1016/j.alcohol.2015.12.002

Matcovitch-Natan O, Winter DR, Giladi A, Vargas Aguilar S, Spinrad A, Sarrazin S, Ben-Yehuda H, David E, Zelada González F, Perrin P, Keren-Shaul H, Gury M, Lara-Astaiso D, Thaiss CA, Cohen M, Bahar Halpern K, Baruch K, Deczkowska A, Lorenzo-Vivas

E et al (2016) Microglia development follows a stepwise program to regulate brain homeostasis. Science 353(6301):aad 8670. https://doi.org/10.1126/science.aad8670

Mattson SN, Bernes GA, Doyle LR (2019) Fetal alcohol spectrum disorders: a review of the neurobehavioral deficits associated with prenatal alcohol exposure. Alcohol Clin Exp Res 43(6):1046–1062. https://doi.org/10.1111/acer.14040

May PA, Gossage JP (2011) Maternal risk factors for fetal alcohol spectrum disorders. Alcohol Res Health 34(1):15–26

McCormick CM, Mathews IZ (2007) HPA function in adolescence: role of sex hormones in its regulation and the enduring consequences of exposure to stressors. Pharmacol Biochem Behav 86(2):220–233. https://doi.org/10.1016/j.pbb.2006.07.012

Meltzer A, Van De Water J (2017) The role of the immune system in autism spectrum disorder. Neuropsychopharmacology 42(1):284–298. https://doi.org/10.1038/npp.2016.158

Mendiola AS, Cardona AE (2018) The IL-1β phenomena in neuroinflammatory diseases. J Neural Transm 125(5):781–795. https://doi.org/10.1007/s00702-017-1732-9

Meyer U, Feldon J, Yee BK (2009) A review of the fetal brain cytokine imbalance hypothesis of schizophrenia. Schizophr Bull 35(5):959–972. https://doi.org/10.1093/schbul/sbn022

Meyer U, Murray PJ, Urwyler A, Yee BK, Schedlowski M, Feldon J (2008) Adult behavioral and pharmacological dysfunctions following disruption of the fetal brain balance between pro-inflammatory and IL-10-mediated anti-inflammatory signaling. Mol Psychiatry 13(2):208–221. https://doi.org/10.1038/sj.mp.4002042

Miller AH, Raison CL (2016) The role of inflammation in depression: from evolutionary imperative to modern treatment target. Nat Rev Immunol 16(1):22–34. https://doi.org/10.1038/nri.2015.5

Montesinos J, Pascual M, Pla A, Maldonado C, Rodríguez-Arias M, Miñarro J, Guerri C (2015) TLR4 elimination prevents synaptic and myelin alterations and long-term cognitive dysfunctions in adolescent mice with intermittent ethanol treatment. Brain Behav Immun 45:233–244. https://doi.org/10.1016/j.bbi.2014.11.015

Mooney SM, Varlinskaya EI (2011) Acute prenatal exposure to ethanol and social behavior: effects of age, sex, and timing of exposure. Behav Brain Res 216(1):358–364. https://doi.org/10.1016/j.bbr.2010.08.014

Moore KW, De Waal Malefyt R, Coffman RL, O'Garra A (2001) Interleukin-10 and the interleukin-10 receptor. Annu Rev Immunol 19(1):683–765. https://doi.org/10.1146/annurev.immunol.19.1.683

Morimoto K, Nakajima K (2019) Role of the immune system in the development of the central nervous system. Front Neurosci 13:916. https://doi.org/10.3389/fnins.2019.00916

Neniskyte U, Vilalta A, Brown GC (2014) Tumour necrosis factor alpha-induced neuronal loss is mediated by microglial phagocytosis. FEBS Lett 588(17):2952–2956. https://doi.org/10.1016/j.febslet.2014.05.046

Neufeld KM, Kang N, Bienenstock J, Foster JA (2011a) Reduced anxiety-like behavior and central neurochemical change in germ-free mice: behavior in germ-free mice. Neurogastroenterol Motility 23(3):255–e119. https://doi.org/10.1111/j.1365-2982.2010.01620.x

Neufeld K-AM, Kang N, Bienenstock J, Foster JA (2011b) Effects of intestinal microbiota on anxiety-like behavior. Comm Integra Biol 4(4):492–494. https://doi.org/10.4161/cib.15702

Noor S, Milligan ED (2018) Lifelong impacts of moderate prenatal alcohol exposure on neuroimmune function. Front Immunol 9:1107. https://doi.org/10.3389/fimmu.2018.01107

Noor S, Sanchez JJ, Vanderwall AG, Sun MS, Maxwell JR, Davies S, Jantzie LL, Petersen TR, Savage DD, Milligan ED (2017) Prenatal alcohol exposure potentiates chronic neuropathic pain, spinal glial and immune cell activation and alters sciatic nerve and DRG cytokine levels. Brain Behav Immun 61:80–95. https://doi.org/10.1016/j.bbi.2016.12.016

Norman AL, Crocker N, Mattson SN, Riley EP (2009) Neuroimaging and fetal alcohol spectrum disorders. Dev Disabil Res Rev 15(3):209–217. https://doi.org/10.1002/ddrr.72

O'Connor MJ, Paley B (2009) Psychiatric conditions associated with prenatal alcohol exposure. Dev Disabil Res Rev 15(3):225–234. https://doi.org/10.1002/ddrr.74

O'Leary CM, Nassar N, Zubrick SR, Kurinczuk JJ, Stanley F, Bower C (2010) Evidence of a complex association between dose, pattern and timing of prenatal alcohol exposure and child behaviour problems: prenatal alcohol exposure and child behaviour. Addiction 105(1):74–86. https://doi.org/10.1111/j.1360-0443.2009.02756.x

Olson HC, Streissguth AP, Sampson PD, Barr HM, Bookstein FL, Thiede K (1997) Association of prenatal alcohol exposure with behavioral and learning problems in early adolescence. J Am Acad Child Adolesc Psychiatry 36(9):1187–1194. https://doi.org/10.1097/00004583-199709000-00010

Onore C, Careaga M, Ashwood P (2012) The role of immune dysfunction in the pathophysiology of autism. Brain Behav Immun 26(3):383–392. https://doi.org/10.1016/j.bbi.2011.08.007

Osborne BF, Turano A, Schwarz JM (2018) Sex differences in the neuroimmune system. Curr Opin Behav Sci 23:118–123. https://doi.org/10.1016/j.cobeha.2018.05.007

Ouyang W, Rutz S, Crellin NK, Valdez PA, Hymowitz SG (2011) Regulation and functions of the IL-10 family of cytokines in inflammation and disease. Annu Rev Immunol 29(1):71–109. https://doi.org/10.1146/annurev-immunol-031210-101312

Park MJ, Sohrabji F (2016) The histone deacetylase inhibitor, sodium butyrate, exhibits neuroprotective effects for ischemic stroke in middle-aged female rats. J Neuroinflammation 13(1):300. https://doi.org/10.1186/s12974-016-0765-6

Pascual M, Montesinos J, Montagud-Romero S, Forteza J, Rodríguez-Arias M, Miñarro J, Guerri C (2017) TLR4 response mediates ethanol-induced neurodevelopment alterations in a model of fetal alcohol spectrum disorders. J Neuroinflammation 14(1):145. https://doi.org/10.1186/s12974-017-0918-2

Pei J, Denys K, Hughes J, Rasmussen C (2011) Mental health issues in fetal alcohol spectrum disorder. J Ment Health 20(5):473–483. https://doi.org/10.3109/09638237.2011.577113

Peirce JM, Alviña K (2019) The role of inflammation and the gut microbiome in depression and anxiety. J Neurosci Res 97(10):1223–1241. https://doi.org/10.1002/jnr.24476

Peng L, Li Z-R, Green RS, Holzmanr IR, Lin J (2009) Butyrate enhances the intestinal barrier by facilitating tight junction assembly via activation of AMP-activated protein kinase in caco-2 cell monolayers. J Nutr 139(9):1619–1625. https://doi.org/10.3945/jn.109.104638

Perry VH, Holmes C (2014) Microglial priming in neurodegenerative disease. Nat Rev Neurol 10(4):217. https://doi.org/10.1038/nrneurol.2014.38

Prehn-Kristensen A, Zimmermann A, Tittmann L, Lieb W, Schreiber S, Baving L, Fischer A (2018) Reduced microbiome alpha diversity in young patients with ADHD. PLoS One 13(7):e0200728. https://doi.org/10.1371/journal.pone.0200728

Prieto GA, Tong L, Smith ED, Cotman CW (2019) TNFα and IL-1β but not IL-18 suppresses hippocampal long-term potentiation directly at the synapse. Neurochem Res 44(1):49–60. https://doi.org/10.1007/s11064-018-2517-8

Probert L (2015) TNF and its receptors in the CNS: the essential, the desirable and the deleterious effects. Neuroscience 302:2–22. https://doi.org/10.1016/j.neuroscience.2015.06.038

Purchiaroni F, Tortora A, Gabrielli M, Bertucci F, Gigante G, Ianiro G, Ojetti V, Scarpellini E, Gasbarrini A (2013) The role of intestinal microbiota and the immune system. Eur Rev Med Pharmacol Sci 17(3):323–333

Raffaele S, Lombardi M, Verderio C, Fumagalli M (2020) TNF production and release from microglia via extracellular vesicles: impact on brain functions. Cells 9(10):2145. https://doi.org/10.3390/cells9102145

Raineki C, Chew L, Mok P, Ellis L, Weinberg J (2016) Short- and long-term effects of stress during adolescence on emotionality and HPA function of animals exposed to alcohol prenatally. Psychoneuroendocrinology 74:13–23. https://doi.org/10.1016/j.psyneuen.2016.08.015

Raineki C, Bodnar TS, Holman PJ, Baglot SL, Lan N, Weinberg J (2017) Effects of early-life adversity on immune function are mediated by prenatal environment: role of prenatal alcohol exposure. Brain Behav Immun 66:210–220. https://doi.org/10.1016/j.bbi.2017.07.001

Rasmussen C (2005) Executive functioning and working memory in fetal alcohol Spectrum disorder. Alcoholism: Clin Exp Res 29(8):1359–1367. https://doi.org/10.1097/01.alc.0000175040.91007.d0

Reid N, Akison LK, Hoy W, Moritz KM (2019) Adverse health outcomes associated with fetal alcohol exposure: a systematic review focused on cardio–renal outcomes. J Stud Alcohol Drugs 80(5):515–523. https://doi.org/10.15288/jsad.2019.80.515

Ren Z, Wang X, Xu M, Frank JA, Luo J (2019) Minocycline attenuates ethanol-induced cell death and microglial activation in the developing spinal cord. Alcohol 79:25–35. https://doi.org/10.1016/j.alcohol.2018.12.002

Ritsu M, Kawakami K, Kanno E, Tanno H, Ishii K, Imai Y, Maruyama R, Tachi M (2017) Critical role of tumor necrosis factor-α in the early process of wound healing in skin. J Dermatol Dermatol Surg 21(1):14–19. https://doi.org/10.1016/j.jdds.2016.09.001

Rothaug M, Becker-Pauly C, Rose-John S (2016) The role of interleukin-6 signaling in nervous tissue. Biochimica et Biophysica Acta (BBA)–molecular. Cell Res 1863(6):1218–1227. https://doi.org/10.1016/j.bbamcr.2016.03.018

Ruffaner-Hanson CD, Fernandez-Oropeza AK, Sun MS, Caldwell KK, Allan AM, Savage DD, Valenzuela CF, Noor S, Milligan ED (2023) Prenatal alcohol exposure alters mRNA expression for stress peptides, glucocorticoid receptor function and immune factors in acutely stressed neonatal brain. Front Neurosci 17:1203557. https://doi.org/10.3389/fnins.2023.1203557

Sachser N, Zimmermann TD, Hennessy MB, Kaiser S (2020) Sensitive phases in the development of rodent social behavior. Curr Opin Behav Sci 36:63–70. https://doi.org/10.1016/j.cobeha.2020.07.014

Sanchez JJ, Noor S, Davies S, Savage D, Milligan ED (2017) Prenatal alcohol exposure is a risk factor for adult neuropathic pain via aberrant neuroimmune function. J Neuroinflammation 14(1):254. https://doi.org/10.1186/s12974-017-1030-3

Sato K (2015) Effects of microglia on neurogenesis. Glia 63(8):1394–1405. https://doi.org/10.1002/glia.22858

Schalbetter SM, von Arx AS, Cruz-Ochoa N, Dawson K, Ivanov A, Mueller FS, Lin H-Y, Amport R, Mildenberger W, Mattei D, Beule D, Földy C, Greter M, Notter T, Meyer U (2022) Adolescence is a sensitive period for prefrontal microglia to act on cognitive development. Sci Adv 8(9):eabi6672. https://doi.org/10.1126/sciadv.abi6672

Schambra UB, Lewis CN, Harrison TA (2017) Deficits in spatial learning and memory in adult mice following acute, low or moderate levels of prenatal ethanol exposure during gastrulation or neurulation. Neurotoxicol Teratol 62:42–54. https://doi.org/10.1016/j.ntt.2017.05.001

Schiepers OJG, Wichers MC, Maes M (2004) Cytokines and major depression. Prog Neuro-Psychopharmacol Biol Psychiatry 29:201–217. https://doi.org/10.1016/j.pnpbp.2004.11.003

Schirmer M, Smeekens SP, Vlamakis H, Jaeger M, Oosting M, Franzosa EA, Ter Horst R, Jansen T, Jacobs L, Bonder MJ, Kurilshikov A, Fu J, Joosten LAB, Zhernakova A, Huttenhower C, Wijmenga C, Netea MG, Xavier RJ (2016) Linking the human gut microbiome to inflammatory cytokine production capacity. Cell 167(4):1125–1136.e8. https://doi.org/10.1016/j.cell.2016.10.020

Schneider H, Pitossi F, Balschun D, Wagner A, Del Rey A, Besedovsky HO (1998) A neuromodulatory role of interleukin-1β in the hippocampus. Proc Natl Acad Sci 95(13):7778–7783. https://doi.org/10.1073/pnas.95.13.7778

Schneider M (2013) Adolescence as a vulnerable period to alter rodent behavior. Cell Tissue Res 354(1):99–106. https://doi.org/10.1007/s00441-013-1581-2

Schwarz JM, Sholar PW, Bilbo SD (2012) Sex differences in microglial colonization of the developing rat brain. J Neurochem 120(6):948–963. https://doi.org/10.1111/j.1471-4159.2011.07630.x

Scott GA, Terstege DJ, Vu AP, Law S, Evans A, Epp JR (2020) Disrupted neurogenesis in germ-free mice: effects of age and sex. Front Cell Develop Biol 8:407. https://doi.org/10.3389/fcell.2020.00407

Segain J-P, Ferrier L, Bonnet C, Blottière HM, Galmiche J-P (2000) Butyrate inhibits inflammatory responses through NFkB inhibition: implications for Crohn's disease. Gut 47(3):397–403. https://doi.org/10.1136/gut.47.3.397

Selenica M-LB, Alvarez JA, Nash KR, Lee DC, Cao C, Lin X, Reid P, Mouton PR, Morgan D, Gordon MN (2013) Diverse activation of microglia by chemokine (C-C motif) ligand 2 overexpression in brain. J Neuroinflammation 10(1):856. https://doi.org/10.1186/1742-2094-10-86

Semple BD, Blomgren K, Gimlin K, Ferriero DM, Noble-Haeusslein LJ (2013) Brain development in rodents and humans: identifying benchmarks of maturation and vulnerability to injury across species. Prog Neurobiol 106–107:1–16. https://doi.org/10.1016/j.pneurobio.2013.04.001

Semple BD, Kossmann T, Morganti-Kossmann MC (2010) Role of chemokines in CNS health and pathology: a focus on the CCL2/CCR2 and CXCL8/CXCR2 networks. J Cereb Blood Flow Metab 30(3):459–473. https://doi.org/10.1038/jcbfm.2009.240

Shimizu Y, Sakata-Haga H, Saikawa Y, Hatta T (2023) Influence of immune system abnormalities caused by maternal immune activation in the postnatal period. Cells 12(5):741. https://doi.org/10.3390/cells12050741

Shohami E (1999) Dual role of tumor necrosis factor alpha in brain injury. Cytokine Growth Factor Rev 10(2):119–130. https://doi.org/10.1016/S1359-6101(99)00008-8

Sierra A, Encinas JM, Deudero JJP, Chancey JH, Enikolopov G, Overstreet-Wadiche LS, Tsirka SE, Maletic-Savatic M (2010) Microglia shape adult hippocampal neurogenesis through apoptosis-coupled phagocytosis. Cell Stem Cell 7(4):483–495. https://doi.org/10.1016/j.stem.2010.08.014

Sittipo P, Choi J, Lee S, Lee YK (2022) The function of gut microbiota in immune-related neurological disorders: a review. J Neuroinflammation 19(1):154. https://doi.org/10.1186/s12974-022-02510-1

Sowell ER, Mattson SN, Kan E, Thompson PM, Riley EP, Toga AW (2008) Abnormal cortical thickness and brain-behavior correlation patterns in individuals with heavy prenatal alcohol exposure. Cereb Cortex 18(1):136–144. https://doi.org/10.1093/cercor/bhm039

Spear LP (2000) The adolescent brain and age-related behavioral manifestations. Neurosci Biobehav Rev 24(4):417–463. https://doi.org/10.1016/S0149-7634(00)00014-2

Suzuki T, Yoshinaga N, Tanabe S (2011) Interleukin-6 (IL-6) regulates claudin-2 expression and tight junction permeability in intestinal epithelium. J Biol Chem 286(36):31263–31271. https://doi.org/10.1074/jbc.M111.238147

Szelényi J (2001) Cytokines and the central nervous system. Brain Res Bull 54(4):329–338. https://doi.org/10.1016/S0361-9230(01)00428-2

Tang J, Bair M, Descalzi G (2021) Reactive astrocytes: critical players in the development of chronic pain. Front Psych 12. https://doi.org/10.3389/fpsyt.2021.682056

Tay TL, Béchade C, D'Andrea I, St-Pierre M-K, Henry MS, Roumier A, Tremblay M-E (2018) Microglia gone rogue: impacts on psychiatric disorders across the lifespan. Front Mol Neurosci 10:421. https://doi.org/10.3389/fnmol.2017.00421

Tay TL, Savage JC, Hui CW, Bisht K, Tremblay M (2017) Microglia across the lifespan: from origin to function in brain development, plasticity and cognition. J Physiol 595(6):1929–1945. https://doi.org/10.1113/JP272134

Terasaki LS, Schwarz JM (2016) Effects of moderate prenatal alcohol exposure during early gestation in rats on inflammation across the maternal-fetal-immune interface and later-life immune function in the offspring. J Neuroimmune Pharmacol 11(4):680–692. https://doi.org/10.1007/s11481-016-9691-8

Terasaki L, Schwarz J (2017) Impact of prenatal and subsequent adult alcohol exposure on pro-inflammatory cytokine expression in brain regions necessary for simple recognition memory. Brain Sci 7(10):125. https://doi.org/10.3390/brainsci7100125

The Human Microbiome Project Consortium (2012) Structure, function and diversity of the healthy human microbiome. Nature 486(7402):207–214. https://doi.org/10.1038/nature11234

Tian L, Ma L, Kaarela T, Li Z (2012) Neuroimmune crosstalk in the central nervous system and its significance for neurological diseases. J Neuroinflammation 9(1):594. https://doi.org/10.1186/1742-2094-9-155

Tonelli LH, Postolache TT (2005) Tumor necrosis factor alpha, interleukin-1 beta, interleukin-6 and major histocompatibility complex molecules in the normal brain and after peripheral immune challenge. Neurol Res 27(7):679–684. https://doi.org/10.1179/016164105X49463

Topper LA, Baculis BC, Valenzuela CF (2015) Exposure of neonatal rats to alcohol has differential effects on neuroinflammation and neuronal survival in the cerebellum and hippocampus. J Neuroinflammation 12(1):160. https://doi.org/10.1186/s12974-015-0382-9

Topper LA, Valenzuela CF (2014) Effect of repeated alcohol exposure during the third trimester-equivalent on messenger RNA levels for interleukin-1β, chemokine (C–C motif) ligand 2, and interleukin 10 in the developing rat brain after injection of lipopolysaccharide. Alcohol 48(8):773–780. https://doi.org/10.1016/j.alcohol.2014.09.032

Turner MD, Nedjai B, Hurst T, Pennington DJ (2014) Cytokines and chemokines: at the crossroads of cell signaling and inflammatory disease. Biochimica et Biophysica Acta (BBA)–molecular. Cell Res 1843(11):2563–2582. https://doi.org/10.1016/j.bbamcr.2014.05.014

Valenzuela CF, Morton RA, Diaz MR, Topper L (2012) Does moderate drinking harm the fetal brain? Insights from animal models. Trends Neurosci 35(5):284–292. https://doi.org/10.1016/j.tins.2012.01.006

Van Der Hee B, Wells JM (2021) Microbial regulation of host physiology by short-chain fatty acids. Trends Microbiol 29(8):700–712. https://doi.org/10.1016/j.tim.2021.02.001

Vella VR, Luft C, Wong KE, Ainsworth-Cruickshank G, Holman PJ, Parfrey L, Weinberg J, Bodnar TS, Raineki C (2023) Ontogenetic changes in immune system function and gut microbiota following prenatal alcohol exposure in male and female rats. Alcohol Clin & Exp Res 47(1):539. https://doi.org/10.1111/acer.15071

Vinolo MAR, Rodrigues HG, Nachbar RT, Curi R (2011) Regulation of inflammation by short chain fatty acids. Nutrients 3(10):858–876. https://doi.org/10.3390/nu3100858

Wagner JL, Zhou FC, Goodlett CR (2014) Effects of one- and three-day binge alcohol exposure in neonatal C57BL/6 mice on spatial learning and memory in adolescence and adulthood. Alcohol 48(2):99–111. https://doi.org/10.1016/j.alcohol.2013.12.001

Wang P, Liu B-Y, Wu M-M, Wei X-Y, Sheng S, You S-W, Shang L-X, Kuang F (2019) Moderate prenatal alcohol exposure suppresses the TLR4-mediated innate immune response in the hippocampus of young rats. Neurosci Lett 699:77–83. https://doi.org/10.1016/j.neulet.2019.01.049

Weström B, Arévalo Sureda E, Pierzynowska K, Pierzynowski SG, Pérez-Cano F-J (2020) The immature gut barrier and its importance in establishing immunity in newborn mammals. Front Immunol 11:1153. https://doi.org/10.3389/fimmu.2020.01153

Willford JA, Richardson GA, Leech SL, Day NL (2004) Verbal and visuospatial learning and memory function in children with moderate prenatal alcohol exposure. Alcohol Clin Exp Res 28(3):497–507. https://doi.org/10.1097/01.ALC.0000117868.97486.2D

Wong EL, Lutz NM, Hogan VA, Lamantia CE, McMurray HR, Myers JR, Ashton JM, Majewska AK (2018) Developmental alcohol exposure impairs synaptic plasticity without overtly altering microglial function in mouse visual cortex. Brain Behav Immun 67:257–278. https://doi.org/10.1016/j.bbi.2017.09.003

Wong EL, Strohm A, Atlas J, Lamantia C, Majewska AK (2021) Dynamics of microglia and dendritic spines in early adolescent cortex after developmental alcohol exposure. Dev Neurobiol 81(6):786–804. https://doi.org/10.1002/dneu.22843

Wozniak D, Hartman R, Boyle M, Vogt S, Brooks A, Tenkova T, Young C, Olney J, Muglia L (2004) Apoptotic neurodegeneration induced by ethanol in neonatal mice is associated with profound learning/memory deficits in juveniles followed by progressive functional recovery in adults. Neurobiol Dis 17(3):403–414. https://doi.org/10.1016/j.nbd.2004.08.006

Wozniak JR, Mueller BA, Mattson SN, Coles CD, Kable JA, Jones KL, Boys CJ, Lim KO, Riley EP, Sowell ER (2017) Functional connectivity abnormalities and associated cognitive deficits in fetal alcohol spectrum disorders (FASD). Brain Imaging Behav 11(5):1432–1445. https://doi.org/10.1007/s11682-016-9624-4

Young SL, Gallo LA, Brookes DSK, Hayes N, Maloney M, Liddle K, James A, Moritz KM, Reid N (2022) Altered bone and body composition in children and adolescents with confirmed prenatal alcohol exposure. Bone 164:116510. https://doi.org/10.1016/j.bone.2022.116510

Young VB, Schmidt TM (2008) Overview of the Gastrointestinal Microbiota, vol 635, pp 29–40. https://doi.org/10.1007/978-0-387-09550-9_3

Zeisel SH (2004) Nutritional importance of choline for brain development. J Am Coll Nutr 23(sup6):621S–626S. https://doi.org/10.1080/07315724.2004.10719433

Zeiss CJ (2021) Comparative milestones in rodent and human postnatal central nervous system development. Toxicol Pathol 49(8):1368–1373. https://doi.org/10.1177/01926233211046933

Zhang X, Lan N, Bach P, Nordstokke D, Yu W, Ellis L, Meadows GG, Weinberg J (2012) Prenatal alcohol exposure alters the course and severity of adjuvant-induced arthritis in female rats. Brain Behav Immun 26(3):439–450. https://doi.org/10.1016/j.bbi.2011.11.005

The Neurobiology of Learning and Memory in Rodent Models of Fetal Alcohol Spectrum Disorders

3

Benjamin J. Clark, Gabriela Acosta, Lilliana Sanchez, and Kehiry Trejo Rico

Abstract

Exposure to ethanol during gestation can lead to the onset of Fetal Alcohol Spectrum Disorders, which describes a range of neurodevelopmental and behavioral dysfunctions that include impairments in learning and memory and can have serious repercussions for scholastic performance during adolescence. The neurobiological basis of learning and memory dysfunction in Fetal Alcohol Spectrum Disorders has been frequently linked to the hippocampal formation, which is due in part to the fact that some hippocampal neurons, called place cells, fire action potentials correlated with an animal's spatial location as well as other features of memory episodes. The goal of this chapter is to provide an overview of research investigating developmental alcohol exposure in rodent models and the impact on learning and memory, hippocampal circuitry, and neural representations of learning and memory. We conclude by highlighting areas in which more concentrated behavioral and neurobiological study is needed to expand and develop rodent models of memory dysfunction in Fetal Alcohol Spectrum Disorders.

B. J. Clark (✉) · G. Acosta · L. Sanchez · K. T. Rico
Department of Psychology, University of New Mexico, Albuquerque, United States
e-mail: bnjclark@unm.edu

Keywords

Fetal alcohol · Spatial · Learning · Memory · Object memory · Exploration · Rodent

3.1 Introduction

Exposure to ethanol during gestation can lead to the onset of Fetal Alcohol Spectrum Disorders (FASD), which describes a range of physical dysmorphology and neurobiological, cognitive, and behavioral deficits in offspring (Popova et al. 2023). In clinical work and in animal models of developmental alcohol exposure, cognitive abnormalities are frequently linked to deficits in learning and memory, which can have serious repercussions for scholastic performance during adolescence and can persist into adulthood (Marquardt and Brigman 2016; Valenzuela et al. 2012). The neurobiological basis of learning and memory dysfunction in FASD has been linked to the hippocampal formation and associated limbic regions (Berman and Hannigan 2000; Marquardt and Brigman 2016; Valenzuela et al. 2012). This is due, in part, to the discovery of hippocampal place cells which are neurons that fire action potentials correlated with the animal's location in the environment (Moser et al. 2017; O'Keefe and Nadel 1978). While these cells are speculated to have a prominent role in "mapping" the spatial extent of the environment, it is now firmly estab-

C. F. Valenzuela et al. (eds.), *Effects of Alcohol on the Brain across the Lifespan*, Advances in Experimental Medicine and Biology 1473, https://doi.org/10.1007/978-3-031-81908-7_3

lished that place cells can modify their firing characteristics in relation to other features (objects, items, actions, etc.) associated with a particular experience, which may reflect a broader role for the hippocampus in learning and memory such as in episodic memory (Eichenbaum and Cohen 2014).

The goal of this chapter is to provide an overview of research investigating the developmental impact of alcohol exposure on learning and memory, hippocampal circuitry, and place cell representations. Although there has been considerable research studying the relationship between developmental alcohol exposure and learning and memory in a wide range of model systems, rodents represent the most commonly used model in FASD research. This preference is supported by research suggesting consistencies between human patients and rodent models in the relationship between blood alcohol concentrations and impact on neural systems underlying cognition (Driscoll et al. 1990; Marquardt and Brigman 2016). In addition, rodent models have enabled investigation of the impact of timing of exposure and dosage on multiple levels of analyses, including behavioral, neural circuits, and neurochemical systems. It is important to clarify that rodents have a gestational period that is much shorter than humans (18–23 days), with significant neural development occurring after birth. This is in contrast to the human gestation period which is comprised of three trimesters occurring prenatally before birth. The development of the rodent nervous system can be divided into three trimesters, with the first- and second-trimester equivalent occurring prenatally (before birth) from gestational day 1 to 20, and the third-trimester equivalent occurring postnatally (after birth) from postnatal day 1 to 10 (Marquardt and Brigman 2016; Patten et al. 2014; West 1987).

In this chapter, we therefore focus our review of the literature on rodent models of FASD. First, we summarize the effects of developmental alcohol from early studies to more recent developments that span spatial learning and memory, visual-spatial memory, as well as "non-spatial" functions such as the learning and retention of higher-order sensory representations of objects and items relevant to memory episodes. We conclude with a summary of research testing hypotheses regarding the relationship between developmental alcohol exposure and the neural systems underlying learning and memory. Although learning and memory involve a broad network of cortical-thalamic-hippocampal systems, we center our discussion on the hippocampus with a particular emphasis on recent work investigating a possible role of hippocampal place cell representations in the memory impairments observed in FASD. Our general goal here is to provide an overview of a memory phenotype in rodent models of FASD and highlight areas in which more concentrated behavioral and neurobiological study could expand and develop models of memory dysfunction in FASD.

3.2 Learning and Memory in Rodent Models of Developmental Alcohol Exposure

3.2.1 Spatial Learning and Memory

A large body of studies has shown that animals, including humans, navigate from one place to another on the basis of learned cues in the environment but can also navigate based on cues derived from their own movements (e.g., optic flow, proprioceptive, and vestibular cues) (Gallistel 1990). Self-motion cues can be used to estimate the current location/orientation in the environment and enable navigation to goal locations, even in the absence of familiar environmental cues (e.g., in darkness) (Maaswinkel and Whishaw 1999; Mittelstaedt and Mittelstaedt 1980; Whishaw et al. 2001; Wallace et al. 2002). The Morris water task represents one of the most commonly used tasks to assess the impact of developmental alcohol on learning and memory (Morris 1981). The task requires that animals learn to navigate to the location of a submerged platform typically located a short distance away from the pool wall and in cloudy, opaque water. Following task acquisition, memory is assessed by removing the platform and placing the rats back in the pool for a short probe test (e.g., 60 sec). It is generally agreed that rats use a distal cues located along the room walls to localize the platform location (Morris 1981), but some studies report that rats can use the pool wall as a frame of reference, i.e., to estimate the distance of the platform from the pool wall (Hamilton et al. 2008) and that rats may use distinct movement patterns from each starting location to reach the goal (Goncalves-Garcia et al. 2024b). Although less commonly assessed, some studies report that animals can use self-motion cues (likely from transport to the pool) to generate an accurate trajectory (direction) to the platform (Clark et al. 2015; Stackman et al. 2012).

A clear outcome of studies investigating the relationship between developmental alcohol exposure and spatial learning and memory in rodents, as measured in the Morris water task, is that impairments are largely dose-dependent (reviewed in Harvey et al. 2019; Marquardt and Brigman 2016). Specifically, while low to moderate doses (blood ethanol concentration (BEC): <100 mg/dL) tend to produce milder and specific effects in the water task (discussed further below), an overwhelming majority of studies report that high dose prenatal or postnatal alcohol exposure (BEC: >100 mg/dL) produce deficits in both hidden platform learning, retention (platform removal probe tests), and moving platform (working memory) variants of the water task (e.g., Christie et al. 2005; Goodlett et al. 1987; Iqbal et al. 2004; Kim et al. 1997; Schneider and Thomas 2016; Thomas et al. 2007, 2008; Wagner et al. 2014; Wozniak et al. 2004; for a detailed summary of this literature see Fig. 1 of Harvey et al. 2019 and Fig. 4 of Marquardt and Brigman 2016). Water task navigation is generally intact when the platform is cued, even in cases of high-dose exposure, indicating that procedural learning (swimming and other motor skills) remains intact. In short, alcohol-exposed animals are capable of performing the general procedural elements of the task and can learn to use cues directly associated with the platform location, but they are unable to learn to use cues (environmental and/or self-motion) to navigate to a hidden platform.

While high-dose developmental alcohol exposure produces clear deficits in hidden platform navigation and impairs its retention, some features of water task performance can be spared at lower doses. For instance, learning the location of a fixed and stable hidden platform is generally spared in cases of low or moderate prenatal alcohol exposure (Cullen et al. 2014; Hamilton et al. 2014; Rodriguez et al. 2016; Savage et al. 2002, 2010; Sutherland et al. 2000). In general, these studies report that low or moderate prenatal alcohol-treated rats are capable of acquiring the hidden platform task rapidly and reach a similar asymptotic level as control rats, even with slightly different training procedures. For instance, Sutherland et al. (2000) observed similar performance between control and moderate prenatal alcohol-exposed rats in a training procedure involving eight trials per day across five days, while Savage et al. (2002) observed similar hidden platform acquisition in adult moderate prenatal alcohol-exposed rats over 12 training trials given in a single day of testing. An important caveat is that most studies have largely restricted their analyses of navigation to general performance measures, without assessment of the particular strategy used. Thus, it is possible that rats in the moderate alcohol exposure groups may have used alternate "non-spatial" strategies to similarly navigate to the hidden platform, as has been described in other rodent models of memory dysfunction (e.g., Berkowitz et al. 2018). Future work should investigate this possibility.

Although the findings above suggest that the learning of a fixed location, sometimes referred to as spatial reference memory, is unaffected in the water task after low or moderate prenatal alcohol exposure, retention or recall of recent spatial learning is more reliably impaired. Savage et al. (2002, 2010) provided a demonstration of this phenomenon by training adult rats to navigate to a fixed hidden platform in a single session composed of 12 training trials. Four days later, rats were returned to the pool for a retention test in which they were given another 12 trials. While adult control and moderate prenatal alcohol-exposed (gestational days 1–21) rats performed similarly during task acquisition, the alcohol-exposed rats showed greater escape latencies on the first one or two trials of the retention test. On the remaining trials, alcohol-exposed rats performed similarly to control rats. Thus, while moderate prenatal alcohol rats demonstrated a clear ability to learn an accurate trajectory to a fixed hidden platform, their retention across a 4-day interval was significantly impaired. Interestingly, Cullen et al. (2014) reported that a learned hidden platform location could be retained 24 h later by adult rats prenatally exposed to alcohol (gestational days 1–22/23), suggesting that retention deficits for fixed, stable locations are more likely to be observed after post-acquisition intervals greater than 24 h. Further support for this idea comes from a study by Matthews and Simson (1998), who trained adult rats in a hidden platform water task with platform removal tests given either 24 h or three days after training. In general, rats prenatally exposed to a high dose of alcohol (gestational days 8–20) learned to navigate to the hidden platform similarly to the control group, again indicating that learning of a fixed spatial location remains intact. In the probe tests, however, rats in the alcohol condition showed poor retention for the platform position three days after training. Consistent with Cullen et al. (2014), rats showed similar retention to controls when tested after 24 h.

Some studies report that developmental alcohol exposure (both moderate and high-dose exposure) can impair the rapid learning of a new spatial location (Acosta et al. 2024; Christie et al. 2005; Savage et al. 2002, 2010; Sutherland et al. 2000). An early investigation into one-trial learning and retention of spatial information was conducted by Sutherland et al. (2000) who used a moving platform variant of the hidden platform task. The task involved moving the platform to a novel location each day of water task training, with adult rats given at least two daily trials. The first trial represents an initial encoding of the new spatial location and subsequent trials represent retention of the initial encoding. Using this procedure, Sutherland et al. (2000) reported that although adult rats in the control and moderate prenatal alcohol exposure (gestational days 1–20) groups showed improved latencies across the two trials, indicative of rapid acquisition of spatial information on the initial trial, prenatally exposed rats showed less rapid learning across the first two daily training trials. A report by Schneider and Thomas (2016) also found that postnatal exposure to alcohol (postnatal days 4–9) can produce impairments in this water task variant in adult rats. These results, along with similar findings by Savage et al. (2002, 2010), may reflect deficits in spatial working memory, which we define here as a memory for a recent spatial experience that is retained for a time beyond the traditional definition of working memory (Olton et al. 1979; Eichenbaum 2012). In this interpretation, prenatal and postnatal alcohol impairs the capacity to remember information that was obtained in a single experience and to retain this information after the delay period.

Consistent with the spatial working memory hypothesis, some studies have shown that prenatal alcohol exposure can impair behavioral flexibility such that rats express a tendency to re-enter recently visited locations rather than navigate directly to a newly reinforced spatial location. For example, using the water task, Hamilton et al. (2014) tested adult rats in a task variant where animals acquired a hidden platform location across ten blocks (four trials per block), followed by six blocks of training in which the platform was moved to a new location. Although moderate prenatal alcohol exposure (gestational days 1–20) failed to impair task acquisition across the 3 days, alcohol-treated rats showed a tendency to swim to the previous location of the platform. In other words, rats expressed perseverative swimming towards the previous spatial location (also see Rodriguez et al. 2016).

Other assessments of spatial working memory after developmental alcohol exposure have used the radial arm maze, Y-maze, M-maze, or other related maze environments (Acosta et al. 2024; for reviews see Harvey et al. 2019; Marquardt and Brigman 2016). In the radial arm maze, rodents visit a location (a maze arm) for a reward once and then move to the next location until all maze arms have been visited while avoiding returning to previously visited maze arms. In the Y-maze, M-maze, or similar task designs, rodents alternate between at least two arms for reward. Entries into previously visited maze arms are counted as spatial working memory errors. Several studies have shown that high-dose exposure to alcohol, either prenatally or postnatally, can increase spatial working memory errors (e.g., Reyes et al. 1989; Riley et al. 1979; Thomas 2004; Thomas et al. 1996, 1997; Zimmerberg et al. 1989, 1991).

Previous work also indicates that spatial working memory deficits can be observed after moderate prenatal alcohol exposure (Brady et al. 2012; Kenton et al. 2020), but are more likely to be detected if the task design is made more challenging by requiring discrimination between overlapping spatial stimuli (Acosta et al. 2024; Sanchez et al. 2025). Brady et al. (2012) tested this idea using a radial arm maze variant of a spatial working memory task that required mice to first visit a single open arm for reward. After consuming the reward, mice were removed from the maze for a short delay (60 sec) and afterward were returned to the maze in which they were given a choice between the previously visited arm and a second previously unavailable arm. Mice were rewarded only when they visited the new arm. Thus, the challenge for mice was to discriminate between the two maze arms, which varied with respect to their spatial location. In other words, in some tests, the maze arms were physically separated by a small distance, while in other tests, they were separated by a greater distance. The authors hypothesized that the choice between nearby arms would be more challenging as they shared similarities in the distal cues that defined their spatial locations. The hippocampus (in particular, the dentate gyrus) is thought to play a central role in the discrimination of stimuli, including spatial locations (Yassa and Stark 2011). Because prenatal alcohol exposure, even in moderate amounts, can disrupt hippocampal processing, including structural features of the dentate gyrus (discussed further below), the authors hypothesized that prenatal alcohol exposure (gestational days 1–20) would impair performance specifically when the maze arms had a small separation (had greater overlap in distal spatial cues). Consistent with these hypotheses, adult mice prenatally exposed to moderate alcohol made a greater number of spatial working memory errors in the small separation condition but not when there was a large separation between arms. Similar findings have been reported for another radial arm maze task variant in which all of the maze arms were made available to the rats but only two were rewarded (Sanchez et al. 2025). In this latter study, rats were given four daily training trials with the two rewarded arms remaining in a fixed location across ten test days. In general, adult male rats exposed to alcohol prenatally (gestational days 1–20) made more spatial working memory errors and were less likely to use a spatial strategy during training.

Many of the observed spatial learning and memory deficits described above have stimulated continued efforts to describe the spatial learning and memory performance of those diagnosed with FASD (Dodge et al. 2019, 2020; Hamilton et al. 2003; Mattson et al. 2010; Woods et al. 2018). One of the first experiments that adapted a virtual Morris water task was Hamilton et al. (2003). In this study, the authors tested children with Fetal Alcohol Syndrome (FAS) in a tabletop computer-based VR task where they were required to navigate to a hidden location in a virtual swimming pool that was surrounded with several stable orienting cues. Children were given a total of four training trials followed by a probe test in which the platform was removed. In general, the authors reported that FAS children traveled greater distances to the virtual platform location. Importantly, when the hidden platform was made visible, FAS subjects navigated similarly to control subjects, suggesting that the basic procedural components of the task were unimpaired, but use of distal visual cues for accurate navigation was significantly disrupted. Dodge et al. (2019) expanded on these findings by showing that deficits in virtual navigation increased with the severity of exposure and that deficits could be observed in both male and female subjects. Importantly, as in studies in rodent models of FASD, Dodge et al. reported that heavy exposure was associated with impaired learning of a fixed spatial location, while moderately exposed individuals were unimpaired at learning the fixed location. Woods et al. (2018) have also observed impairments in virtual navigation in children with a history of heavy exposure but also reported greater impairments in male children. In addition, Woods et al. reported reduced activation in temporal, parahippocampal, and related limbic brain regions (also see Roediger et al. 2020). Again, numerous studies using animal models have reported disruptions to hippocampal function and structure (discussed further below). Collectively, the findings summarized above illustrate consistencies between the quantity of alcohol exposure and outcomes in tasks designed to assess spatial behavior, and related neural systems, in both human subjects and rodent models.

3.2.2 Visual-Spatial and Object-Location Memory

A number of studies have reported that humans diagnosed with FASD, as well as animal models of FASD, express impairments in visual-spatial and object-location memory (e.g., Kenton et al. 2020; Mattson et al. 2010, 2019; Olson et al. 1998; Sanchez et al. 2019; Uecker and Nadel 1996). While these deficits are likely related to the observed spatial learning and memory impairments described above, and involve overlapping neural systems, visual-spatial processing has a few distinct features and is considered in a separate section here. For instance, in contrast to tests examining spatial navigation where subjects navigate to goal locations, visual-spatial and object-location memory typically involves tests requiring that animals or human participants learn relationships between objects or other stimuli with their respective spatial locations. For instance, Uecker and Nadel (1996) tested children with FAS in a "Memory for 16 Objects" test in which subjects first studied an array of 16 objects on a tabletop and then were asked to recall the objects and their matching locations. The recall tests occurred either immediately or after a 24-h delay. In both the immediate and delayed tests, FAS children failed to identify objects with their original locations but were capable of identifying the objects in the immediate test condition. Further, the authors found that FAS children had difficulties recalling the general spatial arrangement of the objects, suggesting that object-location memory was severely impaired.

Similar object-location deficits have been reported in rodent models of developmental alcohol exposure (summarized in Table 3.1). In a recent study by Sanchez et al. (2019), adult male rats exposed to moderate alcohol throughout gestation (gestational days 1–20) were tested in an object-place paired associate task that required discrimination of objects based on their maze arm location in a radial arm maze. In this design, identical pairs of objects were presented at two spatial locations occupying two maze arms. Thus, rats were reinforced only when they selected an object paired with a specific maze arm, but not when the same object was encountered in the second maze arm. Over the course of ten training days, control rats showed improvement in their accuracy in selecting the correct object. However, the alcohol-treated rats were significantly slower to learn the task and failed to reach a similar level of performance by the end of the ten days of training. Supporting these findings, a recent report by Terasaki and Schwarz (2017) demonstrated impaired performance in a novel object location task after moderate prenatal alcohol (gestational days 10–16). In sum, these findings are consistent with Uecker and Nadel (1996), suggesting that alcohol exposure during gestation, even when moderate, can produce impairments in object-location learning.

Binge-like alcohol exposure during the third-trimester equivalent in rats can also produce deficits in object-location memory. Gursky et al. (2021) tested rats in a spontaneous object exploration paradigm where rats were presented with four objects in an open field and were allowed to explore the objects for 5 min. The task makes use of the rodent's natural proclivity to explore novel objects. After this initial exposure, rats were removed from the open field and the object array was modified such that two of the objects swapped locations. Rats were returned and allowed to explore this new object-location con-

Table 3.1 The effects of developmental alcohol exposure on visual-spatial and object-place learning and memory. *PreAE* prenatal alcohol exposure; *PostAE* postnatal alcohol exposure

Type	Reference	Species	Sex	Method and dose (BEC: mg/dL)	1st	2nd	3rd	Testing age	Task(s)	Results
Visual-spatial and object-place learning and memory	Wilson et al. (2011)	Mice (C57BL/6By)	Male and female	Two s.c. injections (2 h apart) of 20% ethanol at 2.5 g/kg (na)			X (postnatal days 7)	3 months	Object placement memory task	– PostAE impaired
	Jablonski et al. (2013)	Rat (long-Evans)	Male and female	Intragastric intubation (412.0)			X (postnatal days 7–9)	Postnatal day 31	Novel object location recognition	– No group differences
	MacIlvane et al. (2016)	Rat (long-Evans)	Male and female	Intragastric intubation (383.6)			X (postnatal days 5–9)	Postnatal days 65–70	Object in context recognition	– No group differences
	Terasaki and Schwarz (2017)	Rat (Sprague-Dawley)	Male and female	Gavage (70)		X (gestational days 10–16)		Postnatal day 90	Novel object location task	– Female PreAE rats were impaired
	Sanchez et al. (2019)	Rat (long-Evans)	Male	Liquid diet (60.8)	X	X		6 months	Object-place paired-associate task	– PreAE impaired
	Kenton et al. (2020)	Mice (C57BL/6 J)	Male and female	Liquid diet (80–90)	X	X		6–7 weeks	Trial unique non-matching-to-location	– Female PreAE mice were impaired on the variable separation/delay variant
	Gursky et al. (2021)	Rat (long-Evans)	Male	Intragastric intubation (360.54)			X (postnatal days 4–9)	Postnatal days 52–72	Object-in-place	– PostAE impaired

figuration. As expected, control rats explored the objects that changed location more than the objects that did not change position, suggesting that they encoded their spatial configuration during initial exploration. In contrast, adult rats that were given a high dose (BEC: 360.54 mg/dL) exposure to alcohol during postnatal days 4–9 did not discriminate between the moved vs. stable objects. Thus, the authors extended the observations by Sanchez et al. (2019), determining that third-trimester alcohol exposure can also produce deficits in object-location processing (also see Wilson et al. 2011 for consistent findings after postnatal day 7 alcohol exposure).

It is important to note that some studies have failed to identify impairments in similar tasks after postnatal alcohol exposure (Jablonski et al. 2013; MacIlvane et al. 2016). For example, MacIlvane et al. (2016) tested rats postnatally exposed (postnatal days 5–9) to a high dose of alcohol (BEC: 383.6 mg/dL) in an object-in-context task which was comprised of two contexts (context A and context B) each with a unique pair of identical objects. After rats explored the objects in the two contexts, and following a 20-min delay, they were given a test session in the same two contexts but with one object from each context replaced by an object from the opposite context. For example, during the test session in context A, the two objects were a combination of a previously sampled object in context A (matched-to-context object) and a previously sampled object in context B (mismatched-to-context object). This test takes advantage of the tendency of healthy rats to explore the mismatched-to-context object over the matched-to-context object (i.e., the novel object-context relationship). Surprisingly, alcohol-treated rats explored the mismatched-to-context object at above chance levels that were similar to control subjects, suggesting that they learned and retained the associative relationship between the objects and context.

A recent approach by investigators has been to develop visual-spatial tests that allow systematic manipulation of memory demands while also providing potential for translation to human subjects (Bussey et al. 2012; Mattson et al. 2019; Oomen et al. 2013). This approach has been adopted by Brigman and colleagues, who have made use of rodent touch screen paradigms to investigate the impact of developmental alcohol on visual-spatial discrimination (Marquardt et al. 2014; Olguin et al. 2020). For instance, using a touch screen apparatus, Kenton and colleagues tested mice in a trial-unique delayed non-matching to location (TUNL) paradigm where the animals were required to select an illuminated square during a sample phase, followed by a variable delay. After delay, the mice were presented with a sample square and a new square with the requirement that mice select the new square (i.e., the non-match). The physical separation of the two squares and the delay interval between sample and test task phases could be varied between trials, allowing the investigator to systematically manipulate the visual discrimination and memory demands of the task. Importantly, TUNL performance is strongly dependent on intact hippocampi as well as dorsal dentate gyrus function (Josey and Brigman 2015) and dorsal CA1 NMDA subunit expression (Kenton et al. 2018). As noted above, structural features of the hippocampus, including the dentate gyrus and NMDA subunit expression, is sensitive to prenatal alcohol exposure.

An important feature of the Kenton et al. study was that the investigators trained mice to learn two variants of the TUNL task: one in which the task was made increasingly difficult (square separation was decreased and delay was increased over trials) and a second where the same task variables were randomly changed between trials. In the first task variant, adult mice exposed to alcohol prenatally (gestational days 1–20) performed similarly to control mice, suggesting that prenatal alcohol exposure did not impact visual-spatial discrimination as long as the task parameters were introduced incrementally. However, when tested in the challenging TUNL task, characterized by varying delay intervals and square separation, prenatal alcohol mice exhibited a deficit in performance, particularly at longer delay intervals. Interestingly, these differences were most apparent in female prenatal alcohol mice, suggesting that visual-spatial discrimination behaviors are impaired in a sex-specific manner after prenatal alcohol exposure (also see Terasaki and Schwarz 2017).

3.2.3 Object Learning and Memory

A large body of research also suggests a role for hippocampal-cortical processing of high-order non-spatial sensory stimuli, such as the discrimination and recall of images or objects (Burke et al. 2018; Johnson et al. 2019; Maurer et al. 2017). There have been several studies examining memory for objects and items after developmental alcohol exposure. For example, in the study by Uecker and Nadel (1991) described above, the authors used the "Memory for 16 Objects" task to question children's memory for objects either immediately after a study period or after a 24-h delay. The authors reported that when tested immediately, FAS children were capable of remembering a similar number of objects as the control group. However, when tested after the 24-h delay, FAS children remembered fewer objects compared to controls. Thus, as for spatial memory, developmental alcohol seems to also affect the retention of object information.

In rodent models, developmental alcohol exposure has also been shown to impair the learning and retention of object and visual stimuli (summarized in Table 3.2); an observation that has also been reported in other model systems such as monkeys (e.g., Clarren et al. 1992). In a study by Popovic et al. (2006), rats were trained to discriminate between drinking cans with subtle differences in their visual features. For instance, in a "simple" task condition, a different brand of drinking can was rewarded while all other cans were of the same brand. In a second "complex" task variant, a distinct can was used for reward, while all the other cans had different brands. Because the rewarded can changed location within the array, rats were required to recognize the features of the label of the rewarded can and discriminate these features from the other cans without reference to its spatial location. For each variant, rats were given five consecutive training sessions followed by a retention test that was conducted ten days after the final acquisition test. The authors reported that adult rats given prenatal alcohol (BEC: 103.5 mg/dL) throughout gestation (prenatal exposure) or throughout lactation (postnatal exposure) were impaired in both task variants. While the impairments were greater for the complex task, alcohol-exposed rats generally showed the same pattern of impaired behavior in the simple task. Thus, alcohol-treated rats (exposed either prenatally or postnatally) displayed greater errors (selection of incorrect cans) throughout training without ever reaching similar performance levels as control rats and continued to display inaccurate performance during retention tests.

In a variant of a spontaneous object exploration task, Patten et al. (2016) found that prenatal alcohol exposure (BEC: 101.5 mg/dL) can also impair the discrimination of the temporal relationships between objects. In this study, rats were first allowed to explore an identical pair of objects for 5 min, followed by second and third 5-min exploration sessions, each containing a new pair of identical objects. Thus, rats explored three distinct sets of objects in sequence. Shortly after the third session, rats were given a test session where they explored two objects again for 5 min with one object from the first session and another from the third session. Previous work has shown that control rats typically explore the object from the first session to a greater extent, indicating greater memory for the recently experienced objects (i.e., the session 3 object). Consistent with these prior observations, Patten et al. (2016) found that control rats explored the session 1 object greater than the session 3 object, while adult rats with prenatal alcohol exposure (gestational days 3–21) showed no exploration preference. Thus, while control rats showed sensitivity to the temporal order of the two objects, the prenatal alcohol-exposed rats did not. These observations contrast with a second experiment in which Patten et al. (2016) gave rats two consecutive exploration sessions (separated by 5 min) where they were presented with the same object pairs but with the distance between the object pairs changed

Table 3.2 The effects of developmental alcohol exposure on object learning and memory. PreAE: prenatal alcohol exposure; PostAE: postnatal alcohol exposure

					Trimester of exposure					
Type	Reference	Species	Sex	Method and dose (BEC: mg/dL)	1st	2nd	3rd	Testing age	Task(s)	Results
Object learning and memory	Kim et al. (1997)	Rat (Sprague-Dawley)	Male	Liquid diet (145–155)	X	X		4–12 months	Object recognition delayed non-match to sample.	– No group differences
	Girard and Wainwright (2002)	Rat (long-Evans)	Male	Intragastric cannula (328.56)			X (postnatal days 6–9)	Postnatal days 270–284	Visual discrimination in Morris water task	– PostAE impaired task acquisition
	Popovic et al. (2006)	Rat (Wistar)	Male and female	Liquid diet (103.5)	X	X	X	Postnatal days 80–126	Simple object recognition task	– PreAE group impaired
									Complex object recognition task	– PreAE group impaired
	Summers et al. (2008)	Mice (C57BL/6 J)	Male and female	I.P. Injection (500)	X (gestational day 8)			Postnatal day 120	Object recognition	– PreAE group impaired
	Röskam and Koch (2009)	Rat (Wistar)	Male	Two i.p. injections (2 h apart) of 20% ethanol at 2.0 g/kg (na)			X (postnatal days 7)	Postnatal days 90–95	Object recognition	– No group differences
	Jablonski et al. (2013)	Rat (long-Evans)	Male and female	Intragastric intubation (412.0)			X (postnatal days 7–9)	Postnatal day 31	Novel object recognition	– No group differences
	Patten et al. (2016)	Rat (Sprague-Dawley)	Male and female	Liquid diet (101.5)	X	X		Postnatal days 55–70	Metric change objects task	– No group differences
									Temporal order objects task	– PreAE group impaired

	MacIlvane et al. (2016)	Rat (long-Evans)	Male and female	Intragastric intubation (383.6)			X (postnatal days 5–9)	Postnatal days 65–70	Object recognition	– PostAE group impaired only after 20 min delay
	Cantacorps et al. (2017)	Mice (C57BL/6 J)	Male	Liquid diet (79.27–81.57)	X	X	X	Postnatal day 60	Object recognition	– No group differences
	Terasaki and Schwarz (2016)	Rat (Sprague-Dawley)	Male and female	Gavage (70)		X (gestational days 10–16)		Postnatal day 90	Object recognition	– Female PAE rats were impaired
	Terasaki and Schwarz (2017)	Rat (Sprague-Dawley)	Male and female	Gavage (70)		X (gestational days 10–16)		Postnatal day 90	Object recognition	– Male PAE rats were impaired
	Sanchez et al. (2019)	Rat (long-Evans)	Male	Liquid diet (60.8)	X	X		6 months	Object discrimination	– No group differences
	Gursky et al. (2021)	Rat (long-Evans)	Male	Intragastric intubation (360.54)			X (postnatal days 4–9)	Postnatal days 52–72	Object recognition	– No group differences

between the two sessions (in session one: 68 cm between objects; in session two: 34 cm between objects). In general, control animals showed a tendency to explore the objects in the second session more than the first session indicating that they detect the subtle change in the spatial relationship between the objects. However, in contrast to the results of the temporal order task, there were no differences in performance between alcohol-exposed and control rats. Thus, the results of Patten et al. suggest that prenatal alcohol exposure selectively impairs the ability to learn the temporal relationship between objects but not their metric relationships.

It is important to point out that while the studies above have shown clear impairments in processing object-related information (also see Girard and Wainwright 2002; MacIlvane et al. 2016; Summers et al. 2008; Terasaki and Schwarz 2016, 2017), there have been several reports of intact performance in object learning and memory tasks (Cantacorps et al. 2017; Gursky et al. 2021; Jablonski et al. 2013; Kim et al. 1997; Röskam and Koch 2009; Sanchez et al. 2019). For instance, Sanchez et al. (2019) tested rats in an object discrimination task where the animals were rewarded for correctly selecting an object. The task involved rats locomoting toward a pair of objects placed side-by-side at the end of an alleyway. The left-right position of the rewarded objects was counterbalanced across training trials to prevent the animals from learning a simple motor response for reward (e.g., left turn). Using this test, Sanchez et al. (2019) found that control and moderate prenatal alcohol-exposed (gestational days 1–20) adult rats acquired the task across eight days of training and performed at similar asymptotic levels as control rats. Gursky et al. (2021), using a high dose (BEC: 360.54 mg/dL) postnatal binge model of alcohol exposure (postnatal days 4–9), made a similar set of observations while using a novel object recognition task. In this task, rats first spontaneously explored a pair of similar objects located in an open field for 5 min. After initial object exploration, rats were removed from the open field and the object array was modified such that one of the objects was replaced with a new object. Rats were returned and allowed to explore the old and new objects. As expected, control rats showed an exploratory preference for the novel object, and a similar pattern of behavior was observed in the alcohol-exposed group, indicating recognition of the previously explored object. Thus, the findings from Sanchez et al. (2019) and Gursky et al. (2021) provide evidence that object recognition can remain unaffected by after prenatal or postnatal exposure to alcohol.

An important consideration for future research will be to determine the nature of the disparate behavioral results in the studies summarized above. An argument that has been made elsewhere is that this variability may reflect differences in complexity (Marquardt and Brigman 2016; Popovic et al. 2006). For example, learning the temporal relationship between objects (Patten et al. 2016) or learning the associative relationships between objects and places (Gursky et al. 2021; Sanchez et al. 2019) may place a greater load on hippocampal function, while simpler object recognition tasks can be solved by compensatory mechanisms and alternative neural circuits (Burke et al. 2018). An additional interpretation is that variability between studies may be due to uncontrolled features of the objects used. For instance, the diversity of visual features in drinking cans used by Popovic may have been particularly challenging with respect to discrimination and recognition and perhaps placed a greater demand on hippocampal processing. Further, Popovic tested retention over a long post-acquisition interval (ten days after acquisition), which, as in the study by Uecker & Nadel (1991), may be a particularly important variable for exposing object processing deficits after developmental alcohol exposure.

To address these hypotheses, Sanchez et al. (2025) tested moderate prenatal alcohol-exposed adult rats in a rodent variant of a target-lure (mnemonic similarity) discrimination task (Johnson et al. 2017, 2019). The task makes use of LEGO objects that can be systematically varied with respect to their perceptual similarity.

That is, LEGO pieces can be removed or added to make the rewarded object appear more similar or distinct from the non-rewarded object. In this study, the authors tested the ability of adult male prenatal alcohol-exposed rats to discriminate between a rewarded object and randomly presented objects that varied in terms of their similarity to the target (rewarded) object. Four days of testing were performed with at least 72 h between each test. Thus, animals had to remember the rewarded object between testing days thereby increasing the memory demands of task performance. In brief, Sanchez et al. found that prenatal alcohol-exposed (gestational days 1–20) adult rats did not express significant impairments in the discrimination of objects, even when the objects were perceptually similar. Thus, even when the object learning procedures were made more complex and challenging with respect to discrimination and retention, prenatal alcohol-exposed rats could still perform as well as control rats.

3.2.4 Spontaneous Exploratory Behaviors

As described in the studies above, it is well documented that rodents can gather information by exploring features of the environment. There is a large literature showing that rodents express an organized pattern of exploratory behaviors in open environments and that these behaviors are spontaneously generated even in the absence of environmental cues (Dudchenko and Wallace 2018; Thompson et al. 2018). The organization of spatial behaviors tends to be centered around one or two locations, termed home bases, where they exhibit a high number of stops of long duration and other behaviors such as grooming and turning (Eilam and Golani 1989). Home bases are rapidly established within a few minutes of entering the environment and are maintained across long-duration tests (Donaldson et al. 2018). From the home base, rodents make exploratory trips into the rest of the environment that end with a rapid and direct trajectory back to the home base (Wallace and Whishaw 2003).

It is well documented that damage to the hippocampus can lead to locomotor hyperactivity (O'Keefe and Nadel 1978) but also produce significant alterations to the organization of spontaneous exploratory behaviors, including disruptions in stop duration and frequency, home base stability, and deficits in the homeward components of exploratory trips (Clark et al. 2005; Gorny et al. 2002; Lehmann et al. 2007; Hines and Whishaw 2005; Wallace and Whishaw 2003; Whishaw et al. 1994). It has been argued extensively that disruptions to the organization of exploratory behaviors may ultimately impair the acquisition and retention of information gathered from the environment (Nadel 1991; O'Keefe and Nadel 1978; Poulter et al. 2018). That prenatal alcohol exposure can lead to alterations to hippocampal structure and function suggests that spontaneous exploratory behavior may also be altered. A recent study tested this hypothesis by placing adult rats exposed to moderate prenatal alcohol or saccharine (controls) in a lighted open field (Osterlund Oltmanns et al. 2022; also see Table 3.3). As expected, moderate prenatal alcohol (gestational days 1–20) adult rats traveled longer distances (i.e., were hyperactive) and expressed less concentrated stops at home base locations. Further, when rats were locomoting between stops (termed progressions), prenatal alcohol rats exhibited a weaker tendency to scale their peak locomotor speeds with the distance traveled. The authors also observed sex differences with male prenatal alcohol rats showing less prominent home base behaviors (clustering of stops near the home base). In sum, the findings suggest that prenatal alcohol exposure can produce alterations to spontaneous exploratory behaviors.

The organized structure of the exploratory behaviors can be observed in the absence of environmental cues (complete darkness) (Hines and Whishaw 2005; Wallace and Whishaw 2003; Whishaw et al. 2001), suggesting a role for self-motion cues in organizing spontaneous spatial behaviors. Previous work has shown that the hippocampus has a particularly important role in processing self-motion cues for accurate navigation in an environment (Wallace and Whishaw

Table 3.3 The effects of developmental alcohol exposure on exploratory behaviors. PreAE: prenatal alcohol exposure

					Trimester of exposure					
Type	Reference	Species	Sex	Method and dose (BEC: mg/dL)	1st	2nd	3rd	Testing age	Task	Results
Exploratory behaviors	Osterlund Oltmanns et al. (2022)	Rat (long-Evans)	Male and female	Liquid diet (42)	X	X		Postnatal days: 135–137	Open field exploratory behaviors—Lighted conditions	– PreAE rats travelled longer distances – Male PreAE rats had weaker stop clustering – Male PreAE rats had smaller changes in heading during stops – PreAE rats had weaker movement scaling
	Schaeffer et al. (2025)	Rat (long-Evans)	Male and female	Liquid diet (42)	X	X		Postnatal days: 142–148	Open field exploratory behaviors—Darkened conditions	– No group differences

2003; Whishaw et al. 2001). Thus, Schaeffer and colleagues (2025) hypothesized that moderate prenatal alcohol exposure should also disrupt spontaneous exploratory behaviors when animals are tested in a darkened open field environment (Table 3.3). As in the study by Osterlund Oltmanns et al. (2022), adult prenatal alcohol-exposed (gestational days 1–20) and control rats were given a single 30-min exposure to an open field in a testing room that was made completely dark. However, in contrast to Osterlund Oltmanns et al., the authors reported that prenatal alcohol rats organized their exploratory behaviors similarly to control rats under these testing conditions, suggesting that the spontaneous movement organization in relation to self-motion cues likely remains intact even after developmental alcohol exposure. The difference in results between these two studies is in support of the general observation that PAE impairs behaviors in several tasks that require the use of environmental cues for accurate navigation (as described above). Further, hippocampal place cell activity is thought to provide a critical signal in the establishment of maps of environmental space with exploratory behaviors playing an important role in updating these spatial representations (O'Keefe and Nadel 1978). Thus, deficits in spontaneous exploratory behaviors may have a relationship with the reported changes in hippocampal place representations in prenatal alcohol exposure (described below).

3.3 Neural Systems Involved in Memory and Developmental Alcohol Exposure

The behavioral studies summarized above support two general conclusions. First, as discussed by others (see Harvey et al. 2019; Marquardt and Brigman 2016), learning and memory impairments are prominent after high-dose alcohol exposure across both spatial and object learning and memory domains. In contrast, learning and memory impairments are more specific after low- or moderate-dose developmental alcohol exposure. In the Morris water task, moderate prenatal alcohol exposure impairs the retention of a fixed spatial location (reference memory) but does not impair its initial acquisition. In moving platform variants and dry land mazes such as the radial arm maze, deficits in spatial working memory and behavioral flexibility have been reported, suggesting that moderate prenatal alcohol exposure may lead to deficits in the retention of recent experience. Deficits in visual-spatial touch screen tasks also show a related outcome with greater impairments for recent visual-spatial experience, especially after longer delay intervals. Deficits may be more readily exposed if spatial working memory tasks include discrimination between spatial stimuli with a high degree of location overlap.

While retention of spatial information appears to be sensitive to moderate developmental alcohol exposure, some studies have reported deficits when animals are required to detect a spatial change in object location, recall the identity of an object based on its spatial position, recall the temporal relationships between objects, or discriminate between nearby spatial locations or visual stimuli on touch screens. In addition, the manner in which rats with moderate prenatal alcohol exposure spontaneously organize their spatial exploratory behaviors markedly differs from control rats, suggesting that moderate prenatal alcohol exposure can affect how animals explore their environment which may ultimately impact how memory representations are updated by relevant neural circuits. A second general conclusion is that while spatial learning and memory (in navigation and visual-spatial tasks) is clearly disrupted by developmental alcohol exposure, detection of impairments in object learning and retention has yielded less consistent results. Variability in study design (complexity of task and uncontrolled features of stimuli) may account for some of these discrepancies. Regardless, additional studies are needed to further interrogate this form of learning and memory in alcohol-exposed animals.

Although the neural basis of learning and memory impairments after developmental alcohol exposure are poorly understood, considerable attention has been directed toward the hippocampal formation (CA1, CA3, dentate gyrus) (Berman and Hannigan 2000; Harvey et al. 2019; Marquardt and Brigman 2016; Valenzuela et al. 2012). Damage to the hippocampus can produce severe impairments in the encoding of new information, especially in tasks where animals are required to discriminate between spatial locations, objects, or visual-spatial stimuli (as in touch screen paradigms described above) (Johnson et al. 2017; Josey and Brigman 2015; Yassa and Stark 2011). Further, damage to the hippocampus can impair retention or recall of recently acquired spatial and object-based memories (Gilbert and Kesner 2003; Lee and Solivan 2010; Sutherland et al. 2001). The hippocampus contains neurons called place cells that discharge action potentials correlated with an animal's spatial location in an environment (Moser et al. 2017; O'Keefe and Nadel 1978). The firing location and firing rates of place cells can rapidly change, or "remap," in response to encountering new environments but can also maintain stability in their firing locations and rates in unchanging environments. While place cells are thought to serve a role in generating "maps" of environmental space, they can also modify their firing to encode experiences, actions, and other "non-spatial" features (objects, items, etc.) that comprise a particular learning and memory task (Eichenbaum and Cohen 2014).

Consistent with findings that developmental alcohol exposure disrupts the encoding and retention/recall of recently learned spatial locations, an extensive body of research has demonstrated that hippocampal circuits and related neuronal activity are impacted (reviewed in Hannigan and Berman 2000; Valenzuela et al. 2012). In humans with FASD, neuroimaging studies have shown that temporal lobe and parahippocampal activation is reduced (Woods et al. 2018). Reductions in hippocampal volumes have also been reported in subjects with FASD or the presence of prenatal alcohol exposure (Dodge et al. 2020). In non-human primate models of FASD, moderate alcohol exposure prenatally can lead to significant neuronal loss in the CA hippocampal subfields (Burke et al. 2022). In rodent models, moderate prenatal alcohol exposure is known to produce deficits in hippocampal synaptic plasticity, especially between perforant path projections and the hippocampus (Fontaine et al. 2016; Savage et al. 2002, 2010; Sutherland et al. 1997). Moderate prenatal alcohol exposure can also disrupt enrichment-induced neurogenesis in the mouse dentate gyrus (Choi et al. 2005) and disrupt NMDA subunit expression in the mouse dentate gyrus (Brady et al. 2013). Pre- and postnatal exposure can alter the expression of hippocampal parvalbumin-positive interneurons in both male and female rats (Bird et al. 2018; Madden et al. 2020). Limbic thalamic and cortical regions that contribute to the functions of the hippocampus and place cell representations can also be damaged after prenatal or postnatal alcohol exposure (e.g., Bird et al. 2023; Gursky and Klintsova 2021; Licheri et al. 2023; Wozniak et al. 2004).

The expression of hippocampal place cell activity is thought to critically depend on synaptic plasticity and perforant path input, especially from the entorhinal cortex as well as input from other limbic cortical regions affected by developmental alcohol exposure (e.g., Cooper and Mizumori 2001; Ormond and McNaughton 2015; Save et al. 2005; Schlesiger et al. 2015). A recent study from our laboratory (Harvey et al. 2020) addressed whether developmental alcohol would impair place cell representations in a rat model of moderate prenatal alcohol exposure (gestational days 1–20) (results summarized in Table 3.4). In this study, hippocampal place cell activity in CA1 and CA3 subregions was monitored in adult rats while they locomoted randomly for scattered food pellets in a large open field environment or locomoted for reward located at each end of a narrow linear track. First, the study determined that prenatal alcohol did not reduce the overall number of place cells in the hippocampus but significantly altered their location-specific firing characteristics. These altered firing characteristics included reductions in peak firing rates, weaker location-specific firing, less consistency in spiking, particularly in the CA3 subregion, and less stability in their firing locations. These differences in firing characteristics were observed in different shaped mazes (cylindrical and a linear track). Thus, hippocampal place cells in moderately alcohol-exposed rats were generally weaker and less consistent regardless of the testing environment.

While an animal locomotes along a restricted path, such as a narrow maze arm, place cell locations, and firing rates can significantly differ in opposite directions of travel (Moser et al. 2017). In other words, place fields can express distinct firing characteristics (“remapping”) in each direction of travel, suggesting a potential mechanism for generating unique cell activity for subtle differences in the path taken to each end of the track. Given that moderate prenatal alcohol exposure impairs location discrimination in spatial tasks using narrow maze arms or in tasks where animals are required to rapidly learn new spatial locations (e.g., Brady et al. 2012), Harvey et al. (2020) tested the hypothesis that distinct firing by hippocampal place cells would be impaired after moderate prenatal alcohol exposure. In brief, the authors found that prenatal alcohol place cell maps (firing rates and locations) were more similar in the two directions of travel of the linear track compared to those recorded in the control group. Thus, prenatal alcohol place cell activity was less likely to discriminate subtle contextual differences on the linear track. These observations, coupled with the fact that prenatal alcohol place cell activity can exhibit reduced location-specific tuning and instability, appear to provide a potential mechanism for spatial impairments after moderate prenatal alcohol exposure. Consistent with this notion are previous reports

Table 3.4 The effects of developmental alcohol exposure on neural representations of learning and memory. PreAE: prenatal alcohol exposure

					Trimester of exposure					
Type	Reference	Species	Sex	Method and dose (BEC: mg/dL)	1st	2nd	3rd	Testing age	Task	Results
Neural representations of learning and memory	Harvey et al. (2020)	Rat (long-Evans)	Male	Liquid diet (60.8)	X	X		>4 months	In vivo hippocampal place cell recordings—Open field locomotion	– PreAE reduced spatial tuning in CA3 – Lowered peak firing rates, spatial coherence, and stability in CA1/CA3 – PreAE reduced theta rhythmicity in CA1/CA3 and phase precession in CA3
									In vivo hippocampal place cell recordings—linear track locomotion	– Firing characteristics as summarized above – PreAE diminished location and firing rate discrimination in CA1 and CA3, respectively
									In vivo hippocampal place cell recordings—cue rotation	– PreAE increased control by cue in CA1/CA3 – PreAE reduced change in spatial tuning in CA1/CA3

that reductions in location-specificity and stability are associated with learning and memory deficits (Cacucci et al. 2008; Lester et al. 2017). Thus, future studies could be designed to test hypotheses regarding the relationship between disruptions in place cell remapping and learning and memory deficits after developmental alcohol exposure.

An additional feature of hippocampal place cell activity is their role in the post-learning strengthening of memories (e.g., Lee and Wilson 2002; Wilson and McNaughton 1994). Specifically, hippocampal neurons engaged during spatial learning, such as the place cells traversed in a specific environmental location, are "replayed" during "offline" periods when the animal is resting. Further, the sequence by which place cell receptive fields are traversed during behavior is retained during these offline replay events. Patterned activity by hippocampal place cells is known to co-occur with fluctuations, or oscillations, in the hippocampal local field potential. Hippocampal neurons are known to oscillate at slow (theta, ~4-12 Hz) and fast frequencies (gamma, 25-100 Hz and sharp wave-ripples, 150-250 Hz) (Buzsáki 2005, 2015; Colgin 2016). These oscillations are generated by rhythmic fluctuations in the excitability of local circuits and represent periods during which sequential task-related spiking is organized. During active behavior, theta rhythms are the predominant rhythmic activity observed in the hippocampus. However, during immobility and in slow-wave sleep, high-frequency sharp wave-ripples dominate the hippocampal local field potential. Current theory suggests that theta oscillations provide a temporal window by which neuronal spiking can be organized during active spatial behavior while sharp wave ripples initiate co-activation, or "replay," of these same cells during subsequent sleep (Buzsáki 2005, 2015; Colgin 2016; Chadwick et al. 2015; Diba and Buzsaki 2007; Dragoi and Buzsaki 2006; Foster and Wilson 2007; Lee and Wilson 2002). The fact that traversed sequences of place fields are organized in the same order within theta cycles and during sharp wave ripples, and that disruption of these patterns impairs spatial memory (Ego-Stengel and Wilson 2009; Girardeau et al. 2009; Robbe and Buzsáki 2009), supports these hypotheses.

Considering the reported disruptions to hippocampal place cell activity and hippocampal circuitry described after prenatal alcohol exposure, a reasonable hypothesis is that developmental alcohol would lead to similar disruptions to patterned replay and hippocampal rhythmic activity. Although these features have not been explored in relation to post-memory acquisition behavior and sleep, recent work from our laboratory (Harvey et al. 2020) has presented evidence that place cells in rats moderately exposed to alcohol are less likely to spike with the local theta rhythm, possibly pointing to a failure of sequential encoding during active behavior (Table 3.4). Specifically, while locomoting for reward on a linear track and in a cylindrical open field, Harvey et al. reported that fewer place cells exhibited significant theta rhythmicity in the moderate prenatal alcohol group. In addition, of the cells that were significantly modulated by theta, prenatal alcohol place cells tended to discharge at a slower frequency. Harvey et al. (2020) also investigated whether spiking by prenatal alcohol place cells was organized in relation to the local theta rhythm. In healthy animals, spikes from place cells are systematically locked to theta cycles such that each spike shifts across successive cycles in a phenomenon known as theta phase precession (e.g., Buzsáki 2005; Colgin 2016; Diba and Buzsaki 2007). In other words, the sequence of spatial firing fields traversed during locomotion corresponds to the sequence of firing during a theta cycle. Harvey et al. reported that the proportion of place cells demonstrating theta phase precession was reduced in both linear track and open-field behavior in the moderate prenatal alcohol group. Thus, the coupling between place cell firing and theta rhythmicity is impaired after moderate prenatal alcohol exposure, suggesting a disruption to the sequential organization of place cell spiking. Future work is needed to determine whether these impairments in place cell features can explain the learning and memory-related deficits observed after moderate prenatal alcohol exposure. Further, whether place cell dysfunction is a general feature of developmental alcohol exposure, and can be observed in other exposure paradigms where the timing and dosage has been varied, remains to be determined.

3.4 Therapeutic Interventions for Learning and Memory Deficits After Developmental Alcohol Exposure

A number of studies have investigated whether developmental alcohol exposure-related learning and memory deficits and hippocampal dysfunction in FASD can be attenuated by dietary supplementation or pharmacological manipulation. Choline supplementation has received considerable attention as a potential therapeutic target in large part due to findings that it influences synaptic plasticity mechanisms and cognitive function in animal models and in humans (reviewed in Ernst et al. 2022). With respect to hippocampal function, choline is known to have an important role during prenatal and postnatal development in supporting neurogenesis, cell proliferation, reducing apoptosis, and fostering synaptogenesis. Importantly, choline supplementation has been associated with improvements in learning and memory in rodent models of FASD (e.g., Ryan et al. 2008; Schneider and Thomas 2016). For instance, Schneider and Thomas (2016) administered choline during early adulthood (postnatal days 40 to 60) in rats with binge-like postnatal alcohol exposure (postnatal days 4–9) and found that supplementation reduced impairments specifically in a test of spatial working memory where rats were required to rapidly learn and retain a spatial location. Other studies have found that supplementation at earlier developmental time points (prior to postnatal day 30) attenuates the impact of developmental alcohol exposure on other forms of learning and memory, including place learning in the Morris water task, spatial discrimination, and associative learning (Thomas and Tran 2012; Thomas et al. 2000, 2007; Thomas 2004; Wagner and Hunt 2006). Specifically, Ryan et al. (2008) reported that both early (postnatal days 11 to 20) and adolescent (postnatal days 21 to 30) choline supplementation improved retention of a recently learned spatial location in a rat model of binge-like postnatal exposure (days 4–9). Several studies have also confirmed the benefits of choline supplementation after prenatal alcohol exposure (e.g., Thomas et al. 2000; Waddell and Mooney 2017).

Another avenue for therapeutic intervention is the potential use of pharmacological agents that manipulate histamine receptors which are thought to influence glutamate and/or acetylcholine release within hippocampal circuits (Goncalves-Garcia and Hamilton 2023). Savage et al. (2010) assessed the effects of the histamine receptor antagonist, ABT-239, delivered prior to learning, and found significant improvements in place memory in the Morris water task by adult rats exposed to moderate alcohol prenatally (gestational days 1–21). At the level of hippocampal neural circuits, ABT-239 administration improves dentate gyrus long-term potentiation in adult moderate prenatal alcohol-exposed (gestational days 1–21) rats (Varaschin et al. 2010). Similar findings have been made using the histamine H_3 receptor agonist, SAR-152954 (Goncalves-Garcia et al. 2024a).

Lastly, non-pharmacological approaches have also been investigated, including the impact of exercise, handling, or environmental enrichment, which are known to promote learning and memory, hippocampal synaptic plasticity, and neurogenesis, as well as ameliorate some of behavioral and functional deficits that follow developmental alcohol exposure (Christie et al. 2005; Hannigan et al. 2007; Lee and Rabe 1999; Redila and Christie 2005; Thomas et al. 2008). For instance, Christie et al. (2005) used a voluntary exercise procedure in which rats prenatally exposed to alcohol (gestational days 1–22) were given access to a running wheel for 12 days after weaning. Rats were then tested in adulthood in a one-trial learning paradigm in the Morris water task. In brief, Christie et al. found that exercise attenuated prenatal alcohol-induced deficits in the Morris water task while also improving dentate gyrus long-term potentiation. Thus, voluntary alcohol was capable of alleviating hippocampal functional and learning deficits. Thomas et al.

(2008) expanded on these findings using a rat model of postnatal alcohol exposure (postnatal days 4–9) and provided access to a running wheel during adolescence (postnatal days 21 to 51). In this study, the authors reported improvements in acquisition and retention of a place memory in the Morris water task. Together, these studies support the conclusion that voluntary exercise can improve learning and memory and hippocampal functional outcomes after developmental alcohol exposure.

3.5 Conclusions

A major aim of the present chapter was to provide a brief overview of the behavioral consequences of developmental alcohol exposure in rodent models with a particular emphasis on the study of learning and memory. Our summary initially focused on investigations involving spatial learning and memory, which has formed one of the most dominant behavioral assessments, but we also provided an overview of less-studied visual-spatial, object learning and memory, and spontaneous behaviors related to information gathering. Our overview presented clear evidence that developmental alcohol exposure leads to deficits in spatial learning and memory in rodent models, with more subtle and specific impairments identified in rodent models of low and moderate prenatal alcohol exposure. Deficits in visual-spatial, object learning and memory, and spontaneous spatial behaviors have been reported after developmental alcohol exposure, but these avenues of research are not well represented in the rodent model literature. Future work is needed, particularly on the impact of developmental alcohol on object learning and memory tasks, where variability in study design may have the greatest influence. Our summary of the literature using rodent modes suggests that the level of exposure and test features determine the learning and memory phenotype after developmental alcohol exposure.

We also discussed evidence suggesting a prominent role for the hippocampal formation in underlying learning and memory deficits in rodent models of developmental alcohol exposure. Our emphasis was largely on the impact on place cell activity which has been argued to serve a critical role in generating "maps" of an animal's experience during learning and memory. We presented evidence indicating that these hippocampal neural signals are affected by moderate prenatal alcohol exposure, but it is clear that more study is needed. Specifically, in relation to the impact of alcohol on the strengthening of memories within hippocampal networks during offline states such as sleep. There is also a need to understand whether these changes in neural activity can provide a general explanation of learning and memory impairment after developmental alcohol exposure. Although the hippocampus served as the primary emphasis in this chapter, it is clear that more work is needed to investigate alterations in an extended network of brain regions that include the entorhinal cortex, parahippocampal regions, and limbic thalamus. These extended hippocampal regions are significantly affected by developmental alcohol exposure and contain neural signals that have been associated with memory functions. Thus, a broader examination of cortical-thalamic-hippocampal networks in relationship to developmental alcohol exposure will be critical in creating a complete picture of the neural changes related to learning and memory impairments.

Acknowledgments The authors are supported by the National Institute on Alcohol Abuse and Alcoholism of the National Institutes of Health under award numbers R01 AA029700, T32 AA014127 and F31 AA030711.

References

Acosta G, Rico KT, Madden JT, LaCour A, Wang E, Sanchez LM, Davies S, Maestas-Olguin C, Cox KB, Reyna NC, Hogeveen J, Savage DD, Pentkowski NS, Clark BJ (2024) The effects of moderate prenatal alcohol exposure on performance in hippocampalsensitive spatial memory and anxiety tasks by adult male and female rat offspring. Alcohol 121:75–86. https://doi.org/10.1016/j.alcohol.2024.08.002. Epub 2024 Aug 8. PMID: 39122134; PMCID: PMC11637952

Berkowitz LE, Harvey RE, Drake E, Thompson SM, Clark BJ (2018) Progressive impairment of directional and spatially precise trajectories by TgF344-Alzheimer's disease rats in the Morris Water task. Sci Rep 8:16153

Berman RF, Hannigan JH (2000) Effects of prenatal alcohol exposure on the hippocampus: spatial behavior, electrophysiology, and neuroanatomy. Hippocampus 10(1):94–110. https://doi.org/10.1002/(SICI)1098-1063(2000)10:1<94::AID-HIPO11>3.0.CO;2-T. PMID: 10706221

Bird CW, Mayfield SS, Lopez KM, Dunn BR, Feng A, Roberts BT, Almeida RN, Chavez GJ, Valenzuela CF (2023) Binge-like ethanol exposure during the brain growth spurt disrupts the function of retrosplenial cortex-projecting anterior thalamic neurons in adolescent mice. Neuropharmacology 241:109738

Bird CW, Taylor DH, Pinkowski NJ, Chavez GJ, Valenzuela CF (2018) Long-term reductions in the population of GABAergic interneurons in the mouse hippocampus following developmental ethanol exposure. Neuroscience 383:60–73

Brady ML, Allan AM, Caldwell KK (2012) A limited access mouse model of prenatal alcohol exposure that produces long-lasting deficits in hippocampal-dependent learning and memory. A limited access mouse model of pae. Alcohol Clin Exp Res 36:457–466

Brady ML, Diaz MR, Iuso A, Everett JC, Valenzuela CF, Caldwell KK (2013) Moderate prenatal alcohol exposure reduces plasticity and alters NMDA receptor subunit composition in the dentate gyrus. J Neurosci 33(3):1062–1067

Burke MW, Slimani H, Ptito M, Ervin FR, Palmour RM (2022) Dose-related reduction in hippocampal neuronal populations in fetal alcohol exposed Vervet monkeys. Brain Sci 12:1117

Burke SN, Gaynor LS, Barnes CA, Bauer RM, Bizon JL, Roberson ED, Ryan L (2018) Shared functions of Perirhinal and Parahippocampal cortices: implications for cognitive aging. Trends Neurosci 41(6):349–359

Bussey TJ, Holmes A, Lyon L, Mar AC, McAllister KA, Nithianantharajah J, Oomen CA, Saksida LM (2012) New translational assays for preclinical modelling of cognition in schizophrenia: the touchscreen testing method for mice and rats. Neuropharmacology 62:1191–1203

Buzsáki G (2005) Theta rhythm of navigation: link between path integration and landmark navigation, episodic and semantic memory. Hippocampus 15:827–840

Buzsáki G (2015) Hippocampal sharp wave-ripple: a cognitive biomarker for episodic memory and planning. Hippocampus 25:1073–1188

Cacucci F, Yi M, Wills TJ, Chapman P, O'Keefe J (2008) Place cell firing correlates with memory deficits and amyloid plaque burden in Tg2576 Alzheimer mouse model. Proc Natl Acad Sci USA 105:7863–7868

Cantacorps L, Alfonso-Loeches S, Moscoso-Castro M, Cuitavi J, Gracia-Rubio I, López-Arnau R, Escubedo E, Guerri C, Valverde O (2017) Maternal alcohol binge drinking induces persistent neuroinflammation associated with myelin damage and behavioural dysfunctions in offspring mice. Neuropharmacology 123:368–384

Chadwick A, van Rossum MC, Nolan MF (2015) Independent theta phase coding accounts for CA1 population sequences and enables flexible remapping. elife 2(4):e03542

Choi IY, Allan AM, Cunningham LA (2005) Moderate fetal alcohol exposure impairs the neurogenic response to an enriched environment in adult mice. Alcohol Clin Exp Res 29:2053–2062

Christie BR, Swann SE, Fox CJ, Froc D, Lieblich SE, Redila V, Webber A (2005) Voluntary exercise rescues deficits in spatial memory and long-term potentiation in prenatal ethanol-exposed male rats. Eur J Neurosci 21:1719–1726

Clark BJ, Hines DJ, Hamilton DA, Whishaw IQ (2005) Movements of exploration intact in rats with hippocampal lesions. Behav Brain Res 163(1):91–99

Clark BJ, Hong NS, Bettenson DJ, Woolford J, Horwood L, McDonald RJ (2015) Maintained directional navigation across environments in the Morris water task is dependent on vestibular cues. J Exp Psychol Anim Learn Cogn 41:301–308

Clarren SK, Astley SJ, Gunderson VM, Spellman D (1992) Cognitive and behavioral deficits in nonhuman primates associated with very early embryonic binge exposures to ethanol. J Pediatr 121:789–796

Colgin LL (2016) Rhythms of the hippocampal network. Nat Rev Neurosci 17:239–249

Cooper BG, Mizumori SJ (2001) Temporary inactivation of the retrosplenial cortex causes a transient reorganization of spatial coding in the hippocampus. J Neurosci 21:3986–4001

Cullen CL, Burne THJ, Lavidis NA, Moritz KM (2014) Low dose prenatal alcohol exposure does not impair spatial learning and memory in two tests in adult and aged rats. PLoS One 9:e101482

Diba K, Buzsaki G (2007) Forward and reverse hippocampal place-cell sequences during ripples. Nat Neurosci 10:1241–1242

Dodge NC, Thomas KGF, Meintjes EM, Molteno CD, Jacobson JL, Jacobson SW (2019) Spatial navigation in children and Young adults with fetal alcohol Spectrum disorders. Alcohol Clin Exp Res 43:2536–2546

Dodge NC, Thomas KGF, Meintjes EM, Molteno CD, Jacobson JL, Jacobson SW (2020) Reduced hippocampal volumes partially mediate effects of prenatal alcohol exposure on spatial navigation on a virtual Water maze task in children. Alcohol Clin Exp Res 44:844–855

Donaldson T, Jennings KT, Cherep LA, McNeela AM, Depreux FF, Jodelka FM et al (2018) Antisense oligonucleotide therapy rescues disruptions in organization of exploratory movements associated with usher syndrome type 1C in mice. Behav Brain Res 338:76–87

Dragoi G, Buzsaki G (2006) Temporal encoding of place sequences by hippocampal cell assemblies. Neuron 50:145–157

Driscoll CD, Streissguth AP, Riley EP (1990) Prenatal alcohol exposure: comparability of effects in humans and animal models. Neurotoxicol Teratol 12:231–237

Dudchenko PA, Wallace D (2018) Neuroethology of spatial cognition. Curr Biol 28(17):R988–R992
Eichenbaum H (2012) *The cognitive neuroscience of memory: an introduction* (2nd ed). Oxford University Press
Eichenbaum H, Cohen NJ (2014) Can we reconcile the declarative memory and spatial navigation views on hippocampal function? Neuron 83(4):764–770
Eilam D, Golani I (1989) Home base behavior of rats (Rattus norvegicus) exploring a novel environment. Behav Brain Res 34(3):199–211
Ego-Stengel V, Wilson MA (2009) Disruption of ripple-associated hippocampal activity during rest impairs spatial learning in the rat. Hippocampus 20:1–10
Ernst AM, Gimbel BA, de Water E, Eckerle JK, Radke JP, Georgieff MK, Wozniak JR (2022) Prenatal and postnatal choline supplementation in fetal alcohol Spectrum disorder. Nutrients 14:688
Fontaine CJ, Patten AR, Sickmann HM, Helfer JL, Christie BR (2016) Effects of pre-natal alcohol exposure on hippocampal synaptic plasticity: sex, age and methodological considerations. Neurosci Biobehav Rev 64:12–34
Foster DJ, Wilson MA (2007) Hippocampal theta sequences. Hippocampus 17:1093–1099
Gallistel CR (1990) The organization of learning. MIT Press, Cambridge, MA, US
Gilbert PE, Kesner RP (2003) Localization of function within the dorsal hippocampus: the role of the CA3 subregion in paired-associate learning. Behav Neurosci 117(6):1385–1394. https://doi.org/10.1037/0735-7044.117.6.1385. PMID: 14674856
Girard TA, Wainwright PE (2002) Testing the spatial- versus object-learning distinction: water-maze performance of male rats exposed to ethanol during the brain growth spurt. Behav Brain Res 134:493–503
Girardeau G, Benchenane K, Wiener SI, Buzsáki G, Zugaro MB (2009) Selective suppression of hippocampal ripples impairs spatial memory. Nat Neurosci 12:1222–1223
Goncalves-Garcia M, Hamilton DA (2023) Unraveling the complex relationship between prenatal alcohol exposure, hippocampal LTP, and learning and memory. Front Mol Neurosci 16:1326089
Goncalves-Garcia M, Davies S, Savage DD, Hamilton DA (2024a) The histamine H_3 receptor inverse agonist SAR-152954 reverses deficits in long-term potentiation associated with moderate prenatal alcohol exposure. Alcohol 118:45–55
Goncalves-Garcia M, Barto D, Reyna N, Clark BJ, Hamilton DA (2024b) The prominence of action sequences and behavioral similarity in the Morris water task. Perspect Behav Sci. In Press 47:449
Goodlett CR, Kelly SJ, West JR (1987) Early postnatal alcohol exposure that produces high blood alcohol levels impairs development of spatial navigation learning. Psychobiology 15:64–74
Gorny JH, Gorny B, Wallace DG, Whishaw IQ (2002) Fimbria–fornix lesions disrupt the dead reckoning (homing) component of exploratory behavior in mice. Learn Mem 9:387–394
Gursky ZH, Klintsova AY (2021) Changes in representation of thalamic projection neurons within prefrontal-thalamic-hippocampal circuitry in a rat model of third trimester binge drinking. Brain Sci 11:323
Gursky ZH, Savage LM, Klintsova AY (2021) Executive functioning-specific behavioral impairments in a rat model of human third trimester binge drinking implicate prefrontal-thalamo-hippocampal circuitry in fetal alcohol Spectrum disorders. Behav Brain Res 405:113208
Hannigan JH, O'Leary-Moore SK, Berman RF (2007) Postnatal environmental or experiential amelioration of neurobehavioral effects of perinatal alcohol exposure in rats. Neurosci Biobehav Rev 31:202–211
Hamilton DA, Kodituwakku P, Sutherland RJ, Savage DD (2003) Children with fetal alcohol syndrome are impaired at place learning but not cued-navigation in a virtual Morris water task. Behav Brain Res 143:85–94
Hamilton DA, Akers KG, Johnson TE, Rice JP, Candelaria FT, Sutherland RJ, Weisend MP, Redhead ES (2008) The relative influence of place and direction in the Morris water task. J Exp Psychol Anim Behav Process 34:31–53
Hamilton DA, Barto D, Rodriguez CI, Magcalas CM, Fink BC, Rice JP, Bird CW, Davies S, Savage DD (2014) Effects of moderate prenatal ethanol exposure and age on social behavior, spatial response perseveration errors and motor behavior. Behav Brain Res 269:44–54
Harvey RE, Berkowitz LE, Hamilton DA, Clark BJ (2019) The effects of developmental alcohol exposure on the neurobiology of spatial processing. Neurosci Biobehav Rev 107:775–794
Harvey RE, Berkowitz LE, Savage DD, Hamilton DA, Clark BJ (2020) Altered hippocampal place cell representation and theta rhythmicity following moderate prenatal alcohol exposure. Curr Biol 30:3556–3569. e3555
Hines DJ, Whishaw IQ (2005) Home bases formed to visual cues but not to self-movement (dead reckoning) cues in exploring hippocampectomized rats. Eur J Neurosci 22(9):2363–2375
Iqbal U, Dringenberg HC, Brien JF, Reynolds JN (2004) Chronic prenatal ethanol exposure alters hippocampal GABAA receptors and impairs spatial learning in the Guinea pig. Behav Brain Res 150:117–125
Jablonski SA, Schreiber WB, Westbrook SR, Brennan LE, Stanton ME (2013) Determinants of novel object and location recognition during development. Behav Brain Res 256:140–150
Johnson SA, Turner SM, Lubke KN, Cooper TL, Fertal KE, Bizon JL, Maurer AP, Burke SN (2019) Experience-dependent effects of Muscimol-induced hippocampal excitation on mnemonic discrimination. Front Syst Neurosci 12. https://doi.org/10.3389/fnsys.2018.00072
Johnson SA, Turner SM, Santacroce LA, Carty KN, Shafiq L, Bizon JL, Maurer AP, Burke SN (2017) Rodent age-related impairments in discriminating perceptually similar objects parallel those observed in humans. Hippocampus 27(7):759–776

Josey M, Brigman JL (2015) Loss of hippocampal function impairs pattern separation on a mouse touchscreen operant paradigm. Neurobiol Learn Mem 125:85–92

Kenton JA, Castillo R, Holmes A, Brigman JL (2018) Cortico-hippocampal GluN2B is essential for efficient visual-spatial discrimination learning in a touchscreen paradigm. Neurobiol Learn Mem 156:60–67

Kenton JA, Castillo VK, Kehrer PE, Brigman JL (2020) Moderate prenatal alcohol exposure impairs visual-spatial discrimination in a sex-specific manner: effects of testing order and difficulty on learning performance. Alcohol Clin Exp Res 44(10):2008–2018

Kim CK, Kalynchuk LE, Kornecook TJ, Mumby DG, Dadgar NA, Pinel JPJ, Weinberg J (1997) Object-recognition and spatial learning and memory in rats prenatally exposed to ethanol. Behav Neurosci 111(5):985. Retrieved from http://files/239/ContentServer.pdf

Lehmann H, Clark BJ, Whishaw IQ (2007) Similar development of cued and learned home bases in control and hippocampal-damaged rats in an open field exploratory task. Hippocampus 17:370–380

Lee MH, Rabe A (1999) Infantile handling eliminates reversal learning deficit in rats prenatally exposed to alcohol. Alcohol 18:49–53

Lee I, Solivan F (2010) Dentate gyrus is necessary for disambiguating similar object- place representations. Learn Mem 17:252–258

Lee AK, Wilson MA (2002) Memory of sequential experience in the hippocampus during slow wave sleep. Neuron 36:1183–1194

Lester AW, Moffat SD, Wiener JM, Barnes CA, Wolbers T (2017) The aging navigational system. Neuron 95:1019–1035

Licheri V, Jacquez BJ, Castillo VK, Sainz DB, Valenzuela CF, Brigman JL (2023) Long-term effects of low prenatal alcohol exposure on GABAergic interneurons of the murine posterior parietal cortex. Alcohol Clin Exp Res 12:2248–2261

Maaswinkel H, Whishaw IQ (1999) Homing with locale, taxon, and dead reckoning strategies by foraging rats: sensory hierarchy in spatial navigation. Behav Brain Res 99(2):143–152. https://doi.org/10.1016/S0166-4328(98)00100-4

MacIlvane NM, Pochiro JM, Hurwitz NR, Goodfellow MJ, Lindquist DH (2016) Recognition memory is selectively impaired in adult rats exposed to binge-like doses of ethanol during early postnatal life. Alcohol 57:55–63

Madden JT, Thompson SM, Magcalas CM, Wagner JL, Hamilton DA, Savage DD et al (2020) Moderate prenatal alcohol exposure reduces parvalbumin expressing GABAergic interneurons in the dorsal hippocampus of adult male and female rat offspring. Neurosci Lett 718:134700. https://doi.org/10.1016/j.neulet.2019.134700

Marquardt K, Brigman JL (2016) The impact of prenatal alcohol exposure on social, cognitive and affective behavioral domains: insights from rodent models. Alcohol 51:1–15

Marquardt K, Sigdel R, Caldwell K, Brigman JL (2014) Prenatal ethanol exposure impairs executive function in mice into adulthood. Alcohol Clin Exp Res 38:2962–2968

Mattson SN, Roesch SC, Fagerlund Å, Autti-Rämö I, Jones KL, May PA, Adnams CM, Konovalova V, Riley EP (2010) Towards a neurobehavioral profile of fetal alcohol Spectrum disorders. Alcohol Clin Exp Res 34:1640–1650

Mattson SN, Riley EP, Gramling L, Delis DC, Jones KL (1998) Neuropsychological comparison of alcohol-exposed children with or without physical features of fetal alcohol syndrome. Neuropsychology 12:146–153

Mattson SN, Bernes GA, Doyle LR (2019) Fetal alcohol Spectrum disorders: a review of the neurobehavioral deficits associated with prenatal alcohol exposure. Alcohol Clin Exp Res 6:1046–1062

Matthews DB, Simson PE (1998) Prenatal exposure to ethanol disrupts spatial memory: effect of the training–testing delay period. Physiol Behav 64:63–67

Maurer AP, Johnson SA, Hernandez AR, Reasor J, Cossio DM, Fertal KE, Burke SN (2017) Age-related changes in lateral entorhinal and CA3 neuron allocation predict poor performance on object discrimination. Front Syst Neurosci 11. https://doi.org/10.3389/fnsys.2017.00049

Mittelstaedt ML, Mittelstaedt H (1980) Homing by path integration in a mammal. *Die Naturwissenschaften*. Springer 67(11):566–567

Morris RGM (1981) Spatial localization does not require the presence of local cues. Learn Motiv 12(2):239–260

Moser EI, Moser MB, McNaughton BL (2017) Spatial representation in the hippocampal formation: a history. Nat Neurosci. Nature Publishing Group 20(11):1448–1464. https://doi.org/10.1038/nn.4653

Nadel L (1991) The hippocampus and space revisited. Hippocampus 3:221–229

O'Keefe J, Nadel L (1978) The hippocampus as a cognitive map. Clarendon Press, Oxford

Olguin SL, Thompson SM, Young JW, Brigman JL (2020) Moderate prenatal alcohol exposure impairs cognitive control, but not attention, on a rodent touchscreen continuous performance task. Genes Brain Behav e12652

Olson HC, Feldman JJ, Streissguth AP, Sampson PD, Bookstein FL (1998) Neuropsychological deficits in adolescents with fetal alcohol syndrome: clinical findings. Alcohol Clin Exp Res 22:1998–2012

Olton DS, Becker JT, Handelmann GE (1979) Hippocampus, space. Memory 2(3):313–365

Oomen CA, Hvoslef-Eide M, Heath CJ, Mar AC, Horner AE, Bussey TJ, Saksida LM (2013) The touchscreen operant platform for testing working memory and pattern separation in rats and mice. Nat Protoc 8:2006–2021

Ormond J, McNaughton BL (2015) Place field expansion after focal MEC inactivations is consistent with loss of Fourier components and path integrator gain reduction. Proc Natl Acad Sci USA 112:4116–4121

Osterlund Oltmanns JR, Schaeffer EA, Goncalves Garcia M, Donaldson TN, Acosta G, Sanchez LM, Davies S, Savage DD, Wallace DG, Clark BJ (2022) Sexually

dimorphic organization of open field behavior following moderate prenatal alcohol exposure. Alcohol Clin Exp Res 46:861–875
Patten AR, Fontaine CJ, Christie BR (2014) A comparison of the different animal models of fetal alcohol spectrum disorders and their use in studying complex behaviors. Front Pediatr 2:93
Patten AR, Sawchuk S, Wortman RC, Brocardo PS, Gil-Mohapel J, Christie BR (2016) Prenatal ethanol exposure impairs temporal ordering behaviours in young adult rats. Behav Brain Res 299:81–89
Popova S, Charness ME, Burd L, Crawford A, Hoyme HE, Mukherjee RAS, Riley EP, Elliott EJ (2023) Fetal alcohol spectrum disorders. Nature Reviews Disease Primers, 9(1):11
Popovic M, Caballerobleda M, Guerri C (2006) Adult rat's offspring of alcoholic mothers are impaired on spatial learning and object recognition in the can test. Behav Brain Res 174(1):101–111
Poulter S, Hartley T, Lever C (2018) The neurobiology of mammalian navigation. Curr Biol 17:R1023–R1042
Redila VA, Christie BR (2005) Exercise-induced changes in dendritic structure and complexity in the adult hippocampal dentate gyrus. Neuroscience 137:1299–1307
Reyes E, Wolfe J, Savage DD (1989) The effects of prenatal alcohol exposure on radial arm maze performance in adult rats. Physiol Behav 46:45–48
Riley EP, Lochry EA, Shapiro NR, Baldwin J (1979) Response perseveration in rats exposed to alcohol prenatally. Pharmacol Biochem Behav 10:255–259
Robbe D, Buzsáki G (2009) Alteration of theta timescale dynamics of hippocampal place cells by a cannabinoid is associated with memory impairment. J Neurosci 29:12597–12605
Rodriguez CI, Magcalas CM, Barto D, Fink BC, Rice JP, Bird CW, Davies S, Pentkowski NS, Savage DD, Hamilton DA (2016) Effects of sex and housing on social, spatial, and motor behavior in adult rats exposed to moderate levels of alcohol during prenatal development. Behav Brain Res 313:233–243
Roediger DJ, Krueger AM, de Water E, Mueller BA, Boys CA, Hendrickson TJ, Schumacher MJ, Mattson SN, Jones KL, Lim KO, Wozniak JR (2020) Hippocampal subfield abnormalities and memory functioning in children with fetal alcohol Spectrum disorders. Neurotoxicol Teratol 83:106944
Röskam S, Koch M (2009) Effects of neonatal and peripubertal ethanol treatment on various aspects of adult rat behavior and brain anatomy. Int J Dev Neurosci 3:249–256
Ryan SH, Williams JK, Thomas JD (2008) Choline supplementation attenuates learning deficits associated with neonatal alcohol exposure in the rat: effects of varying the timing of choline administration. Brain Res 1237:91–100
Sanchez LM, Acosta G, Cushing SD, Johnson SA, Turner SM, Davies S, Savage DD, Burke SN, Clark BJ (2025) The effects of moderate prenatal alcohol exposure on performance in object and spatial discrimination tasks by adult male rats. Behav Brain Res 478:115324. https://doi.org/10.1016/j.bbr.2024.115324. Epub 2024 Nov 7. PMID: 39521144; PMCID: PMC11606775
Sanchez LM, Goss J, Wagner J, Davies S, Savage DD, Hamilton DA, Clark BJ (2019) Moderate prenatal alcohol exposure impairs performance by adult male rats in an object-place paired-associate task. Behav Brain Res 360:228–234
Savage DD, Becher M, Torre AJ, Sutherland RJ (2002) Dose-dependent effects of prenatal ethanol exposure on synaptic plasticity and learning in mature offspring. Alcohol Clin Exp Res 26:1752–1758
Savage DD, Rosenberg MJ, Wolff CR, Akers KG, El-Emawy A, Staples MC, Varaschin RK, Wright CA, Seidel JL, Caldwell KK et al (2010) Effects of a novel cognition-enhancing agent on fetal ethanol-induced learning deficits. Alcohol Clin Exp Res 34:1793–1802
Save E, Paz-Villagran V, Alexinsky T, Poucet B (2005) Functional interaction between the associative parietal cortex and hippocampal place cell firing in the rat. Eur J Neurosci 2:522–530
Schaeffer EA, LaCour A, Donaldson TN, Linsenbardt DN, Davies S, Savage DD, Wallace DG, Clark BJ (2025) Organization of spontaneous spatial behaviors under dark conditions is unaffected in adult male and female long-Evans rats after moderate prenatal alcohol exposure. Behav Neurosci 139(1):44–52. https://doi.org/10.1037/bne0000589. Epub 2024 Apr 18. PMID: 38635177
Schlesiger MI et al (2015) The medial entorhinal cortex is necessary for temporal organization of hippocampal neuronal activity. Nat Neurosci 18:1123–1132
Schneider RD, Thomas JD (2016) Adolescent choline supplementation attenuates working memory deficits in rats exposed to alcohol during the third trimester equivalent. Alcohol Clin Exp Res 40:897–905
Stackman RW, Lora JC, Williams SB (2012) Directional responding of C57BL/6J mice in the Morris water maze is influenced by visual and vestibular cues and is dependent on the anterior thalamic nuclei. J Neurosci 32:10211–10225
Summers BL, Henry CM, Rofe AM, Coyle (2008) Dietary zinc supplementation during pregnancy prevents spatial and object recognition memory impairments caused by early prenatal ethanol exposure. Behav Brain Res 186(2):230–8. https://doi.org/10.1016/j.bbr.2007.08.011. Epub 2007 Aug 19. PMID: 17884190
Sutherland RJ, McDonald RJ, Savage DD (1997) Prenatal exposure to moderate levels of ethanol can have long-lasting effects on hippocampal synaptic plasticity in adult offspring. Hippocampus 7:232–238
Sutherland RJ, McDonald RJ, Savage DD (2000) Prenatal exposure to moderate levels of ethanol can have long-lasting effects on learning and memory in adult offspring. Psychobiology 28:532–539
Sutherland RJ, Weisend MP, Mumby D, Astur RS, Hanlon FM, Koerner A, Thomas MJ, Wu Y, Moses SN, Cole C, Hamilton DA, Hoesing JM (2001) Retrograde amnesia after hippocampal damage: recent vs. remote memories in two tasks. Hippocampus 11:27–42

Terasaki LS, Schwarz JM (2016) Effects of moderate prenatal alcohol exposure during early gestation in rats on inflammation across the maternal-fetal-immune Interface and later-life immune function in the offspring. J Neuroimmune Pharmacol 4:680–692

Terasaki LS, Schwarz JM (2017) Impact of prenatal and subsequent adult alcohol exposure on pro-inflammatory cytokine expression in brain regions necessary for simple recognition memory. Brain Sci 7:125

Thomas JD, La Fiette MH, Quinn VR, Riley EP (2000) Neonatal choline supplementation ameliorates the effects of prenatal alcohol exposure on a discrimination learning task in rats. Neurotoxicol Teratol 22:703–711

Thomas J (2004) Perinatal choline supplementation attenuates behavioral alterations associated with neonatal alcohol exposure in rats. Neurotoxicol Teratol 26:35–45

Thomas JD, Tran TD (2012) Choline supplementation mitigates trace, but not delay, eyeblink conditioning deficits in rats exposed to alcohol during development. Hippocampus 22:619–630

Thomas JD, Wasserman EA, West JR, Goodlett CR (1996) Behavioral deficits induced by bingelike exposure to alcohol in neonatal rats: importance of developmental timing and number of episodes. Dev Psychobiol 29:433–452

Thomas JD, Weinert SP, Sharif S, Riley EP (1997) MK-801 administration during ethanol withdrawal in neonatal rat pups attenuates ethanol-induced behavioral deficits. Alcohol Clin Exp Res 21:8

Thomas JD, Biane JS, O'Bryan KA, O'Neill TM, Dominguez HD (2007) Choline supplementation following third-trimester-equivalent alcohol exposure attenuates behavioral alterations in rats. Behav Neurosci 121:120–130

Thomas JD, Sather TM, Whinery LA (2008) Voluntary exercise influences behavioral development in rats exposed to alcohol during the neonatal brain growth spurt. Behav Neurosci 122:1264–1273

Thompson SM, Berkowitz LE, Clark BJ (2018) Behavioral and neural subsystems of rodent exploration. Learn Motiv 61:3–15

Uecker A, Nadel L (1996) Spatial locations gone awry: object and spatial memory deficits in children with fetal alcohol syndrome. Neuropsychologia 34:209–223

Valenzuela CF, Morton RA, Diaz MR, Topper L (2012) Does moderate drinking harm the fetal brain? Insights from animal models. Trends in Neurosciences 35(5) 284e292

Varaschin RK, Akers KG, Rosenberg MJ, Hamilton DA, Savage DD (2010) Effects of the cognition-enhancing agent ABT-239 on fetal ethanol-induced deficits in dentate gyrus synaptic plasticity. J Pharmacol Exp Ther 334:191–198

Waddell J, Mooney SM (2017) Choline and working memory training improve cognitive deficits caused by prenatal exposure to ethanol. Nutrients 9:1080

Wagner AF, Hunt PS (2006) Impaired trace fear conditioning following neonatal ethanol: reversal by choline. Behav Neurosci 120:482–487

Wagner JL, Zhou FC, Goodlett CR (2014) Effects of one- and three-day binge alcohol exposure in neonatal C57BL/6 mice on spatial learning and memory in adolescence and adulthood. Alcohol 48:99–111

Wallace DG, Whishaw IQ (2003) NMDA lesions of Ammon's horn and the dentate gyrus disrupt the direct and temporally paced homing displayed by rats exploring a novel environment: evidence for a role of the hippocampus in dead reckoning. Eur J Neurosci 18:513–523

Wallace DG et al (2002) Vestibular information is required for dead reckoning in the rat. J Neurosci 22(22):10009–10017. https://doi.org/10.1523/jneurosci.22-22-10009.2002

West JR (1987) Fetal alcohol-induced brain damage and the problem of determining temporal vulnerability: a review. Alcohol Drug Res 7:423–441

Whishaw IQ, Hines DJ, Wallace DG (2001) Dead reckoning (path integration) requires the hippocampal formation: evidence from spontaneous exploration and spatial learning tasks in light (allothetic) and dark (idiothetic) tests. Behav Brain Res 127(1–2):49–69. https://doi.org/10.1016/S0166-4328(01)00359-X

Whishaw IQ, Cassal JC, Majchrzak M (1994) Short-stops in rats with fimbria-fornix lesions: evidence for change in the mobility gradient. Hippocampus 4:577–582

Wilson M, McNaughton B (1994) Reactivation of hippocampal ensemble memories during sleep. Science 265:676–679

Wilson DA, Peterson J, Basavaraj BS, Saito M (2011) Local and regional network function in behaviorally relevant cortical circuits of adult mice following postnatal alcohol exposure. Alcohol Clin Exp Res 35:1974–1984

Woods KJ, Thomas KGF, Molteno CD, Jacobson JL, Jacobson SW, Meintjes EM (2018) Prenatal alcohol exposure affects brain function during place learning in a virtual environment differently in boys and girls. Brain Behav 8:e01103

Wozniak DF, Hartman RE, Boyle MP, Vogt SK, Brooks AR, Tenkova T, Young C, Olney JW, Muglia LJ (2004) Apoptotic neurodegeneration induced by ethanol in neonatal mice is associated with profound learning/memory deficits in juveniles followed by progressive functional recovery in adults. Neurobiol Dis 17:403–414

Yassa MA, Stark CEL (2011) Pattern separation in the hippocampus. Trends Neurosci 34:515–525

Zimmerberg B, Mattson S, Riley EP (1989) Impaired alternation test performance in adult rats following prenatal alcohol exposure. Pharmacol Biochem Behav 32:293–299

Zimmerberg B, Sukel HL, Stekler JD (1991) Spatial learning of adult rats with fetal alcohol exposure: deficits are sex-dependent. Behav Brain Res 42:49–56

Impact of Developmental Alcohol Exposure on the Thalamus

4

C. Fernando Valenzuela, Natalie M. Reid, Benjamin B. Blanco, Victoria L. Carlson, and Alynna B. Do

Abstract

This chapter comprehensively explores the impact of prenatal alcohol (ethanol) exposure (PAE) on the thalamus, integrating findings from animal models and human studies spanning various developmental stages. Animal model investigations, encompassing first and second trimester-equivalent exposures and the critical third trimester, where the brain growth spurt starts, reveal specific alterations in thalamic structures and circuits, emphasizing the specificity of damage to corticothalamic loops. The ventrobasal thalamic nucleus exhibits a unique response to PAE, involving intricate interactions with postnatal neurogenesis and neurotrophin responsiveness. Third trimester-equivalent exposure consistently induces apoptotic neurodegeneration in various thalamic nuclei, highlighting the heightened susceptibility of the visual thalamus, particularly the lateral geniculate nucleus, during critical developmental periods. The nucleus reuniens, vital for cognitive processes, was shown to be significantly affected by alcohol exposure during this period. Investigations into the trigeminal/somatosensory system activity revealed disruptions in glucose utilization and increased neuronal activity in the thalamus. Research on binge-like alcohol exposure during the brain growth spurt demonstrates lasting modifications in action-potential properties and synaptic currents in thalamic neurons projecting to the retrosplenial cortex. Human studies, employing advanced techniques like super-resolution fetal MRI and functional MRI, underscore the PAE-induced structural and functional consequences in the thalamus and its connections, spanning from fetal development to adulthood. The complex effects of PAE on thalamic structure and function vary across developmental stages, emphasizing the importance of considering factors such as age and concurrent exposures. The development of higher-resolution imaging tools is essential for assessing the impact of PAE on the structure and function of individual thalamic nuclei in humans.

Keywords

Fetal · Embryonic · Prenatal · Pregnancy · Alcohol · Thalamus · Animal model · Human · Imaging · Structure · Morphology · Function · Development

C. F. Valenzuela (✉) · N. M. Reid · B. B. Blanco
V. L. Carlson · A. B. Do
Department of Neurosciences and New Mexico Alcohol Research Center, School of Medicine, University of New Mexico Health Sciences Center, Albuquerque, NM, USA
e-mail: fvalenzuela@salud.unm.edu

C. F. Valenzuela et al. (eds.), *Effects of Alcohol on the Brain across the Lifespan*, Advances in Experimental Medicine and Biology 1473, https://doi.org/10.1007/978-3-031-81908-7_4

This chapter focuses on the impact of prenatal alcohol (ethanol) exposure (PAE) on the thalamus, which orchestrates complex interactions within neural networks, shaping fundamental aspects of perception, attention, and cognition. Understanding PAE's effects on thalamic structure and function is paramount in elucidating the pathophysiology of the spectrum of disorders caused by alcohol consumption during pregnancy, known as fetal alcohol spectrum disorder (FASD). This chapter aims to navigate the intricate landscape of thalamic development and dysfunction in the context of PAE, synthesizing insights garnered from animal models and human studies alike to offer a comprehensive understanding of PAE's effects on this critical component of the brain's neural network.

4.1 An Overview of Fetal Alcohol Spectrum Disorder

Fetal alcohol spectrum disorder (FASD) encompasses a wide range of neurobehavioral impairments with or without craniofacial abnormalities that occur after PAE. FASD serves as an overarching designation for a cluster of disorders, which includes fetal alcohol syndrome (FAS), partial fetal alcohol syndrome (pFAS), alcohol-related neurodevelopmental disorder, and alcohol-related birth defects (Popova et al. 2023). The harmful effects of PAE depend on factors such as amounts/dosages, duration of exposure, time of exposure, and other considerations (Popova et al. 2023). FASD can cause lifelong neuropsychological and behavioral impairments; common neuropsychological impairments seen in FASD include executive and cognitive functional deficits, impaired reasoning and comprehension, which can lead to difficulties in school, and predispose individuals to mental illness (Popova et al. 2023). On the other hand, individuals with FASD exhibit many notable strengths in social relationships, prosocial behaviors, specific skills (tactile strengths, artistic abilities, athleticism), and resilience, with the latter being particularly important considering the high incidence of trauma or maltreatment experienced within this population (Kautz-Turnbull et al. 2022).

Fetal alcohol syndrome (FAS) is characterized by distinct physical attributes, such as growth restrictions, and facial dysmorphology, which includes features like short palpebral fissure, a thin vermillion border, and an indistinct philtrum. It is further characterized by growth restrictions and the presence of global cognitive and intellectual deficits. Partial FAS diagnosis may be assigned to individuals exhibiting some diagnostic criteria of FAS without presenting all features, such as growth impairment and decreased head circumference. Alcohol-related neurodevelopmental disorder lacks physical anomalies but spans a spectrum of cognitive impairment, ranging from mild to severe. Finally, alcohol-related birth defects are diagnosed in cases with a history of PAE and include systemic malformations (e.g., musculoskeletal or cardiovascular).

Globally, FASD exhibits a prevalence of 0.77%, while in Canada, USA, and Europe, this figure ranges from 2 to 5%. Several factors contribute to the occurrence of FASD, including unknown pregnancy status, lack of prenatal care, maternal alcohol use disorder, psychiatric disorders, history of abuse, having had previous children with FASD, and poverty (Popova et al. 2023). FASDs are generally underdiagnosed, a multifactorial issue stemming from societal stigma surrounding maternal alcohol use, leading to reduced reporting, overlapping diagnoses with conditions like attention-deficit hyperactivity disorder, reliance on facial dysmorphology typically associated with the most extreme side of the spectrum (i.e., FAS), and the overall complexity of diagnosis due to a broad spectrum of impairments and presentations (Wozniak et al. 2019).

Alcohol can traverse the placenta, impacting fetal development through diverse mechanisms, including oxidative injury, apoptotic neurodegeneration, and disruptions of synaptogenesis, neuronal migration, neurotransmission, and synaptic plasticity. Individuals with FASD exhibit abnormal brain structures and functional connectivity in neuroimaging studies. The scope of neuropathological findings, however, varies based on factors such as the doses and timing of PAE, genetics, and concurrent substance use. Genetic factors contribute to neuropathological findings, exemplified by alleles for alcohol dehydrogenase that accelerate alcohol metabolism, thereby diminishing its effects on fetal development (Wozniak et al. 2019). PAE exerts significant deleterious effects on numerous brain regions, with the thalamus displaying particular sensitivity to its actions.

4.2 The Thalamus: An Overview

The human thalamus is located centrally in the brain and has a variety of essential physiological functions. The thalamus, constituting the largest diencephalic structure, consists of various nuclei, each playing distinct functions. These roles range from functioning as a sensory and motor relay structure to regulating complex tasks such as consciousness and alertness. Traditionally, the prevailing view of the thalamus is that it is a relay station. However, in recent years, this notion has been questioned, with studies suggesting thalamic involvement with more advanced cognitive functions, including attention, memory, executive function, and learning (Worden et al. 2021). Moreover, its considerable metabolic requirements and extensive connections with the neocortex defy the characterization of it as a simple relay station. Although the precise role of the thalamus remains to be fully understood, its undeniable significance is firmly established as it plays pivotal roles in various fundamental aspects of brain function.

4.2.1 Development

Thalamic development is intricate, involving dynamic processes such as patterning and cell proliferation, which lay the foundation for the stepwise generation of neurons, each with specific spatial locations within the thalamus. Another crucial element in thalamic development is the regulation of thalamocortical axons and their impact on cortical development. Additionally, corticothalamic projections influence thalamic development. All of these processes are potential targets of PAE. However, our understanding of the development of specific thalamic nuclei remains limited, and further research is needed to elucidate the mechanisms by which PAE affects this process.

The initial thalamic complex is derived from the diencephalon and includes the prethalamus, the mid-diencephalic organizer, and the dorsal thalamus. After neural tube formation in the embryo, the formation of the prethalamus and thalamus is induced by various transcription factors and the mid-diencephalic organizer (Scholpp and Lumsden 2010). The mid-diencephalic organizer functions to release necessary cell signaling molecules, such as sonic hedgehog protein that induces thalamic neuronal development and differentiation. Later on, the mid-diencephalic organizer matures into the zona limitans intrathalamica, which becomes a signaling center as well as a border between the prethalamus and thalamus (Nakagawa 2019).

Sonic hedgehog, wingless-type MMTV integration site family, and fibroblast growth factors orchestrate the regionalization and differentiation of the thalamus (Nakagawa 2019). Beyond extrinsic cues, transcription factors like Pax6 contribute to normal thalamic size. Examination of clonal cell lineages has yielded valuable insights into the mechanisms underlying thalamic neurogenesis and gliogenesis, which populate it during a brief period of embryonic development. Shi et al. (2017) demonstrated that thalamic radial glial progenitors give rise to distinct cohorts of neuronal progeny with specific spatial configurations and nuclear arrangements correlating with functionality. An anteriorly located clonal cluster primarily contributes cells to nuclei associated with cognitive functions (e.g., anterodorsal and anteroventral), whereas the medial ventral posterior group of clonal cells contributes predominantly to the formation of sensory/motor nuclei (e.g., ventral posterolateral, ventral lateral) (Fig. 4.1). In rodents, thalamic

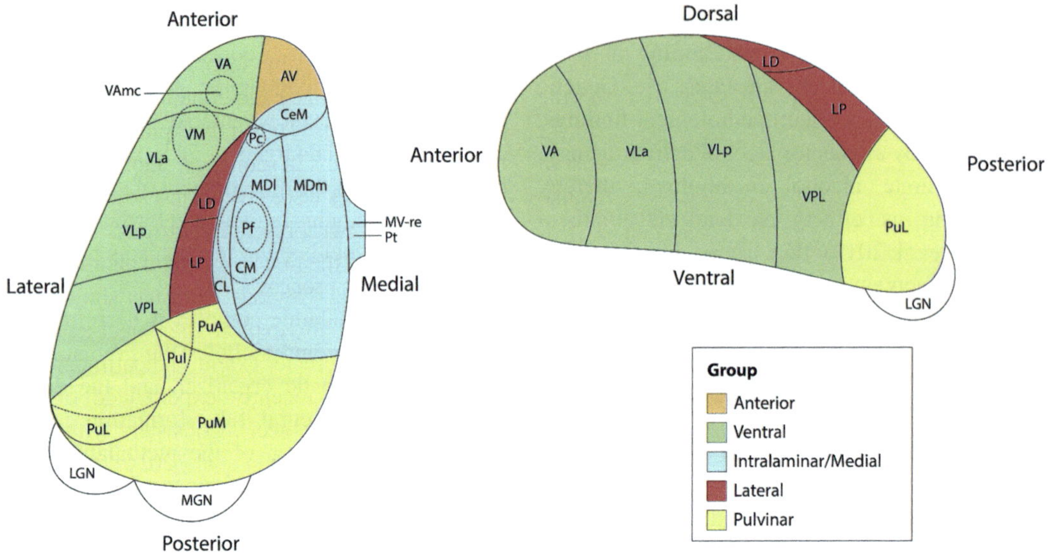

Fig. 4.1 Illustrative representation of thalamic nuclei. The left panel shows a transverse view of the left thalamus from the top. Dashed lines mark the location of nuclei situated more profound within the thalamus. In the right panel, a sagittal perspective from the side is presented. Thalamic nuclei groups are color-coded as indicated. *AV* anteroventral; *CeM* central medial; *CL* central lateral; *CM* centromedian; *LD* laterodorsal; *LGN* lateral geniculate nucleus; *LP* lateral posterior; *MGN* medial geniculate nucleus; *MDm* mediodorsal medial; *MDl* mediodorsal lateral; *MV-re* reuniens; *Pc* paracentral; *Pf* parafascicular; *Pt* paratenial; *PuA* pulvinar anterior; *PuI* pulvinar inferior; *PuL* pulvinar lateral; *PuM* pulvinar medial; *VA* ventral anterior; *VAmc* ventral anterior magnocellular; *VLa* ventral lateral anterior; *VLp* ventral lateral posterior; *VM* ventral medial; *VPL* ventral posterolateral. This figure is reprinted with permission from Elsevier, and it was initially published in Weeland et al. (2022)

nuclei predominantly consist of excitatory neurons, except for the lateral geniculate nucleus (LGN), which also includes inhibitory neurons. In contrast, the thalamic reticular nucleus exclusively contains GABAergic neurons, which deliver inhibitory input to other thalamic nuclei. The caudal progenitor domain of the developing thalamus generates all thalamic excitatory neurons (Vue et al. 2007). In contrast, the rostral progenitor domain in the developing thalamus and prethalamus generates LGN and thalamic reticular nuclear neurons (Vue et al. 2007; Inamura et al. 2011).

In the process of thalamic development, afferent input from both subcortical and cortical sources plays a crucial role, impacting gene expression and connectivity (Nakagawa 2019). Various studies, including those involving enucleation and subcortical input ablation in neonatal mice, have demonstrated the significant influence of these inputs on thalamic organization. Alterations in gene expression patterns and aberrant axonal projections in response to these manipulations highlight the importance of afferent input in shaping thalamic development. Additionally, cross-modal changes in gene expression and calcium waves suggest that modifying specific subcortical sensory inputs can have broader effects across sensory modalities and influence different nuclei. Transcription factors like Gbx2 and FOXP2 exhibit dynamic expression patterns, and their loss-of-function mutations impact thalamic nuclei size, indicating potential roles in specifying nuclear identities in postmitotic neurons.

The initial cortical neurogenetic and migration processes operate largely without dependence on subcortical input. Nevertheless, interactions between thalamocortical projections and the cortex commence before the conclusion of cortical neurogenesis and neuronal migration (Molnar and Kwan 2024). These interactions play a role in conveying early signals for forming neuronal circuits within specific cortical areas.

Thalamocortical projections play crucial roles in cortical development and, as discussed above, reciprocal corticothalamic projections influence thalamic nuclear hierarchical development. Thalamocortical axons establish a topographically organized innervation of the cortex, facilitating the refinement of cortical area formation through excitatory inputs (Molnar and Kwan 2024). The development of thalamocortical projections involves both extrinsic and intrinsic mechanisms (Nakagawa 2019). Studies manipulating thalamocortical input in mice demonstrate the instructive role of thalamocortical input in determining the organization of neocortical subregions. Alterations in the embryonic thalamus, such as changes in sonic hedgehog signaling or genetic manipulations, result in corresponding shifts in principal sensory nuclei and cortical area characteristics. The absence of thalamic input leads to disrupted area borders and altered gene expression patterns. Thalamocortical axons also regulate cortical neurogenesis, with evidence suggesting a role in promoting the division of specific neural progenitor cell populations, influencing cortical area size.

4.2.2 Structure and Function

The thalamus is located above the midbrain (mesencephalon), and its location allows for multi-directional nerve connections to the cerebral cortex. Anatomically speaking, it constitutes the upper and lateral boundaries of the third ventricle, with its dorsal surface contributing to the structure of the lateral ventricle. Laterally, the thalamus is bordered by the posterior arms of the internal capsule. The thalamus is composed mostly of gray matter with some areas of white matter. The internal medullary lamina is a white matter structure that divides the thalamic nuclei into its anterior, medial, and lateral complexes. There are approximately 50–60 thalamic nuclei (Kosif 2016) that project to different cortical areas. These nuclei are anatomically divided into six larger groups (with their respective sub-nuclei provided in parentheses) (Fig. 4.1):

1. Anterior nuclear complex (anteroventral, anteromedial, anterodorsal, and laterodorsal nuclei)
2. Medial-midline group (mediodorsal magnocellular, mediodorsal parvocellular, paratenial, paraventricular, and nucleus reuniens nuclei)
3. Lateral group (ventral anterior lateral posterior, ventral lateral, ventral medial, ventral posterior, ventral posterolateral, ventral posteromedial, ventral posterior lateral, ventral posterior inferior, and lateral posterior nuclei)
4. Intralaminar group (centromedian, central lateral, paracentral, and parafascicular nuclei)
5. Posterior group (pulvinar complex, pulvinar inferior/medial nucleus)
6. Metathalamus (lateral geniculate, medial geniculate, and reticular nuclei)

The thalamus is predominantly composed of nuclei (e.g., anterior, medial, and lateral groups, as well as the lateral and medial geniculate), and their projections to the cortex exhibit specific localization patterns compared to reticular and intralaminar nuclei (Sheridan and Tadi 2024). These relay nuclei, categorized as lateral, medial, and anterior, play a significant role in clinical contexts, with notable nuclei such as ventral posterolateral, ventral posteromedial, lateral geniculate, medial geniculate, and ventral lateral belonging to the lateral nuclear group. In contrast, the reticular nucleus, surrounding the lateral thalamus, provides GABAergic inhibitory projections to the thalamus, regulating its activity. The intralaminar nuclei, receiving inputs from the basal ganglia, also contribute projections to the cortex.

Functionally, the thalamus serves three primary roles encompassing sensory, motor, and cognitive processing. Known for its involvement in sensory processing, distinct nuclei play key roles in transmitting sensory information. The LGN relays visual data to the visual cortex, while the medial geniculate conveys auditory information to the auditory cortex. Additionally, the ventral posterolateral nucleus receives pain, temperature, and touch information from the spinal spinothalamic tract and the lemniscal pathway, while the ventral posteromedial nucleus processes facial sensation from the trigeminal nerve. Beyond sensory functions, the motor thalamus, composed of the ventral anterior, ventral lateral, and ventral medial thalamic nuclei, plays a crucial role in motor processing. The motor nuclei receive cerebellar and basal ganglia afferent input, contributing to motor planning, execution, and control. Particularly, the motor thalamus is vital for real-time movement control, facilitated by inputs from the reticular nucleus and disinhibitory signals from the globus pallidus, enabling voluntary movements (Torrico and Munakomi 2024). Collectively, these circuitries within the thalamus significantly contribute to motor control, coordination, and execution of movements. The thalamus also supports different cognitive functions including memory, attention, and consciousness. The anterior thalamic nuclei are part of the limbic memory system that involves the hippocampal formation, retrosplenial cortex, and mammillary nuclei (Aggleton and O'Mara 2022). Midline and intralaminar nuclei have been associated with and have a role in awareness. These nuclei do not produce awareness per se but provide necessary arousal for subsequent awareness (Kosif 2016). Other recent studies have explored the role of the mediodorsal nucleus in learning and memory for complex spatial configurations (Chao et al. 2022). The paraventricular thalamus has also been found to be important for balancing risk and reward (Worden et al. 2021).

4.2.3 Clinical Implications

Given the thalamus' sensory and motor functions, certain brain lesions will lead to specific deficits in these domains. After a thalamic stroke, the affected patients will have an absence of sensation and tingling on the side opposite the thalamic lesion. Traditionally, thalamic strokes are thought of as purely sensory strokes, which can be associated with severe chronic pain (Torrico and Munakomi 2024). Thalamic pathology is also implicated in seizures with studies showing involvement in both generalized and frontal types (Martin-Lopez et al. 2017). Additionally, the thalamus is implicated in several neurological and psychiatric disorders. Most prominently among these is its role in alcohol use disorder and Korsakoff's syndrome (Segobin and Pitel 2021). The most prominent sign of Korsakoff's syndrome is severe anterograde and retrograde amnesia. Two brain networks, the Papez and frontocerebellar circuits, are affected in Korsakoff's, and the thalamus contributes to both of these circuits. Crucially, preclinical and clinical investigations highlight the central involvement of the thalamus in the pathophysiology of FASD, as elaborated upon in the sections below.

4.3 Effects of Alcohol on Thalamus Morphology in Animal Models of FASD

The effects of alcohol on thalamic morphology have been characterized in rodents (mice and rats) and non-human primates. It is crucial to note the substantial difference in the rodent gestational period compared to humans (18–23 days for mice/rats). Notably, a considerable portion of brain development in rodents unfolds postnatally, distinguishing them from the human developmental timeline. For rats and mice, the first and second trimester-equivalent periods spans gestational days 1–10 and 10–20, respectively, and the third trimester equivalent (start of the brain growth spurt) takes place postnatally (from postnatal day 1 to 10) (Patten et al. 2014). Drinking patterns during pregnancy vary considerably, ranging from early alcohol ingestion until pregnancy is recognized to consumption throughout all trimesters of gestation (Ethen et al. 2009). Therefore, research has explored the impact of alcohol on the thalamus throughout all of these developmental stages, as discussed below.

4.3.1 First and Second Trimester-Equivalent Exposure

Several studies examined the effect of PAE during the rodent equivalent to the first two trimesters of human pregnancy. Pioneering studies from Granato and collaborators shed light on the PAE-induced thalamo-cortico-thalamic loop alterations in adult rats. Minciacchi et al. (1993) focused on the organization of thalamocortical connections in adult rats exposed to 2.4 g/kg of alcohol/day via gastric intubation on gestational days 14–19 (blood ethanol concentration (BEC) = not determined). Anterograde tracing experiments revealed a significant thinning of the thalamic-recipient zone in the sensorimotor cortex of PAE rats. Additionally, aberrant thalamocortical terminations in layer 5a were observed. However, neurons of origin for corticothalamic projections appeared quantitatively comparable between control and alcohol-exposed cases. In a subsequent study that used the same exposure paradigm, Granato et al. (1995) discovered permanent modifications in the thalamocortical circuits of PAE-exposed adult rats, with alterations noted at the level of axon terminals. The effects of PAE were nuclei-specific, including the anterior intralaminar and midline nuclei, along with impaired crossed corticothalamic fibers. Interestingly, the damage within the sensorimotor cortex displayed a gradient, being more severe in lateral fields. Cells labeled in the thalamus and layer 5 of the sensorimotor cortex demonstrated typical values of areal density. In contrast, corticothalamic neurons in layer 6, particularly in the lateral agranular sensorimotor field, showed reduced areal density values compared to controls. The study of Santarelli et al. (1995) further extended these studies by exploring the preventive effects of administration of acetyl-L-carnitine (75 mg/kg/day in drinking water concurrently with alcohol on gestational days 14–19 and during lactation). The researchers found that acetyl-L-carnitine treatment reduced the alcohol-induced changes in thalamocortical circuits. Collectively, the findings of these studies highlight the specificity of damage in different regions, the thinning of cortical zones, and the potential preventive effects of certain interventions, such as acetyl-L-carnitine.

A related study examined the impact of PAE on vibrissal somatosensory cortical barrel network development, which is modulated by the serotonin-containing thalamocortical afferents (Powrozek and Zhou 2005). The investigators used C57BL/6 mice exposed to alcohol from embryonic day 7 to 18 (liquid diet as the sole nutritional source; peak maternal BEC = 104 mg/dl on embryonic day 14). On postnatal day 7, cortices were analyzed for thalamocortical fibers and mature neurons. Thalamocortical terminals displaying 5-HT immunoreactivity delineate a barrel field pattern related to vibrissae in layer 4. These thalamocortical fibers exhibit a transient expression of the serotonin transporter and 5-HT1B receptors that presynaptically inhibit thalamocortical transmission. The results show that, despite there being no decrease in the overall brain weight or volume, PAE leads to a significant reduction in both the area of the posterior medial barrel subfield (containing serotonin-positive thalamocortical fibers) and the size of individual barrels within that defined area. These effects may lead to changes in whisker discriminatory sensitivity, serving as a model for studying compromised sensory function following PAE.

White et al. (2015) investigated the impact of binge PAE during the first trimester equivalent on the reciprocal part of the circuit (i.e., corticothalamic) in Swiss Webster mice. Alcohol was intraperitoneally administered to pregnant dams (2.9 g/kg initially followed by 1.45 g/kg 2 h later; embryonic days 11.5, 12.5, and 13.5; peak maternal BECs were ~ 250–290 mg/dl). It was found that cortical neurons that project to the thalamus, labeled with bromodeoxyuridine and T-box brain transcription factor 1, were located in the deep cortical layers and their numbers were unaffected by PAE. These results suggest that alcohol has differential effects on thalamocortical vs. corticothalamic projections and emphasize the significance of diversifying rodent strains in modeling FASD to advance understanding of its complex etiology. Future studies should determine if alterations in cortical neurons drive thalamic neuron damage or vice versa.

Studies have also provided valuable insights into the effects of PAE on the ventrobasal thalamic nucleus and its interactions with the trigeminal-somatosensory pathway. Mooney and Miller (1999) employed stereological methods to assess the impact of PAE on the number of neurons in the ventrobasal thalamic nucleus (liquid diet containing 6.7% alcohol on gestational days 11–21, which was gradually introduced between gestational days 6–10; peak BEC = ~150 mg/dl). Surprisingly, they found no significant effect on ventrobasal thalamic nucleus volume or neuron count. This contradicted the numerical matching hypothesis, which predicted a reduction in ventrobasal third-order thalamic neurons based on observed reductions in trigeminal-somatosensory second and fourth-order neurons. A subsequent study by the same investigative team delved into the molecular mechanisms underlying these effects (Mooney and Miller 2001). Employing the same PAE paradigm, the researchers demonstrated that neuronal death is absent in the ventrobasal nucleus but does occur in the somatosensory cortex. The somatosensory cortex exhibited reduced bcl-2 expression, altering the bcl-2/bax ratio, while the thalamus remained unaffected. These findings highlighted the differential effects of PAE on death-related proteins in the cortex and thalamus during critical periods of neuronal development.

The long-term consequences of PAE on postnatal neurogenesis in the ventrobasal thalamic nucleus were explored in a subsequent study (Mooney and Miller 2001). The exposure paradigm was the same as described above. The study revealed alterations in the pattern of change in neuronal number over time, suggesting a lasting impact on the rate of neuronal acquisition. The rate of cell proliferation was higher in alcohol-exposed animals, indicating a complex interplay of factors influencing postnatal neurogenesis. In a previous study, it was shown that rats exhibit a decrease in ventrobasal thalamic nucleus neurons during the first postnatal week followed by a 2.5-fold increase over 3 weeks, likely due to changes in neurogenesis (Mooney and Miller 2007). Interestingly, neurotrophins influence cycling cell numbers in the ventrobasal thalamic nucleus, where alcohol alters neurotrophin levels, as well as those of their corresponding receptors (Mooney and Miller 2011). Alcohol increased NGF expression but decreased levels of the low-affinity receptor p75. These findings suggest that postnatal cell proliferation in the ventrobasal thalamic nucleus is neurotrophin-responsive and influenced by alcohol exposure. Taken together, these studies indicate that the developing ventrobasal thalamic nucleus has a relatively unique response to PAE. Alcohol exposure during prenatal development does not affect cell proliferation, but it has a delayed impact on cells proliferating after birth. Despite this, the total number of neurons in the developing ventrobasal thalamic nucleus remains unaffected by alcohol. One potential justification for this phenomenon is that the ventrobasal thalamic nucleus may display compensatory responses to alcohol-induced damage via postnatal neurogenesis.

Neuroimaging techniques have been used to evaluate the effects of PAE on thalamic structure. Zhang et al. (2019) explored the impact of PAE during the initial stages of pregnancy on brain structure using a C57BL/6 J mouse model (10% alcohol in drinking water from 0.5 to 8.5 gestational days; BEC = not determined). The study assessed the brains of male offspring at adolescence and adulthood using diffusion tensor imaging (DTI). The results indicate that PAE had no substantial impact on the organization of white matter in various brain regions, including the thalamus, internal capsule, and corpus callosum. Whole brain volume and volumes of specific brain regions (e.g., neocortex, cerebellum, and caudate putamen) were not significantly affected (thalamus volume was not measured). However, a subsequent study by Nguyen et al. (2020) that used a similar exposure paradigm did detect changes in thalamic volume. BECs in this study were variable (67% of the mice had a BEC of 0–2 mg/dl; 28% had a BEC of 2–50 mg/dl, and 5% had a BEC of 50–86 mg/dl). PAE resulted in significant volume changes, including enlargement of the thalamus and hypothalamus at postnatal day 28. Although this study did not detect changes in DTI, histological studies revealed that mice exposed to PAE showed decreased neurofilament-positive staining in the thalamus and striatum.

A lack of an effect of early PAE on brain volumes was also observed in a study by Fish et al. (2016) where alcohol was administered to pregnant mice during gestational day 8.0 (2.8 g/kg via intraperitoneal injection; peak maternal BEC = 380 mg/dl), and the effects were assessed in adolescent offspring. Although there were no significant changes in brain volume, shape differences were observed in the striatum, corpus callosum, cerebellum, and hypothalamus (but not the thalamus). However, some effects on thalamic structure were detected in a primate FASD model. Specifically, Wang et al. (2020) exposed rhesus macaques to PAE (oral self-administration of 1.5 g/kg alcohol per day from pre-pregnancy to the first 60 days of a 168-day gestational term; BECs = 20–120 mg/dl; (Jimenez et al. 2019)) and characterized its effect on brain structure. Interestingly, fetal MRI at gestation day 135 found negative correlations between maternal BEC and brainstem and thalamic volume. Overall, these neuroimaging studies suggest that the impact PAE on thalamic morphology is contingent upon the timing and dosage of alcohol administration, along with the specific route through which it is delivered.

4.3.2 Third Trimester-Equivalent Exposure

4.3.2.1 Induction of Apoptotic Neurodegeneration

Studies conducted by various research groups have consistently demonstrated that exposure to binge-like doses of alcohol on postnatal day 7 induces apoptotic neurodegeneration in the thalamus (subcutaneous injection; 5 g/kg; peak BEC = 500 mg/dl). Ikonomidou et al. (2000) observed a substantial increase in the density of degenerating neurons, with a 12-fold and 60-fold rise in the mediodorsal and laterodorsal thalamus of rats, respectively, within 24 h of alcohol treatment. Intriguingly, certain neuronal populations, such as those in the mediodorsal and ventral thalamus, exhibited a significant response to alcohol starting on embryonic day 19, reaching its peak at P0, and diminishing rapidly thereafter. Conversely, other neuronal populations, including the laterodorsal and anteroventral thalamus, displayed a response beginning on embryonic day 19, peaking at postnatal day 3, and gradually returning to baseline levels by postnatal day 14 (Ikonomidou et al. 2000). Olney et al. (2002) and Wozniak et al. (2004) observed comparable outcomes in mice administered alcohol on postnatal day 7, as detailed earlier. Notably, the anterior thalamic nucleus exhibited significant neuroapoptotic neurodegeneration, with the most substantial impact observed in the anterodorsal thalamic nucleus. This region displayed a 60% decrease in total volume on postnatal days 14, 30, and 90. Apoptotic neuronal damage was also evident across all regions of the extended hippocampal circuit. Alcohol exposure resulted in marked deficits in spatial learning and memory at 1 month of age. However, there was a notable improvement in function during subsequent development, indicating a recovery of cognitive abilities.

Other laboratories have confirmed the findings of these studies using the same murine exposure paradigm. Ieraci and Herrera (2006) detected an increase in apoptotic neurodegeneration in the laterodorsal thalamic nucleus and determined that nicotinamide administration after alcohol exposure reduced caspase-3 activation and apoptosis, preventing behavioral abnormalities induced by alcohol. Ullah et al. (2013, 2011) found that alcohol exposure at postnatal day 7 led to apoptosis and extensive neuronal loss in the cortex and thalamus, an effect that was reduced by co-treatment with pyruvate. Furumiya and Hashimoto (2011) found that postnatal day 7 alcohol exposure triggered apoptosis across multiple brain regions, including the thalamus, and caused persistent impairments in spatial learning. Smiley et al. (2023) determined that the same exposure paradigm causes a strong neuroapoptotic response in the anterior thalamic nucleus and in numerous regions intricately linked to these nuclei (e.g., mammillary bodies, dorsal subiculum, and the midline cerebral cortex's cingulate, prefrontal, and retrosplenial regions). In the anterior dorsal thalamic nucleus, the investigators detected a normal developmental neuron loss, resulting in nearly a 50% neuronal number reduction from postnatal day 7 to postnatal day 70. Interestingly, this process persisted in animals from the PAE group, even after a substantial alcohol-induced deficit at postnatal day 7. After correcting for this normal age-related neuronal decrease, it was determined that alcohol caused a reduction in neuronal density of 35% at postnatal day 7 and 54% at postnatal day 70.

Interestingly, the effect of alcohol on anterior thalamic nucleus neuronal survival depends on the alcohol exposure paradigm. Bird et al. (2023) subjected transgenic mice expressing fluorescently tagged GABAergic interneurons (VGAT-Venus mice) to alcohol vapor exposure on postnatal day 7 (4 h. in chamber; peak pup BEC = ~400 mg/dl). This paradigm activated caspase 3 in the reticular thalamic, anteroventral, and anterodorsal nuclei at postnatal day 7 (8-hr. post-alcohol vapor exposure). However, in contrast to the subcutaneous injection studies, a noteworthy absence of a significant neuronal density reduction was noted in these nuclei of young adult mice. A similar finding was reported in another thalamic nucleus (ventrolateral) by Livy et al. (2001) in rats (at postnatal day 10) subjected to binge-like alcohol exposure during the first, second, and/or third trimester equivalent via intragastric gavage between postnatal days 4–9; BEC = 266–398 mg/dl. Specifically, the only significant finding was that rats in both the pair-fed and alcohol groups, when exposed during the ventrolateral nucleus neurogenesis period (embryonic days 14–15), showed decreases in both volume and cell number. The disparities between these studies could be linked to the slower rises in blood alcohol concentrations in vapor- or gavage-exposed pups when compared to those receiving subcutaneous injections.

A study by Farber et al. (2010) demonstrated comparable effects in the primate brain to those observed in rodents, providing evidence that binge-like alcohol exposure in the final trimester of pregnancy might induce parallel effects in the human brain. Cynomolgus macaques (full gestation = 160–165 days) were injected with alcohol intravenously (~2 g/kg), succeeded by a maintenance dose (0.2 g/kg/hr. for 6 h) to uphold the BEC within the 300–400 mg/dl range. Fetal brains were collected 1 h after the last alcohol maintenance dose. Exposure during earlier gestational ages (days 105–135) primarily affected the caudate/putamen, cerebellum, entorhinal cortex, inferior colliculus, nucleus accumbens, subiculum, and thalamus. Conversely, at later gestational ages (days 140–155), some cerebral cortex regions were primarily affected. The most severe damage was observed in the caudate/putamen, subiculum, inferior colliculus, thalamus, and cerebellum at gestational days 105–135.

4.3.2.2 Effects on the Lateral Geniculate Nucleus

A number of studies collectively shed light on the detrimental effects of alcohol exposure during various developmental stages on the visual system, particularly emphasizing the impact on the LGN of the thalamus. The LGN relays visual information to the primary visual cortex. Olney et al. (2002) demonstrated the induction of caspase-3 activation in the LGN in response to binge-like alcohol intoxication (subcutaneous injection; 5 g/kg; peak BEC = 500 mg/dl). In a follow-up study, the investigators demonstrated that retinal ganglion cells and neurons in the LGN, superior colliculus, and visual cortex displayed a high susceptibility to the apoptogenic effects of alcohol (Tenkova et al. 2003). The critical period of heightened sensitivity extended from postnatal days 1 to 4 for ganglion cells and from postnatal days 4 to 7 for other neuronal components of the pathway. Notably, a blood alcohol level near 120 mg/dL was the threshold to trigger apoptosis in visual neurons, underscoring the acute vulnerability of the visual thalamus to alcohol-induced apoptosis.

Dursun et al. (2011) investigated the consequences of early postnatal alcohol exposure on the morphology of retinal ganglion cells and the dorsolateral LGN. The study included transgenic mice (on a C57BL/6 background), both male and female, expressing yellow fluorescent protein regulated by a Thy-1 (thymus cell antigen 1) promoter. Alcohol was administered to mouse pups at 3 g/kg/day through intragastric intubation from postnatal days (PDs) 3 to 20. The findings revealed that alcohol exposure induced significant alterations in retinal ganglion cell morphology (reduction in soma area and total dendritic field area, and increase in dendritic tortuosity and mean branch angle), and a notable decrease in the numbers of neurons in both the ganglion cell layer and the LGN.

In the study of Papia et al. (2010) pregnant vervet monkeys (full-term gestation = 165 days) drank ~2.5 g/kg/day of alcohol four times/week during the last trimester of pregnancy (alcohol consumption started at gestational days 72–112 and lasted for 7–13 days; BEC = 61–132 mg/dl). The investigators measured neuronal and glial cell numbers in the parvocellular and magnocellular layers of the LGN in offspring (ages 1–35 days). There were no significant effects of PAE on layer volume or cell numbers in the LGN. However, a noteworthy finding emerged in the reduced soma size of magnocellular neurons in PAE subjects.

Collectively, these studies underscore the vulnerability of the visual system, particularly the LGN, to the adverse effects of alcohol exposure during critical developmental periods. The findings emphasize the importance of considering the timing, duration, and pattern of alcohol exposure in understanding its impact on neural structures involved in visual processing. Considering that visual alterations are common in individuals with FASD (Ayoub et al. 2023), further research in this area is crucial for developing strategies to mitigate the detrimental effects of alcohol on the developing LGN and other components of the visual system.

4.3.2.3 Effects on the Nucleus Reuniens

This thalamic nucleus is a key player in a complex network that involves hippocampal and cortical structures, acting as a foundational substrate for cognitive processes. It serves as a vital component of corticothalamic circuits, facilitating medial prefrontal cortex-hippocampus communication. Impairment of the nucleus reuniens has been linked to various neuropsychiatric disorders (Dolleman-van der Weel et al. 2019). Pioneering studies from the Klintsova laboratory have demonstrated that this nucleus is an important target of developmental alcohol exposure. Gursky et al. (2019) investigated the lasting neuroanatomical effects of early postnatal alcohol exposure in the nucleus reuniens of female rats. Using a dose of 5.25 g/kg/day of alcohol via intragastric intubation during postnatal days 4–9 (BEC = 378 mg/dl), the investigators observed a significant decrease (21%) in total neuron number and a reduction in total volume (18%) exclusively in this nucleus at postnatal day 72. These alterations were not observed in the rhomboid nucleus or in non-neuronal cell numbers. Gursky and Klintsova (2022) subsequently demonstrated that moderate doses of alcohol during postnatal days 4–9 (3.00 g/kg/day via intragastric intubation; BEC = 144 mg/dl) caused significant neuronal and non-neuronal cell loss, leading to reduced nucleus reuniens volume. Furthermore, Gursky et al. (2020) determined that a single alcohol exposure during postnatal day 7 on the nucleus reuniens (5.25 g/kg via intragastric intubation; BEC = 364–381 mg/dl) is sufficient to permanently damage it. This exposure paradigm induced apoptotic cell death on day 7 and caused short-term cell loss on day 11. At postnatal day 72, rats exposed to alcohol exhibited permanent cell loss, including neuronal number and volume reductions in the nucleus reuniens.

A subsequent investigation conducted by the same team demonstrated that high exposure to alcohol (5.25 g/kg/day; postnatal days 4–9) selectively modifies the ratio of nucleus reuniens neurons concurrently innervating both the medial prefrontal cortex and ventral hippocampus in adulthood (Gursky et al. 2021). Notably, this alcohol exposure paradigm also reduces the cumulative length of axon terminals originating from the medial prefrontal cortex in the nucleus reuniens of adult female rats (Smith et al. 2022). These studies were groundbreaking in revealing that alcohol exposure during the third trimester equivalent of human pregnancy induces enduring alterations in reciprocal connectivity between the nucleus reuniens and cortical regions. These changes could contribute to documented deficits in executive function in rats exposed to this alcohol paradigm. Specifically, behavioral assessments in adulthood unveiled alcohol-induced impairments in object-in-place preference and rule-switching performance on a plus maze task (Gursky et al. 2021). These findings support the idea that executive function impairments resulting from alcohol exposure during late gestation may stem from damage to medial prefrontal cortex- thalamic nucleus reuniens-CA1 hippocampus circuitry.

4.4 Effects of Alcohol on Thalamus Function in Animal Models of FASD

Relatively few studies have examined the consequences of developmental alcohol exposure on thalamic function. These studies have used biochemical, imaging, and electrophysiological techniques in rat and mouse models of FASD.

4.4.1 First and Second Trimester-Equivalent Exposure

Ledig et al. (1988), tested the impact of maternal alcohol consumption during pregnancy and lactation on brain gamma-aminobutyric acid (GABA) levels in 3-week-old rat pups. The dams drank 8 g alcohol/kg/day (BECs = 100 mg/dl) starting 2 months before mating, and then during pregnancy and lactation. The results indicated a notable decrease in GABA levels in the cerebellum, hippocampus, pons, and thalamus. Conversely, a significant increase in GABA levels was noted in the frontal cortex, olfactory bulbs, anterior colliculus, and amygdala.

Miller and Dow-Edwards (1993) investigated the impact of PAE on activity in the trigeminal/somatosensory system induced by sensory input. Using mature offspring of dams fed either a liquid diet containing 6.7% alcohol on gestational days 11–21 (gradually introduced between gestational days 6–10; peak BEC = ~150 mg/dl) or a control diet, the researchers examined the effects of whisker stimulation-evoked glucose utilization in the trigeminal principal sensory nucleus, the ventrobasal thalamus, and the somatosensory cortex. The results showed that whisker stimulation increased glucose utilization to a lesser extent in the ventrobasal thalamic complex of alcohol-treated rats compared to controls.

Mitchell and Snyder-Keller (2003) investigated the expression of the immediate-early gene, c-fos, and cleaved caspase-3. Pregnant rats were subjected to alcohol exposure (liquid diet containing 4.5% alcohol, gestational days 5–22; BEC = not determined but estimated to be ~80 mg/dl), cocaine binge (three injections of 15 mg/kg subcutaneously 1 hr. apart on embryonic day 22), or a combination of both. PAE elevated Fos immunoreactivity levels in the medial thalamus, among other brain regions, at the 0-, 3-, and 24-h time points after birth. Enhanced cleaved caspase-3 expression in the medial thalamus was observed at 0 and 24 h post-birth in the alcohol group. The authors concluded that prenatal exposure to cocaine and alcohol increases neuronal activity and activates apoptotic pathways in the developing brain in a time- and region-specific manner.

Rodriguez et al. (2016) examined the impact of light PAE on functional brain connectivity in adult rats (drinking-in-the-dark limited access paradigm; 5% alcohol in saccharin throughout pregnancy; peak BEC = 60 mg/dl). The investigators used group-independent component analysis applied to resting-state functional MRI data to explore the global effects of PAE on neural function. The findings unveiled 17 components distributed across various brain regions, including the cerebellum, cortex, hippocampus, midbrain, and thalamus. PAE males exhibited diminished connectivity among cortical, midbrain, and thalamic components. Conversely, PAE males showed heightened connectivity among cerebellar, cortical, hippocampal, midbrain, and striatal components. On the other hand, females exhibited increased connectivity between cerebellar, hippocampal, striatal, and thalamic components. The results of this study indicate that even mild PAE can disrupt thalamic functional network connectivity in a sex-dependent manner.

4.4.2 Third Trimester-Equivalent Exposure

Ke et al. (2011) investigated the impact of alcohol on endoplasmic reticulum stress in C57BL/6 mice injected with alcohol at postnatal day 7

(5 g/kg via subcutaneous injection; pup BEC = 338 mg/dl). The study revealed that alcohol exposure increases the expression of endoplasmic reticulum stress-inducible proteins and activates associated signaling pathways, including mesencephalic astrocyte-derived neurotrophic factor. The induction of endoplasmic reticulum stress occurs within 4 h of alcohol injection, and some endoplasmic reticulum stress-inducible proteins remain elevated 24 h later. The distribution of alcohol-induced endoplasmic reticulum stress is observed in neocortical, hippocampal, and thalamic neurons.

Work from the Valenzuela laboratory assessed the effects of binge-like alcohol exposure during the brain growth spurt on retrosplenial cortex-projecting anterior thalamic neurons in adolescent mice (Bird et al. 2023). As mentioned above in sect. 4.3.2.1., this study used transgenic mice expressing fluorescently tagged GABAergic interneurons that were exposed to alcohol vapor for 4 h on postnatal day 7 (peak pup BEC = ~400 mg/dl). Electrophysiological recordings from retrogradely labeled neurons indicated significant effects of alcohol treatment on action-potential properties and synaptic currents. Specifically, we observed notable treatment-induced effects on both the instantaneous action potential frequency and overshoot, with sex by treatment interactions evident for threshold and overshoot in neurons of the anterodorsal thalamic nucleus. Additionally, a sex by treatment interaction surfaced for action potential number in anteroventral neurons. Apart from changes in excitability, we identified enduring modifications in the spontaneous excitatory and inhibitory synaptic transmission within anterodorsal nucleus neurons projecting to the retrosplenial cortex, but not in those in the anteroventral nucleus. These changes include a reduction in both the frequency and amplitude of spontaneous excitatory postsynaptic currents and an elevation in the total charge of spontaneous inhibitory postsynaptic currents. Given that the anterior thalamus is a key component of the limbic memory system, these findings suggest damage to this thalamic region could underlie enduring cognitive deficits associated with developmental alcohol exposure.

4.5 Effects of PAE on Thalamic Morphology in Humans

The approaches that have been commonly used to assess the impact of PAE on human thalamic abnormalities involve postmortem investigations and neuroimaging methodologies. Postmortem studies enable a comprehensive examination of brain anatomy and pathology, while non-invasive neuroimaging techniques like structural MRI, tensor-based morphometry (TBM), and DTI provide insights into structural changes over time in vivo. These methodologies have played a pivotal role in advancing the understanding of the effect of PAE on thalamic structure.

4.5.1 Postmortem Studies

In 1993, Coulter et al. (1993) conducted a case study involving a comprehensive neuropathologic evaluation of an infant with fetal alcohol effects who succumbed to acute respiratory distress at 2 months of age. The investigation centered on a girl born at 36–37 weeks to a mother who acknowledged binge alcohol drinking and occasional marijuana use during the first trimester. The infant exhibited numerous abnormalities commonly associated with FAS, notably severe microcephaly. Frontal lobes were fused in the anterior portion coupled with a lack of olfactory bulbs and tracts. Fusion of the thalamus and caudate nuclei extended to the level of the pineal gland, accompanied by optic nerve and tract hypoplasia, and evidence of an obstructed pituitary gland due to an enlarged and bulbous hypothalamus. These findings collectively contributed to a complex cerebral anomaly resembling incomplete holoprosencephaly and septo-optic dysplasia, underscoring the profound impact of PAE on brain development.

Jarmasz et al. (2017) conducted a retrospective survey of autopsies that met specific inclusion criteria. Thalamic abnormalities were identified in two cases. The first case involved a 19-year-old mother with a history of physical abuse and alcohol consumption during pregnancy. The baby, delivered by emergency cesarean section, died immediately. Postmortem examination revealed several abnormalities, including asymmetric cerebellum, posterior corpus callosum partial agenesis, and microscopic glial heterotopia in the meninges ventral to the thalamus and posterior to the midbrain. The second case featured a 3.5-year-old girl clinically diagnosed with FAS. Her mother had a history of regularly sniffing gasoline during pregnancy, and detailed birth information was unavailable as the girl was in foster care. The girl exhibited severe cognitive delay and succumbed to pneumonia. Postmortem findings included occipital microgyria with laminar necrosis, microcephaly, and a cavitated infarct in the thalamus.

4.5.2 Neuroimaging Studies

Several studies have investigated the impact of PAE on thalamic structure across different developmental stages. These investigations have utilized advanced imaging techniques, including super-resolution fetal MRI with tissue segmentation, high-resolution MRI, DTI, tractography, cranial sonography, tensor-based morphometry, and quantitative susceptibility mapping for indirect measurements of brain iron content. The following discussion will provide an overview of these studies, beginning with those conducted during early developmental stages and progressing to those focusing on later stages of development.

4.5.2.1 Fetal Development and Infancy

Stuempflen et al. (2023) used high-resolution fetal MRI to identify regional effects of PAE on human brain structure. Gestational age was 27 weeks and maternal age 30–31 years. On average, individuals in the PAE group reported weekly consumption of one to three standard drinks and acknowledged engaging in at least one episode of binge drinking, defined as consuming four or more drinks on a single occasion during their pregnancy. Fetuses with PAE exhibited significantly smaller volumes of the periventricular zone and larger volumes of the corpus callosum. The models for the remaining structures, including the deep gray nuclei (basal ganglia and thalamus) and total brain volume, did not reveal a significant impact of PAE. Results indicate that even minor PAE (1–3 standard drinks per week) affects the structure of certain fetal brain regions (but not the thalamus).

A previous study by Taylor et al. (2015) utilized DTI and tractography to examine the impact of PAE on white matter development in newborns (36–44 weeks postconception). The mothers in the PAE group consumed quantities of alcohol ranging from 6.2 to 14.0 standard drinks on each occasion, consistent with heavy binge drinking. Three primary white matter bundles were evaluated: (1) transcallosal tracts responsible for commissural left-right hemispheric connections and corona radiata, (2) left and right projection fibers connecting the cortex with the spinal cord, brainstem, and thalamus, and (3) left and right association cortico-cortical fibers. The axial diffusivity, particularly in medial and inferior white matter regions associated with early myelination, showed a strong correlation with maternal drinking. Specifically, increased PAE was linked to decreased axial diffusivity across all the callosal network. Within the left and right projection fibers, elevated PAE was found to be associated with lower axial diffusivity in bilateral tractographic connections within three regions: the corticospinal tracts and the superior and posterior thalamic radiations. Similarly, in the association networks, higher PAE levels were correlated with reduced axial diffusivity bilaterally in three regions: the uncinate fasciculus and superior and inferior longitudinal fasciculi. These results indicate that axial diffusivity should be considered in the examination of how PAE affects early white matter development.

Donald et al. (2016a) characterized the impact of PAE on neonatal gray matter volume. It was a sub-study of a large birth cohort study that aimed to explore the brain structure of infants who were exposed to alcohol in utero. There was variation in the number of drinks per occasion, with a significant proportion consuming 2–3 drinks per occasion 1–3 times per week. While alcohol consumption was most prevalent in the first trimester, it also occurred in the second and third trimesters. Two to four weeks old infants underwent unsedated MRI. The study showed that there was a significant decrease in the total gray matter volume in the PAE group compared to controls, indicating a broad impact on the overall brain structure of these infants. Upon correction of the gray matter volume changes, specific subcortical gray matter regions demonstrated significantly decreased volumes. The specific areas included: the left hippocampus, which indicated an early effect of PAE on areas critical for memory and spatial navigation; both the right and left amygdala, centers responsible for emotional regulation as well as memory processing; and the left thalamus, highlighting impaired sensory and motor signal relay as well as the active regulation of sleep, alertness, and cognition.

The results from these studies suggest that the impact of PAE on thalamic structure may not manifest immediately and becomes apparent during the neonatal period. Crucially, these effects are influenced by the dosage and drinking patterns, as well as the characteristics of the studied population, including socioeconomic status. These combined findings emphasize the importance of taking multiple factors into account to gain a comprehensive understanding of how PAE affects thalamic structure during early developmental stages.

4.5.2.2 Childhood, Adolescence, and Young Adulthood

An early study by Hughes et al. (1991) investigated the presence and significance of linear areas of echogenicity in the thalami and basal ganglia of neonates that was observed via cranial sonography. Initially, this abnormality had been described in patients with congenital infections and in patients with trisomy 13; it was originally associated with a necrotizing vasculopathy. In their study of 25 patients, only 4 of them had an isolated cytomegalovirus infection and there were no reported cases of the other infections. Among the other diagnoses encountered was FAS. These findings highlighted the diversity of conditions that can present with thalamic linear areas of echogenicity and indicated that patients with these sonographic abnormalities warrant comprehensive screening. Over the past three decades, cerebral ultrasound has emerged as a reliable method for studying the neonatal brain. It offers a cost-effective, non-invasive approach to neuroimaging that is readily available in most hospitals. Initially used to detect major abnormalities like intraventricular hemorrhage, it has expanded its diagnostic capabilities through different acoustic windows and ongoing technological advancements (e.g., higher transducer frequencies and 3D imaging) (Dudink and Jeanne Steggerda 2020). These enhancements enable the identification of various lesion patterns in several brain regions, including the thalamus. Therefore, it is important to further investigate the potential utility of this technique in identifying PAE-induced alterations in this and other brain regions.

Lebel et al. (2008) utilized DTI with diffusion tractography and two-dimensional region-of-interest analysis to investigate how PAE affects white and deep gray matter in various brain regions of 24 children (5–13 years old). Among the diagnosed cases, two children were identified with FAS, while the remaining individuals received diagnoses of other forms of FASD. Detailed information regarding the quantity of alcohol consumed during pregnancy and the potential use of other substances was not accessible. The findings of Lebel et al. supported prior DTI studies, showing microstructural abnormalities in the corpus callosum, right temporal white matter, and bilateral posterior cingulate white matter in children with FASD. Furthermore, they found that the volume of white matter was marginally more reduced than gray matter in individuals with FASD. In children with FASD compared to controls, there were observed reductions in fractional anisotropy in the bilateral inferior and superior longitudinal fasciculi, left thalamus, right cingulum, and splenium of the corpus callosum. Additionally, there was an increase in mean diffusivity in the right thalamus, right putamen, right corticospinal tract, left inferior longitudinal fasciculus, globus pallidus, and bilateral inferior fronto-occipital fasciculus. These changes in fractional anisotropy values may imply alterations in structural order in the thalamus and putamen.

A study by Nardelli et al. (2011) explored the link between specific structural brain injuries and cognitive, motor, and behavioral challenges using high-resolution MRI. The study included children and adolescents (6 to 17 years old) with FASD compared to age/sex-matched controls and investigated whether these differences persist across different developmental stages. Approximately half of the subjects were likely exposed during the first trimester to 1–2 drinks once a week and the other half was likely exposed throughout pregnancy or to binge drinking at any point during gestation. Results revealed a reduction in the volumes of the amygdala, caudate, globus pallidus, hippocampus, putamen, and thalamus. Notably, volume reductions in the globus pallidus, hippocampus, and thalamus were consistent across all age subgroups (6–9, 10–13, and 14–17 years old), while the caudate and putamen showed smaller volumes in FASD only within the two youngest subgroups, and the amygdala was smaller in the two oldest subgroups. These findings highlight that PAE induces age-dependent volume reductions in deep gray matter. Interestingly, the investigators did not find differences in brain volumes between those exposed to moderate vs. high alcohol levels.

A study by Sowell et al. (2010) explored the impact of prenatal exposure to methamphetamine, alcohol, and methamphetamine plus alcohol on brain structure in 9–11-year-old children. The investigators used high-resolution MRI and tensor-based morphometry to examine regional variations in brain substructure volumes by initially aligning all brain images globally to a common brain template. Subsequently, localized deformations are applied to modify each subject's anatomy, ensuring alignment with the overall group-averaged template. Inclusion criteria for PAE included exposure to ≥ 4 drinks/occasion at least once per week or ≥ 14 drinks/week. Individuals with prenatal drug exposure exhibited changes in numerous brain regions, encompassing the anterior and posterior cingulate, dorsal and ventral frontal and parietal cortices, inferior temporal, medial temporal, striatum, and thalamus, as well as the cerebellar and posterior callosal white matter. For the most part, these effects predominantly manifest bilaterally across all cortical and subcortical regions. In addition, the investigators found that those individuals exposed prenatally to drugs of abuse are prone to exhibit lower Full-Scale Intelligence Quotient scores and diminished volumes in the thalamus and medial occipital regions compared to those without substantial histories of prenatal drug exposure. The results of this study highlight the enduring negative effects of PAE on brain morphology during childhood and adolescence. Notably, it introduces new findings indicating that children with prenatal methamphetamine exposure experience distinct alterations in brain morphology beyond those associated with PAE alone. While there are certain regional variations, both exposure groups exhibit a comparable pattern of diminished volumes in subcortical structures such as the striatum and thalamus, as well as in lateral cortical regions, including the temporoparietal lobes and the right frontal pole. This implies that overlapping brain systems are impacted.

Meintjes et al. (2014) also used tensor-based morphometry methods to analyze structural brain imaging data from children with FASD and matched controls (9.5–11.0 years). Two cohorts of women were enlisted: (1) heavy drinkers, defined as those consuming ≥ 14 drinks/week and/or participating in binge drinking (≥ 5 drinks/occasion), and (2) controls, consisting of 14 individuals who abstained from drinking and 2 who consumed minimal alcohol during pregnancy. In general agreement with the study of Sowell et al. (2010), the investigators found that the degree of PAE was linked to a substantial correlation with widespread reductions in brain tissue across regions but not deformations. These reductions occurred in various brain regions, including the dorsolateral frontal cortex, inferior occipital lobe, midbrain, precuneus, superior cerebellum, superior parietal lobule, thalamus, ventromedial frontal lobe. However, after factoring out overall brain size, no significant associations with PAE were detected, suggesting that the observed effects were linked to differences in brain size. The results obtained from independent component analysis validated the regions of structural deformation associated with PAE identified in the tensor-based morphometry analysis. Specifically, three out of the four regions carrying the most substantial weight—the supero-medial cerebellum/inferior surface of the occipital lobe, the precuneus, and the thalamus/midbrain—coincided with areas that exhibited reductions in brain tissue volume due to PAE in the tensor-based morphometry analysis. These findings suggest that employing a data-driven independent component analysis of MRI scans could potentially serve as an indicator of PAE, particularly in cases where the diagnosis is more difficult in cases that lack facial dysmorphology.

Nakhid et al. (2022) conducted a study that focused on the potential alteration of brain iron levels associated with PAE and the implication that these alterations can have on cognitive, behavioral, and mental health outcomes. They studied 20 children and adolescents aged 7.5–15 years, all of whom had confirmed PAE and were compared to 44 normal developing controls. Twelve participants reached the exposure threshold specified in the Canadian FASD diagnostic guidelines ($\geq$ 7 drinks/week or $\geq$ 2 binge episodes of at least 4 drinks at some point during pregnancy), and eight participants with PAE had verified exposure but the quantity was unknown. The study employed quantitative susceptibility mapping, a novel MRI technique that quantifies magnetic susceptibility differences within tissue, allowing to indirectly measure variations in tissue composition, including iron content. Postmortem studies validate quantitative susceptibility mapping's reliability in measuring ferritin in subcortical gray matter. Before corrections for multiple comparisons, the PAE group exhibited higher susceptibility in the thalamus. However, there were no significant group differences in magnetic susceptibility after correcting for multiple comparisons in the thalamus and other regions of interest (amygdala, caudate, globus pallidus, hippocampus, nucleus accumbens, and putamen). Interestingly, the study also explored whether a relationship between thalamic susceptibility and intelligence quotient existed; however, there was no significant association found between them. It is important to highlight that, unlike other studies, the group exposed to PAE did not show a decrease in volume in the thalamus but reductions were observed in the left putamen, bilateral pallidum, and bilateral caudate. The observed variations may be attributed to the relatively limited sample size of the study or disparities in the severity of PAE compared to other studies.

Boateng et al. (2023) set out to investigate structural brain differences in subjects ranging from 3–21 years old with FASD, examining PAE in each FASD subtype. PAE was characterized by documented consumption of $\geq$ 6 drinks/week for $\geq$ 2 weeks during pregnancy, $\geq$ 3 drinks/occasion on $\geq$ 2 occasions during pregnancy, a history of intoxication or treatment for an alcohol-related disorder, and a positive alcohol screening test. Previous studies have indicated individuals with partial FAS or FAS have the most prominent effect on structural brain development when compared to alcohol-related neurodevelopmental disorder and control groups. In agreement with these findings, the investigators found no significant difference in overall brain volume between individuals with alcohol-related neurodevelopmental disorder and the control group, but the partial FAS/FAS subtypes had a significantly smaller brain volume than both the alcohol-related neurodevelopmental disorder and control groups. Children with FASD exhibited reduced volume in each of the specific structures examined (corpus callosum, cerebellum, thalamus, caudate, putamen, pallidum, and hippocampus). When accounting for estimated total intracranial volume, there were additional decreases in volume beyond what would be anticipated based solely on the degree of microcephaly in the basal ganglia, cerebellum, hippocampus, and thalamus. The authors conclude that while overall brain volume appeared similar between children with PAE/alcohol-related neurodevelopmental disorder and controls, analysis of specific brain structures revealed comparable levels of damage in both FASD subtypes. Unlike prior research, which linked dysmorphology to more severe structural abnormalities, this study suggests that some structures are equally sensitive to PAE even without facial dysmorphology. The detected differences among FASD subtypes could potentially elucidate the similarities in performance between individuals with alcohol-related neurodevelopmental disorder and children with FAS/partial FAS on specific cognitive tasks.

In summary, the results of these studies illuminate the complex interplay between PAE and structural changes in the developing thalamus. Together, these investigations underscore the intricate impact of PAE on thalamic structure, emphasizing the necessity for careful consideration of variables like age, concurrent exposures, and subtype differences to grasp its diverse effects on thalamic development fully.

4.6 Effects of PAE on Thalamic Function in Humans

The investigation into the effects of PAE on the thalamus in humans has predominantly focused on structural aspects, with fewer studies delving into the impact on thalamic function during various developmental stages. Employing techniques such as functional MRI, positron emission tomography, magnetic resonance spectroscopy, and magnetoencephalography, researchers have strived to comprehend the intricate dynamics of thalamic function influenced by in-utero alcohol exposure.

4.6.1 Studies with Infants

Studies involving infants have provided crucial insights into the early consequences of PAE. These studies underscore the need for longitudinal studies to comprehensively understand connectivity disturbances associated with in-utero alcohol exposure, particularly in older children. Donald et al. (2016b) studied PAE's effect on interhemispheric functional brain connectivity in neonates, offering preliminary insights into the establishment of intrinsic connectivity during early infancy. Mothers in the PAE group were initially screened using the Alcohol, Smoking, and Substance Involvement Screening followed by interviews to ensure that they had a positive history of moderate to severe alcohol use during any trimester of pregnancy. Employing resting-state functional MRI on a cohort of 60 subjects aged 2 to 4 weeks, which included 13 with PAE and 14 age-matched controls, the study utilized a multivariate model to examine connectivity among sensorimotor intrinsic functional networks. Seed-based analyses were also conducted to identify group differences in interhemispheric connectivity of intrinsic motor networks. Six distinct networks in the brain were identified: (1) anterior motor (including the supplementary motor area and the superior frontal gyrus), (2) posterior motor (including the postcentral gyrus, supplementary motor area, paracentral lobule and left thalamus), (3) left somatosensory (situated within the left postcentral gyrus), (4) right somatosensory (situated within the right, and to a lesser extent, the left postcentral gyrus), (5) a bilateral network manifested in the striatum, displaying heightened activation in both the left and right putamen, and (6) a diffuse thalamic network incorporating regions like the brainstem, hippocampus, amygdala, and globus pallidus. The research unveiled heightened connectivity between the brainstem and the anterior motor network in infants affected by PAE, as well as increased connectivity between the striatum and both anterior and posterior motor networks. However, there was preliminary evidence of attenuated interhemispheric connectivity between the left and right somatosensory networks in PAE infants. Further exploration of hemispheric connectivity in motor networks did not yield significant effects of PAE. The clinical significance of internetwork differences was also examined, with only connectivity between the striatum and the anterior motor network (but not the thalamus) showing a moderating effect on the association with clinical deficits.

Grewen et al. (2015) investigated the impact of prenatal marijuana exposure with or without concurrent exposure to other substances, including alcohol, on early brain development. Maternal Timeline Followback interviews were used to evaluate prenatal drug exposure. Examining resting-state functional connectivity in sleeping infants aged 2–6 weeks, the research included neonates with prenatal marijuana exposure in combination with other substances (nicotine, alcohol, opiates, and/or selective serotonin reuptake inhibitors) denoted as +MJ ($n = 20$), those exposed to the same substances without marijuana (−MJ, $n = 23$), and drug-free controls ($n = 20$). A significant percentage of both the +MJ and −MJ groups disclosed prenatal cigarette smoking (87%) and alcohol consumption (30%) during at least one trimester of pregnancy, while relatively lower rates of prenatal antidepressant and opiate use were reported in both groups. Both the +MJ and −MJ groups, in comparison to controls, exhibited hyper-connectivity of the left amygdala seed with the orbital frontal cortex and hypo-connectivity of the posterior thalamus seed with the hippocampus. These findings suggest a potential susceptibility to various substances of abuse, including alcohol, within these circuits.

It should be noted that a study by Candelaria-Cook et al. (2022) explored the impact of PAE on resting-state alpha peak frequency in infants. Because this study also examined PAE's effects on children and adolescents, it is discussed at the end of sect. 4.6.2.

4.6.2 Studies with Children, Adolescents, and Young Adults

Woods et al. (2018) investigated the influence of PAE on place learning assessed virtually, examining neural correlates of deficits in 57 children (41 alcohol-exposed; 16 controls; mean age = 9.4 years; 29 boys). The characteristics of the women recruited for this study are the same as those described in sect. 4.5.2.2 (Meintjes et al. 2014). Using functional MRI, the study explored the effects of PAE during place learning in a computer-generated environment. Results indicated that PAE in boys correlated with decreased performance and diminished activation in various brain regions, including the parahippocampal gyrus and thalamus. Conversely, PAE did not significantly affect performance or brain region activation in girls. The study underscored sex-specific differences in navigation strategies, emphasizing that boys tend to use an allocentric strategy, while girls tend to rely more on landmarks. The noted dose-dependent impairment in place learning among heavily exposed boys could be associated with suboptimal activation of the parahippocampal gyrus, a region linked to allocentric navigation. In contrast, the lack of effects in girls indicates that navigational strategies relying on landmarks may be less vulnerable to the influence of PAE.

Ware et al. (2021) examined whether functional connectivity patterns moderate cognition in children (8–16 years old) with PAE compared to non-exposed controls. Standardized attention and executive functioning tasks were administered alongside resting-state functional MRI. Results revealed that PAE was associated with diminished functional connectivity between specific regions, such as the left temporoparietal junction and left ventral frontal cortex. Conversely, PAE was linked to elevated functional connectivity between certain regions, including the left thalamus and dorsal frontal cortex. Crucially, in children with PAE, functional connectivity patterns were indicative of negative predictions for cognitive performance, whereas in controls, they were associated with positive predictions. These results propose that elevated intra-network connectivity coupled with diminished internetwork functional connectivity could signify inefficient network specialization and impaired long-range connectivity within attention network regions following PAE. This might contribute to attention and executive dysfunction in affected children.

A study evaluated the structural and functional integrity of the brains in individuals with FAS (Clark et al. 2000). The sample comprised 19 subjects with an age range of 16–30 years and a mean age of 20.6 years. MRI was utilized to examine brain structure, while positron emission tomography with [^{18}F]-fluorodeoxyglucose assessed brain function. The MRI revealed only one abnormality, a thin corpus callosum, which was associated with the subject displaying the lowest intelligence quotient. Regional cerebral metabolic rate reductions were observed in five brain regions, including the thalamus and basal ganglia. These findings suggest a continuum of neuropathology in FAS, indicating that in cases with relatively less severe intellectual deficits, the underlying cause may be functional rather than structural.

Fagerlund et al. (2006) investigated the presence of brain metabolic alterations in adolescents (14–21 years old) with and without FASD using proton nuclei magnetic resonance spectroscopy, a non-invasive imaging method that allows a direct assessment of the biochemical composition and metabolite irregularities in specific brain regions. Individuals in the PAE group were diagnosed with FAS (n = 3), pFAS (n = 3), and alcohol-related neurodevelopmental disorder (n = 4). Ten matching controls were also studied. In the spectrum obtained through this technique, the predominant resonances emanate from N-acetylaspartate, compounds containing choline, and creatine and phosphocreatine. N-acetylaspartate serves as an indicator of neuronal/axonal density and viability. Choline-containing compounds play a role in membrane synthesis and degradation, while creatine reflects high-energy phosphate metabolism. Ten individuals with confirmed heavy PAE and FASD diagnoses (three with FAS, three with partial pFAS, and four with alcohol-related neurobehavioral disorder) were compared to a control group. The study revealed lower N-acetylaspartate/choline ratio and/or N-acetylaspartate/creatine ratio in various brain regions, including the cerebellar dentate nucleus, corpus callosum, frontal white matter, parietal and frontal cortices, and thalamus, suggesting lasting or permanent changes in brain metabolism due to PAE. Interestingly, an increase in the glial markers choline and creatine was observed, while the neuronal marker N-acetylaspartate remained unchanged, indicating that these metabolic alterations primarily involve glial cells rather than neurons.

In a study conducted by Pinner et al. (2020), the objective was to enhance the early detection of FASD by investigating the interrelation between brain function, measured through magnetoencephalography, and brain structure (axonal density, diameter, and myelination in white matter), assessed by fractional anisotropy from DTI, in adolescents aged 12–21 years. Maternal alcohol consumption was verified by legal records (e.g., documentation of drinking while intoxicated), maternal interview, and/or several accounts of maternal drinking during pregnancy. The researchers employed joint independent component analysis to examine the connection between magnetoencephalography and functional anisotropy and its correlation with behavior. Five components were identified, with one particular component (#15) showing alterations induced by PAE. Specifically, associations were noted with fractional anisotropy values in various brain regions implicated in the motor pathway, including the anterior portions of the right hemisphere corpus callosum, anterior right thalamic radiation (connecting thalamus to prefrontal cortex), posterior limb of the left internal capsule, short-range association fibers in both the right and left hemispheres, the superior region of the left corona radiata, and the anterior portion of the right corona radiata. Notably, fractional anisotropy values were higher in healthy controls compared to individuals with FASD, and increased magnetoencephalography amplitudes were correlated with higher fractional anisotropy values. These findings suggest that integrating magnetoencephalography and fractional anisotropy data reveals unique associations between brain structure and function, potentially aiding in the differentiation of adolescents with FASD from typically developing controls.

Stephen et al. (2021) studied the effects of PAE on corticothalamic connectivity in children (8–13 years old) with and without FASD. A multidisciplinary team assessed all children and established the FASD diagnosis by consensus. The study investigated the relationship between alpha power and corticothalamic connectivity. Studies support the involvement of lateral geniculate thalamic pacemaker neurons and thalamocortical projections in the generation of resting alpha oscillations (Valdes-Hernandez et al. 2010; Hughes et al. 2004; Silva et al. 1991). Alpha oscillations are associated with various cognitive functions including memory, attention, and language (Ippolito et al. 2022). The study employed rest magnetoencephalography with eyes open or closed, along with MRI for structural and DTI data. Magnetoencephalography spectral analysis was conducted for both sensor and source data, and mean fractional anisotropy in corticothalamic regions of interest was estimated. The FASD group exhibited reduced mean fractional anisotropy in three corticothalamic regions of interest (posterior limb of internal capsule-left hemisphere, posterior limb of internal capsule-right hemisphere, anterior limb of the internal capsule-left hemisphere) and this was significantly correlated with alpha power at both sensor and source levels. In conclusion, this research underscores differences in corticothalamic connectivity between children with FASD and those without, with this connectivity being linked to resting alpha power. These results provide additional insights into the underlying mechanisms influencing altered behavior in children with FASD. Future investigations should explore whether resting alpha oscillations impact attention capabilities in this population.

Candelaria-Cook et al. (2022) further explored the impact of PAE on resting-state alpha peak frequency in infants, children, and adolescents (6 months to 17 years old), providing novel insights into the developmental trajectory of this measure. As mentioned above, resting alpha oscillations are influenced by corticothalamic tracts and thalamic pacemaker neurons. Infants' PAE status was determined either through maternal self-report or by detecting positive alcohol metabolite biomarkers. Utilizing resting-state magnetoencephalography data, alpha peak frequency was determined from parietal/occipital regions. The findings revealed that alpha peak frequency increases as age increases in both groups but was significantly reduced in subjects with PAE/FASD compared to controls across developmental stages. This finding is consistent with the results of a study with 8–12-year-old children with FASD who exhibited reduced peak alpha power (Candelaria-Cook et al. 2021). In a targeted subgroup of participants, the study explored the relationship between alpha peak frequency and brain structure. Among typically developing children, a positive correlation emerged, indicating that higher alpha peak frequencies were associated with increased cerebral white matter volume. However, no such correlation was observed within the group affected by PAE and FASD. Additionally, both the control and PAE/FASD groups displayed a significant negative correlation, linking higher alpha peak frequencies with reduced total gray matter volume. These findings underscore the persistent adverse effects of PAE on processes dependent on normal thalamic functioning.

In summary, the combined results of these functional studies underscore that PAE disrupts functional thalamic connectivity, impacting cognition and highlighting a continuum of neuropathology throughout development. Overall, the findings provide a limited understanding of how thalamic function is affected by PAE and stress the need for additional studies.

4.7 Conclusion and Future Directions

The collective findings from structural and functional studies with animal models underscore the pivotal role of the thalamus in the pathophysiology of FASD. Studies with animal models on the impact of PAE on thalamic structures and circuits have yielded valuable insights. The studies encompassed a range of developmental stages, focusing on both the first and second trimester-equivalent exposures, as well as the critical third trimester-equivalent period. Studies examining PAE during the rodent equivalent of the first two trimesters of human pregnancy revealed multifaceted alterations in the thalamo-cortico-thalamic loop, emphasizing the specificity of damage in different regions. The examination of the vibrissal somatosensory cortical barrel network provided a model for studying compromised sensory function following PAE. Additionally, research on binge alcohol exposure during the first trimester equivalent highlighted differential effects on thalamocortical vs. corticothalamic projections, underlining the complexity of PAE's impact on thalamic circuits. Investigations into the ventrobasal thalamic nucleus shed light on its unique response to PAE, involving alterations in postnatal neurogenesis and neurotrophin responsiveness. The third trimester-equivalent exposure studies delved into apoptotic neurodegeneration, unveiling intricate timelines of vulnerability across thalamic nuclei. The findings showcased the differential effects of alcohol on various thalamic regions, with implications for cognitive processes. The vulnerability of the anterior thalamic nuclei was further characterized in our laboratory, showing that third trimester-equivalent alcohol exposure induces significant changes in the function of retrosplenial cortex-projecting neurons that are part of the limbic memory system. Additionally, the nucleus reuniens emerged as a crucial thalamic nucleus affected by PAE that could underlie deficits in executive function. While structural studies have contributed valuable information, there is a need for additional studies on PAE's effects on thalamic function at different developmental stages. The comprehensive investigation into the effects of alcohol on thalamic morphology and function in animal models of FASD will continue to provide valuable insights into the intricate consequences of PAE.

Human studies have provided strong evidence of thalamic involvement in the neurophysiological deficits associated with FASD. Notably, findings from super-resolution fetal MRI reveal that even minor PAE affects specific fetal brain regions, such as the corpus callosum and periventricular zone. Transitioning to the neonatal period, investigations uncover significant decreases in total gray matter and specific subcortical regions due to PAE. Cranial sonography emerges as a valuable tool that should be utilized more, expanding diagnostic capabilities for PAE-induced alterations in the thalamus and other brain regions. Moving into childhood, adolescence, and young adulthood, studies unveil enduring negative effects on thalamic morphology, with regionally pervasive brain tissue reductions linked to the degree of PAE. While emphasizing the sensitivity of specific structures to PAE even without facial dysmorphology, these collective findings underscore the intricate interplay between PAE and thalamic structural changes, necessitating consideration of variables like age, concurrent exposures, and subtype differences for a comprehensive understanding across developmental stages. However, due to the limited resolution of MRI, the field must turn to postmortem human brain research to unravel the neurobiological foundations of volumetric MRI observations in small brain regions such as the thalamic nuclei. Similarly, postmortem studies could provide insights by considering findings from MRI investigations, refining hypotheses, and focusing on specific brain areas. Only through the integration of these methodologies can we comprehensively elucidate the pathophysiology of FASD. There is a strong need to increase the number of brain samples of individuals with FASD in brain banks around the world.

Significant progress has been made in the exploration of PAE's effects on thalamic function, employing advanced techniques like functional MRI, positron emission tomography, magnetic resonance spectroscopy, and magnetoencephalography. Studies with infants have illuminated early consequences, revealing alterations in interhemispheric functional brain connectivity, particularly in relation to motor networks, and the impact of prenatal marijuana exposure on early brain functional circuitry. Investigations with children, adolescents, and young adults shed light on diverse facets, including place learning deficits associated with decreased activation in multiple brain regions, altered functional connectivity patterns influencing cognition, and lasting metabolic changes in various brain regions. The above-presented findings underscore the need for comprehensive understanding in addressing the intricate dynamics of thalamic function affected by in-utero alcohol exposure across different developmental stages, with potential implications for clinical interventions and targeted support. The intersection of structural and functional investigations reveals a limited understanding of how PAE induces regionally distinct alterations in thalamic circuits, therefore influencing cognitive processes across developmental stages.

Moreover, the existing body of research underscores the critical need for better imaging tools to enhance precision in mapping thalamic structure and function. For instance, it was recently discovered that the development of the human thalamus follows a lateral-to-medial pattern. Additionally, thalamocortical connectivity exhibits a posterior-to-anterior gradient, with thalamofrontal connectivity emerging later than other connections within the thalamocortical network (Zheng et al. 2023). Future imaging studies should address whether PAE affects these processes, leading to an abnormal thalamic developmental trajectory.

Moving forward, expanding human investigations, especially in older age groups, will be imperative for a comprehensive understanding of FASD neuropathology. It is also important to determine if alterations in the sensorimotor functions of the thalamus indirectly impact cognition. The intricate dynamics of thalamic function affected by PAE call for continued interdisciplinary efforts, incorporating advanced imaging methodologies and integrating insights from both animal models and human studies. Such an approach is pivotal not only for advancing our understanding of FASD but also for informing targeted therapeutic interventions and support strategies tailored to the unique challenges faced by individuals affected by PAE. As mentioned earlier, the thalamus is not just a passive relay station but an active player in adaptive processes. Since thalamic neurons have the capacity for synaptic plasticity even in the mature brain, it is crucial to explore whether their stimulation—for example, via targeted neurorehabilitation or transcranial focused ultrasound—can offer therapeutic benefits for individuals with FASD (Kim et al. 2023; Munivenkatappa and Agrawal 2016).

Acknowledgments Supported by NIH grants R01 AA015614 and P50 AA022534 (CFV). The authors declare no conflict of interest. While preparing this work, Grammarly and ChatGPT were used to improve language and readability. After using these tools, the authors reviewed and edited the content as needed and take full responsibility for the publication's content.

References

Aggleton JP, O'Mara SM (2022) The anterior thalamic nuclei: core components of a tripartite episodic memory system. Nat Rev Neurosci 23(8):505–516

Ayoub L, Aring E, Gyllencreutz E, Landgren V, Svensson L, Landgren M et al (2023) Visual and ocular findings in children with fetal alcohol spectrum disorders (FASD): validating the FASD eye code in a clinical setting. BMJ Open Ophthalmol 8(1):e001215

Bird CW, Mayfield SS, Lopez KM, Dunn BR, Feng A, Roberts BT et al (2023) Binge-like ethanol exposure during the brain growth spurt disrupts the function of retrosplenial cortex-projecting anterior thalamic neurons in adolescent mice. Neuropharmacology 241:109738

Boateng T, Beauchamp K, Torres F, Ruffaner-Hanson CD, Pinner JFL, Vakamudi K et al (2023) Brain structural differences in children with fetal alcohol spectrum disorder and its subtypes. Front Neurosci 17:1152038

Candelaria-Cook FT, Schendel ME, Flynn L, Cerros C, Kodituwakku P, Bakhireva LN et al (2022) Decreased resting-state alpha peak frequency in children and adolescents with fetal alcohol spectrum disorders or prenatal alcohol exposure. Dev Cogn Neurosci 57:101137

Candelaria-Cook FT, Schendel ME, Flynn L, Hill DE, Stephen JM (2021) Altered resting-state neural oscillations and spectral power in children with fetal alcohol Spectrum disorder. Alcohol Clin Exp Res 45(1):117–130

Chao OY, Nikolaus S, Yang YM, Huston JP (2022) Neuronal circuitry for recognition memory of object and place in rodent models. Neurosci Biobehav Rev 141:104855

Clark CM, Li D, Conry J, Conry R, Loock C (2000) Structural and functional brain integrity of fetal alcohol syndrome in nonretarded cases. Pediatrics 105(5):1096–1099

Coulter CL, Leech RW, Schaefer GB, Scheithauer BW, Brumback RA (1993) Midline cerebral dysgenesis, dysfunction of the hypothalamic-pituitary axis, and fetal alcohol effects. Arch Neurol 50(7):771–775

Dolleman-van der Weel MJ, Griffin AL, Ito HT, Shapiro ML, Witter MP, Vertes RP et al (2019) The nucleus reuniens of the thalamus sits at the nexus of a hippocampus and medial prefrontal cortex circuit enabling memory and behavior. Learn Mem 26(7):191–205

Donald KA, Fouche JP, Roos A, Koen N, Howells FM, Riley EP et al (2016a) Alcohol exposure in utero is associated with decreased gray matter volume in neonates. Metab Brain Dis 31(1):81–91

Donald KA, Ipser JC, Howells FM, Roos A, Fouche JP, Riley EP et al (2016b) Interhemispheric functional brain connectivity in neonates with prenatal alcohol exposure: preliminary findings. Alcohol Clin Exp Res 40(1):113–121

Dudink J, Jeanne Steggerda S (2020) Horsch S, eur USbg. State-of-the-art neonatal cerebral ultrasound: technique and reporting. Pediatr Res 87(Suppl 1):3–12

Dursun I, Jakubowska-Dogru E, van der List D, Liets LC, Coombs JL, Berman RF (2011) Effects of early postnatal exposure to ethanol on retinal ganglion cell morphology and numbers of neurons in the dorsolateral geniculate in mice. Alcohol Clin Exp Res 35(11):2063–2074

Ethen MK, Ramadhani TA, Scheuerle AE, Canfield MA, Wyszynski DF, Druschel CM et al (2009) Alcohol consumption by women before and during pregnancy. Matern Child Health J 13(2):274–285

Fagerlund A, Heikkinen S, Autti-Ramo I, Korkman M, Timonen M, Kuusi T et al (2006) Brain metabolic alterations in adolescents and young adults with fetal alcohol spectrum disorders. Alcohol Clin Exp Res 30(12):2097–2104

Farber NB, Creeley CE, Olney JW (2010) Alcohol-induced neuroapoptosis in the fetal macaque brain. Neurobiol Dis 40(1):200–206

Fish EW, Holloway HT, Rumple A, Baker LK, Wieczorek LA, Moy SS et al (2016) Acute alcohol exposure during neurulation: Behavioral and brain structural consequences in adolescent C57BL/6J mice. Behav Brain Res 311:70–80

Furumiya J, Hashimoto Y (2011) Effects of ethanol exposure on spatial learning in mice during synaptogenesis. Nihon Arukoru Yakubutsu Igakkai Zasshi 46(2):250–259

Granato A, Santarelli M, Sbriccoli A, Minciacchi D (1995) Multifaceted alterations of the thalamo-cortico-thalamic loop in adult rats prenatally exposed to ethanol. Anat Embryol (Berl) 191(1):11–23

Grewen K, Salzwedel AP, Gao W (2015) Functional connectivity disruption in neonates with prenatal marijuana exposure. Front Hum Neurosci 9:601

Gursky ZH, Klintsova AY (2022) Rat model of late gestational alcohol exposure produces similar life-Long changes in thalamic nucleus Reuniens following moderate- versus high-dose insult. Alcohol Alcohol 57(4):413–420

Gursky ZH, Savage LM, Klintsova AY (2019) Nucleus reuniens of the midline thalamus of a rat is specifically damaged after early postnatal alcohol exposure. Neuroreport 30(10):748–752

Gursky ZH, Savage LM, Klintsova AY (2021) Executive functioning-specific behavioral impairments in a rat model of human third trimester binge drinking implicate prefrontal-thalamo-hippocampal circuitry in fetal alcohol Spectrum disorders. Behav Brain Res 405:113208

Gursky ZH, Spillman EC, Klintsova AY (2020) Single-day postnatal alcohol exposure induces apoptotic cell death and causes long-term neuron loss in rodent thalamic nucleus Reuniens. Neuroscience 435:124–134

Hughes SW, Lorincz M, Cope DW, Blethyn KL, Kekesi KA, Parri HR et al (2004) Synchronized oscillations at alpha and theta frequencies in the lateral geniculate nucleus. Neuron 42(2):253–268

Hughes P, Weinberger E, Shaw DW (1991) Linear areas of echogenicity in the thalami and basal ganglia of neonates: an expanded association. Work in progress Radiology 179(1):103–105

Ieraci A, Herrera DG (2006) Nicotinamide protects against ethanol-induced apoptotic neurodegeneration in the developing mouse brain. PLoS Med 3(4):e101

Ikonomidou C, Bittigau P, Ishimaru MJ, Wozniak DF, Koch C, Genz K et al (2000) Ethanol-induced apoptotic neurodegeneration and fetal alcohol syndrome. Science 287(5455):1056–1060

Inamura N, Ono K, Takebayashi H, Zalc B, Ikenaka K (2011) Olig2 lineage cells generate GABAergic neurons in the prethalamic nuclei, including the zona incerta, ventral lateral geniculate nucleus and reticular thalamic nucleus. Dev Neurosci 33(2):118–129

Ippolito G, Bertaccini R, Tarasi L, Di Gregorio F, Trajkovic J, Battaglia S et al (2022) The role of alpha oscillations among the Main neuropsychiatric disorders in the adult and developing human brain: evidence from the last 10 years of research. Biomedicines 10(12)

Jarmasz JS, Basalah DA, Chudley AE, Del Bigio MR (2017) Human brain abnormalities associated with prenatal alcohol exposure and fetal alcohol Spectrum disorder. J Neuropathol Exp Neurol 76(9):813–833

Jimenez VA, Wang X, Newman N, Walter NAR, Gonzales S, Lo JO et al (2019) Detecting neurodevelopmental effects of early-gestation ethanol exposure: a nonhuman primate model of ethanol drinking during pregnancy. Alcohol Clin Exp Res 43(2):250–261

Kautz-Turnbull C, Adams TR, Petrenko CLM (2022) The strengths and positive influences of children with fetal alcohol Spectrum disorders. Am J Intellect Dev Disabil 127(5):355–368

Ke Z, Wang X, Liu Y, Fan Z, Chen G, Xu M et al (2011) Ethanol induces endoplasmic reticulum stress in the developing brain. Alcohol Clin Exp Res 35(9):1574–1583

Kim HC, Lee W, Weisholtz DS, Yoo SS (2023) Transcranial focused ultrasound stimulation of cortical and thalamic somatosensory areas in human. PLoS One 18(7):e0288654

Kosif R (2016) The thalamus: a review of its functional anatomy. Medical Research Archives 4(8)

Lebel C, Rasmussen C, Wyper K, Walker L, Andrew G, Yager J et al (2008) Brain diffusion abnormalities in children with fetal alcohol spectrum disorder. Alcohol Clin Exp Res 32(10):1732–1740

Ledig M, Ciesielski L, Simler S, Lorentz JG, Mandel P (1988) Effect of pre- and postnatal alcohol consumption on GABA levels of various brain regions in the rat offspring. Alcohol Alcohol 23(1):63–67

Livy DJ, Maier S, West JR (2001) Fetal alcohol exposure and temporal vulnerability: effects of binge-like alcohol exposure on the ventrolateral nucleus of the thalamus. Alcohol Clin Exp Res 25(5):774–780

Martin-Lopez D, Jimenez-Jimenez D, Cabanes-Martinez L, Selway RP, Valentin A, Alarcon G (2017) The role of thalamus versus cortex in epilepsy: evidence from human ictal Centromedian recordings in patients

assessed for deep brain stimulation. Int J Neural Syst 27(7):1750010
Meintjes EM, Narr KL, van der Kouwe AJ, Molteno CD, Pirnia T, Gutman B et al (2014) A tensor-based morphometry analysis of regional differences in brain volume in relation to prenatal alcohol exposure. Neuroimage Clin 5:152–160
Miller MW, Dow-Edwards DL (1993) Vibrissal stimulation affects glucose utilization in the trigeminal/somatosensory system of normal rats and rats prenatally exposed to ethanol. J Comp Neurol 335(2):283–284
Minciacchi D, Granato A, Santarelli M, Sbriccoli A (1993) Modifications of thalamo-cortical circuitry in rats prenatally exposed to ethanol. Neuroreport 4(4):415–418
Mitchell ES, Snyder-Keller A (2003) C-fos and cleaved caspase-3 expression after perinatal exposure to ethanol, cocaine, or the combination of both drugs. Brain Res Dev Brain Res 147(1–2):107–117
Molnar Z, Kwan KY. Development and evolution of Thalamocortical connectivity. Cold Spring Harb Perspect Biol. 2024;16(1)
Mooney SM, Miller MW (1999) Effects of prenatal exposure to ethanol on systems matching: the number of neurons in the ventrobasal thalamic nucleus of the mature rat. Brain Res Dev Brain Res 117(1):121–125
Mooney SM, Miller MW (2001) Effects of prenatal exposure to ethanol on the expression of bcl-2, bax and caspase 3 in the developing rat cerebral cortex and thalamus. Brain Res 911(1):71–81
Mooney SM, Miller MW (2007) Postnatal generation of neurons in the ventrobasal nucleus of the rat thalamus. J Neurosci 27(19):5023–5032
Mooney SM, Miller MW (2011) Role of neurotrophins on postnatal neurogenesis in the thalamus: prenatal exposure to ethanol. Neuroscience 179:256–266
Munivenkatappa A, Agrawal A (2016) Role of thalamus in recovery of traumatic brain injury. J Neurosci Rural Pract 7(Suppl 1):S76–SS9
Nakagawa Y (2019) Development of the thalamus: from early patterning to regulation of cortical functions. Wiley Interdiscip Rev Dev Biol 8(5):e345
Nakhid D, McMorris C, Sun H, Gibbard WB, Tortorelli C, Lebel C (2022) Brain volume and magnetic susceptibility differences in children and adolescents with prenatal alcohol exposure. Alcohol Clin Exp Res 46(10):1797–1807
Nardelli A, Lebel C, Rasmussen C, Andrew G, Beaulieu C (2011) Extensive deep gray matter volume reductions in children and adolescents with fetal alcohol spectrum disorders. Alcohol Clin Exp Res 35(8):1404–1417
Nguyen VT, Tieng QM, Mardon K, Zhang C, Chong S, Galloway GJ et al (2020) Magnetic resonance imaging and micro-computed tomography reveal brain morphological abnormalities in a mouse model of early moderate prenatal ethanol exposure. Neurotoxicol Teratol 77:106849
Olney JW, Tenkova T, Dikranian K, Muglia LJ, Jermakowicz WJ, D'Sa C et al (2002) Ethanol-induced caspase-3 activation in the in vivo developing mouse brain. Neurobiol Dis 9(2):205–219
Papia MF, Burke MW, Zangenehpour S, Palmour RM, Ervin FR, Ptito M (2010) Reduced soma size of the M-neurons in the lateral geniculate nucleus following foetal alcohol exposure in non-human primates. Exp Brain Res 205(2):263–271
Patten AR, Fontaine CJ, Christie BR (2014) A comparison of the different animal models of fetal alcohol spectrum disorders and their use in studying complex behaviors. Front Pediatr 2:93
Pinner JFL, Coffman BA, Stephen JM (2020) Covariation between brain function (MEG) and structure (DTI) differentiates adolescents with fetal alcohol Spectrum disorder from typically developing controls. Neuroscience 449:74–87
Popova S, Charness ME, Burd L, Crawford A, Hoyme HE, Mukherjee RAS et al (2023) Fetal alcohol spectrum disorders. Nat Rev Dis Primers 9(1):11
Powrozek TA, Zhou FC (2005) Effects of prenatal alcohol exposure on the development of the vibrissal somatosensory cortical barrel network. Brain Res Dev Brain Res 155(2):135–146
Rodriguez CI, Davies S, Calhoun V, Savage DD, Hamilton DA (2016) Moderate prenatal alcohol exposure alters functional connectivity in the adult rat brain. Alcohol Clin Exp Res 40(10):2134–2146
Santarelli M, Granato A, Sbriccoli A, Gobbi G, Janiri L, Minciacchi D (1995) Alterations of the thalamocortical system in rats prenatally exposed to ethanol are prevented by concurrent administration of acetyl-L-carnitine. Brain Res 698(1–2):241–247
Scholpp S, Lumsden A (2010) Building a bridal chamber: development of the thalamus. Trends Neurosci 33(8):373–380
Segobin S, Pitel AL (2021) The specificity of thalamic alterations in Korsakoff's syndrome: implications for the study of amnesia. Neurosci Biobehav Rev 130:292–300
Sheridan N, Tadi P (2024) Neuroanatomy, Thalamic Nuclei. StatPearls. Treasure Island (FL) ineligible companies. Disclosure: Prasanna Tadi declares no relevant financial relationships with ineligible companies
Shi W, Xianyu A, Han Z, Tang X, Li Z, Zhong H et al (2017) Ontogenetic establishment of order-specific nuclear organization in the mammalian thalamus. Nat Neurosci 20(4):516–528
Silva LR, Amitai Y, Connors BW (1991) Intrinsic oscillations of neocortex generated by layer 5 pyramidal neurons. Science 251(4992):432–435
Smiley JF, Bleiwas C, Marino BM, Vaddi P, Canals-Baker S, Wilson DA et al (2023) Estimates of total neuron number show that neonatal ethanol causes immediate and lasting neuron loss in cortical and subcortical areas. Front Neurosci 17:1186529
Smith IF, Gursky ZH, Klintsova AY (2022) Representation of prefrontal axonal efferents in the thalamic nucleus reuniens in a rodent model of fetal alcohol exposure during third trimester. Front Behav Neurosci 16:993601

Sowell ER, Leow AD, Bookheimer SY, Smith LM, O'Connor MJ, Kan E et al (2010) Differentiating prenatal exposure to methamphetamine and alcohol versus alcohol and not methamphetamine using tensor-based brain morphometry and discriminant analysis. J Neurosci 30(11):3876–3885

Stephen JM, Hill DE, Candelaria-Cook FT (2021) Examining the effects of prenatal alcohol exposure on corticothalamic connectivity: a multimodal neuroimaging study in children. Dev Cogn Neurosci 52:101019

Stuempflen M, Schwartz E, Diogo MC, Glatter S, Pfeiler B, Kienast P et al (2023) Fetal MRI based brain atlas analysis detects initial in utero effects of prenatal alcohol exposure. Cereb Cortex 33(11):6852–6861

Taylor PA, Jacobson SW, van der Kouwe A, Molteno CD, Chen G, Wintermark P et al (2015) A DTI-based tractography study of effects on brain structure associated with prenatal alcohol exposure in newborns. Hum Brain Mapp 36(1):170–186

Tenkova T, Young C, Dikranian K, Labruyere J, Olney JW (2003) Ethanol-induced apoptosis in the developing visual system during synaptogenesis. Invest Ophthalmol Vis Sci 44(7):2809–2817

Torrico TJ, Munakomi S (2024) Neuroanatomy, Thalamus. StatPearls. Treasure Island (FL) ineligible companies. Disclosure: Sunil Munakomi declares no relevant financial relationships with ineligible companies

Ullah N, Naseer MI, Ullah I, Kim TH, Lee HY, Kim MO (2013) Neuroprotective profile of pyruvate against ethanol-induced neurodegeneration in developing mice brain. Neurol Sci 34(12):2137–2143

Ullah N, Naseer MI, Ullah I, Lee HY, Koh PO, Kim MO (2011) Protective effect of pyruvate against ethanol-induced apoptotic neurodegeneration in the developing rat brain. Neuropharmacology 61(8):1248–1255

Valdes-Hernandez PA, Ojeda-Gonzalez A, Martinez-Montes E, Lage-Castellanos A, Virues-Alba T, Valdes-Urrutia L et al (2010) White matter architecture rather than cortical surface area correlates with the EEG alpha rhythm. NeuroImage 49(3):2328–2339

Vue TY, Aaker J, Taniguchi A, Kazemzadeh C, Skidmore JM, Martin DM et al (2007) Characterization of progenitor domains in the developing mouse thalamus. J Comp Neurol 505(1):73–91

Wang X, Cuzon Carlson VC, Studholme C, Newman N, Ford MM, Grant KA et al (2020) In utero MRI identifies consequences of early-gestation alcohol drinking on fetal brain development in rhesus macaques. Proc Natl Acad Sci USA 117(18):10035–10044

Ware AL, Long X, Lebel C (2021) Functional connectivity of the attention networks is altered and relates to neuropsychological outcomes in children with prenatal alcohol exposure. Dev Cogn Neurosci 48:100951

Weeland CJ, Vriend C, van der Werf Y, Huyser C, Hillegers M, Tiemeier H et al (2022) Thalamic subregions and obsessive-compulsive symptoms in 2,500 children from the general population. J Am Acad Child Adolesc Psychiatry 61(2):321–330

White SA, Weber JN, Howard CD, Favero CB (2015) Effects of binge ethanol exposure during first-trimester equivalent on corticothalamic neurons in Swiss Webster outbred mice. Neuroreport 26(18):1083–1088

Woods KJ, Thomas KGF, Molteno CD, Jacobson JL, Jacobson SW, Meintjes EM (2018) Prenatal alcohol exposure affects brain function during place learning in a virtual environment differently in boys and girls. Brain Behav 8(11):e01103

Worden R, Bennett MS, Neacsu V (2021) The thalamus as a blackboard for perception and planning. Front Behav Neurosci 15:633872

Wozniak DF, Hartman RE, Boyle MP, Vogt SK, Brooks AR, Tenkova T et al (2004) Apoptotic neurodegeneration induced by ethanol in neonatal mice is associated with profound learning/memory deficits in juveniles followed by progressive functional recovery in adults. Neurobiol Dis 17(3):403–414

Wozniak JR, Riley EP, Charness ME (2019) Clinical presentation, diagnosis, and management of fetal alcohol spectrum disorder. Lancet Neurol 18(8):760–770

Zhang CR, Kurniawan ND, Yamada L, Fleming W, Kaminen-Ahola N, Ahola A et al (2019) Early gestational ethanol exposure in mice: effects on brain structure, energy metabolism and adiposity in adult offspring. Alcohol 75:1–10

Zheng W, Zhao L, Zhao Z, Liu T, Hu B, Wu D (2023) Spatiotemporal developmental gradient of thalamic morphology, microstructure, and connectivity from the third trimester to early infancy. J Neurosci 43(4):559–570

Adult Hippocampal Neurogenesis as a Therapeutic Target in Fetal Alcohol Spectrum Disorder

5

Lee Anna Cunningham, Elif Tunc-Ozcan, and Arasely M. Rodriguez

Abstract

This review is focused on adult hippocampal neurogenesis as a potential therapeutic target in fetal alcohol spectrum disorder (FASD). Adult hippocampal neurogenesis refers to the production of new hippocampal dentate granule cells (DGCs) from a replenishable pool of neural stem and progenitor cells throughout life. Adult-generated DGCs have been shown to exert a profound influence on hippocampal network activity in experimental animals and have been implicated in the regulation of many hippocampal-dependent behaviors and emotional states, including certain forms of learning and memory, anxiety, mood, and stress resilience. While adult hippocampal neurogenesis in humans remains controversial, many studies support its existence and impact on hippocampal function in human health and disease. Here, we review mechanisms of adult hippocampal neurogenesis under physiological conditions, as described primarily in rodent brain, its impact on network activity and behavior, and the negative effects of developmental alcohol exposure on this process. We then explore hippocampal neurogenesis as a potential target for FASD therapy using pharmacological and neurophysiological approaches known to stimulate adult hippocampal neurogenesis, currently available for clinical use in FASD patients.

Keywords

Adult hippocampal neurogenesis · Prenatal alcohol exposure · FASD · Gestational alcohol exposure · Antidepressants · Transcranial stimulation · Dentate gyrus

5.1 Introduction

Fetal alcohol spectrum disorder (FASD) covers a spectrum of neurological outcomes, spanning from severe cognitive impairment due to high prenatal alcohol exposure to milder behavioral issues resulting from moderate alcohol exposure during pregnancy (May et al. 2014). FASD is the foremost environmentally caused intellectual and neurodevelopmental disability, with an estimated annual economic burden as high as $4 billion in the United States (Lupton et al. 2004). The prevalence of FASD among school-age children within the United States has been estimated at ~2.5% (May et al. 2014). According to a recent Centers for Disease Control and Prevention (CDC) report, approximately 1 in 10 women consume alcohol

L. A. Cunningham (✉) · E. Tunc-Ozcan
A. M. Rodriguez
Department of Neurosciences, University of New Mexico Health Sciences Center,
Albuquerque, New Mexico, USA
e-mail: leeanna@salud.unm.edu

C. F. Valenzuela et al. (eds.), *Effects of Alcohol on the Brain across the Lifespan*, Advances in Experimental Medicine and Biology 1473, https://doi.org/10.1007/978-3-031-81908-7_5

during pregnancy at some point (Tan et al. 2015), yet there are limited interventions available to address the cognitive and neurobehavioral issues linked with FASD (Kodituwakku and Kodituwakku 2011). Both clinical and preclinical studies on FASD demonstrate that even moderate prenatal alcohol exposure can lead to impairment of brain function (Petrelli et al. 2018). Notably, over 80% of women who report drinking during pregnancy consume moderate levels of alcohol (1–3 drinks per day), whereas less than 20% engage in binge drinking (> 4 drinks per day). While prenatal alcohol exposure affects many brain regions, structural and functional deficits in the hippocampus are widely documented (Berman and Hannigan 2000; Autti-Ramo et al. 2002; Willoughby et al. 2008; Nardelli et al. 2011).

Preclinical research in rodent models of FASD has demonstrated impairment of adult hippocampal neurogenesis in offspring exposed to alcohol during various stages of brain development. Persistent reductions in postnatal neurogenesis occur across a range of developmental exposure times, some of which are designed to mimic exposures during the first and second trimester of human brain development, corresponding to the entire gestational period in mice and rats, or third trimester-equivalent exposures, where alcohol is administered during the first 2 weeks after birth (Choi et al. 2005; Ieraci and Herrera 2007; Gil-Mohapel et al. 2010a, 2011; Klintsova et al. 2007, 2012; Kajimoto et al. 2013, 2016; Gustus et al. 2020). Because adult hippocampal neurogenesis has been shown in many experimental studies to regulate aspects of hippocampal function that are often impaired in FASD (including learning, mood, anxiety, and stress), it is important to explore this unique form of brain plasticity as a therapeutic target that could mitigate some of these cognitive and neurobehavioral problems (Streissguth and O'Malley 2000; O'Connor and Paley 2009; Hellemans et al. 2010; Pei et al. 2011). Here we summarize current knowledge regarding adult hippocampal neurogenesis and its role in hippocampal plasticity, followed by a review of developmental alcohol's effects on these processes. We then explore potential pharmacological and neurophysiological approaches to remediate neurogenesis deficits, including use of a commonly prescribed selective serotonin reuptake inhibitor (SSRI) and non-invasive transcranial stimulation strategies. We focus on these therapeutics because they both stimulate neurogenesis in experimental animals, and they have been utilized with some benefit or are currently being tested in clinical FASD.

5.2 Adult Hippocampal Neurogenesis

Neurogenesis refers to the creation of new neurons from neural stem cells (NSCs). Neurogenesis reaches its highest levels during early development when new neurons are rapidly formed to construct the neural networks essential for brain function. Neurogenesis then ceases in most regions of the mammalian brain following brain development. The rate of hippocampal neurogenesis peaks during mid-gestation in humans, and during the first postnatal week in laboratory rats and mice (Bond et al. 2022). Within the hippocampal dentate gyrus, NSCs transition from a rapidly dividing to a largely quiescent state, with peak proliferation at postnatal day 3 (Bond et al. 2020). However, quiescent NSCs persist within the subgranular zone (SGZ) of the hippocampal dentate gyrus and continuously generate low numbers of new dentate granule cells (DGCs) throughout life at a much slower tempo and a more protracted maturational timeline than what occurs during developmental neurogenesis. The continuous production of new DGCs that integrate into the existing circuitry can profoundly modify hippocampal function long after developmental neurogenesis has ceased (Kempermann and Gage 1999; Ming and Song 2011; Bonaguidi et al. 2012; Kempermann et al. 2015).

The generation of DGCs in the adult hippocampus is a dynamically regulated stepwise process. Radial glial cells, type 1 cells, are the largely quiescent NSCs located in the dentate SGZ. Once activated, these NSCs give rise to neural progenitors, type 2 cells. These cells have self-renewal capacity, but at some point, they undergo asymmetric cell division and generate neuroblasts, type 3 cells, which rapidly exit the cell cycle and migrate a short distance into the dentate granule cell layer, where they mature to become functional DGCs that integrate into pre-existing hippocampal circuitry (Bonaguidi et al. 2012; Kempermann et al. 2015). The production, survival, and integration of newborn neurons is a highly regulated process that is dependent upon the expression and action of many factors within the neurogenic niche as well as the activity of the hippocampal circuitry itself. Indeed, many more immature neurons are produced that survive to maturity as functionally integrated newborn DGCs (Biebl et al. 2000; Dranovsky et al. 2011; Al-Onaizi et al. 2020; Babcock et al. 2021; Amanollahi et al. 2023). Adult-born DGCs substantially remodel the pre-existing hippocampal circuitry (Braun and Jessberger 2014), and are thought to play a unique role in hippocampal circuitry as they traverse through an immature state in which they have physiologically distinct properties from mature neurons (Ge et al. 2007, 2008). For example, young adult-generated DGCs are more excitable than mature granule neurons that are generated during development, and exhibit a lowered threshold for long-term potentiation (LTP). Many studies in experimental animals have demonstrated that these immature DGCs (4–6 weeks old in mouse and rat) play critical roles in different forms of hippocampal-dependent functions (Ming and Song 2011; Rodriguez-Iglesias et al. 2019; Li et al. 2023), including memory consolidation (Kumar et al. 2020), spatial/temporal pattern separation (Clelland et al. 2009), affect regulation (Snyder et al. 2011), social recognition (Cope et al. 2020), stress-induced anxiety (Planchez et al. 2021), and social avoidance (Lagace et al. 2010). It has been postulated that adult-generated DGCs may be recruited to networks that represent the environment in which they mature, distinct from those DGCs generated early during development (Tashiro et al. 2007; Aimone et al. 2009, 2014).

While ongoing debate surrounds the extent and duration of adult hippocampal neurogenesis in the adult human brain (Kempermann 2014; Sorrells et al. 2018; Lucassen et al. 2019), mounting evidence suggests that hippocampal neurogenesis may continue throughout life in human (Eriksson et al. 1998; Knoth et al. 2010; Spalding et al. 2013; Boldrini et al. 2018; Moreno-Jimenez et al. 2019; Tobin et al. 2019). It has been estimated that hundreds of new neurons are integrated into the hippocampus daily in middle-aged adults (Spalding et al. 2013; Lucassen et al. 2019). These estimates are concurrent across distinct studies utilizing carbon dating of genomic neuronal DNA (Spalding et al. 2013), and stereological analysis of hippocampal neurogenesis in postmortem tissue from individuals free of psychiatric treatments or cognitive impairment (Boldrini et al. 2018). Even a relatively small addition of these new neurons could significantly influence hippocampal circuits relevant to human behavior due to their unique electrophysiological and synaptic properties (Aimone and Gage 2011; Snyder and Cameron 2012; Snyder 2019).

5.3 Functional Roles of Adult Hippocampal Neurogenesis

Increased numbers of young adult-born DGCs are associated with improved hippocampal function, while decreased numbers are associated with deficient hippocampal function (Bonaguidi et al. 2012; Braun and Jessberger 2014; Kempermann et al. 2015). For example, inhibiting the proliferation of SGZ neural progenitors and subsequent production of new DGCs with irradiation or genetic methods lead to changes in affective behaviors in animal models and abolishes the effect of antidepressants (Santarelli et al. 2003; Wang et al. 2008; Surget et al. 2011; Brooker et al. 2017). Inhibiting the activity of adult-born immature neurons promotes stress susceptibility while increasing the numbers or activity of adult-born neurons confers stress resilience (Hill et al. 2015; Anacker et al. 2018; Tunc-Ozcan et al. 2019; Segi-Nishida and Suzuki 2022). Further, chronic stress reduces adult neurogenesis, which contributes to depression and anxiety-like symptoms in rodents (Tunc-Ozcan et al. 2019, 2021) and primates (Wu et al. 2014), and decreases hippocampal volume in humans (Wang et al. 2010; Boldrini et al. 2013). Recently, it has been shown that adult-born neurons directly inhibit the activity of mature granule cells in the dentate gyrus, which resulted in increased resilience against chronic stress (Anacker et al. 2018). Also, the survival of adult-born immature neurons is increased by living in an enriched environment in mice previously exposed to social defeat stress (Schloesser et al. 2010). Silencing adult-generated DGCs via transgenic tools blocks the stress resilience afforded by enriched environment (Schloesser et al. 2010; Pereira-Caixeta et al. 2017).

Immature neurons can bidirectionally manipulate hippocampal network activity based on incoming synaptic information, as lateral entorhinal inputs reduce the activity of mature neurons, while medial entorhinal inputs increase it (Luna et al. 2019). In addition, adult-born immature neurons form synapses with CA3 pyramidal neurons and interneurons both in the dentate gyrus and CA3, which gives them the ability for disynaptic feedback inhibition on mature granule cells (Toni et al. 2008; Temprana et al. 2015). Interestingly, immature neurons transform the synapses of existing mature granule cells, which lead to rewiring of the hippocampal circuitry (Toni and Schinder 2015). These properties provide great flexibility to adult-born neurons and partly explain their involvement in learning and memory functions as well as pattern separation.

Adult-born neurons contribute to various forms of learning and memory, such as spatial navigation, contextual fear conditioning, and pattern discrimination. Factors that reduce adult neurogenesis impair spatial learning (Snyder et al. 2005; Dupret et al. 2008; Garthe et al. 2009), while increasing the number of adult-born neurons improves spatial learning (Sahay et al. 2011). Eliminating adult-born neurons leads to various learning and memory deficits, measured by hippocampal-dependent behavioral tests, such as the Morris water maze, Barnes maze, and radial arm (Winocur et al. 2006; Jessberger et al. 2009; Kitamura et al. 2009). Also, silencing adult-born neurons impairs the acquisition and recall of contextual fear conditioning, which depends on the functional hippocampus (Huckleberry et al. 2018). Deficits in fear conditioning have been repeatedly accompanied by reduced numbers or complete ablation of adult-born neurons (Tronel et al. 2012). On the other hand, running increases the number of adult-born neurons while improving spatial learning and memory (Sahay et al. 2011; Vivar et al. 2023). Likewise, living in an enriched environment makes mice learn more flexibly in a spatial learning and memory task, due to increased adult neurogenesis (Garthe et al. 2009).

Pattern separation is the discrimination of highly similar contexts or events that occur close to each other in time. Pattern separation is crucial to avoid interference between memories and is regulated by adult hippocampal neurogenesis. In the context fear discrimination test, which measures the freezing behavior of animals in a chamber where a mild foot shock is delivered compared to a very similar chamber with no foot shock, animals with impaired adult neurogenesis perform worse than controls (Tronel et al. 2012). However, in the same task, increasing adult neurogenesis by enhancing the survival of adult-born neurons improves pattern separation, as measured by increased freezing in the shock context compared to no shock context (Sahay et al. 2011). Also, increased neurogenesis as a result of exercise leads to better spatial discrimination in a touch screen test (Creer et al. 2010). Nevertheless, stress-induced impairments in adult neurogenesis lead to deficits in pattern separation, which is proposed to be one of the underlying mechanisms for affective disorders and cognitive decline, especially post-traumatic stress disorder (Gandy et al. 2017; Wu et al. 2021).

5.4 Adult Hippocampal Neurogenesis and FASD

The significance of studying adult hippocampal neurogenesis as a potential therapeutic target in FASD is primarily drawn from animal studies indicating lasting reductions in neurogenesis after exposure to alcohol during development (Choi et al. 2005; Ieraci and Herrera 2007; Gil-Mohapel et al. 2010a, b, 2011; 2014; Klintsova et al. 2007, 2012; Kajimoto et al. 2013, 2016;

Gustus et al. 2020). In addition, adult hippocampal neurogenesis regulates many hippocampal-dependent behaviors that are also disrupted in clinical FASD (Streissguth and O'Malley 2000; O'Connor and Paley 2009; Hellemans et al. 2010).

Using a continuous access model of gestational alcohol exposure in mice and bromodeoxyuridine (BrdU) labeling of hippocampal progenitors, early studies first demonstrated that moderate prenatal alcohol exposure (PAE) (peak maternal blood alcohol of 80–120 mg%) had no effect on adult hippocampal neurogenesis when mice were housed under standard conditions. However, in this moderate exposure paradigm, offspring were unable to mount a neurogenic response to housing under enriched living conditions, which is one of the most potent drivers of adult hippocampal neurogenesis (Choi et al. 2005). Control mice displayed a two- to threefold increase in neurogenesis under conditions of social and physical enrichment, whereas PAE mice displayed no neurogenic response. Impairment of enrichment-mediated neurogenesis was not due to impaired progenitor proliferation or size of the progenitor pool, but appeared to be due to impaired survival of newly generated DGCs in response to enrichment. Using inducible transgenic reporter mice to fate map the neurogenic lineage in this PAE model, subsequent studies provided evidence to indicate decreased survival and integration of newly generated early postmitotic DGCs coupled with impaired pattern discrimination learning (Kajimoto et al. 2013). Although the numbers of adult-generated DGCs in PAE mice were significantly diminished compared to controls under enriched conditions, patch-clamp recordings of retrovirally birthdated DGCs demonstrated a marked increase in the frequency of spontaneous excitatory input onto newborn DGCs in PAE suggesting a compensatory response and alterations in circuit function (Kajimoto et al. 2016). Interestingly, PAE blunted enrichment-mediated increases in dendritic complexity within both developmentally generated as well as adult-generated DGCs, indicating an overall impairment of hippocampal structural and functional plasticity (Kajimoto et al. 2013; Gustus et al. 2020). These studies imply that exposure to even moderate levels of alcohol during gestation may limit the ability of individuals to optimally adapt to enriched social and physical experiences, and possibly limit their ability to respond optimally to certain types of behavioral training therapy. This is important, since moderate drinking is the most common pattern of drinking during pregnancy in women. Importantly, moderate alcohol exposure during the third trimester equivalent using vapor chamber delivery to achieve pup blood alcohol levels of ~160 mg% did not impair the neurogenic response to enrichment in mice (Gustus et al. 2019), even though this dosing paradigm resulted in long-term deficits in white matter integrity (Newville et al. 2017, 2022).

Subsequent studies examined the effects of binge-like exposure to ethanol during the early postnatal period in rats, corresponding to the human third trimester equivalent (Klintsova et al. 2007). Using daily intragastric gavage to attain pup blood alcohol concentrations of 315–409 mg% at postnatal age 4–10, these investigators demonstrated a marked reduction in the number of adult-generated DGCs under standard housing conditions, as assessed using BrdU to label proliferating progenitors in adulthood (Klintsova et al. 2007). This group later showed that the deficit in adult hippocampal neurogenesis in response to high dose alcohol during the early postnatal period in rat could be restored to baseline with exercise and social enrichment (Hamilton et al. 2012). Going one step further, Gil-Mohapel et al., (Gil-Mohapel et al. 2011) investigated the impact of developmental alcohol during all three trimester equivalents in rats (i.e., from conception to postnatal day 10), where peak blood alcohol concentrations were ~ 140–194 mg% during gestation (maternal) and 210 mg% (pup) by postnatal day 10. These investigators found no effect on the rate of adult hippocampal neurogenesis under standard housing conditions at this alcohol dosing, but did observe a statistically significant reduction in the expression of an immature neuronal maturation marker, NeuroD, in young adult rats, suggesting dysregulated maturation of adult-generated

DGCs. Using BrdU labeling methods, these investigators later demonstrated an impact of moderate gestational exposure on hippocampal neurogenesis in aged (~1 year old), but not adolescent, rats reared under standard housing conditions (Gil-Mohapel et al. 2014). In contrast, a single very high dose exposure to alcohol at postnatal day 7 (subcutaneous injection of 5.5 gm/kg ethanol) reduced the pool of hippocampal progenitors, blunted progenitor proliferation, and decreased the total number of neuroblasts in young adult mouse hippocampus (4 months of age) under standard housing conditions (Ieraci and Herrera 2007).

Taken together these studies suggest that high dose (binge-level) but not moderate dose alcohol exposure during the early postnatal period in mice and rats (i.e., human third trimester equivalent) markedly impairs adult hippocampal neurogenesis in mice and rats even under standard housing conditions. In contrast, exposure to moderate levels of alcohol throughout gestation (i.e., first and second trimester exposures) has no effect on baseline rates of adult hippocampal neurogenesis under standard housing conditions, but severely impairs the neurogenic response to behavioral challenge of living in an enriched environment and may speed the decline of neurogenesis in aged mice. Although species-specific differences in hippocampal neurogenesis across the lifespan make it challenging to compare across species (Bond et al. 2022), time course analysis of human and animal neurogenesis with respect to lifespan suggests that rodent models may offer insights into human childhood neurogenesis (Snyder 2019). Given that cognitive and mental health issues in FASD emerge during brain development, the persistent impairment of postnatal hippocampal neurogenesis in animal models aligns with clinical observations of reduced hippocampal volume and dysfunctional temporal lobe networks in FASD (Riikonen et al. 1999; Autti-Rämö 2002; Sowell et al. 2007; Willoughby et al. 2008). Due to lack of in vivo neuroimaging markers, it remains unknown whether adult hippocampal neurogenesis is impaired in FASD; however, postmortem analysis has identified impaired hippocampal neurogenesis associated with human neurodegenerative disease (Moreno-Jimenez et al. 2019; Terreros-Roncal et al. 2021).

5.5 Antidepressants and Neurogenesis

Antidepressants are commonly prescribed in adults and children with FASD to deal with comorbid depression and anxiety (Caldwell et al. 2008; Weyrauch et al. 2017; Andreu-Fernández et al. 2024). Senturias et al. (2022) analyzed data from the 2017 IBM Watson Health MarketScan Multistate Medicaid and Commercial Claims focusing on claims for children ages 0 to 17 with a FASD diagnosis, to review psychotropic medication use. They found that antidepressants were the third most common psychotropic drug covered by private insurance and the fifth most common psychotropic drug covered by Medicaid. Wrath et al. (2022) utilized Canada's national FASD database to examine medication prescription patterns. Antidepressants were commonly prescribed to the FASD population, especially if they had salient facial features. Importantly they found that rates of antidepressant use in the FASD population differed with age; patients older than 25 were prescribed antidepressants more commonly than younger cohorts. This shift of antidepressant use may be due to the black box warning of some antidepressants in populations below 25. However, it can also indicate a change in symptomology in FASD patients with age. Females were also more commonly prescribed antidepressants than males (Wrath et al. 2022). The most common class of antidepressant prescribed in the FASD population are selective serotonin reuptake inhibitors (SSRIs) (Durr et al. 2021). Despite their common use in FASD populations, the effectiveness of antidepressants in the FASD population has not been comprehensively addressed. Mela conducted a systematic review of articles addressing the utility of psychotropic drugs on FASD patients (Mela et al. 2018; Mela et al. 2020; Durr et al. 2021). Coe et al. (2001) mentioned that there was a positive response to the antidepressant sertraline (Coe et al. 2001).

Ritfeld et al. (2024) noted the Coe paper as the only paper to discuss effectiveness of antidepressants in patients with FASD.

The cellular and molecular mechanisms underlying the behavioral effects of antidepressants are not well defined. However, adult hippocampal neurogenesis has been implicated repeatedly in antidepressant action, behavioral responses to stress, and the pathophysiology of mood disorders (Snyder 2019). Abnormal hippocampal structure and function are common in patients with mood disorders. Chronic antidepressant treatment improves these abnormalities and prevents hippocampal volume reduction in patients (Boldrini et al. 2013). Also chronic antidepressant treatment increases the number of immature neurons in the SGZ of the dentate gyrus in animal models (Sheline et al. 2003; Malberg 2004; Wang et al. 2008; Zhao et al. 2008; Surget et al. 2011; Boldrini et al. 2013; Clarke et al. 2017). In contrast, reducing the number of adult-born immature DGCs leads to changes in affective behaviors (Snyder et al. 2011) and abolishes the antidepressant effect (Santarelli et al. 2003; Surget et al. 2008, 2011; Wang et al. 2008; Brooker et al. 2017). Further, environmental enrichment and exercise have antidepressant effects and improve mood together with a rise in neurogenesis (Schloesser et al. 2010; Sahay et al. 2011; Vivar et al. 2023).

It has recently been shown that increased numbers of immature DGCs mediate behavioral effects of multiple classes of antidepressants in mice via a shared signaling pathway (Tunc-Ozcan et al. 2021). Furthermore, multiple classes of antidepressants, including tricyclic antidepressants (Han et al. 2011), monoamine oxidase inhibitors (Morais et al. 2014), selective serotonin reuptake inhibitors (SSRIs) (Tunc-Ozcan et al. 2019), and serotonin-noradrenaline reuptake inhibitors (SNRIs) (Belovicova et al. 2017) increase adult-hippocampal neurogenesis.

These antidepressants manifest effects several weeks after beginning the treatment. Noticeably, adult-born immature neurons form synaptic outputs roughly around the first 3 weeks of their birth (Carli et al. 2021). Then, it takes 4 to 6 weeks for adult-born immature neurons to fully incorporate into the dentate gyrus circuit (Snyder et al. 2001; Toni et al. 2007, 2008) and contribute to behavioral changes (Denny et al. 2012; Braun and Jessberger 2014; Brooker et al. 2017). These observations led to the neurogenic hypothesis of antidepressant action, which suggests that the delay of therapeutic responses to antidepressants reflect the time required for adult-born immature neurons to incorporate into the hippocampal circuitry and initiate effective physiological and behavioral changes (Malberg and Schechter 2005; Park 2019; Carli et al. 2021; Wu et al. 2021).

Chronic antidepressant treatment also increases the activity of immature neurons, which was demonstrated using chemogenetic technology in transgenic mice to selectively manipulate the activity of adult-born immature dentate gyrus neurons without affecting the number of newborn cells (Tunc-Ozcan et al. 2019). Remarkably, activating immature DGCs is sufficient to alleviate depression-like behavior and reverse the adverse effects of unpredictable chronic mild stress (Tunc-Ozcan et al. 2019; Rawat et al. 2022). The rapid-acting effects of ketamine, a newly approved antidepressant, are mediated by such an increase in the neuronal activity of adult-generated DGCs, while its sustained effects are facilitated by an increase in the numbers of immature neurons (Rawat et al. 2022, 2024). On the contrary, chronic silencing of immature neurons via chemogenetic tools prevents antidepressant effects and leads to increased depression- and anxiety-like behaviors in mice (Tunc-Ozcan et al. 2019).

These results point to the importance of immature neurons to hippocampal network activity and computations (Alvarez et al. 2016; Drew et al. 2016; Luna et al. 2019) and how this can influence antidepressant action. Since newborn neurons form synapses more readily, are more excitable, and have more synaptic plasticity (Cope and Gould 2019), the complex effects of antidepressants and experiences that have antidepressant-like effects rely on both increased numbers of immature dentate gyrus neurons and their activity plus changes in their synaptic connectivity in parallel to changes in hippocampal circuitry.

Whether SSRI administration can restore adult hippocampal neurogenesis in rodent models of FASD has not been explored. Although SSRIs enhance neural plasticity through a number of mechanisms, studies outlined above demonstrate that stimulation of neurogenesis is a key requirement for its antidepressant action. Notably, previous work in the field of developmental toxicology demonstrated that fluoxetine, a first-line antidepressant, ameliorates depression induced by developmental arsenic exposure via a neurogenic mechanism (Tyler et al. 2014).

5.6 Transcranial Stimulation Strategies

Adult hippocampal neurogenesis is an activity-dependent process that is tightly linked to excitation (Deisseroth et al. 2004; Song et al. 2016). The survival and incorporation of adult-generated DGCs into the existing circuitry requires activation of existing DGCs through activity-dependent synaptic input during a critical period in their maturation (Ge et al. 2007; Kheirbek et al. 2012). Stimulation of hippocampal inputs to newborn DGCs via deep brain stimulation or chemogenetic activation of the entorhinal cortex promotes neurogenesis (Stone et al. 2011; Yun et al. 2018, 2023a); facilitates spatial memory (Stone et al. 2011), behavioral pattern separation and cognitive flexibility (Yun et al. 2023b); and induces antidepressive-like behavior that relies on the generation of new DGCs (Yun et al. 2018). Potential mechanisms by which developmental alcohol impairs adult hippocampal neurogenesis include altered neurophysiological function of the existing hippocampal circuitry and/or inability of newly generated DGCs to respond appropriately to activity-dependent integration. Prior studies have demonstrated that dendritic branching in response to experience is significantly impaired in both developmentally generated and adult-generated DGCs in mice exposed to alcohol throughout gestation (Kajimoto et al. 2016). Indeed, broad impairment of neuronal plasticity following gestational alcohol has been well-documented in preclinical studies of FASD suggesting that disrupted plasticity is a primary contributor to cognitive deficits associated with FASD (Patten et al. 2013; Lantz et al. 2015). Stimulation strategies that target enhanced plasticity may therefore also target adult hippocampal neurogenesis in clinical FASD.

Techniques to stimulate the function of the hippocampal circuit and adult newborn dentate granule cells include electric convulsive therapy, deep brain stimulation, transcranial magnetic stimulation (TMS), and transcranial direct current stimulation (tDCS). Among these, tDCS is a readily accessible, non-invasive, and safe brain stimulation technique that shows promise as a neuromodulatory therapy that is being explored for various neurological and psychiatric conditions (Bikson et al. 2016). Depending on the polarity of the stimulation, tDCS can have a depolarizing or hyperpolarizing effect (Antal et al. 2004). Importantly, tDCS has been shown to have neuroplastic effects (Pelletier and Cicchetti 2015; Das et al. 2016; Barbati et al. 2022). tDCS can increase long-term potentiation in the hippocampus of mice by transiently increasing intracellular calcium, which in turn increases BDNF exon 1 expression through pCREB binding (Podda et al. 2016). This tDCS treatment improved hippocampal-dependent spatial navigation and recognition in the Morris water maze and novel object recognition task (Podda et al. 2016). Furthermore, tDCS has been shown to effect structural plasticity. Paciello et al. (2018) observed an increase in spine density in the auditory cortex after tDCS in this area. Increased spine density has also been demonstrated in the motor cortex after tDCS stimulation, which was accompanied by enhanced forelimb strength in mice (Barbati et al. 2022).

More recent studies have demonstrated that multisession tDCS enhances hippocampal neurogenesis and context discrimination in mice. Yu et al. (2017), applied anodal tDCS on the skull over the hippocampus in mice daily for 10 days and observed increased proliferation of dentate NSCs and increased survival of newly generated DGCs (Yu et al. 2017). In addition, this study demonstrated that this tDCS protocol also enhanced performance on a contextual fear discrimination task that was reliant on adult hippocampal neurogenesis. These results suggest an important role for neurogenesis as a mediator of cognitive improvement following tDCS therapy.

In humans, tDCS can enhance synaptic plasticity in an activity-dependent manner and can promote cognitive and motor functions (Saucedo Marquez et al. 2013; Jones et al. 2017; Pisoni et al. 2018), and represents a potential therapeutic approach to treat clinical FASD. Boroda et al., (2019) describe a randomized controlled clinical trial of tDCS therapy targeting the prefrontal cortex, coupled with cognitive training in children with FASD (Boroda et al. 2020). This study demonstrated that tDCS was well-tolerated in children and that children who underwent tDCS paired with concurrent cognitive training did better than sham controls when evaluated for attention; however, there were no differences in working memory or on a trail-making task. In humans with depression, tDCS that targeted the dorso-lateral prefrontal cortex increased gray matter in the target and also in areas connected to the target, the bilateral posterior cingulate cortex, the subgenual anterior cingulate cortex and the right hippocampus, thalamus and left caudate brain regions (Jog et al. 2023). This shows that tDCS can be used to target depression, a common ailment in the FASD population, and can also affect circuitry. The effects of tDCS are not constrained to the area of interest. tDCS has been investigated as a complimentary technique to facial emotion-recognition training, emotion and empathy training, and social interaction modeling training in adults with autism spectrum disorder (Wilson et al. 2021). In these studies, tDCS subjects displayed increased verbal fluency in emotional words and emotional quotients scores.

TMS is a similar type of non-invasive stimulation therapy that targets brain regions using a magnetic field generated over the scalp to boost neural activity, with some clinical benefit in children with attention deficit hyperactivity disorder (ADHD) (Gomez et al. 2014) and autism spectrum disorder (Gomez et al. 2017). Repetitive TMS (rTMS) has been approved by the FDA for use in major depressive disorder. In mice, rTMS has been shown to increase neurogenesis in radiation-induced brain injury, with reduction in anxiety-like behavior as measured by the open field test and the open plus maze; these mice also showed increased spatial navigation through the Morris water maze (Qin et al. 2024). Interestingly, Qin et al. (2024) showed that blocking BDNF was sufficient to block the rTMS improvements. BDNF is important to neuroplasticity. Ramírez-Rodríguez et al. (2023)saw that rTMS increased neurogenesis in chronically stressed mice, with improvement in depression-like behaviors measured through coat state, open field test, and forced swim test (Ramírez-Rodríguez et al. 2023). The antidepressive effects of rTMS was further observed by Zuo et al. (2020). They observed that rTMS in stress-induced depression improved depression-like behaviors and this was accompanied by increases in hippocampal neurogenesis and proteins involved in neuroplasticity (i.e., BDNF, trkB, p-trkB) (Zuo et al. 2020). Interestingly, a combination therapy of fluoxetine and rTMS on stress-induced depression in mice showed additive improvement on depressive-like behavior (Ramírez-Rodríguez et al. 2023). Although there were additive improvements, there were similar levels of increased neurogenesis in rTMS, fluoxetine, and combo-treated mice compared to control-stressed mice (Ramírez-Rodríguez et al. 2023). A proof-of-concept study on rTMS in FASD children was conducted by Melder et al. (2023). Although well-tolerated, there was no improvement in attention, social-emotional regulation, and executive functions (Melder et al. 2023). However, Melder et al. (2023) suggest that their rTMS intervention protocol was insufficient as it had the lowest number of sessions in total and per week compared to other effective interventions.

Taken together, these studies suggest that transcranial stimulation may hold promise in the treatment of FASD. Although evidence indicates that transcranial stimulation upregulates hippocampal neurogenesis in experimental rodents, it will be important to test this in preclinical models of FASD to determine whether neurogenic deficits can be reversed.

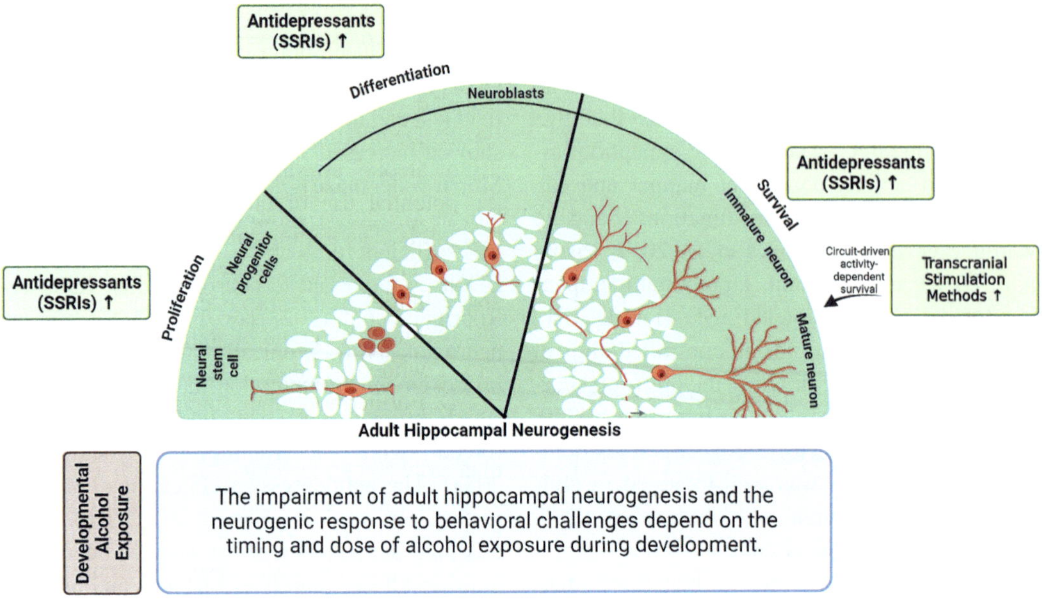

Fig. 5.1 Impact of developmental alcohol on adult hippocampal neurogenesis and potential therapeutic benefits of serotonin reuptake inhibitors (SSRIs) and transcranial stimulation

5.7 Summary and Conclusion

In summary, adult hippocampal neurogenesis is a unique form of neural plasticity that regulates many hippocampal functions that are deficient in FASD. Although it is currently unknown whether adult neurogenesis is impaired in clinical FASD, ample preclinical work has shown detrimental effects of developmental alcohol on adult hippocampal neurogenesis and the neurogenic response to experience. Potential approaches to ameliorate these deficits include antidepressants and transcranial stimulation techniques, which are both available and being tested in clinical FASD (Fig. 5.1). While these approaches are promising, it is important to test their mechanisms of action in preclinical models, including their ability to target hippocampal function and adult hippocampal neurogenesis in preclinical models of FASD.

References

Aimone JB, Gage FH (2011) Modeling new neuron function: a history of using computational neuroscience to study adult neurogenesis. Eur J Neurosci 33:1160–1169. https://doi.org/10.1111/j.1460-9568.2011.07615.x

Aimone JB, Wiles J, Gage FH (2009) Computational influence of adult neurogenesis on memory encoding. Neuron 61:187–202. https://doi.org/10.1016/j.neuron.2008.11.026

Aimone JB, Li Y, Lee SW, Clemenson GD, Deng W, Gage FH (2014) Regulation and function of adult neurogenesis: from genes to cognition. Physiol Rev 94:991–1026. https://doi.org/10.1152/physrev.00004.2014

Al-Onaizi M, Al-Khalifah A, Qasem D, ElAli A (2020) Role of microglia in modulating adult neurogenesis in health and neurodegeneration. Int J Mol Sci 21. https://doi.org/10.3390/ijms21186875

Alvarez DD, Giacomini D, Yang SM, Trinchero MF, Temprana SG, Buttner KA, Beltramone N, Schinder AF (2016) A disynaptic feedback network activated by experience promotes the integration of new granule cells. Science 354:459–465. https://doi.org/10.1126/science.aaf2156

Amanollahi M, Jameie M, Heidari A, Rezaei N (2023) The dialogue between Neuroinflammation and adult neurogenesis: mechanisms involved and alterations in neurological diseases. Mol Neurobiol 60:923–959. https://doi.org/10.1007/s12035-022-03102-z

Anacker C, Luna VM, Stevens GS, Millette A, Shores R, Jimenez JC, Chen B, Hen R (2018) Hippocampal neurogenesis confers stress resilience by inhibiting the ventral dentate gyrus. Nature 559:98–102. https://doi.org/10.1038/s41586-018-0262-4

Andreu-Fernández V, La Maida N, Marquina M, Mirahi A, García-Algar O, Pichini S, Minutillo A (2024) Novel interventions on comorbidities in patients with fetal alcohol Spectrum disorder (FASD): an integrative review. Biomedicines 12:496. https://doi.org/10.3390/biomedicines12030496

Antal A, Varga ET, Kincses TZ, Nitsche MA, Paulus W (2004) Oscillatory brain activity and transcranial direct current stimulation in humans. Neuroreport 15:1307–1310. https://doi.org/10.1097/01.wnr.0000127460.08361.84

Autti-Ramo I, Autti T, Korkman M, Kettunen S, Salonen O, Valanne L (2002) MRI findings in children with school problems who had been exposed prenatally to alcohol. Dev Med Child Neurol 44:98–106

Autti-Rämö I (2002) Foetal alcohol syndrome–a multifaceted condition. Dev Med Child Neurol 44:141–144

Babcock KR, Page JS, Fallon JR, Webb AE (2021) Adult hippocampal neurogenesis in aging and Alzheimer's disease. Stem Cell Reports 16:681–693. https://doi.org/10.1016/j.stemcr.2021.01.019

Barbati SA, Podda MV, Grassi C (2022) Tuning brain networks: the emerging role of transcranial direct current stimulation on structural plasticity. Front Cell Neurosci 16. https://doi.org/10.3389/fncel.2022.945777

Belovicova K, Bogi E, Koprdova R, Ujhazy E, Mach M, Dubovicky M (2017) Effects of venlafaxine and chronic unpredictable stress on behavior and hippocampal neurogenesis of rat dams. Neuro Endocrinol Lett 38:19–26

Berman RF, Hannigan JH (2000) Effects of prenatal alcohol exposure on the hippocampus: spatial behavior, electrophysiology, and neuroanatomy. Hippocampus 10:94–110. https://doi.org/10.1002/(SICI)1098-1063(2000)10:1<94::AID-HIPO11>3.0.CO;2-T

Biebl M, Cooper CM, Winkler J, Kuhn HG (2000) Analysis of neurogenesis and programmed cell death reveals a self-renewing capacity in the adult rat brain. Neurosci Lett 291:17–20. https://doi.org/10.1016/s0304-3940(00)01368-9

Bikson M et al (2016) Safety of transcranial direct current stimulation: evidence based update 2016. Brain Stimul 9:641–661. https://doi.org/10.1016/j.brs.2016.06.004

Boldrini M, Santiago AN, Hen R, Dwork AJ, Rosoklija GB, Tamir H, Arango V, John Mann J (2013) Hippocampal granule neuron number and dentate gyrus volume in antidepressant-treated and untreated major depression. Neuropsychopharmacology 38:1068–1077. https://doi.org/10.1038/npp.2013.5

Boldrini M, Fulmore CA, Tartt AN, Simeon LR, Pavlova I, Poposka V, Rosoklija GB, Stankov A, Arango V, Dwork AJ, Hen R, Mann JJ (2018) Human hippocampal neurogenesis persists throughout aging. Cell Stem Cell 22:589–599 e585. https://doi.org/10.1016/j.stem.2018.03.015

Bonaguidi MA, Song J, Ming GL, Song H (2012) A unifying hypothesis on mammalian neural stem cell properties in the adult hippocampus. Curr Opin Neurobiol 22:754–761. https://doi.org/10.1016/j.conb.2012.03.013

Bond AM, Ming GL, Song H (2022) What is the relationship between hippocampal neurogenesis across different stages of the lifespan? Front Neurosci 16:891713. https://doi.org/10.3389/fnins.2022.891713

Bond AM, Berg DA, Lee S, Garcia-Epelboim AS, Adusumilli VS, Ming GL, Song H (2020) Differential timing and coordination of neurogenesis and Astrogenesis in developing mouse hippocampal subregions. Brain Sci 10. https://doi.org/10.3390/brainsci10120909

Boroda E, Krueger AM, Bansal P, Schumacher MJ, Roy AV, Boys CJ, Lim KO, Wozniak JR (2020) A randomized controlled trial of transcranial direct-current stimulation and cognitive training in children with fetal alcohol spectrum disorder. Brain Stimulation: Basic, Translational, and Clinical Research in Neuromodulation 13:1059–1068. https://doi.org/10.1016/j.brs.2020.04.015

Braun SM, Jessberger S (2014) Adult neurogenesis: mechanisms and functional significance. Development 141:1983–1986. https://doi.org/10.1242/dev.104596

Brooker SM, Gobeske KT, Chen J, Peng CY, Kessler JA (2017) Hippocampal bone morphogenetic protein signaling mediates behavioral effects of antidepressant treatment. Mol Psychiatry 22:910–919. https://doi.org/10.1038/mp.2016.160

Caldwell KK, Sheema S, Paz RD, Samudio-Ruiz SL, Laughlin MH, Spence NE, Roehlk MJ, Alcon SN, Allan AM (2008) Fetal alcohol spectrum disorder-associated depression: evidence for reductions in the levels of brain-derived neurotrophic factor in a mouse model. Pharmacol Biochem Behav 90:614–624. https://doi.org/10.1016/j.pbb.2008.05.004

Carli M, Aringhieri S, Kolachalam S, Longoni B, Grenno G, Rossi M, Gemignani A, Fornai F, Maggio R, Scarselli M (2021) Is adult hippocampal neurogenesis really relevant for the treatment of psychiatric disorders? Curr Neuropharmacol 19:1640–1660. https://doi.org/10.2174/1570159X18666200818194948

Choi IY, Allan AM, Cunningham LA (2005) Moderate fetal alcohol exposure impairs the neurogenic response to an enriched environment in adult mice. Alcohol Clin Exp Res 29:2053–2062

Clarke M, Razmjou S, Prowse N, Dwyer Z, Litteljohn D, Pentz R, Anisman H, Hayley S (2017) Ketamine modulates hippocampal neurogenesis and pro-inflammatory cytokines but not stressor induced neurochemical changes. Neuropharmacology 112:210–220. https://doi.org/10.1016/j.neuropharm.2016.04.021

Clelland CD, Choi M, Romberg C, Clemenson GD Jr, Fragniere A, Tyers P, Jessberger S, Saksida LM, Barker RA, Gage FH, Bussey TJ (2009) A functional role for adult hippocampal neurogenesis in spatial pattern separation. Science 325:210–213. https://doi.org/10.1126/science.1173215

Coe JH, Sidders J, Riley K, Waltermire JA, Hagerman RJ (2001) A survey of medication responses in children and adolescents with fetal alcohol syndrome.

Cope EC, Gould E (2019) Adult neurogenesis, glia, and the extracellular matrix. Cell Stem Cell 24:690–705. https://doi.org/10.1016/j.stem.2019.03.023

Cope EC, Waters RC, Diethorn EJ, Pagliai KA, Dias CG, Tsuda M, Cameron HA, Gould E (2020) Adult-born neurons in the hippocampus are essential for social memory maintenance. eNeuro 7:ENEURO.0182. https://doi.org/10.1523/ENEURO.0182-20.2020

Creer DJ, Romberg C, Saksida LM, van Praag H, Bussey TJ (2010) Running enhances spatial pattern separation in mice. Proc Natl Acad Sci USA 107:2367–2372. https://doi.org/10.1073/pnas.0911725107

Das S, Holland P, Frens MA, Donchin O (2016) Impact of transcranial direct current stimulation (tDCS) on neuronal functions. Front Neurosci 10. https://doi.org/10.3389/fnins.2016.00550

Deisseroth K, Singla S, Toda H, Monje M, Palmer TD, Malenka RC (2004) Excitation-neurogenesis coupling in adult neural stem/progenitor cells. Neuron 42:535–552

Denny CA, Burghardt NS, Schachter DM, Hen R, Drew MR (2012) 4- to 6-week-old adult-born hippocampal neurons influence novelty-evoked exploration and contextual fear conditioning. Hippocampus 22:1188–1201. https://doi.org/10.1002/hipo.20964

Dranovsky A, Picchini AM, Moadel T, Sisti AC, Yamada A, Kimura S, Leonardo ED, Hen R (2011) Experience dictates stem cell fate in the adult hippocampus. Neuron 70:908–923. https://doi.org/10.1016/j.neuron.2011.05.022

Drew LJ, Kheirbek MA, Luna VM, Denny CA, Cloidt MA, Wu MV, Jain S, Scharfman HE, Hen R (2016) Activation of local inhibitory circuits in the dentate gyrus by adult-born neurons. Hippocampus 26:763–778. https://doi.org/10.1002/hipo.22557

Dupret D, Revest JM, Koehl M, Ichas F, De Giorgi F, Costet P, Abrous DN, Piazza PV (2008) Spatial relational memory requires hippocampal adult neurogenesis. PLoS One 3:e1959. https://doi.org/10.1371/journal.pone.0001959

Durr MR, Petryk S, Mela M, DesRoches A, Wekerle M, Newaz S (2021) Utilization of psychotropic medications in children with FASD: a retrospective review. BMC Pediatr 21:512. https://doi.org/10.1186/s12887-021-02986-5

Eriksson PS, Perfilieva E, Björk-Eriksson T, Alborn A-M, Nordborg C, Peterson DA, Gage FH (1998) Neurogenesis in the adult human hippocampus. Nat Med 4:1313

Gandy K, Kim S, Sharp C, Dindo L, Maletic-Savatic M, Calarge C (2017) Pattern separation: a potential marker of impaired hippocampal adult neurogenesis in major depressive disorder. Front Neurosci 11:571. https://doi.org/10.3389/fnins.2017.00571

Garthe A, Behr J, Kempermann G (2009) Adult-generated hippocampal neurons allow the flexible use of spatially precise learning strategies. PLoS One 4:e5464. https://doi.org/10.1371/journal.pone.0005464

Ge S, Sailor KA, Ming GL, Song H (2008) Synaptic integration and plasticity of new neurons in the adult hippocampus. J Physiol 586:3759–3765. https://doi.org/10.1113/jphysiol.2008.155655

Ge S, Yang CH, Hsu KS, Ming GL, Song H (2007) A critical period for enhanced synaptic plasticity in newly generated neurons of the adult brain. Neuron 54:559–566. https://doi.org/10.1016/j.neuron.2007.05.002

Gil-Mohapel J, Boehme F, Kainer L, Christie BR (2010a) Hippocampal cell loss and neurogenesis after fetal alcohol exposure: insights from different rodent models. Brain Res Rev 64:283–303. https://doi.org/10.1016/j.brainresrev.2010.04.011

Gil-Mohapel J, Simpson JM, Titterness AK, Christie BR (2010b) Characterization of the neurogenesis quiescent zone in the rodent brain: effects of age and exercise. Eur J Neurosci 31:797–807. https://doi.org/10.1111/j.1460-9568.2010.07132.x

Gil-Mohapel J, Boehme F, Patten A, Cox A, Kainer L, Giles E, Brocardo PS, Christie BR (2011) Altered adult hippocampal neuronal maturation in a rat model of fetal alcohol syndrome. Brain Res 1384:29–41. https://doi.org/10.1016/j.brainres.2011.01.116

Gil-Mohapel J, Titterness AK, Patten AR, Taylor S, Ratzlaff A, Ratzlaff T, Helfer J, Christie BR (2014) Prenatal ethanol exposure differentially affects hippocampal neurogenesis in the adolescent and aged brain. Neuroscience 273:174–188. https://doi.org/10.1016/j.neuroscience.2014.05.012

Gomez L, Vidal B, Morales L, Baez M, Maragoto C, Galvizu R, Vera H, Cabrera I, Zaldivar M, Sanchez A (2014) Low frequency repetitive transcranial magnetic stimulation in children with attention deficit/hyperactivity disorder. Preliminary results. Brain Stimul 7:760–762. https://doi.org/10.1016/j.brs.2014.06.001

Gomez L, Vidal B, Maragoto C, Morales LM, Berrillo S, Vera Cuesta H, Baez M, Denis M, Marin T, Cabrera Y, Sanchez A, Alarcon C, Selguera M, Llanez Y, Dieguez L, Robinson M (2017) Non-invasive brain stimulation for children with autism Spectrum disorders: a short-term outcome study. Behav Sci (Basel) 7. https://doi.org/10.3390/bs7030063

Gustus K, Li L, Newville J, Cunningham LA (2020) Functional and structural correlates of impaired enrichment-mediated adult hippocampal neurogenesis in a mouse model of prenatal alcohol exposure. Brain Plast 6:67–82. https://doi.org/10.3233/BPL-200112

Gustus K, Lozano E, Newville J, Li L, Valenzuela CF, Cunningham LA (2019) Resistance of postnatal hippocampal neurogenesis to alcohol toxicity in a third trimester-equivalent mouse model of gestational alcohol exposure. Alcohol Clin Exp Res 43:2504. https://doi.org/10.1111/acer.14207

Hamilton GF, Boschen KE, Goodlett CR, Greenough WT, Klintsova AY (2012) Housing in environmental complexity following wheel running augments survival of newly generated hippocampal neurons in a rat model of binge alcohol exposure during the third trimester equivalent. Alcohol Clin Exp Res 36:1196–1204. https://doi.org/10.1111/j.1530-0277.2011.01726.x

Han X, Tong J, Zhang J, Farahvar A, Wang E, Yang J, Samadani U, Smith DH, Huang JH (2011) Imipramine treatment improves cognitive outcome associated with enhanced hippocampal neurogenesis after traumatic brain injury in mice. J Neurotrauma 28:995–1007. https://doi.org/10.1089/neu.2010.1563

Hellemans KG, Sliwowska JH, Verma P, Weinberg J (2010) Prenatal alcohol exposure: fetal programming and later life vulnerability to stress, depression and anxiety disorders. Neurosci Biobehav Rev 34:791–807. https://doi.org/10.1016/j.neubiorev.2009.06.004

Hill AS, Sahay A, Hen R (2015) Increasing adult hippocampal neurogenesis is sufficient to reduce anxiety and depression-like behaviors. Neuropsychopharmacology 40:2368–2378. https://doi.org/10.1038/npp.2015.85

Huckleberry KA, Shue F, Copeland T, Chitwood RA, Yin W, Drew MR (2018) Dorsal and ventral hippocampal adult-born neurons contribute to context fear memory. Neuropsychopharmacology 43:2487–2496. https://doi.org/10.1038/s41386-018-0109-6

Ieraci A, Herrera DG (2007) Single alcohol exposure in early life damages hippocampal stem/progenitor cells and reduces adult neurogenesis. Neurobiol Dis 26:597–605. https://doi.org/10.1016/j.nbd.2007.02.011

Jessberger S, Clark RE, Broadbent NJ, Clemenson GD Jr, Consiglio A, Lie DC, Squire LR, Gage FH (2009) Dentate gyrus-specific knockdown of adult neurogenesis impairs spatial and object recognition memory in adult rats. Learn Mem 16:147–154. https://doi.org/10.1101/lm.1172609

Jog MA, Anderson C, Kubicki A, Boucher M, Leaver A, Hellemann G, Iacoboni M, Woods R, Narr K (2023) Transcranial direct current stimulation (tDCS) in depression induces structural plasticity. Sci Rep 13:2841. https://doi.org/10.1038/s41598-023-29792-6

Jones KT, Peterson DJ, Blacker KJ, Berryhill ME (2017) Frontoparietal neurostimulation modulates working memory training benefits and oscillatory synchronization. Brain Res 1667:28–40. https://doi.org/10.1016/j.brainres.2017.05.005

Kajimoto K, Allan A, Cunningham LA (2013) Fate analysis of adult hippocampal progenitors in a murine model of fetal alcohol spectrum disorder (FASD). PLoS One 8:e73788. https://doi.org/10.1371/journal.pone.0073788

Kajimoto K, Valenzuela CF, Allan AM, Ge S, Gu Y, Cunningham LA (2016) Prenatal alcohol exposure alters synaptic activity of adult hippocampal dentate granule cells under conditions of enriched environment. Hippocampus 26:1078–1087. https://doi.org/10.1002/hipo.22588

Kempermann G (2014) Off the beaten track: new neurons in the adult human striatum. Cell 156:870–871. https://doi.org/10.1016/j.cell.2014.02.027

Kempermann G, Gage FH (1999) New nerve cells for the adult brain. Sci Am 280:48–53. https://doi.org/10.1038/scientificamerican0599-48

Kempermann G, Song H, Gage FH (2015) Neurogenesis in the adult hippocampus. Cold Spring Harb Perspect Biol 7:a018812. https://doi.org/10.1101/cshperspect.a018812

Kheirbek MA, Tannenholz L, Hen R (2012) NR2B-dependent plasticity of adult-born granule cells is necessary for context discrimination. J Neurosci 32:8696–8702. https://doi.org/10.1523/JNEUROSCI.1692-12.2012

Kitamura T, Saitoh Y, Takashima N, Murayama A, Niibori Y, Ageta H, Sekiguchi M, Sugiyama H, Inokuchi K (2009) Adult neurogenesis modulates the hippocampus-dependent period of associative fear memory. Cell 139:814–827. https://doi.org/10.1016/j.cell.2009.10.020

Klintsova AY, Hamilton GF, Boschen KE (2012) Long-term consequences of developmental alcohol exposure on brain structure and function: therapeutic benefits of physical activity. Brain Sci 3:1–38. https://doi.org/10.3390/brainsci3010001

Klintsova AY, Helfer JL, Calizo LH, Dong WK, Goodlett CR, Greenough WT (2007) Persistent impairment of hippocampal neurogenesis in young adult rats following early postnatal alcohol exposure. Alcohol Clin Exp Res 31:2073–2082. https://doi.org/10.1111/j.1530-0277.2007.00528.x

Knoth R, Singec I, Ditter M, Pantazis G, Capetian P, Meyer RP, Horvat V, Volk B, Kempermann G (2010) Murine features of neurogenesis in the human hippocampus across the lifespan from 0 to 100 years. PLoS One 5:e8809. https://doi.org/10.1371/journal.pone.0008809

Kodituwakku PW, Kodituwakku EL (2011) From research to practice: an integrative framework for the development of interventions for children with fetal alcohol spectrum disorders. Neuropsychol Rev 21:204–223. https://doi.org/10.1007/s11065-011-9170-1

Kumar D et al (2020) Sparse activity of hippocampal adult-born neurons during REM sleep is necessary for memory consolidation. Neuron 107:552–565 e510. https://doi.org/10.1016/j.neuron.2020.05.008

Lagace DC, Donovan MH, DeCarolis NA, Farnbauch LA, Malhotra S, Berton O, Nestler EJ, Krishnan V, Eisch AJ (2010) Adult hippocampal neurogenesis is functionally important for stress-induced social avoidance. Proc Natl Acad Sci USA 107:4436–4441. https://doi.org/10.1073/pnas.0910072107

Lantz CL, Sipe GO, Wong EL, Majewska AK, Medina AE (2015) Effects of developmental alcohol exposure on potentiation and depression of visual cortex responses. Alcohol Clin Exp Res 39:1434–1442. https://doi.org/10.1111/acer.12775

Li YD, Luo YJ, Song J (2023) Optimizing memory performance and emotional states: multi-level enhancement of adult hippocampal neurogenesis. Curr Opin Neurobiol 79:102693. https://doi.org/10.1016/j.conb.2023.102693

Lucassen PJ, Toni N, Kempermann G, Frisen J, Gage FH, Swaab DF (2019) Limits to human neurogenesis-really? Mol Psychiatry 25:2207. https://doi.org/10.1038/s41380-018-0337-5

Luna VM, Anacker C, Burghardt NS, Khandaker H, Andreu V, Millette A, Leary P, Ravenelle R, Jimenez JC, Mastrodonato A, Denny CA, Fenton AA, Scharfman HE, Hen R (2019) Adult-born hippocampal neurons bidirectionally modulate entorhinal inputs into the dentate gyrus. Science 364:578–583. https://doi.org/10.1126/science.aat8789

Lupton C, Burd L, Harwood R (2004) Cost of fetal alcohol spectrum disorders. Am J Med Genet C Semin

Med Genet 127C:42–50. https://doi.org/10.1002/ajmg.c.30015
Malberg JE (2004) Implications of adult hippocampal neurogenesis in antidepressant action. J Psychiatry Neurosci 29:196–205
Malberg JE, Schechter LE (2005) Increasing hippocampal neurogenesis: a novel mechanism for antidepressant drugs. Curr Pharm Des 11:145–155. https://doi.org/10.2174/1381612053382223
May PA, Baete A, Russo J, Elliott AJ, Blankenship J, Kalberg WO, Buckley D, Brooks M, Hasken J, Abdul-Rahman O (2014) Prevalence and characteristics of fetal alcohol spectrum disorders. Pediatrics 134:855–866
Mela M, Okpalauwaekwe U, Anderson T, Eng J, Nomani S, Ahmed A, Barr AM (2018) The utility of psychotropic drugs on patients with fetal alcohol Spectrum disorder (FASD): a systematic review. Psyc Clin Psychopharmacol 28:436–445. https://doi.org/10.1080/24750573.2018.1458429
Mela M, Hanlon-Dearman A, Ahmed A, Rich SD, Densmore R, Reid D, Barr AM, Osser D, Anderson T, Suberu B (2020) Treatment algorithm for the use of psychopharmacological agents in individuals prenatally exposed to alcohol and/or with diagnosis of fetal alcohol spectrum disorder (FASD). J Popul Ther Clin Pharmacol 27:e1–e13
Melder A, Wittmann E, Bulubas L, Dornheim B, Kerber K, Vogelmann U, Campana M, Hubert J, Schmidt V, Heinen F, Padberg F, Landgraf MN (2023) Transcranial magnetic stimulation as a feasible, non-invasive, neuromodulatory intervention in fetal alcohol spectrum disorders. A very first proof of concept. Eur J Paediatr Neurol 47:131–142. https://doi.org/10.1016/j.ejpn.2023.10.004
Ming GL, Song H (2011) Adult neurogenesis in the mammalian brain: significant answers and significant questions. Neuron 70:687–702. https://doi.org/10.1016/j.neuron.2011.05.001
Morais M, Santos PA, Mateus-Pinheiro A, Patricio P, Pinto L, Sousa N, Pedroso P, Almeida S, Filipe A, Bessa JM (2014) The effects of chronic stress on hippocampal adult neurogenesis and dendritic plasticity are reversed by selective MAO-A inhibition. J Psychopharmacol 28:1178–1183. https://doi.org/10.1177/0269881114553646
Moreno-Jimenez EP, Flor-Garcia M, Terreros-Roncal J, Rabano A, Cafini F, Pallas-Bazarra N, Avila J, Llorens-Martin M (2019) Adult hippocampal neurogenesis is abundant in neurologically healthy subjects and drops sharply in patients with Alzheimer's disease. Nat Med 25:554–560. https://doi.org/10.1038/s41591-019-0375-9
Nardelli A, Lebel C, Rasmussen C, Andrew G, Beaulieu C (2011) Extensive deep gray matter volume reductions in children and adolescents with fetal alcohol spectrum disorders. Alcohol Clin Exp Res 35:1404–1417. https://doi.org/10.1111/j.1530-0277.2011.01476.x
Newville J, Valenzuela CF, Li L, Jantzie LL, Cunningham LA (2017) Acute oligodendrocyte loss with persistent white matter injury in a third trimester equivalent mouse model of fetal alcohol spectrum disorder. Glia 65:1317–1332. https://doi.org/10.1002/glia.23164
Newville J, Howard TA, Chavez GJ, Valenzuela CF, Cunningham LA (2022) Persistent myelin abnormalities in a third trimester-equivalent mouse model of fetal alcohol spectrum disorder. Alcohol Clin Exp Res 46:77–86. https://doi.org/10.1111/acer.14752
O'Connor MJ, Paley B (2009) Psychiatric conditions associated with prenatal alcohol exposure. Dev Disabil Res Rev 15:225–234. https://doi.org/10.1002/ddrr.74
Paciello F, Podda MV, Rolesi R, Cocco S, Petrosini L, Troiani D, Fetoni AR, Paludetti G, Grassi C (2018) Anodal transcranial direct current stimulation affects auditory cortex plasticity in normal-hearing and noise-exposed rats. Brain Stimul 11:1008–1023. https://doi.org/10.1016/j.brs.2018.05.017
Park SC (2019) Neurogenesis and antidepressant action. Cell Tissue Res 377:95–106. https://doi.org/10.1007/s00441-019-03043-5
Patten AR, Gil-Mohapel J, Wortman RC, Noonan A, Brocardo PS, Christie BR (2013) Effects of ethanol exposure during distinct periods of brain development on hippocampal synaptic plasticity. Brain Sci 3:1076–1094. https://doi.org/10.3390/brainsci3031076
Pei J, Denys K, Hughes J, Rasmussen C (2011) Mental health issues in fetal alcohol spectrum disorder. J Ment Health 20:438–448. https://doi.org/10.3109/09638237.2011.577113
Pelletier SJ, Cicchetti F (2015) Cellular and molecular mechanisms of action of transcranial direct current stimulation: evidence from in vitro and in vivo models. Int J Neuropsychopharmacol 18:pyu047–pyu047. https://doi.org/10.1093/ijnp/pyu047
Pereira-Caixeta AR, Guarnieri LO, Pena RR, Dias TL, Pereira GS (2017) Neurogenesis inhibition prevents enriched environment to prolong and strengthen social recognition memory, but not to increase BDNF expression. Mol Neurobiol 54:3309–3316. https://doi.org/10.1007/s12035-016-9922-2
Petrelli B, Weinberg J, Hicks GG (2018) Effects of prenatal alcohol exposure (PAE): insights into FASD using PAE mouse models. Biochem Cell Biol 96:131. https://doi.org/10.1139/bcb-2017-0280
Pisoni A, Mattavelli G, Papagno C, Rosanova M, Casali AG, Romero Lauro LJ (2018) Cognitive enhancement induced by anodal tDCS drives circuit-specific cortical plasticity. Cereb Cortex 28:1132–1140. https://doi.org/10.1093/cercor/bhx021
Planchez B, Lagunas N, Le Guisquet AM, Legrand M, Surget A, Hen R, Belzung C (2021) Increasing adult hippocampal neurogenesis promotes resilience in a mouse model of depression. Cells 10. https://doi.org/10.3390/cells10050972
Podda MV, Cocco S, Mastrodonato A, Fusco S, Leone L, Barbati SA, Colussi C, Ripoli C, Grassi C (2016) Anodal transcranial direct current stimulation boosts synaptic plasticity and memory in mice via epigen-

etic regulation of Bdnf expression. Sci Rep 6:22180. https://doi.org/10.1038/srep22180

Qin T, Guo L, Wang X, Zhou G, Liu L, Zhang Z, Ding G (2024) Repetitive transcranial magnetic stimulation ameliorates cognitive deficits in mice with radiation-induced brain injury by attenuating microglial pyroptosis and promoting neurogenesis via BDNF pathway. Cell Commun Signal 22:216. https://doi.org/10.1186/s12964-024-01591-0

Ramírez-Rodríguez GB, Meneses San-Juan D, Rico-Becerra AI, González-Olvera JJ, Reyes-Galindo V (2023) Repetitive transcranial magnetic stimulation and fluoxetine reverse depressive-like behavior but with differential effects on Olig2-positive cells in chronically stressed mice. Neuropharmacology 236:109567. https://doi.org/10.1016/j.neuropharm.2023.109567

Rawat R, Tunc-Ozcan E, McGuire TL, Peng CY, Kessler JA (2022) Ketamine activates adult-born immature granule neurons to rapidly alleviate depression-like behaviors in mice. Nat Commun 13:2650. https://doi.org/10.1038/s41467-022-30386-5

Rawat R, Tunc-Ozcan E, Dunlop S, Tsai YH, Li F, Bertossi R, Peng CY, Kessler JA (2024) Ketamine's rapid and sustained antidepressant effects are driven by distinct mechanisms. Cell Mol Life Sci 81:105. https://doi.org/10.1007/s00018-024-05121-6

Riikonen R, Salonen I, Partanen K, Verho S (1999) Brain perfusion SPECT and MRI in foetal alcohol syndrome. Dev Med Child Neurol 41:652–659. https://doi.org/10.1017/S0012162299001358

Ritfeld GJ, Kable JA, Holton JE, Coles CD (2024) Effectiveness of psychotropic medications in children with prenatal alcohol and drug exposures: a case series and model of care. Child Psychiatry Hum Dev 55:744–753. https://doi.org/10.1007/s10578-022-01451-3

Rodriguez-Iglesias N, Sierra A, Valero J (2019) Rewiring of memory circuits: connecting adult newborn neurons with the help of microglia. Front Cell Dev Biol 7:24. https://doi.org/10.3389/fcell.2019.00024

Sahay A, Scobie KN, Hill AS, O'Carroll CM, Kheirbek MA, Burghardt NS, Fenton AA, Dranovsky A, Hen R (2011) Increasing adult hippocampal neurogenesis is sufficient to improve pattern separation. Nature 472:466–470. https://doi.org/10.1038/nature09817

Santarelli L, Saxe M, Gross C, Surget A, Battaglia F, Dulawa S, Weisstaub N, Lee J, Duman R, Arancio O, Belzung C, Hen R (2003) Requirement of hippocampal neurogenesis for the behavioral effects of antidepressants. Science 301:805–809. https://doi.org/10.1126/science.1083328

Saucedo Marquez CM, Zhang X, Swinnen SP, Meesen R, Wenderoth N (2013) Task-specific effect of transcranial direct current stimulation on motor learning. Front Hum Neurosci 7:333. https://doi.org/10.3389/fnhum.2013.00333

Schloesser RJ, Lehmann M, Martinowich K, Manji HK, Herkenham M (2010) Environmental enrichment requires adult neurogenesis to facilitate the recovery from psychosocial stress. Mol Psychiatry 15:1152–1163. https://doi.org/10.1038/mp.2010.34

Segi-Nishida E, Suzuki K (2022) Regulation of adult-born and mature neurons in stress response and antidepressant action in the dentate gyrus of the hippocampus. Neurosci Res. https://doi.org/10.1016/j.neures.2022.08.010

Senturias Y, Ali MM, West K (2022) Psychotropic medication utilization among children diagnosed with fetal alcohol Spectrum disorder. Pediatrics 150:e2022056797. https://doi.org/10.1542/peds.2022-056797

Sheline YI, Gado MH, Kraemer HC (2003) Untreated depression and hippocampal volume loss. Am J Psychiatry 160:1516–1518. https://doi.org/10.1176/appi.ajp.160.8.1516

Snyder JS (2019) Recalibrating the relevance of adult neurogenesis. Trends Neurosci 42:164–178. https://doi.org/10.1016/j.tins.2018.12.001

Snyder JS, Cameron HA (2012) Could adult hippocampal neurogenesis be relevant for human behavior? Behav Brain Res 227:384–390. https://doi.org/10.1016/j.bbr.2011.06.024

Snyder JS, Kee N, Wojtowicz JM (2001) Effects of adult neurogenesis on synaptic plasticity in the rat dentate gyrus. J Neurophysiol 85:2423–2431. https://doi.org/10.1152/jn.2001.85.6.2423

Snyder JS, Hong NS, McDonald RJ, Wojtowicz JM (2005) A role for adult neurogenesis in spatial long-term memory. Neuroscience 130:843–852. https://doi.org/10.1016/j.neuroscience.2004.10.009

Snyder JS, Soumier A, Brewer M, Pickel J, Cameron HA (2011) Adult hippocampal neurogenesis buffers stress responses and depressive behaviour. Nature 476:458–461. https://doi.org/10.1038/nature10287

Song J, Olsen RH, Sun J, Ming GL, Song H (2016) Neuronal circuitry mechanisms regulating adult mammalian neurogenesis. Cold Spring Harb Perspect Biol 8. https://doi.org/10.1101/cshperspect.a018937

Sorrells SF, Paredes MF, Cebrian-Silla A, Sandoval K, Qi D, Kelley KW, James D, Mayer S, Chang J, Auguste KI, Chang EF, Gutierrez AJ, Kriegstein AR, Mathern GW, Oldham MC, Huang EJ, Garcia-Verdugo JM, Yang Z, Alvarez-Buylla A (2018) Human hippocampal neurogenesis drops sharply in children to undetectable levels in adults. Nature 555:377. https://doi.org/10.1038/nature25975

Sowell ER, Lu LH, O'Hare ED, McCourt ST, Mattson SN, O'Connor MJ, Bookheimer SY (2007) Functional magnetic resonance imaging of verbal learning in children with heavy prenatal alcohol exposure. Neuroreport 18:635–639. https://doi.org/10.1097/WNR.0b013e3280bad8dc

Spalding KL, Bergmann O, Alkass K, Bernard S, Salehpour M, Huttner HB, Bostrom E, Westerlund I, Vial C, Buchholz BA, Possnert G, Mash DC, Druid H, Frisen J (2013) Dynamics of hippocampal neurogenesis in adult humans. Cell 153:1219–1227. https://doi.org/10.1016/j.cell.2013.05.002

Stone SS, Teixeira CM, Devito LM, Zaslavsky K, Josselyn SA, Lozano AM, Frankland PW (2011) Stimulation of entorhinal cortex promotes adult neurogenesis and facilitates spatial memory. J Neurosci Off J Soc Neurosci 31:13469–13484. https://doi.org/10.1523/JNEUROSCI.3100-11.2011

Streissguth AP, O'Malley K (2000) Neuropsychiatric implications and long-term consequences of fetal alcohol spectrum disorders. Semin Clin Neuropsychiatry 5:177–190

Surget A, Saxe M, Leman S, Ibarguen-Vargas Y, Chalon S, Griebel G, Hen R, Belzung C (2008) Drug-dependent requirement of hippocampal neurogenesis in a model of depression and of antidepressant reversal. Biol Psychiatry 64:293–301. https://doi.org/10.1016/j.biopsych.2008.02.022

Surget A, Tanti A, Leonardo ED, Laugeray A, Rainer Q, Touma C, Palme R, Griebel G, Ibarguen-Vargas Y, Hen R, Belzung C (2011) Antidepressants recruit new neurons to improve stress response regulation. Mol Psychiatry 16:1177–1188. https://doi.org/10.1038/mp.2011.48

Tan CH, Denny CH, Cheal NE, Sniezek JE, Kanny D (2015) Alcohol use and binge drinking among women of childbearing age–United States, 2011-2013. MMWR Morb Mortal Wkly Rep 64:1042–1046. https://doi.org/10.15585/mmwr.mm6437a3

Tashiro A, Makino H, Gage FH (2007) Experience-specific functional modification of the dentate gyrus through adult neurogenesis: a critical period during an immature stage. J Neurosci 27:3252–3259. https://doi.org/10.1523/JNEUROSCI.4941-06.2007

Temprana SG, Mongiat LA, Yang SM, Trinchero MF, Alvarez DD, Kropff E, Giacomini D, Beltramone N, Lanuza GM, Schinder AF (2015) Delayed coupling to feedback inhibition during a critical period for the integration of adult-born granule cells. Neuron 85:116–130. https://doi.org/10.1016/j.neuron.2014.11.023

Terreros-Roncal J, Moreno-Jimenez EP, Flor-Garcia M, Rodriguez-Moreno CB, Trinchero MF, Cafini F, Rabano A, Llorens-Martin M (2021) Impact of neurodegenerative diseases on human adult hippocampal neurogenesis. Science 374:1106–1113. https://doi.org/10.1126/science.abl5163

Tobin MK, Musaraca K, Disouky A, Shetti A, Bheri A, Honer WG, Kim N, Dawe RJ, Bennett DA, Arfanakis K, Lazarov O (2019) Human hippocampal neurogenesis persists in aged adults and Alzheimer's disease patients. Cell Stem Cell 24:974–982 e973. https://doi.org/10.1016/j.stem.2019.05.003

Toni N, Schinder AF (2015) Maturation and functional integration of new granule cells into the adult hippocampus. Cold Spring Harb Perspect Biol 8:a018903. https://doi.org/10.1101/cshperspect.a018903

Toni N, Laplagne DA, Zhao C, Lombardi G, Ribak CE, Gage FH, Schinder AF (2008) Neurons born in the adult dentate gyrus form functional synapses with target cells. Nat Neurosci 11:901–907. https://doi.org/10.1038/nn.2156

Toni N, Teng EM, Bushong EA, Aimone JB, Zhao C, Consiglio A, van Praag H, Martone ME, Ellisman MH, Gage FH (2007) Synapse formation on neurons born in the adult hippocampus. Nat Neurosci 10:727–734. https://doi.org/10.1038/nn1908

Tronel S, Belnoue L, Grosjean N, Revest JM, Piazza PV, Koehl M, Abrous DN (2012) Adult-born neurons are necessary for extended contextual discrimination. Hippocampus 22:292–298. https://doi.org/10.1002/hipo.20895

Tunc-Ozcan E, Peng CY, Zhu Y, Dunlop SR, Contractor A, Kessler JA (2019) Activating newborn neurons suppresses depression and anxiety-like behaviors. Nat Commun 10:3768. https://doi.org/10.1038/s41467-019-11641-8

Tunc-Ozcan E, Brooker SM, Bonds JA, Tsai YH, Rawat R, McGuire TL, Peng CY, Kessler JA (2021) Hippocampal BMP signaling as a common pathway for antidepressant action. Cell Mol Life Sci 79:31. https://doi.org/10.1007/s00018-021-04026-y

Tyler CR, Solomon BR, Ulibarri AL, Allan AM (2014) Fluoxetine treatment ameliorates depression induced by perinatal arsenic exposure via a neurogenic mechanism. Neurotoxicology 44:98–109. https://doi.org/10.1016/j.neuro.2014.06.006

Vivar C, Peterson B, Pinto A, Janke E, van Praag H (2023) Running throughout middle-age keeps old adult-born neurons wired. eNeuro 10:ENEURO.0084. https://doi.org/10.1523/ENEURO.0084-23.2023

Wang JW, David DJ, Monckton JE, Battaglia F, Hen R (2008) Chronic fluoxetine stimulates maturation and synaptic plasticity of adult-born hippocampal granule cells. J Neurosci 28:1374–1384. https://doi.org/10.1523/JNEUROSCI.3632-07.2008

Wang Z, Neylan TC, Mueller SG, Lenoci M, Truran D, Marmar CR, Weiner MW, Schuff N (2010) Magnetic resonance imaging of hippocampal subfields in posttraumatic stress disorder. Arch Gen Psychiatry 67:296–303. https://doi.org/10.1001/archgenpsychiatry.2009.205

Weyrauch D, Schwartz M, Hart B, Klug MG, Burd L (2017) Comorbid mental disorders in fetal alcohol Spectrum disorders: a systematic review. J Dev Behav Pediatr 38:283. https://doi.org/10.1097/DBP.0000000000000440

Willoughby KA, Sheard ED, Nash K, Rovet J (2008) Effects of prenatal alcohol exposure on hippocampal volume, verbal learning, and verbal and spatial recall in late childhood. J Int Neuropsychol Soc 14:1022–1033. https://doi.org/10.1017/S1355617708081368

Wilson JE, Trumbo MC, Tesche CD (2021) Transcranial direct current stimulation (tDCS) improves empathy and recognition of facial emotions conveying threat in adults with autism Spectrum disorder (ASD). Neuro Regulation 8:87–87. https://doi.org/10.15540/nr.8.2.87

Winocur G, Wojtowicz JM, Sekeres M, Snyder JS, Wang S (2006) Inhibition of neurogenesis interferes with hippocampus-dependent memory function.

Hippocampus 16:296–304. https://doi.org/10.1002/hipo.20163

Wrath AJ, Mela M, Le T, Durr M, DesRoches A, Nomani S, Anderson T, Unsworth K, Cook J (2022) Psychotropic medication usage in individuals with fetal alcohol Spectrum disorders (FASD) and psychiatric co-morbidities in Canada. J Fetal Alcohol Spectrum Dis 4:e70–e90. https://doi.org/10.22374/jfasd.v4iSP1.21

Wu MV, Shamy JL, Bedi G, Choi CW, Wall MM, Arango V, Boldrini M, Foltin RW, Hen R (2014) Impact of social status and antidepressant treatment on neurogenesis in the baboon hippocampus. Neuropsychopharmacology 39:1861–1871. https://doi.org/10.1038/npp.2014.33

Wu Z, Xiao L, Wang H, Wang G (2021) Neurogenic hypothesis of positive psychology in stress-induced depression: adult hippocampal neurogenesis, neuroinflammation, and stress resilience. Int Immunopharmacol 97:107653. https://doi.org/10.1016/j.intimp.2021.107653

Yu F, Shukla DK, Armstrong RC, Marion CM, Radomski KL, Selwyn RG, Dardzinski BJ (2017) Repetitive model of mild traumatic brain injury produces cortical abnormalities detectable by magnetic resonance diffusion imaging, histopathology, and behavior. J Neurotrauma 34:1364–1381. https://doi.org/10.1089/neu.2016.4569

Yun S, Soler I, Tran F, Haas HA, Shi R, Bancroft GL, Suarez M, de Santis CR, Reynolds RP, Eisch AJ (2023a) Behavioral pattern separation and cognitive flexibility are enhanced in a mouse model of increased lateral entorhinal cortex-dentate gyrus circuit activity. bioRxiv. https://doi.org/10.1101/2023.01.26.525756

Yun S, Soler I, Tran FH, Haas HA, Shi R, Bancroft GL, Suarez M, de Santis CR, Reynolds RP, Eisch AJ (2023b) Behavioral pattern separation and cognitive flexibility are enhanced in a mouse model of increased lateral entorhinal cortex-dentate gyrus circuit activity. Front Behav Neurosci 17:1151877. https://doi.org/10.3389/fnbeh.2023.1151877

Yun S, Reynolds RP, Petrof I, White A, Rivera PD, Segev A, Gibson AD, Suarez M, DeSalle MJ, Ito N, Mukherjee S, Richardson DR, Kang CE, Ahrens-Nicklas RC, Soler I, Chetkovich DM, Kourrich S, Coulter DA, Eisch AJ (2018) Stimulation of entorhinal cortex-dentate gyrus circuitry is antidepressive. Nat Med 24:658–666. https://doi.org/10.1038/s41591-018-0002-1

Zhao C, Deng W, Gage FH (2008) Mechanisms and functional implications of adult neurogenesis. Cell 132:645–660. https://doi.org/10.1016/j.cell.2008.01.033

Zuo C, Cao H, Ding F, Zhao J, Huang Y, Li G, Huang S, Jiang H, Jiang Y, Wang F (2020) Neuroprotective efficacy of different levels of high-frequency repetitive transcranial magnetic stimulation in mice with CUMS-induced depression: involvement of the p11/BDNF/Homer1a signaling pathway. J Psychiatr Res 125:152–163. https://doi.org/10.1016/j.jpsychires.2020.03.018

Impact of Developmental Alcohol Exposure on Sleep Physiology

6

Valentina Licheri and Jonathan L. Brigman

Abstract

The present chapter summarizes the clinical and preclinical findings collected to date, showing the impact of developmental alcohol exposure on sleep physiology. Sleep is a complex physiological process that plays a pivotal role in maintaining overall health and well-being via its involvement in regulating physiological, cognitive, and emotional functions. Clinical studies consistently report a high prevalence of sleep disturbances in children and adolescents diagnosed with fetal alcohol spectrum disorders (FASDs), including short sleep duration, sleep anxiety, bedtime resistance, increased sleep fragmentation, and parasomnias. It is established that alcohol consumption during gestation impairs brain development, leading to structural and functional alterations that may affect sleep architecture. In addition, clinical investigations have found a significant correlation between sleep-wake cycle disruptions and cognitive impairments after developmental alcohol exposure, and sleep disturbances are increasingly recognized as a substantial problem among FASD patients. However, the molecular mechanisms underlying these disturbances are poorly understood. Surprisingly, few studies with animal models of FASDs have characterized the effect of developmental ethanol exposure on sleep physiology, and these have focused on high doses. This chapter provides an overview of the current knowledge, reports the sleep disturbances in FASD patients, and then summarizes the gap in understanding the molecular and physiological mechanisms.

Keywords

Prenatal alcohol exposure · Sleep disturbances · Sleep physiology · FASDs · NREM · REM

This chapter explores the impact of developmental alcohol exposure on sleep processes. Alterations in sleep duration and quality have been widely reported in people with FASD, and clinical investigations have found a significant correlation between sleep-wake cycle disruptions and cognitive impairments after developmental alcohol exposure. This chapter describes the present understanding of sleep alterations observed in children and adolescents diagnosed with FASD, first providing a brief overview of the sleep phases and sleep disturbances, and then

V. Licheri (✉) · J. L. Brigman
Department of Neurosciences and New Mexico Alcohol Research Center, School of Medicine, University of New Mexico Health Science Center, Albuquerque, NM, USA
e-mail: vlicheri@salud.unm.edu

C. F. Valenzuela et al. (eds.), *Effects of Alcohol on the Brain across the Lifespan*, Advances in Experimental Medicine and Biology 1473, https://doi.org/10.1007/978-3-031-81908-7_6

summarizing the clinical evidence collected to date. Finally, we review the preclinical studies performed in rodent models to pinpoint the molecular mechanism underlying the sleep alterations. Furthermore, we briefly outline the sleep disturbances observed in adult drinkers to highlight the current knowledge about the relationship between sleep physiology and alcohol consumption, as well as to emphasize the need for the development of innovative therapeutic strategies for FASD and alcohol use disorder (AUD).

6.1 Fetal Alcohol Spectrum Disorders Prevalence and Impact

Fetal alcohol spectrum disorders (FASDs) represent a global health issue due to their high prevalence. The Global Epidemiological Data Report estimates the global prevalence of children and youth diagnosed with FASDs at ~7.7 per 1000 population (Lange et al. 2017). The highest prevalence was observed in the World Health Organization (WHO) European Region at ~19.8 per 1000 population, while the prevalence for the WHO Eastern Mediterranean Region is 0.1 per 1000. Interestingly, the country with the highest estimated prevalence is South Africa (111.1 per 1000 population) (Lange et al. 2017; Popova et al. 2017, 2023). The estimated FASD prevalence in the United States is 1–5% of the population (May et al. 2018). The term FASDs encompasses the group of cognitive and behavioral difficulties observed in children, adolescents, and adults who were exposed to alcohol before birth (Center for Substance Abuse 2014). Persons diagnosed with FASD exhibit deficits in executive function (Mattson et al. 2019; Khoury et al. 2015; Kodituwakku 2009), learning and memory (Crocker et al. 2011; Mattson and Roebuck 2002), and attention (Coles et al. 2002; Mattson et al. 2006). Parents of children with FASD commonly report emotional and behavioral problems, suggesting difficulties in emotional processing (Petrenko et al. 2017). Sleep disturbances are a clinically important symptom of FASD that impacts daytime activities (Wengel et al. 2011). Multiple clinical investigations have observed higher sleep-wake cycle disruptions in children with underlying neurodevelopmental disorders and that decreased sleep quality was correlated with cognitive impairments, including difficulties in cognitive flexibility, memory and learning, attention, and verbal creativity (Jan et al. 2008; Stores 2001). In the context of FASD, the prevalence of children exhibiting sleep alterations, including short sleep duration, sleep fragmentation, sleepwalking, and night awakening, has been estimated to be as high as 80% (Troese et al. 2008; Jan et al. 2010). Despite having the highest prevalence of symptoms reported in clinical studies, sleep disturbances in patients with FASD are still under-investigated and poorly understood, in part due to challenges in early diagnosis (Benz et al. 2009; Ipsiroglu et al. 2013). Physicians often fail to ask about possible alterations in sleep physiology and parents can fail to report these (Bertrand 2009; Ipsiroglu et al. 2013). To date, clinical investigations reporting objective sleep assessments in children with FASD are limited, with few studies using sleep measures such as polysomnography and actigraphy (Kamara and Beauchaine 2020; Pesonen et al. 2009; Wengel et al. 2011; Goril et al. 2016). Surprisingly, there are also a limited number of studies investigating the sleep patterns and architecture in animal models of FASD. Although sleep disturbances represent a serious and significant problem in the FASD community, our current knowledge is not adequate to fully address the issue. Better characterization of the sleep patterns following developmental alcohol exposure may also help explain the cognitive and emotional difficulties commonly seen in children, adolescents, and adults with FASD.

6.2 Sleep Physiology

Sleep is a complicated physiological process characterized by an active state of unconsciousness, during which the brain is reactive primarily to internal stimuli (Brinkman et al. 2024). In addition, sleep plays a pivotal role in multiple biological and physiological processes including metabolic regulation (Schmidt 2014), metabolite clearance from the brain (Xie et al. 2013), brain plasticity, memory processing (Abel et al. 2013; Rasch and Born 2013; Tononi and Cirelli 2014), and maintenance of synaptic homeostasis (Tononi and Cirelli 2003, 2006). Sleep is regulated by two systems: the circadian system and sleep-wake homeostasis. The circadian system controls and synchronizes biological rhythms, including the sleep and wake cycle, hormonal secretion (melatonin and cortisol), and body temperature cycle (Borbély and Achermann 1999; Bathory and Tomopoulos 2017). The activity of the circadian system is regulated by the suprachiasmatic nucleus (SCN) in the hypothalamus (Colwell 2011; Hastings et al. 2019). Furthermore, the sleep-wake cycle is governed by sleep-wake homeostasis, a process that regulates and maintains the balance between wakefulness and sleep (Deboer 2015).

6.2.1 Sleep Phases

Even though sleep is commonly thought of as a state of rest, it is a dynamic process characterized by significant changes in brain wave activity with distinguishable electrophysiological properties. It is organized into two major phases: non-rapid eye movement (NREM) sleep and rapid eye movement (REM) sleep which are cycled throughout the dark phase. NREM sleep can be further segmented into three sub-phases numbered N1, N2, and N3; N3 is also called slow-wave sleep (SWS) (Patel et al. 2024; Brinkman et al. 2024). The NREM sleep phase represents 75–80% of total sleep duration, while the REM covers the remaining 20–25% (Institute of Medicine Committee on Sleep and Research 2006).

Each phase and sub-phase of sleep is associated with distinctive electrophysiological properties. N1 represents the transition from wakefulness to sleep, during which electroencephalogram (EEG) recordings exhibit alpha waves (frequency 8–12 Hz) (Berry et al. 2017; Feriante and Araujo 2024). The brain activity recorded during N2 shows the presence of sleep spindles (brief oscillation of 12–14 Hz) and K-complexes (0.5–2 Hz) (Berry et al. 2017); there are several studies supporting the hypothesis that sleep in this phase is crucial for memory consolidation since it has been found a significant increase in density of sleep spindles in learners compared to the control group (Gais et al. 2002; Feriante and Araujo 2024). N3 is characterized by the presence of delta waves (0.5–2 Hz) (Berry et al. 2017, Feriante and Araujo 2024). REM sleep is well-acknowledged as the phase of dreaming, and its EEG spectral analysis exhibits theta waves (1–3 Hz). REM is also involved in memory processing and consolidation (Boyce et al. 2017; Perogamvros and Schwartz 2012), and in regulating emotional states (Walker and Van Der Helm 2009). Interestingly, NREM and REM sleep phases are involved in different memory processes, with the NREM sleep phase being required to consolidate declarative memory, while the REM sleep phase is thought to be critical for procedural memory (Rasch and Born 2013).

The neuroanatomical underpinnings of sleep are very complex, with various brain regions (see Fig. 6.1) involved in the control and regulation of the sleep-wake cycle, including the basal forebrain, hypothalamus, brainstem, and thalamocortical circuits (Liu and Dan 2019; Wilson et al. 2024; Sulaman et al. 2023). The wake phase is regulated by noradrenergic neurons expressed in the locus coeruleus (LC), serotonergic neurons in the raphe nuclei (RN), and orexinergic neurons of the lateral hypothalamus (LH) (Sulaman et al. 2023; Oh et al. 2019). In contrast, the neuronal populations involved in the sleep transition include GABAergic neurons expressed in the ventrolateral preoptic nuclei (VLPO), median preoptic nuclei (MnPO), brainstem, and the melanin-concentrating hormone-producing (MCH) neurons in the diencephalon (Oh et al. 2019). Interestingly, the basal forebrain plays a crucial role in regulating both sleep and wakefulness. The cholinergic neurons, expressed in this brain area, are active during the wake and REM sleep phases (Lee et al. 2005).

Sleep phases are mediated by specific neuronal populations. NREM sleep is regulated by the GABAergic/galaninergic neurons expressed in the ventrolateral area (VLPO) (Sulaman et al.

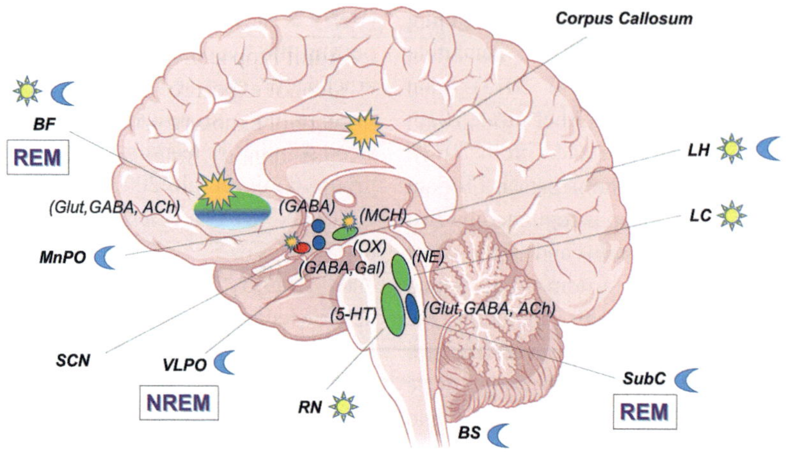

Fig. 6.1 Schematic overview of the key brain regions and neuronal population involved in sleep-wake regulation, sleep phases, and sleep disturbances following developmental alcohol exposure. ☀ = Brain regions involved in the wake phase: locus coeruleus (LC), raphe nuclei (RN), and lateral hypothalamus (LH). ☾= Brain regions involved in the sleep phase: ventrolateral preoptic nucleus (VLPO; NREM phase), subcoeruleus nucleus (SubC; REM phase); median preoptic nuclei (MnPO), melanin-concentrating hormone-producing (MCH) neurons expressed in the diencephalon, brainstem (BS). ☀☾= Brain regions involved in both wake and sleep: basal forebrain (BF) and LH modulate. ✸= Brain areas most affected by developmental alcohol exposure: corpus callosum, LH, BF, and suprachiasmatic nucleus (SCN). Neurotransmitter systems involved in sleep processes: *Glut* glutamatergic neurons; *GABA* GABAergic neurons; *Ach* cholinergic neurons; *OX* orexinergic neurons; *Gal* galaninergic neurons; *5-HT* serotonergic neurons; *NE* noradrenergic neurons

2023). Cortical interneurons, specifically parvalbumin- and somatostatin-expressing interneurons, are involved in the propagation of slow waves during NREM, while the thalamus, specifically GABAergic neurons in the thalamic reticular nucleus, regulates delta wave activity and sleep spindles (Venner et al. 2016; Sulaman et al. 2023).

In contrast, glutamatergic, GABAergic, and cholinergic neurons in the subcoeruleus nucleus are implicated in controlling the REM sleep phase (Sulaman et al. 2023). To further complicate the picture, the transition from NREM to the REM sleep phase is regulated by monoaminergic and orexinergic neurons expressed in the pons, while the GABAergic neurons in the lateral hypothalamus are involved in modulating the REM sleep phase (Sulaman et al. 2023).

Even from this brief overview, it is clear that sleep is a complex process regulated by the orchestrated activity of several brain regions and neurotransmitters. Sleep also changes dynamically during development and is critical for brain maturation during early life, childhood, and adolescence.

6.2.2 Sleep Changes Across Development

Sleep plays a pivotal role in several physiological processes, allowing for optimal physical and mental health maintenance from early life until adulthood. Recent findings highlight the relationship between sleep and brain maturation, characterized by changes in sleep patterns and sleep architecture across ages. During the early stage of life (0–2 years of age), the total sleep duration is typically between 14 and 20 h per day, and sleep patterns are biphasic, including a midday nap and overnight sleep (Iglowstein et al. 2003; Lokhandwala and Spencer 2022). Starting in early childhood (3–5 years), there is a transition from the biphasic to the monophasic pattern, characterized by overnight sleep, that persists until adulthood (Galland et al. 2012). Importantly, the primary function of sleep before 2–3 years of age is to support neural reorganization and learning (Cao et al. 2020). In contrast, after the age of 2–3 years, sleep is involved in maintaining overall health through repair and waste clearance (Cao et al. 2020). The changes observed in sleep architecture across development are characterized by a significant decrease in REM sleep, the primary sleep phase (80% of total sleep) during infancy (Carskadon and Dement 2005), which is unsurprising, given its prominent role in brain maturation (Cao et al. 2020). NREM phases also change during maturation, with the time spent in N3 decreasing in adolescence, while the N2 phase is characterized by sigma activity (10–16 Hz) increasing across childhood (Mcclain et al. 2016). These significant changes in EEG spectral profile reflect the specific role played by each sleep phase in brain development. Extensive literature supports the hypothesis that REM sleep is the phase associated with maturational processes, particularly in the development of the visual cortex and motor system (Dumoulin Bridi et al. 2015; Blumberg et al. 2013). Although the role of NREM sleep in brain development has not been extensively investigated, several hypotheses suggest that it may be involved in refining neural networks through synaptic downscaling and pruning (Knoop et al. 2021).

6.3 Sleep Disorders

The term "sleep disorders" incorporates a wide range of conditions impacting the duration and quality of sleep, leading to impairments in daytime activities. Inadequate sleep can have detrimental effects on overall health and quality of life (Karna et al. 2024). The International Classification of Sleep Disorders (Third Edition, ICSD-3) classifies sleep disorders into several main categories (American Academy of Sleep Medicine 2023):

- *Insomnia*: this sleep disorder is characterized by difficulty falling asleep and/or staying asleep.
- *Sleep-disordered breathing*: this group of disorders is characterized by abnormal and difficult respiration during sleep, including obstructive sleep apnea (OSA), central sleep apnea, sleep-related hypoventilation, and sleep-related hypoxemia (Burman 2017).
- *Central disorders of hypersomnolence*: sleep disorders characterized by excessive daytime sleepiness not due to sleep disruption during the night and/or misaligned circadian rhythms, such as hypersomnia and narcolepsy.
- *Circadian rhythm sleep-wake disorders (CRSD)*: a group of sleep disorders due to misaligned circadian rhythms with the external environment.
- *Parasomnias*: this group of sleep disorders includes abnormal behaviors that occur during the sleep phase without affecting the sleep architecture, including sleepwalking, sleep terrors, sleep talking, sleep-related eating disorders, and nightmare disorders.
- *Sleep-related movement disorders*: a category of sleep disorders characterized by abnormal physical and involuntary movement during the sleep phase, including restless leg syndrome (RLS) and periodic limb movement disorder (PLMD).

6.4 Adult Alcohol Consumption and Sleep

6.4.1 Sleep Disturbances Observed in Adult Consumers

While the focus of this review is on the impact of *developmental* alcohol exposure on sleep, it should be noted that there is a strong, although complex, relationship between alcohol consumption and sleep disturbances. Both clinical and preclinical studies have been conducted to investigate the association between alcohol consumption and sleep disorders among the adult population (for review see He et al. 2019). Clinical investigations have found that 10–40% of the general population use alcohol as a sleep aid (Sharma et al. 2022; Dorrian and Skinner 2012; Roehrs and Roth 2012). To understand the acute effects of alcohol on sleep physiology, studies conducted on healthy adults (no alcohol use disorder) have demonstrated that drinking alcohol (blood alcohol concentration (BAC) between 50–80 mg/dl) before bedtime significantly increases the duration of NREM sleep and decreases REM sleep at the beginning of the night (Sharma et al. 2022; Thakkar et al. 2015; Williams et al. 1983; Feige et al. 2006). Meanwhile, sleep physiology in the second half of the night is characterized by fragmented sleep, episodes of awakening, and increased REM density (Feige et al. 2006; Arnedt et al. 2011; Sharma et al. 2022). Interestingly, this "biphasic effect" of alcohol on sleep physiology seems to be dose-related, as at the beginning of the night, when the value of BAC is higher, alcohol promotes sleep, but during the latter part of the night when the BAC starts to decrease, alcohol disrupts the sleep (Sharma et al. 2022; Prinz et al. 1980).

Studies conducted to investigate the chronic effects of alcohol consumption on sleep have demonstrated several sleep disturbances, including sleep short duration, insomnia, and difficulty falling asleep (Chakravorty et al. 2016; He et al. 2019). Recently, alcohol consumption has also been shown to be related to breathing sleep disorders such as snoring and obstructive sleep apnea (Kolla et al. 2018). Furthermore, several studies have investigated the impact of alcohol withdrawal on sleep physiology. Sleep during acute withdrawal (abstinent for days and weeks) is characterized by insomnia, shorter NREM sleep phase, and decreased total sleep time (Sharma et al. 2022). Interestingly, the effects of long alcohol withdrawal on REM sleep are controversial, as one study found a significant increase in the REM sleep phase after 1 year of abstinence (Drummond et al. 1998) while other studies reported no significant alteration in the REM sleep phase (Brower et al. 2001; Williams et al. 1983). In summary, the consumption of alcohol during adulthood significantly affects the sleep architecture characterized by insomnia and shorter total sleep duration.

6.4.2 Sleep Disturbances Observed in Animal Studies

Preclinical rodent studies have also been conducted to better understand the sleep disturbances induced by a single binge alcohol drinking and chronic exposure. Intriguingly, one study using a single session of "drinking in the dark" (DID) paradigm combined with sleep recordings in C57BL/6 J found a significant increase in the NREM sleep phase during the first 4 h after drinking (Sharma et al. 2018). In contrast, 12 h post-binge, the authors observed a significant reduction of the NREM sleep phase (Sharma et al. 2018) suggesting that alcohol consumption may impact sleep homeostasis. Human studies suggest that sleep induction is the most common behavioral effect frequently observed during the first few hours after alcohol consumption. Similarly, preclinical studies performed to investigate the impact of chronic alcohol consumption on sleep physiology in rodents also found a significant correlation between the amount of alcohol consumed and the time spent awake, and a negative correlation with the NREM sleep phase affecting the quality of sleep (Sharma et al. 2022).

Interestingly, rodent studies performed to characterize the effect of alcohol withdrawal on sleep-wake physiology revealed a significant increase in the wake phase, and a significant reduction in the NREM sleep phase (Sharma et al. 2020). Surprisingly, as in human studies, the effects of withdrawal on the REM sleep phase in preclinical studies are also variable. In a few studies, the authors observed an increased REM sleep phase (Kubota et al. 2002; Mendelson et al. 1978), while other studies demonstrated a significant reduction (Rouhani et al. 1998; Gitlow et al. 1973).

In summary, clinical and preclinical studies suggest that acute alcohol consumption positively modulates sleep phase during the first half of the sleep duration, increasing the quality and quantity of NREM sleep during this period. In contrast, acute alcohol consumption leads to short and fragmented sleep during the second half of the night. Conversely, insomnia and difficulty falling asleep are common sleep disturbances observed in chronic alcohol consumers.

6.5 Developmental Alcohol Exposure and Sleep

It is well known that persons with FASD, in particular children and adolescents, show more sleep disorders as compared to typically developing children (Hayes et al. 2020; Chandler-Mather et al. 2021). As previously mentioned, sleep disruptions were reported in some of the first studies describing the effect induced by alcohol consumption during pregnancy (Abel 1984), and reports from those with FASD, parents, and caregivers consistently report sleep disturbances as a major characteristic of FASD.

6.5.1 Sleep Disorders Observed in FASD Patients

Clinical and preclinical studies conducted to date have extensively documented that developmental alcohol exposure during the prenatal and postnatal periods leads to significant alterations in sleep patterns and architecture. According to reports from parents or caregivers of FASD children, the most frequent sleep disorders reported are insomnias, specifically difficulty falling asleep (Gerstner et al. 2023; Hayes et al. 2020), shortened periods of sleep episodes (Inkelis and Thomas 2018), decrease in the total hours of sleep (Hayes et al. 2020; Benson et al. 2023), and sleep fragmentation (Troese et al. 2008; Goril et al. 2016).

A prospective study conducted in 2018–2020, that examined children diagnosed with FASD (median age 8 years) and typically developing children (median age 10 years), combined screening and objective methods to measure possible alterations in sleep and revealed many interesting findings (Dylag et al. 2021). In the first phase, a caregiver scored sleep problems with the Child Sleep Habit Questionnaire (CSHQ). This questionnaire is recognized to be a well-established tool for screening pediatric issues, specifically sleep disturbances (Owens et al. 2000; Lewandowski et al. 2011; Sen and Spruyt 2020). Briefly, the caregiver answers 33 questions covering several aspects of sleep such as bedtime resistance, sleep duration, parasomnias, and sleep-distorted breathing observed during the last 7 days before administering the questionnaire (Dylag et al. 2021). The second phase of the investigation included only children in the FASD group that scored above 41 points with CSHQ, and utilized lab-attended polysomnography (PSG) measurements (Berry et al. 2017), including EEG, electrooculogram, chin and tibial electromyogram, electrocardiogram, ventilator monitoring, breathing effort, snoring, and body position. The data revealed frequent night-waking episodes, parasomnias, and daytime sleepiness in the FASD group compared to the control group (Dylag et al. 2021). Furthermore, sleep architecture was more altered in the FASD group, with longer time spent in the N1 sleep stage, shorter time spent in the N3 sleep stage, and shorter REM sleep. Interestingly, these significant changes in sleep architecture are accompanied by higher hypopnea and central apnea indices (Dylag et al. 2021). This data set aligns with a previous study on children diagnosed with FASD (N = 33, 4.1–12.1 years), which scored sleep behavior via CSHQ filled out by caregivers. Similarly, this study assessed the participants with high CSHQ scores (range: 46–72) with polysomnography. While only five children underwent PSG, as the CSQH reported high scores in sleep onset and maintenance, such as bedtime resistance, sleep anxiety, sleep duration, and night waking, the study reported mild sleep-disordered breathing, including obstructive apnea and hypopnea also accompanied by fragmented sleep (Chen et al. 2012). Despite the small sample size, these findings demonstrate that the results from sleep assessment using PSG are aligned with those obtained from questionnaires, indicating that the single use of sleep questionnaires may be sufficient to characterize sleep behavior in FASD patients.

A study that utilized an online questionnaire consisting of three simple questions investigating difficulty falling asleep, staying asleep, and frequent waking during the night in children aged 5–17 years diagnosed with FASD, found that the most common sleep disturbance observed was difficulty falling asleep (56.4%), then difficulty staying asleep (44.8%) and waking early (29.4%) (Hayes et al. 2020). Similar findings were corroborated by a longitudinal study investigating the relation between sleep problems and distinct patterns of maternal alcohol consumption during pregnancy in children from 2 to 9 years old (Chandler-Mather et al. 2021). In that study, the patterns of maternal alcohol consumption during the pregnancy were categorized as: (1) abstinent (mothers reported no alcohol consumption across the entire pregnancy); (2) occasional (occasional frequency no more than 1–2 standard drinks per occasion across the entire pregnancy); (3) low (less than 7 drinks per week across the pregnancy and no more than 1–2 drinks per occasion); (4) heavy (more than 7 drinks per week across the pregnancy, or more than 5 drinks drank more than 2 occasions/week across the pregnancy, or more than 11 drinks per occasion regardless of frequency of consumption). Sleep problems were again assessed through a questionnaire (four questions) filled out by the primary caregiver (almost always the biological mother) at different ranges of children's age (2–3 years, 6–7 years, 8–9 years). Low consumption did not alter sleep patterns, while heavy consumption significantly increased the incidence of sleep problems. Interestingly, children in the range of age 6–7 years old did not show any significant impairment in sleep in occasional and low alcohol consumption. In contrast, heavy consumption significantly impaired the sleep patterns. Interestingly, no considerable sleep problems were observed in any category of alcohol consumption in the range of 8–9 years (Chandler-Mather et al. 2021).

More recently, a clinical study using a structured naturalistic observation-based methodology provided strong findings supporting the power and validity of observational paradigms to evaluate sleep-wake behaviors in children and adolescents diagnosed with FASD. This observational study analyzed retrospective data from assessments performed between 2011 and 2014 at the Sleep/Wake-Behavior Clinic (British Columbia Children's Hospital), in 40 pediatric FASD patients aged 1.8–17.5 years old (Ipsiroglu et al. 2019). Consulting the available logs/diaries, the authors found that 98% of patients experienced difficulties falling asleep and insomnia. In addition, the sleep was characterized by hypermotor events during the nighttime, and sleep-disordered breathing and sleep disturbances were confirmed by video recordings from the original study (Ipsiroglu et al. 2015). The subjects exhibited fragmented and non-restorative sleep accompanied by foot and limb movements (Ipsiroglu et al. 2011, 2015), confirming the observation reported by caregivers. Analysis of the polysomnographic recordings obtained from a small sample of patients also found alterations in sleep architecture, specifically prolonged REM latency. This analysis also confirmed the sleep-disordered breathing reported by caregivers as recordings of apnea-hypopnea in these subjects (Ipsiroglu et al. 2019). Taken together, data from this observational study confirm that self-report of sleep disruptions correlates highly with video scoring and PSG recording of sleep quality, that caregivers accurately report sleep-disordered breathing in those with FASD, and further underscore that children with FASD commonly exhibit dysregulated sleep-wake behaviors.

In summary, alcohol consumption during the gestational period induces long-lasting impairments in sleep physiology. As the majority of FASD studies have been performed in humans, this set of clinical investigations has limitations due to the complexity of correlating outcomes with crucial variables such as volume and timing of alcohol consumption and maternal diet, suggesting that preclinical studies may be able to fill a gap in understanding how developmental alcohol exposure alters sleep physiology.

6.5.2 Preclinical Studies About Sleep Disorders

The clinical evidence showing the impact of alcohol consumption during pregnancy on sleep physiology has also been investigated in preclinical models using a variety of developmental alcohol exposure paradigms. Before reviewing these studies, it is first important to understand the different developmental alcohol exposure paradigms used to study the effects of alcohol on the developing brain in rodents.

6.5.2.1 Developmental Alcohol Exposure Rodent Models

In the context of FASD research, rodents represent the most employed animal model as they offer important advantages such as short gestational period and large litter size. Specifically, mice are the most commonly used due to the availability of transgenic models and their physiological and genetic similarities to humans (Almeida et al. 2020). Despite the similarities in brain structure and physiology observed between rodents and humans, it is also important to consider their limitations: the equivalent of the third-trimester development in human gestation occurs after birth in rodents. In addition, mice metabolize alcohol at much higher rates than in humans. Together, these two differences have led to multiple approaches being developed to model human development and reach relevant physiological levels of exposure. Several prenatal alcohol exposure (PAE) paradigms differ in the route of administration, including intraperitoneal injection, liquid diet, voluntary drinking, oral and intragastric gavage, and vapor alcohol chambers (Almeida et al. 2020). The choice of route administration significantly impacts the pattern of alcohol exposure and the amount of alcohol consumed. Furthermore, investigating the detrimental effects of alcohol on the developing brain allows the researchers to correlate the outcomes observed in cognitive impairments, behavior, and sleep with the amount of alcohol consumed, the dose administered, and the timing of exposure during the gestational period.

6.5.2.2 Rodent PAE and Sleep Disturbances

While mice are a commonly used animal model in preclinical FASD work, most studies examining sleep have been done in rats. For example, Stone and colleagues examined the effects of PAE on sleep in young adulthood (Stone et al. 1996). Briefly, pregnant rats were fed an alcohol diet (35% ethanol-derived calories) from gestational day 0 through birth. Sleep stage analysis, measured via EEG recordings in young adult female offspring (6 months old), revealed a significant decrease in REM sleep measured during the light phase compared to the control group (Stone et al. 1996). Interestingly, no significant difference was found in NREM sleep, suggesting that the developmental alcohol exposure paradigm may selectively modulate REM sleep in young subjects. Despite the interesting results, EEG recordings were performed over limited period (3 h during the light phase) and sleep was only evaluated in female offspring.

Another recent preclinical study examined the sleep behavior in rats prenatally exposed to an alcohol diet (36% ethanol-derived calories). Specifically, the sleep behavior in juvenile (postnatal days 23–25) and adolescent (postnatal days 35–36) periods on pairs of rats (paired by prenatal group and in keeping with their social nature) through video recordings (Ipsiroglu et al. 2019). Interestingly, the results show overall changes in sleep behavior with age, regardless of alcohol exposure. The authors observed that the overall number of arousal and movements decreased across age while the duration of sleep bouts increased. The analysis revealed that PAE males spent more time awake and showed longer sleep latencies at both time points. The study also found a significant increase in position changes and longer transition bouts, indicating a non-restorative sleep (Ipsiroglu et al. 2019) in PAE rats directly in line with clinical observations reported in human studies.

Other studies have combined prenatal exposure with EEG sleep recordings in rats using more discrete exposures. Offspring from pregnant rats were exposed to 95% ethanol solution via intraperitoneal injection on gestational day 7. Subsequently, EEG recordings combined with electromyography were performed at 3 months of age (Sylvester et al. 2000). The results showed a significant reduction of time spent in the REM sleep phase in PAE females compared to saline control females. The percentage of REM sleep in the PAE group was half that of the control group, and fewer episodes of REM sleep were observed. Interestingly, the analysis of sleep behavior in male rats did not show significant differences

(Sylvester et al. 2000). The study further examined REM during 6 h segments across the 24-h recording time and found that control rats had higher levels of REM sleep percentage during the light phase, suggesting that PAE rats had difficulties initiating REM sleep in this period (Sylvester et al. 2000). Taken together, the results discussed in these preclinical studies have demonstrated that developmental alcohol exposure across the first and second trimesters impaired sleep physiology in young/adult offspring. Furthermore, the sleep phase most affected is the REM sleep. Similar results were seen in rat pups that were exposed to alcohol throughout gestation. In this model, pregnant rat dams were exposed to alcohol (7–12% alcohol, liquid diet) during the entire gestation period, and the sleep-wake behavior of pups (from 7 to 20 postnatal days) was assessed using a movement-sensitive mattress. The results revealed that PAE male and female pups exposed had a significant decrease in REM sleep and spent more time in the awake phase compared to controls (Hilakivi 1986; Hilakivi et al. 1987). These results suggest that a diverse array of PAE models targeting the equivalent of the first and second trimesters in rats significantly impacts the sleep-wake cycle in infancy and throughout the lifespan, inducing significant alterations in the REM sleep phase that may be sexually dimorphic.

6.5.2.3 Rodent Perinatal Alcohol Exposure and Sleep Disturbances

While PAE has been the focus of most preclinical sleep studies modeling FASD, a handful of studies have also examined the impact of perinatal alcohol exposure targeting third trimester-equivalent exposure by exposing neonates to alcohol via intragastric or subcutaneous injections. Volgin and Kubin used an intragastric alcohol exposure (2.6 g/kg of alcohol twice daily) during postnatal days 4–9 and investigated the sleep behavior in adult male rat offspring. EEG recordings and electromyograms showed that rats exposed to alcohol had reduced sleep duration and a shorter REM sleep phase compared to the controls during the active phase (Volgin and Kubin 2012). Although the statistical analysis was not significant, the NREM sleep phase analysis revealed a reduced duration in the alcohol group compared to the control, while the sleep-wake cycle during the rest phase did not show any differences between the two groups (Volgin and Kubin 2012).

In contrast to PAE approaches, mice have been utilized to examine the effects of perinatal alcohol exposure on sleep. For example, in a series of studies from the Wilson lab, female and male pup mice were injected subcutaneously with alcohol (2.5 g/kg) twice on postnatal day 7 at 2-h intervals. Then, telemetry recordings were used to characterize the NREM sleep phase in adult mice (3 months old) (Wilson et al. 2016; Lewin et al. 2018; Apuzzo et al. 2020). Alcohol-exposed mice exhibited shorter NREM sleep and episodes of sleep fragmentation compared to saline-injected controls. Furthermore, the authors found that alcohol exposure affects the N2 phase considering the reduction of sleep spindle density and the N3 phase due to a decrease of delta oscillations (Wilson et al. 2016, Lewin et al. 2018, Apuzzo et al. 2020). These findings were later replicated using the same binge alcohol exposure paradigm as alcohol-exposed mice showed a significant reduction in the total duration of NREM sleep and NREM sleep bout duration (Shah et al. 2023).

Taken together, these studies suggest that PAE negatively modulates the REM sleep phase, while perinatal binge-alcohol exposure specifically alters the NREM sleep phase during adult life. These differences may be due to the specific developmental time point when alcohol is delivered, as it is known that gestational alcohol exposure prevalently affects neurogenesis and neuronal migration, processes that occur across embryonic days 10–17 (Miyoshi and Fishell 2011), while neonatal exposure induces cell loss, given that the cortical neurons and interneurons have reached their final position in all cortical layers (Miyoshi et al. 2010; Rymar and Sadikot 2007). These findings also confirm that developmental alcohol exposure disrupts the physiology of the sleep-wake cycle as observed in human studies and suggest that sleep disturbances depend on several factors, such as the dose and timing of alcohol exposure during gestation, involving specific and different molecular pathways. However, the mechanism by which developmental exposure alters these processes is still unclear.

6.5.3 Potential Mechanisms Underlying Sleep Disturbances in FASD

The mechanisms underlying sleep disturbances in FASD patients and developmental alcohol exposure rodent models are still being investigated. Prenatal alcohol exposure may induce structural defects in the central nervous system, but alterations of neurotransmission may also contribute to the sleep problems described in the previous sections.

Numerous neuroimaging studies have characterized structural abnormalities in FASD populations and have found that one of the most affected brain regions is the corpus callosum (Anna Dyląg et al. 2016; Donald et al. 2015), either in reduced size, altered shape, or complete absence (Fraize et al. 2023; Yang et al. 2012). Importantly, agenesis of the corpus callosum has been shown to be associated with a decrease in REM sleep duration (Yang et al. 2012) and it has been recently found that the total callosal resection is associated with reduction in the propagation of non-REM slow waves (Bernardi et al. 2021). Taken together, these findings suggest that the alterations in the sleep architecture extensively reported in human and rodent models may be due to altered callosal integrity and connectivity.

Other morphological changes seen in FASD may be related to altered sleep patterns. For example, studies in persons with FASD also report a higher incidence of apneic/hypopneic events during sleep which may be related to physical features of FASD such as micrognathia or high-arched palate (Del Campo and Jones 2017) that may induce an airway obstruction (Dylag et al. 2021), although these associations have not been fully explored.

Sleep disturbances in FASD may also be mediated by changes in neurotransmitter systems that are involved in the sleep-wake process (see Fig. 6.1). The long-term effects of developmental alcohol exposure on GABAergic transmission are well-characterized with multiple preclinical studies demonstrating that PAE or perinatal alcohol exposure with high or low-moderate doses significantly reduced the number of GABAergic interneurons and cholinergic interneurons (Kenton et al. 2020; Smiley et al. 2015; Smiley et al. 2021). For example, binge alcohol exposure to pups at postnatal day 7 induced a significant reduction of cholinergic neurons and parvalbumin interneurons in the basal forebrain (Smiley et al. 2021). Considering the role played by the basal forebrain in the regulation of the sleep-wake cycle, the loss of the cholinergic interneurons could explain non-REM sleep fragmentation and slow-wave oscillation impairments reported in several developmental alcohol-exposure rodent models. In addition, the reduction of orexinergic neurons expressed in the hypothalamus and in the anterior cingulate cortex (Smiley et al. 2021; Olateju et al. 2017), observed in offspring exposed to alcohol during the gestation and the perinatal period, may explain the difficulty falling asleep reported in human and rodent studies. In addition, preclinical investigations conducted using a low-moderate gestational alcohol exposure paradigm, where dams were exposed to 10% of alcohol during the entire gestational period, have found a significant alteration in the expression of GABAergic interneurons in orbitofrontal cortex and posterior parietal cortex in adult offspring (Kenton et al. 2020; Licheri et al. 2023). These findings may explain the altered NREM sleep phase observed in several studies, given that the cortical interneurons played a pivotal role in this specific phase of sleep.

Another critical brain region involved in the control of sleep is the suprachiasmatic nucleus and there is an extensive literature reporting significant changes in the function and structure of this brain region following developmental alcohol exposure. Briefly, gestational alcohol exposure during gestational days 10–21 (35% ethanol-derived calories) significantly affected circadian function in adult rats (Chen et al. 2006). The disruptions in the circadian system may be responsible for the insomnias frequently reported by the caregivers in CSHQ. Furthermore, the alterations of the circadian rhythm could explain the abnormal levels of melatonin observed in FASD children (Goril et al. 2016).

Findings from both clinical and preclinical studies suggest that there may be multiple factors in developmental alcohol exposure impact on sleep. These include morphological changes to brain structures, such as the corpus callosum and suprachiasmatic nucleus, and dysmorphology that may alter respiratory pathways during sleep. In addition, preclinical studies have shown that PAE and perinatal exposure robustly alter several of the neurotransmitter systems known to regulate specific phases and sub-phases of sleep.

6.6 Conclusions

The current literature, including clinical and preclinical findings, clearly demonstrates the negative impact of developmental alcohol exposure on sleep. However, further studies are needed to pinpoint and characterize the specific molecular pathways involved in sleep impairment and identify potential therapeutic targets for sleep disturbances.

6.6.1 Limitations of Current Approaches

There continues to be a disconnect between clinical and preclinical research approaches in examining sleep following developmental alcohol exposure. While the assessment of sleep physiology in children and adolescents with FASD is based on observational studies, several investigations have extensively demonstrated that sleep questionnaires provide valuable data about sleep impairments. However, the patients' caregivers typically fill out the sleep questionnaires, who may not report other physical signs of sleep disturbances, such as breathing abnormalities. Considering these limitations, future clinical studies must combine objective measurements (polysomnography and actigraphy) with observational records. In contrast, preclinical studies typically use objective measurements such as video monitoring and EEG but primarily focus on high doses of alcohol exposure. Both gestational and neonatal paradigms used to assess sleep physiology in rodent models of FASD result in blood alcohol concentrations ranging from 167 to 500 mg/dl. While these studies provide important information on establishing that PAE and perinatal exposure impact sleep, it would be valuable to investigate whether low and moderate doses modulate sleep patterns in adolescent and adult offspring. This is particularly important considering that cognitive and executive function difficulties that correspond with those seen in FASD are commonly observed also in low-moderate developmental alcohol exposure paradigms.

Given the recent focus on the importance of sex differences in both alcohol research in general and FASD specifically, it is important to note that in the current literature on sleep disturbances and developmental alcohol exposure, gender/sex differences are not explicitly addressed. Investigating and characterizing these differences is crucial for improving the current research. Understanding whether developmental alcohol exposure affects men and women differently could significantly improve the development of personalized treatments.

6.6.2 Summary and Future Directions

Despite some limitations of the studies performed to date, it is clear that sleep impairments in the FASD population represent a significant concern due to their potential impact on daytime activities, school performance, and overall health. Findings from clinical studies repeatedly confirm that sleep disturbance is a significant component of FASD. Further, studies combining observation, caregiver and/or self-report, and direct measurement of sleep quality confirm that data from questionnaires correlate highly with direct measurement of insomnia, sleep-wake behaviors, and sleep-disordered breathing in those with FASD. Preclinical findings have demonstrated that developmental alcohol exposure impacts both NREM and REM sleep phases and that these may be timing-dependent. High-dose perinatal alcohol exposure consistently alters the NREM sleep phase, characterized by short duration and fragmentation, while PAE alters the REM sleep phase. It is clear from this review that additional studies are needed to test possible mechanisms involved in these specific altered sleep patterns.

As noted throughout this review, there are still relatively few studies focused on the relationship between developmental alcohol exposure and sleep disturbances. This may be due to several challenges that need to be overcome. As mentioned in the clinical study section, the diagnosis of sleep disturbances is still a big challenge given the complexity of symptoms. Furthermore, the assessment of sleep physiology using questionnaires alone may underestimate the diagnosis, while using polysomnography may not be feasible in some cases. While still limited, rodent studies may fill a critical gap as they allow for careful control of exposure levels and environmental conditions and allow for careful monitoring and characterization of sleep behaviors. As increasingly translational methods and new innovative tools are developed, it is hoped that preclinical studies can provide new insights into the relationship between prenatal alcohol exposure and sleep.

References

Abel EL (1984) Prenatal effects of alcohol. Drug Alcohol Depend 14:1–10

Abel T, Havekes R, Saletin JM, Walker MP (2013) Sleep, plasticity and memory from molecules to whole-brain networks. Curr Biol 23:R774–R788

Almeida L, Andreu-Fernández V, Navarro-Tapia E, Aras-López R, Serra-Delgado M, Martínez L, García-Algar O, Gómez-Roig MD (2020) Murine models for the study of fetal alcohol Spectrum disorders: an overview. Front Pediatr 8:359

Anna Dyląg K, Sikora-Sporek A, Bańdo B, Boroń-Zyss J, Drożdż D, Dumnicka P, Przybyszewska K, Sporek M, Walocha JW, Wojciechowski W, Urbanik A (2016) Magnetic resonance imaging (Mri) findings among children with fetal alcohol syndrome (Fas), partial fetal alcohol syndrome (pfas) and alcohol related neurodevelopmental disorders (Arnd). Przegl Lek 73:605–609

Apuzzo J, Saito M, Wilson DA (2020) Post-exposure environment modulates long-term developmental ethanol effects on behavior, neuroanatomy, and cortical oscillations. Brain Res 1748:147128

Arnedt JT, Rohsenow DJ, Almeida AB, Hunt SK, Gokhale M, Gottlieb DJ, Howland J (2011) Sleep following alcohol intoxication in healthy, young adults: effects of sex and family history of alcoholism. Alcohol Clin Exp Res 35:870–878

Bathory E, Tomopoulos S (2017) Sleep regulation, physiology and development, sleep duration and patterns, and sleep hygiene in infants, toddlers, and preschool-age children. Curr Probl Pediatr Adolesc Health Care 47:29–42

Benson AA, Mughal R, Dimitriou D, Halstead EJ (2023) Towards a distinct sleep and Behavioural profile of fetal alcohol Spectrum disorder (Fasd): a comparison between Fasd, autism and typically developing children. J Integr Neurosci 22:77

Benz J, Rasmussen C, Andrew G (2009) Diagnosing fetal alcohol spectrum disorder: history, challenges and future directions. Paediatr Child Health 14:231–237

Bernardi G, Avvenuti G, Cataldi J, Lattanzi S, Ricciardi E, Polonara G, Silvestrini M, Siclari F, Fabri M, Bellesi M (2021) Role of corpus callosum in sleep spindle synchronization and coupling with slow waves. Brain Commun 3:fcab108

Berry RB, Brooks R, Gamaldo C, Harding SM, Lloyd RM, Quan SF, Troester MT, Vaughn BV (2017) Aasm scoring manual updates for 2017 (version 2.4). J Clin Sleep Med 13:665–666

Bertrand J (2009) Interventions for children with fetal alcohol spectrum disorders (Fasds): overview of findings for five innovative research projects. Res Dev Disabil 30:986–1006

Blumberg MS, Marques HG, Iida F (2013) Twitching in sensorimotor development from sleeping rats to robots. Curr Biol 23:R532–R537

Borbély AA, Achermann P (1999) Sleep homeostasis and models of sleep regulation. J Biol Rhythm 14:557–568

Boyce R, Williams S, Adamantidis A (2017) Rem sleep and memory. Curr Opin Neurobiol 44:167–177

Brinkman JE, Reddy V, Sharma S (2024) Physiology of sleep. *StatPearls*. Treasure Island (Fl) ineligible companies. Disclosure: Vamsi Reddy declares no relevant financial relationships with ineligible companies. In: Disclosure: Sandeep Sharma declares no relevant financial relationships with ineligible companies.: StatPearls publishing copyright © 2024. StatPearls Publishing Llc

Brower KJ, Aldrich MS, Robinson EA, Zucker RA, Greden JF (2001) Insomnia, self-medication, and relapse to alcoholism. Am J Psychiatry 158:399–404

Burman D (2017) Sleep Disorders: Sleep-Related Breathing Disorders. Fp Essent 460:11–21

Cao J, Herman AB, West GB, Poe G, Savage VM (2020) Unraveling why we sleep: quantitative analysis reveals abrupt transition from neural reorganization to repair in early development. Sci Adv 6

Carskadon MA, Dement WC (2005) Normal human sleep: an overview. Principles and practice of sleep medicine 4:13–23

Center for Substance Abuse, P (2014) Samhsa/Csat Treatment Improvement Protocols. Addressing Fetal Alcohol Spectrum Disorders (Fasd). Rockville (Md): Substance Abuse and Mental Health Services Administration (Us)

Chakravorty S, Chaudhary NS, Brower KJ (2016) Alcohol dependence and its relationship with insomnia and other sleep disorders. Alcohol Clin Exp Res 40:2271–2282

Chandler-Mather N, Occhipinti S, Donovan C, Shelton D, Dawe S (2021) An investigation of the link between prenatal alcohol exposure and sleep problems across childhood. Drug Alcohol Depend 218:108412

Chen CP, Kuhn P, Advis JP, Sarkar DK (2006) Prenatal ethanol exposure alters the expression of period genes governing the circadian function of beta-endorphin neurons in the hypothalamus. J Neurochem 97:1026–1033

Chen ML, Olson HC, Picciano JF, Starr JR, Owens J (2012) Sleep problems in children with fetal alcohol spectrum disorders. J Clin Sleep Med 8:421–429

Coles CD, Platzman KA, Lynch ME, Freides D (2002) Auditory and visual sustained attention in adolescents prenatally exposed to alcohol. Alcohol Clin Exp Res 26:263–271

Colwell CS (2011) Linking neural activity and molecular oscillations in the Scn. Nat Rev Neurosci 12:553–569

Crocker N, Vaurio L, Riley EP, Mattson SN (2011) Comparison of verbal learning and memory in children with heavy prenatal alcohol exposure or attention-deficit/hyperactivity disorder. Alcohol Clin Exp Res 35:1114–1121

Deboer T (2015) Behavioral and electrophysiological correlates of sleep and sleep homeostasis. Curr Top Behav Neurosci 25:1–24

Del Campo M, Jones KL (2017) A review of the physical features of the fetal alcohol spectrum disorders. Eur J Med Genet 60:55–64

Donald KA, Eastman E, Howells FM, Adnams C, Riley EP, Woods RP, Narr KL, Stein DJ (2015) Neuroimaging effects of prenatal alcohol exposure on the developing human brain: a magnetic resonance imaging review. Acta Neuropsychiatr 27:251–269

Dorrian J, Skinner N (2012) Alcohol consumption patterns of shiftworkers compared with dayworkers. Chronobiol Int 29:610–618

Drummond SP, Gillin JC, Smith TL, Demodena A (1998) The sleep of abstinent pure primary alcoholic patients: natural course and relationship to relapse. Alcohol Clin Exp Res 22:1796–1802

Dumoulin Bridi MC, Aton SJ, Seibt J, Renouard L, Coleman T, Frank MG (2015) Rapid eye movement sleep promotes cortical plasticity in the developing brain. Sci Adv 1:e1500105

Dylag KA, Bando B, Baran Z, Dumnicka P, Kowalska K, Kulaga P, Przybyszewska K, Radlinski J, Roozen S, Curfs L (2021) Sleep problems among children with fetal alcohol Spectrum disorders (Fasd)- an explorative study. Ital J Pediatr 47:113

Feige B, Gann H, Brueck R, Hornyak M, Litsch S, Hohagen F, Riemann D (2006) Effects of alcohol on polysomnographically recorded sleep in healthy subjects. Alcohol Clin Exp Res 30:1527–1537

Feriante J, Araujo JF (2024) Physiology, rem sleep. StatPearls. Treasure Island (Fl) ineligible companies. Disclosure: John Araujo declares no relevant financial relationships with ineligible companies.: StatPearls publishing copyright © 2024, StatPearls Publishing Llc

Fraize J, Convert G, Leprince Y, Sylvestre-Marconville F, Kerdreux E, Auzias G, Lefèvre J, Delorme R, Elmaleh-Bergès M, Hertz-Pannier L, Germanaud D (2023) Mapping corpus callosum surface reduction in fetal alcohol spectrum disorders with sulci and connectivity-based parcellation. Front Neurosci 17:1188367

Gais S, Mölle M, Helms K, Born J (2002) Learning-dependent increases in sleep spindle density. J Neurosci 22:6830–6834

Galland BC, Taylor BJ, Elder DE, Herbison P (2012) Normal sleep patterns in infants and children: a systematic review of observational studies. Sleep Med Rev 16:213–222

Gerstner T, Saevareid HI, Johnsen ÅR, Løhaugen G, Skranes J (2023) Sleep disturbances in Norwegian children with fetal alcohol spectrum disorders (Fasd) with and without a diagnosis of attention-deficit hyperactivity disorder or epilepsy. Alcohol Clin Exp Res (Hoboken) 47:589–599

Gitlow SE, Bentkover SH, Dziedzic SW, Khazan N (1973) Persistence of abnormal rem sleep response to ethanol as a result of previous ethanol ingestion. Psychopharmacologia 33:135–140

Goril S, Zalai D, Scott L, Shapiro CM (2016) Sleep and melatonin secretion abnormalities in children and adolescents with fetal alcohol spectrum disorders. Sleep Med 23:59–64

Hastings MH, Maywood ES, Brancaccio M (2019) The mammalian circadian timing system and the suprachiasmatic nucleus as its pacemaker. Biology (Basel) 8

Hayes N, Moritz KM, Reid N (2020) Parent-reported sleep problems in school-aged children with fetal alcohol spectrum disorder: association with child behaviour, caregiver, and family functioning. Sleep Med 74:307–314

He S, Hasler BP, Chakravorty S (2019) Alcohol and sleep-related problems. Curr Opin Psychol 30:117–122

Hilakivi L (1986) Effects of prenatal alcohol exposure on neonatal sleep-wake behaviour and adult alcohol consumption in rats. Acta Pharmacol Toxicol (Copenh) 59:36–42

Hilakivi L, Tuomisto L, Hilakivi I, Kiianmaa K, Hellevuo K, Hyytiä P (1987) Effect of prenatal alcohol exposure on neonatal sleep-wake behaviour and adult alcohol consumption in the Aa and Ana rat lines. Alcohol Alcohol 22:231–240

Iglowstein I, Jenni OG, Molinari L, Largo RH (2003) Sleep duration from infancy to adolescence: reference values and generational trends. Pediatrics 111:302–307

Inkelis SM, Thomas JD (2018) Sleep in infants and children with prenatal alcohol exposure. Alcohol Clin Exp Res 42:1390

Institute Of Medicine Committee On Sleep, M. & Research (2006) The National Academies Collection: reports funded by National Institutes of Health. In: Colten HR, Altevogt BM (eds) Sleep disorders and sleep deprivation: an unmet public health problem. National Academies Press (Us), Washington (Dc). Copyright © 2006, National Academy of Sciences

Ipsiroglu O, Black A, Vilela-Barbosa A, Vatikiotis-Bateson E, Mckellin W (2011) Tl-090 describing restless leg syndrome (Rls) as a cause of circadian rhythm sleep disorders (Crsd): video studies in the home setting: from ethnography to quantitative analyses. Sleep Med 12:S82–S83

Ipsiroglu OS, Hung YH, Chan F, Ross ML, Veer D, Soo S, Ho G, Berger M, Mcallister G, Garn H, Kloesch G, Barbosa AV, Stockler S, Mckellin W, Vatikiotis-Bateson E (2015) "diagnosis by behavioral observation" home-videosomnography–a rigorous ethnographic approach to sleep of children with neurodevelopmental conditions. Front Psych 6:39

Ipsiroglu OS, Mckellin WH, Carey N, Loock C (2013) "They silently live in terror…" why sleep problems and night-time related quality-of-life are missed in children with a fetal alcohol spectrum disorder. Soc Sci Med 79:76–83

Ipsiroglu OS, Wind K, Hung YA, Berger M, Chan F, Yu W, Stockler S, Weinberg J (2019) Prenatal alcohol exposure and sleep-wake behaviors: exploratory and naturalistic observations in the clinical setting and in an animal model. Sleep Med 54:101–112

Jan JE, Asante KO, Conry JL, Fast DK, Bax MC, Ipsiroglu OS, Bredberg E, Loock CA, Wasdell MB (2010) Sleep health issues for children with Fasd: clinical considerations. Int J Pediatr 2010:1

Jan JE, Owens JA, Weiss MD, Johnson KP, Wasdell MB, Freeman RD, Ipsiroglu OS (2008) Sleep hygiene for children with neurodevelopmental disabilities. Pediatrics 122:1343–1350

Kamara D, Beauchaine TP (2020) A review of sleep disturbances among infants and children with neurodevelopmental disorders. Rev J Autism Dev Disord 7:278–294

Karna B, Sankari A, Tatikonda G (2024) Sleep disorder. StatPearls. Treasure Island (Fl) ineligible companies. Disclosure: Abdulghani Sankari declares no relevant financial relationships with ineligible companies. In: Disclosure: Geethika Tatikonda declares no relevant financial relationships with ineligible companies. StatPearls Publishing. Copyright © 2024, StatPearls Publishing LLC

Kenton JA, Ontiveros T, Bird CW, Valenzuela CF, Brigman JL (2020) Moderate prenatal alcohol exposure alters the number and function of Gabaergic interneurons in the murine orbitofrontal cortex. Alcohol 88:33–41

Khoury JE, Milligan K, Girard TA (2015) Executive functioning in children and adolescents prenatally exposed to alcohol: a meta-analytic review. Neuropsychol Rev 25:149–170

Knoop MS, De Groot ER, Dudink J (2021) Current ideas about the roles of rapid eye movement and non-rapid eye movement sleep in brain development. Acta Paediatr 110:36–44

Kodituwakku PW (2009) Neurocognitive profile in children with fetal alcohol spectrum disorders. Dev Disabil Res Rev 15:218–224

Kolla BP, Foroughi M, Saeidifard F, Chakravorty S, Wang Z, Mansukhani MP (2018) The impact of alcohol on breathing parameters during sleep: a systematic review and meta-analysis. Sleep Med Rev 42:59–67

Kubota T, De A, Brown RA, Simasko SM, Krueger JM (2002) Diurnal effects of acute and chronic administration of ethanol on sleep in rats. Alcohol Clin Exp Res 26:1153–1161

Lange S, Probst C, Gmel G, Rehm J, Burd L, Popova S (2017) Global prevalence of fetal alcohol Spectrum disorder among children and youth: a systematic review and meta-analysis. JAMA Pediatr 171:948–956

Lee MG, Hassani OK, Alonso A, Jones BE (2005) Cholinergic basal forebrain neurons burst with theta during waking and paradoxical sleep. J Neurosci 25:4365–4369

Lewandowski AS, Toliver-Sokol M, Palermo TM (2011) Evidence-based review of subjective pediatric sleep measures. J Pediatr Psychol 36:780–793

Lewin M, Ilina M, Betz J, Masiello K, Hui M, Wilson DA, Saito M (2018) Developmental ethanol-induced sleep fragmentation, behavioral hyperactivity, cognitive impairment and Parvalbumin cell loss are prevented by lithium co-treatment. Neuroscience 369:269–277

Licheri V, Jacquez BJ, Castillo VK, Sainz DB, Valenzuela CF, Brigman JL (2023) Long-term effects of low prenatal alcohol exposure on Gabaergic interneurons of the murine posterior parietal cortex. Alcohol Clin Exp Res (Hoboken) 47:2248–2261

Liu D, Dan Y (2019) A motor theory of sleep-wake control: arousal-action circuit. Annu Rev Neurosci 42:27–46

Lokhandwala S, Spencer RMC (2022) Relations between sleep patterns early in life and brain development: a review. Dev Cogn Neurosci 56:101130

Mattson SN, Bernes GA, Doyle LR (2019) Fetal alcohol Spectrum disorders: a review of the neurobehavioral deficits associated with prenatal alcohol exposure. Alcohol Clin Exp Res 43:1046–1062

Mattson SN, Calarco KE, Lang AR (2006) Focused and shifting attention in children with heavy prenatal alcohol exposure. Neuropsychology 20:361–369

Mattson SN, Roebuck TM (2002) Acquisition and retention of verbal and nonverbal information in children with heavy prenatal alcohol exposure. Alcohol Clin Exp Res 26:875–882

May PA, Chambers CD, Kalberg WO, Zellner J, Feldman H, Buckley D, Kopald D, Hasken JM, Xu R, Honerkamp-Smith G, Taras H, Manning MA, Robinson LK, Adam MP, Abdul-Rahman O, Vaux K, Jewett T, Elliott AJ, Kable JA, Akshoomoff N, Falk D, Arroyo JA, Hereld D, Riley EP, Charness ME, Coles CD, Warren KR, Jones KL, Hoyme HE (2018) Prevalence of fetal alcohol Spectrum disorders in 4 Us communities. JAMA 319:474–482

Mcclain IJ, Lustenberger C, Achermann P, Lassonde JM, Kurth S, Lebourgeois MK (2016) Developmental changes in sleep spindle characteristics and sigma power across early childhood. Neural Plast 2016:3670951

Medicine AAOS (2023) International classification of sleep disorders: Icsd-3-text revision. American Academy of Sleep Medicine

Mendelson WB, Majchrowicz E, Mirmirani N, Dawson S, Gillin JC, Wyatt RJ (1978) Sleep during chronic ethanol administration and withdrawal in rats. J Stud Alcohol 39:1213–1223

Miyoshi G, Fishell G (2011) Gabaergic interneuron lineages selectively sort into specific cortical layers during early postnatal development. Cereb Cortex 21:845–852

Miyoshi G, Hjerling-Leffler J, Karayannis T, Sousa VH, Butt SJ, Battiste J, Johnson JE, Machold RP, Fishell G (2010) Genetic fate mapping reveals that the caudal ganglionic eminence produces a large and diverse population of superficial cortical interneurons. J Neurosci 30:1582–1594

Oh J, Petersen C, Walsh CM, Bittencourt JC, Neylan TC, Grinberg LT (2019) The role of co-neurotransmitters in sleep and wake regulation. Mol Psychiatry 24:1284–1295

Olateju OI, Bhagwandin A, Ihunwo AO, Manger PR (2017) Changes in the cholinergic, Catecholaminergic, Orexinergic and serotonergic structures forming part of the sleep Systems of Adult Mice Exposed to intrauterine alcohol. Front Neuroanat 11:110

Owens JA, Spirito A, Mcguinn M (2000) The Children's sleep habits questionnaire (Cshq): psychometric properties of a survey instrument for school-aged children. Sleep 23:1043–1051

Patel AK, Reddy V, Shumway KR, Araujo JF (2024) Physiology, sleep stages. StatPearls. Treasure Island (Fl) ineligible companies. Disclosure: Vamsi Reddy declares no relevant financial relationships with ineligible companies. Disclosure: Karlie Shumway declares no relevant financial relationships with ineligible companies. Disclosure: John Araujo declares no relevant financial relationships with ineligible companies.: StatPearls publishing. Copyright © 2024, StatPearls Publishing LLC

Perogamvros L, Schwartz S (2012) The roles of the reward system in sleep and dreaming. Neurosci Biobehav Rev 36:1934–1951

Pesonen AK, Räikkönen K, Matthews K, Heinonen K, Paavonen JE, Lahti J, Komsi N, Lemola S, Järvenpää AL, Kajantie E, Strandberg T (2009) Prenatal origins of poor sleep in children. Sleep 32:1086–1092

Petrenko CLM, Pandolfino ME, Quamma J, Carmichael Olson H (2017) Emotional understanding in school-aged children with fetal alcohol Spectrum disorders: a promising target for intervention. J Popul Ther Clin Pharmacol 24:e21–e31

Popova S, Charness ME, Burd L, Crawford A, Hoyme HE, Mukherjee RAS, Riley EP, Elliott EJ (2023) Fetal alcohol spectrum disorders. Nat Rev Dis Primers 9:11

Popova S, Lange S, Probst C, Gmel G, Rehm J (2017) Estimation of national, regional, and global prevalence of alcohol use during pregnancy and fetal alcohol syndrome: a systematic review and meta-analysis. Lancet Glob Health 5:e290–e299

Prinz PN, Roehrs TA, Vitaliano PP, Linnoila M, Weitzman ED (1980) Effect of alcohol on sleep and nighttime plasma growth hormone and cortisol concentrations. J Clin Endocrinol Metab 51:759–764

Rasch B, Born J (2013) About sleep's role in memory. Physiol Rev 93:681–766

Roehrs T, Roth T (2012) Insomnia pharmacotherapy. Neurotherapeutics 9:728–738

Rouhani S, Dall'ava-Santucci J, Bajenaru O, Emmanouilidis E, Tran G, Manicom R, Dinh-Xuan AT, Poenaru S (1998) Effects of muscimol or homotaurine on sleep-wake states in alcohol-dependent rats during withdrawal. Pharmacol Biochem Behav 59:955–960

Rymar VV, Sadikot AF (2007) Laminar fate of cortical Gabaergic interneurons is dependent on both birthdate and phenotype. J Comp Neurol 501:369–380

Schmidt MH (2014) The energy allocation function of sleep: a unifying theory of sleep, torpor, and continuous wakefulness. Neurosci Biobehav Rev 47:122–153

Sen T, Spruyt K (2020) Pediatric sleep tools: an updated literature review. Front Psych 11:317

Shah P, Kaneria A, Fleming G, Williams CRO, Sullivan RM, Lemon CH, Smiley J, Saito M, Wilson DA (2023) Homeostatic Nrem sleep and salience network function in adult mice exposed to ethanol during development. Front Neurosci 17:1267542

Sharma R, Parikh M, Mishra V, Zuniga A, Sahota P, Thakkar M (2022) Sleep, sleep homeostasis and arousal disturbances in alcoholism. Brain Res Bull 182:30–43

Sharma R, Sahota P, Thakkar MM (2018) A single episode of binge alcohol drinking causes sleep disturbance, disrupts sleep homeostasis, and down-regulates equilibrative nucleoside transporter 1. J Neurochem 146:304–321

Sharma R, Sahota P, Thakkar MM (2020) Chronic alcohol exposure reduces acetylated histones in the sleep-wake regulatory brain regions to cause insomnia during withdrawal. Neuropharmacology 180:108332

Smiley JF, Bleiwas C, Canals-Baker S, Williams SZ, Sears R, Teixeira CM, Wilson DA, Saito M (2021) Neonatal ethanol causes profound reduction of cholinergic cell number in the basal forebrain of adult animals. Alcohol 97:1–11

Smiley JF, Saito M, Bleiwas C, Masiello K, Ardekani B, Guilfoyle DN, Gerum S, Wilson DA, Vadasz C (2015) Selective reduction of cerebral cortex Gaba neurons in a late gestation model of fetal alcohol spectrum disorder. Alcohol 49:571–580

Stone WS, Altman HJ, Hall J, Arankowsky-Sandoval G, Parekh P, Gold PE (1996) Prenatal exposure to alcohol in adult rats: relationships between sleep and memory deficits, and effects of glucose administration on memory. Brain Res 742:98–106

Stores G (2001) Sleep-wake function in children with neurodevelopmental and psychiatric disorders. Semin Pediatr Neurol 8:188–197

Sulaman BA, Wang S, Tyan J, Eban-Rothschild A (2023) Neuro-orchestration of sleep and wakefulness. Nat Neurosci 26:196–212

Sylvester L, Kapron CM, Smith C (2000) In utero ethanol exposure decreases rapid eye movement sleep in female Sprague-Dawley rat offspring. Neurosci Lett 289:13–16

Thakkar MM, Sharma R, Sahota P (2015) Alcohol disrupts sleep homeostasis. Alcohol 49:299–310

Tononi G, Cirelli C (2003) Sleep and synaptic homeostasis: a hypothesis. Brain Res Bull 62:143–150

Tononi G, Cirelli C (2006) Sleep function and synaptic homeostasis. Sleep Med Rev 10:49–62

Tononi G, Cirelli C (2014) Sleep and the price of plasticity: from synaptic and cellular homeostasis to memory consolidation and integration. Neuron 81:12–34

Troese M, Fukumizu M, Sallinen BJ, Gilles AA, Wellman JD, Paul JA, Brown ER, Hayes MJ (2008) Sleep fragmentation and evidence for sleep debt in alcohol-exposed infants. Early Hum Dev 84:577–585

Venner A, Anaclet C, Broadhurst RY, Saper CB, Fuller PM (2016) A novel population of wake-promoting Gabaergic neurons in the ventral lateral hypothalamus. Curr Biol 26:2137–2143

Volgin DV, Kubin L (2012) Reduced sleep and impaired sleep initiation in adult male rats exposed to alcohol during early postnatal period. Behav Brain Res 234:38–42

Walker MP, Van Der Helm E (2009) Overnight therapy? The role of sleep in emotional brain processing. Psychol Bull 135:731–748

Wengel T, Hanlon-Dearman AC, Fjeldsted B (2011) Sleep and sensory characteristics in young children with fetal alcohol spectrum disorder. J Dev Behav Pediatr 32:384–392

Williams DL, Maclean AW, Cairns J (1983) Dose-response effects of ethanol on the sleep of young women. J Stud Alcohol 44:515–523

Wilson DA, Masiello K, Lewin MP, Hui M, Smiley JF, Saito M (2016) Developmental ethanol exposure-induced sleep fragmentation predicts adult cognitive impairment. Neuroscience 322:18–27

Wilson DA, Sullivan RM, Smiley JF, Saito M, Raineki C (2024) Developmental alcohol exposure is exhausting: sleep and the enduring consequences of alcohol exposure during development. Neurosci Biobehav Rev 158:105567

Xie L, Kang H, Xu Q, Chen MJ, Liao Y, Thiyagarajan M, O'donnell J, Christensen DJ, Nicholson C, Iliff JJ, Takano T, Deane R, Nedergaard M (2013) Sleep drives metabolite clearance from the adult brain. Science 342:373–377

Yang Y, Phillips OR, Kan E, Sulik KK, Mattson SN, Riley EP, Jones KL, Adnams CM, May PA, O'connor MJ, Narr KL, Sowell ER (2012) Callosal thickness reductions relate to facial dysmorphology in fetal alcohol spectrum disorders. Alcohol Clin Exp Res 36:798–806

Recent Advances in the Role of Non-coding RNAs in Fetal Alcohol Spectrum Disorders

7

Ariana N. Pritha, Andrea A. Pasmay, and Shahani Noor

Abstract

Despite numerous preclinical studies modeling fetal alcohol spectrum disorder (FASD)-associated neurodevelopmental deficits to date, a comprehensive molecular landscape dictating these deficits remains poorly understood. Non-coding RNAs constitute a substantial layer of epigenetic regulation of gene expression at the transcriptional, post-transcriptional, translational, and post-translational levels. Yet, little is known about the differential expression of non-coding RNAs in the context of prenatal alcohol exposure (PAE) that are mechanistically linked with FASD-related neurobehavior deficits. This chapter reviews our current knowledge from preclinical studies in non-coding RNA-mediated molecular mechanisms that may underlie FASD pathophysiology. This chapter also summarizes relevant clinical evidence and current efforts in utilizing these non-coding RNA molecules as biomarkers of PAE-associated deficits impacting central nervous system (CNS) function. Unraveling the diverse roles of various species of non-coding RNAs is critical to enhancing our comprehension of these intricate molecular pathways. Understanding these pathways would likely contribute to identifying critical molecular target(s) for developing efficient treatment strategies and prognostic and diagnostic markers fostering advancements in treating and managing FASD-related CNS dysfunction.

Keywords

Prenatal alcohol exposure · Developmental alcohol exposure · Neuroimmune · Preclinical models · circRNA · miRNA · lncRNA · Blood biomarker

7.1 Introduction

Exposure to alcohol during pregnancy, termed prenatal alcohol exposure (PAE), can result in a spectrum of adverse outcomes, including congenital anomalies, facial structural irregularities, impaired growth, and disruptions in the central nervous system (CNS), collectively referred to as fetal alcohol spectrum disorders (FASDs) (Price et al. 2017; Mattson et al. 2019; Popova et al. 2016). FASD is associated with cognitive (i.e., intellectual ability, learning, and memory) and behavioral (e.g., mood, attention, and impulse control) deficits and sensory dysfunction (Mattson et al. 2019; Franklin et al. 2008). FASD

A. N. Pritha · A. A. Pasmay · S. Noor (✉)
Department of Neurosciences and New Mexico Alcohol Research Center, School of Medicine, University of New Mexico Health Sciences Center, Albuquerque, NM, USA
e-mail: snoor@salud.unm.edu

C. F. Valenzuela et al. (eds.), *Effects of Alcohol on the Brain across the Lifespan*, Advances in Experimental Medicine and Biology 1473, https://doi.org/10.1007/978-3-031-81908-7_7

occurs worldwide, and it is considered the most preventable cause of mental disability in the Western world and a leading cause of intellectual disabilities, indicating it is a public health concern of paramount importance. Although in the last few decades, FASD has been well-described regarding the long-term physical health problems that impact the CNS (Panczakiewicz et al. 2016; Reid et al. 2021), the pathophysiological mechanism(s) of these long-lasting effects of PAE affecting the CNS are still under investigation.

As our nervous system development continues postnatally, substantial structural remodeling and functional connectivity influence cognition and affective processing (Vijayakumar et al. 2018), judgment, decision-making, and impulse control (Giedd and Rapoport 2010). Developmental alcohol exposure disrupts normal brain development and dysregulates genes related to neurogenesis and gliogenesis (Schaffner et al. 2020; Salem et al. 2021a; Arzua et al. 2021). In recent years, many preclinical studies with developmental alcohol exposure paradigms recapitulated the human FASD phenotypes. Utilizing these preclinical models, significant progress has been made in comprehending the differential regulation of protein-coding genes and how they influence various PAE-related CNS deficits. These studies identified numerous immediate and enduring alternations of gene expression changes as immediate and long-term consequences of PAE. In addition to genes specific to neurodevelopment, specific reprogramming of genes related to the immune system, the neuroendocrine system, and metabolism has been documented in numerous studies (Zhang et al. 2022b; Sibuh et al. 2023; Liu et al. 2021b). These genes are often sex-, developmental age-, and tissue region-specific, indicating the presence of specific modulators dictating these gene expression changes (Kleiber et al. 2013).

The effects of alcohol are transgenerational, and enduring epigenetic modifications have been documented in PAE clinical studies and preclinical studies as the root cause of FASD-related deficits (Lussier et al. 2017). "Epigenetic" generally refers to the modulatory effects of various environmental factors on the genome structure via DNA and histone modification, allowing chromatin condensation or relaxation to suppress or allow gene expression, respectively. Non-coding RNAs (ncRNAs) have emerged as a substantial layer of epigenetic control that plays critical transcriptional and post-transcriptional processes modulating gene expression (Mercer and Mattick 2013; Kaikkonen et al. 2011; Zhang et al. 2019a). A rapidly growing research area has focused on investigating non-coding RNA molecules in FASD (Mahnke et al. 2018). Prior studies reported PAE-induced aberrant expression of small non-coding RNAs such as microRNAs (miRNAs) and linear long non-coding RNAs (lncRNAs) (Balaraman et al. 2013; Seyednejad and Sartor 2022; Pappalardo-Carter et al. 2013). Additionally, non-coding RNAs of intermediate size, such as small nucleolar RNAs, have been described in the context of PAE (Laufer et al. 2013) and extensively reviewed (Mahnke et al. 2018). More recent investigations into the role of a novel class of long non-coding RNA, circular RNAs (circRNAs), have revealed a complex landscape of dysregulation that may contribute to the pathophysiology of FASD (Paudel et al. 2020; Papageorgiou et al. 2023; Noor et al. 2023a). However, the detailed molecular network of these ncRNAs underlying FASD-related dysfunction is not fully understood.

Our current knowledge of the mechanistic details of key contributory roles of ncRNAs impacting neurodevelopment and CNS function across the lifespan is limited. Several studies have revealed a direct link between these non-coding RNAs and neurobehavior and molecular alterations related to FASD, including neuronal apoptosis, neural crest cell migration, and brain connectivity (Chen et al. 2015; Tseng et al. 2019; Chen and Kannan 2023). Not limited to critical roles during the early postnatal development, PAE-induced non-coding RNA dysregulation has been reported in the adult brain and spinal cord to modulate gene expression that is critical to neuronal function (Zhang et al. 2015; Laufer et al. 2013; Papageorgiou et al. 2023; Noor et al. 2023a).

Interestingly, animal models of PAE demonstrate that heightened neuroimmune reactivity is a major driver of FASD-related CNS dysfunction (Tominaga and Caterina 2004, Topiwala et al. 2017, Himmelrich et al. 2020; Chap. 2 by Vella et al. in this volume). Several animal studies support the emerging hypothesis that underlying CNS dysfunction from PAE may not manifest until a second immune challenge is encountered (Terasaki and Schwarz 2016; Sanchez et al. 2017). Although several studies have focused on the roles of non-coding RNAs in neuronal survival and function (Salem et al. 2021b; Balaraman et al. 2013; Yuan et al. 2020; Chen et al. 2015), their roles within the context of PAE-induced neuroinflammatory consequences are unknown. Recent studies have now explored this avenue and suggested potential mechanistic roles of non-coding RNAs not only in regulating PAE-induced basal changes in gene expression but also in contributing to how PAE reprograms neuroimmune responses to subsequent exposure to stress during the early postnatal period or to subsequent immune insults during adulthood (Ruffaner-Hanson et al. 2023; Noor et al. 2023a). Additionally, non-coding RNAs are detected in the peripheral blood or other biofluid samples in various neurological conditions, emphasizing the potential of non-coding RNAs as biomarkers for diagnosis and monitoring health outcomes associated with FASD (Tseng et al. 2019; Bao et al. 2014).

The specific roles of these various classes of non-coding RNAs in cellular processes and their interactions with other coding and non-coding biomolecules remain fertile ground for investigation. As we delve deeper into summarizing the non-coding RNA landscape of PAE, we hope these novel molecular insights may pave the way for future diagnostic and therapeutic strategies to address the heterogeneous FASD-related health issues. In this chapter, we reviewed recent progress made in this emerging area of research. We have primarily focused on preclinical studies suggesting mechanistic roles of these non-coding RNAs as modulators of FASD-related CNS dysfunction and clinical and preclinical evidence considering the feasibility of these molecules for future use as biomarkers of PAE-induced CNS dysfunction.

7.2 Micro RNAs (miRNA)

MiRNAs represent small non-coding RNAs, 18–25 nucleotide long molecules. MiRNAs are transcribed as precursor molecules, termed "pri-miRNAs," which are subsequently cleaved by the endoribonucleases Drosha and Dicer (Fig. 7.1). RNA-binding protein, DGCR8, recognizes specific motifs within the pri-miRNAs, while Drosha cleaves the pri-miRNA duplex, giving its characteristic hairpin structure, yielding a pre-miRNA with a two nucleotide 3′ overhang. Pre-miRNAs are subsequently exported to the cytoplasm by exportin 5 (XPO5)/Ran/GTP complex, where they undergo further processing by RNase III endonuclease and Dicer. Dicer removes the terminal loop, resulting in a mature miRNA duplex. MiRNAs can originate from the 5′ end or the 3′ end of the pre-miRNA hairpin. Based on thermodynamic stability and nucleotide composition, a strand will be loaded into Argonaute (AGO) family proteins. The loaded strand serves as the guide strand, while the unloaded strand is degraded. Canonical animal miRNAs are encoded individually or as clusters, with clustered miRNAs transcribed together. Although miRNAs are typically located within the introns of protein-coding genes, there are seldom functional connections between the host gene and the miRNA. MiRNA biogenesis can occur via non-canonical pathways such as Drosha/DGCR8-independent and Dicer-independent pathways. miRNA biogenesis is extensively discussed in these reviews (Mahnke et al. 2018; Treiber et al. 2019; Balaraman et al. 2013). The primary mechanism of how miRNAs inhibit the expression of target genes is via direct binding to target mRNAs; a single miRNA can regulate multiple genes. Their involvement has been reported in numerous normal and pathological processes including cell proliferation, apoptosis, neuroinflammation, and synaptic plasticity (Zhang et al. 2023a; Treiber et al. 2019). MiRNAs have been implicated in neurodevelopmental and neurodegenerative disorders, as well as alcohol use disorders (AUDs) in numerous preclinical and clinical studies (Zhu et al. 2022; Zhang et al. 2023b; Lim et al. 2021)

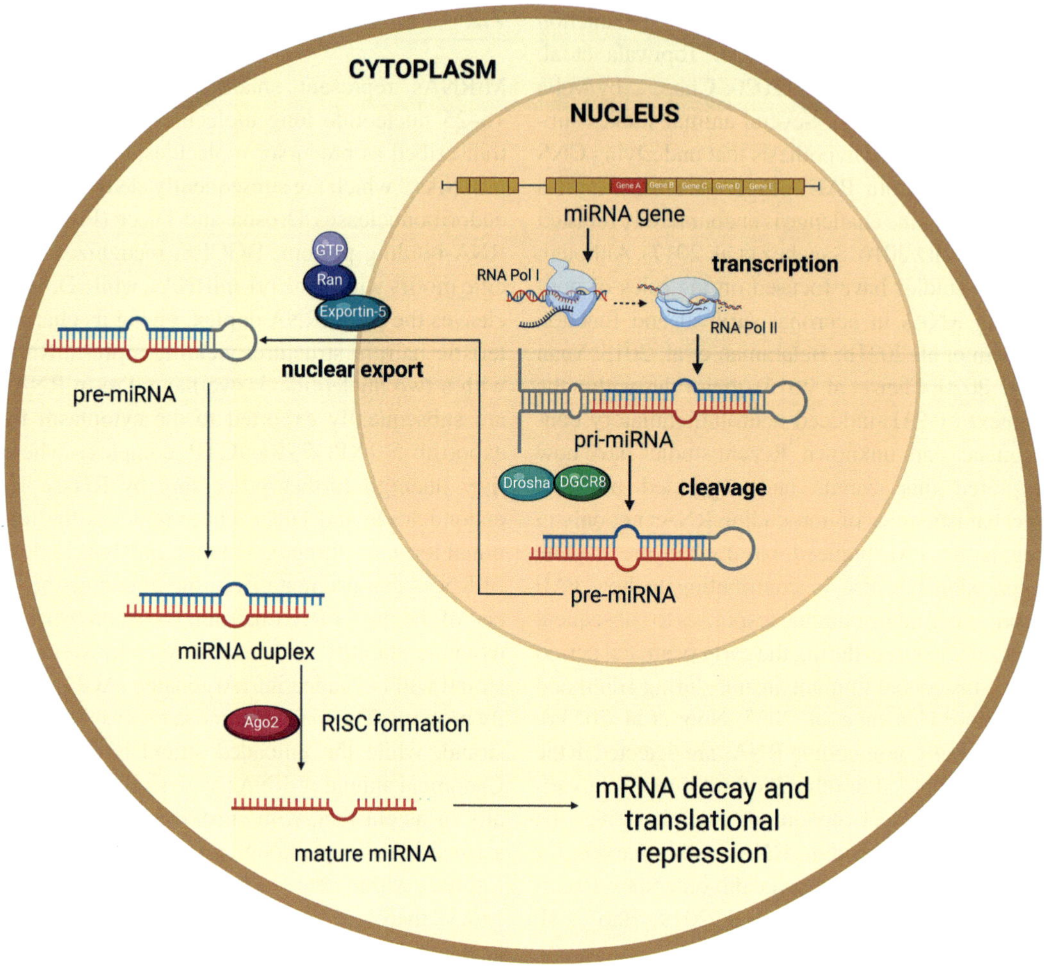

Fig. 7.1 A brief schematic of miRNA biogenesis. This schematic provides a brief overview of miRNA Biogenesis. In the nucleus, pri-miRNA is transcribed by RNA polymerase II and then processed by the Drosha-DGCR8 complex into pre-miRNA. The pre-miRNA is exported to the cytoplasm, and Dicer further processes it into a double-stranded miRNA duplex. One strand of the miRNA duplex is incorporated into RISC, which guides the complex to target mRNAs for silencing through degradation or translational repression. This illustration was created with Biorender.com

7.2.1 MiRNAs Regulate Ethanol-Induced Effects During Embryogenesis and Early Postnatal Development

To date, multiple preclinical and cell culture studies have examined PAE-induced effects on miRNA expression in the context of early postnatal CNS development (Balaraman et al. 2013; Mahnke et al. 2018; Wang et al. 2009). A number of miRNAs, including miR-9, miR-21, miR-153, miR-335a, miR-10a, and miR-10b, were found to be dysregulated in cultured fetal mouse neural stem cells or in the developing zebrafish brain (Wang et al. 2009; Tal et al. 2012; Sathyan et al. 2007). Putative gene targets of deregulated miRNAs are involved in cell cycle control, apoptosis, and cell differentiation, which are the main processes affected by ethanol toxicity.

More recently, several studies have explored key mechanistic details of how these miRNAs

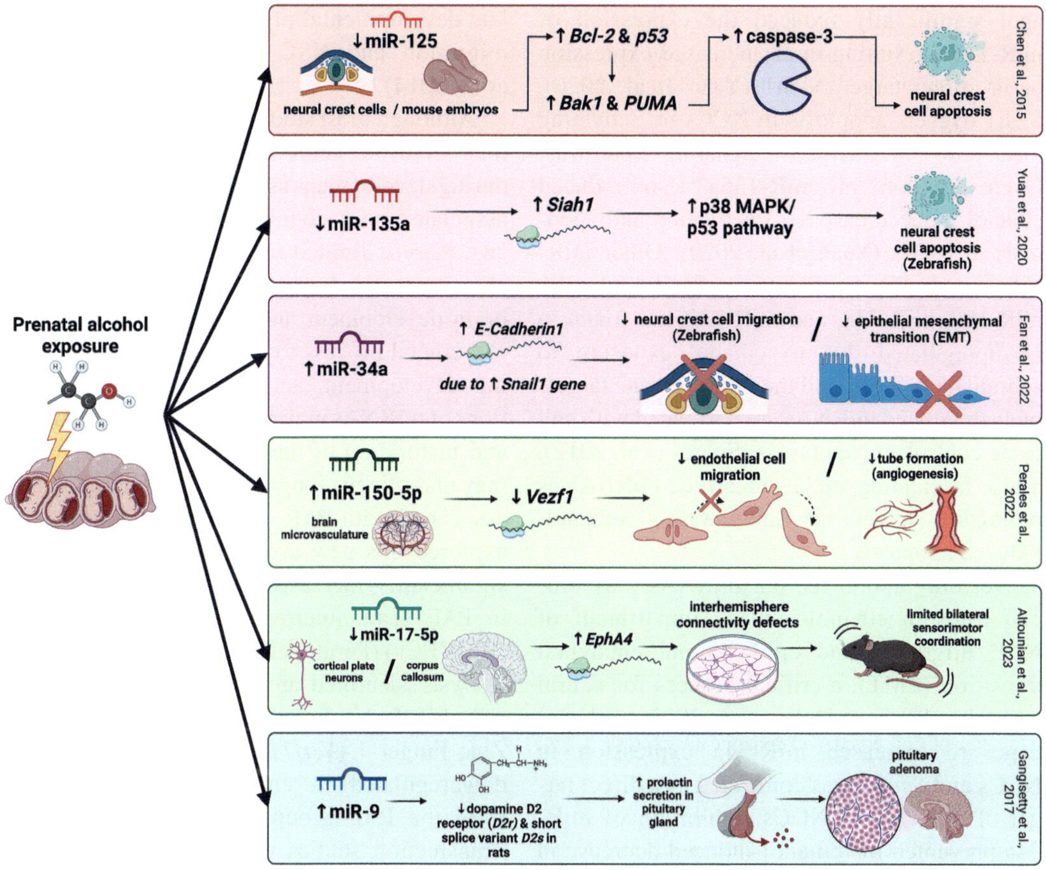

Fig. 7.2 Mechanisms of miRNA impacting PAE-induced effects on the nervous system. Based on current literature, this schematic shows various miRNAs and their critical roles in regulating PAE-induced molecular and functional deficits in the nervous system. Ethanol-induced downregulation of miR-125 and miR-135a promotes apoptosis during neural crest development in zebrafish and mice. Ethanol exposure upregulates miR-34a and reduces epithelial–mesenchymal transition in zebrafish neural crest development. PAE induces upregulation of miR-150-5p and reduces angiogenesis through changes in cell migration in the formation of the brain microvasculature in mice—PAE-induced downregulation of miR-17-5p results in deficits in sensorimotor coordination through the regulation of interhemispheric connectivity. Also, PAE-induced upregulation of miR-9 reduces dopamine receptors, contributing to the development of pituitary adenomas. This illustration was created with Biorender.com

target critical mRNA transcripts impacting early CNS development and function (Fig. 7.2). Ethanol exposure induced apoptosis in neural crest cells (NCCs) via increasing pro-apoptotic effector protein, Bak1 and p53-upregulated modulator of apoptosis (PUMA) protein expression and was associated with a dose-dependent decrease in miR-125b expression in the NCCs of embryos (Chen et al. 2015). Overexpression with miR-125b mimics significantly reduced Bak1 and PUMA protein expression in NCCs, indicating its critical role in ethanol-induced apoptosis (Chen et al. 2015). Additionally, microinjection of miR-125b mimics significantly reduces ethanol-induced caspase-3 activation and growth retardation in cultured mouse embryos exposed to ethanol (Chen et al. 2015). Notably, Fuqiang Yuan et al. utilized the zebrafish model of ethanol exposure and identified another key miRNA, miR-135a, that interacts with Bak1 and PUMA via inhibiting the p38 MAPK/p53 pathway. This study found that ethanol significantly reduced the expression of miR-135a, resulting in an increased expression of its direct target, Siah1 (Yuan et al. 2020). Siah1 triggers apoptosis in NCCs by activating the p38 MAPK/p53 signaling pathway. Overexpression of miR-135a also reduced ethanol-induced caspase-3 activation and apoptosis in NCCs (Yuan et al. 2020). Other miRNAs, including miR-725, miR-30d, let-7k, miR-100, miR-738, and miR-732, are found to be upregulated due to ethanol exposure to zebrafish embryos, and the putative gene targets of dysregulated miRNAs are correlated with cell cycle control and apoptosis (Soares et al. 2012). These fascinating studies identified miRNAs as promising targets in ameliorating ethanol-induced apoptosis.

Not only apoptosis, but miRNAs play crucial roles in ethanol-induced impairment of NCC migration and epithelial–mesenchymal transition (EMT), a critical process for neural crest development (Fan et al. 2022). Ethanol exposure increased miR-34a expression in NCCs and *Snail1* was confirmed as a direct target of miR-34a in NCCs. Inhibition of miR-34a prevented the ethanol-induced decrease in *Snail1* mRNA expression and restored NCC migration, implicating miR-34a in ethanol-induced migration impairment (Fan et al. 2022). Furthermore, downregulation of miR-34a restored ethanol-induced elevation of other critical regulators of EMT, E-cadherin, and Vimentin, indicating a role in modulating epithelial–mesenchymal transition (Fan et al. 2022). Other studies also linked miRNA as a key player in regulating PAE and its potential effects on placental insufficiency and fetal growth. These findings identified a set of 11 miRNAs that may play a role in mediating the effects of prenatal alcohol exposure on fetal growth by disrupting trophoblast invasion and placental EMT, highlighting their potential as therapeutic targets for FASD (Balaraman et al. 2016). Additionally, ethanol exposure during late gestation downregulated several miRNAs, including miR-335 and miR-10b, which regulate developmental processes such as neuronal migration and NSC differentiation (Kleiber et al. 2014).

Although most studies explored miRNAs and their role in neuronal-specific dysregulation during development as the main factor for PAE-associated neurocognitive and behavioral deficits, there is limited research on PAE-mediated changes to the brain microvasculature altering brain development and function. Angiogenesis is a crucial process to vascularize the CNS during development, and ethanol-induced alterations of miRNAs may play a role in the formation and maturation of the microvasculature, which may underlie the long-lasting behavioral deficits associated with PAE. A study by Perales et al. explored this new avenue and found that PAE significantly increased miR-150-5p expression in PAE brain microvascular endothelial cells (BMVECs) (Perales et al. 2022). Bioinformatics analysis identified an evolutionarily conserved miR-150-5p binding site in Vascular Endothelial Zinc Finger 1 (*Vezf1*), which was significantly downregulated in embryonic (E18) cortices from the PAE group and in PAE BMVECs. Transfection studies with miR-150-5p mimics or inhibitors demonstrated that miR-150-5p overexpression significantly reduced *Vezf1* expression, and downstream target endothelin 1 (*Edn1*) expression was also decreased. Ethanol treatment hindered BMVECs' migration and tube formation, which was mitigated by miR-150-5p inhibition or *Vezf1* overexpression. This study identified a new mechanism for how miR-150-5p alteration may impact the development of the cortical microvasculature during PAE and potentially contribute to deficits seen in patients with FASD (Perales et al. 2022).

7.2.2 MiRNA as Regulators of Long-Lasting CNS Dysfunction

Phenomenal studies by Laufer et al. (2013) and Ignacio et al. (2014) identified that PAE-induced miRNA alterations are long lasting and can be detected in adolescent and adult PAE brains consequent to acute and chronic alcohol exposure. These miRNA changes were often specific to the window of gestational exposure, brain regions, and developmental age (Laufer et al. 2013). Importantly, in preclinical models, social enrichment reversed some of these miRNA changes by ethanol, highlighting the critical roles of these miRNAs underlying the etiology of alcohol-induced biological mechanisms dictating neurobehavioral alterations (Ignacio et al. 2014).

Recent studies have explored whether PAE-induced miRNA changes in the brain can act as a master regulator of long-lasting neuromodulation during adulthood. PAE increases s usceptibility to tumorigenesis, including hyperprolactinemia driven by increased prolactin (PRL) secretion from the pituitary gland during adulthood (Jabbar et al. 2018). Lactotrope cell growth and PRL production are primarily regulated by the tonic inhibitory control of the dopaminergic system. Gangisetty et al. explored the role of miR-9 prolactin secretion from the pituitary gland (Gangisetty et al. 2017). Utilizing mimics and anti-miR-9-oligo, this study confirmed a regulatory role of PAE-induced miR-9 in reducing *D2r* and *D2s* expressions and increased PRL production and secretion from the pituitary.

MiRNAs may play a role in functional deficits associated with brain connectivity during adulthood, and impaired interhemispheric connectivity may affect the ability to perform complex tasks, as observed even in non-severe cases of FASD (Altounian et al. 2023). Another study utilized moderate alcohol exposure during prenatal brain wiring to study the effects of PAE on corpus callosum (CC) development and identified a role of miR-17-5p in this process (Altounian et al. 2023). PAE induced aberrant navigation of interhemispheric CC axons, leading to ectopic termination in the contralateral cortex. This study revealed the presence of miR-17-5p in cortical plate neurons and the corpus callosum, with decreased expression in response to ethanol exposure. Overexpression of miR-17-5p rescued the aberrant callosal projections induced by PAE and established ephrin type A receptor 4 (EphA4) as a mediator of the callosal targeting of CC axons. Thus, neuronal miR-17p-mediated regulation of axonal guidance may have implications for interhemispheric cortical connectivity and associated behavioral alterations in FASD.

7.3 Long Non-coding RNAs

Long non-coding RNAs comprise many non-coding RNA species with over 200 base pairs. Most annotated lncRNAs are RNA polymerase II transcribed; hence, they are similar in structure to mRNA and may have cap structures and poly-A tails (Mattick et al. 2023; Zhang et al. 2019c). To date, around 30k lncRNAs have been identified, but their functions have not yet been characterized (Frankish et al. 2019). Through interactions with DNA, RNA, and proteins, lncRNAs can influence chromatin structure and function, regulate the transcription of nearby and distant genes (Fig. 7.3), and impact RNA splicing, stability, and translation (Statello et al. 2021). During prenatal development, lncRNAs can influence the fate determination and differentiation of neural stem cells (NSC) by modulating key signaling pathways involved in neurogenesis (Dong et al. 2015). LncRNAs have been implicated in serving as molecular scaffolds or decoys, engaging with transcription factors or complexes that modify chromatin to regulate the expression of genes responsible for governing the proliferation, differentiation, and migration of NSCs (Aliperti et al. 2021). Additionally, lncRNAs can compete with endogenous RNAs (ceRNAs), effectively sequestering miRNAs or interacting with mRNAs or other non-coding RNAs, such as circular RNAs, and regulate the fate of NSCs (Ratti et al. 2020).

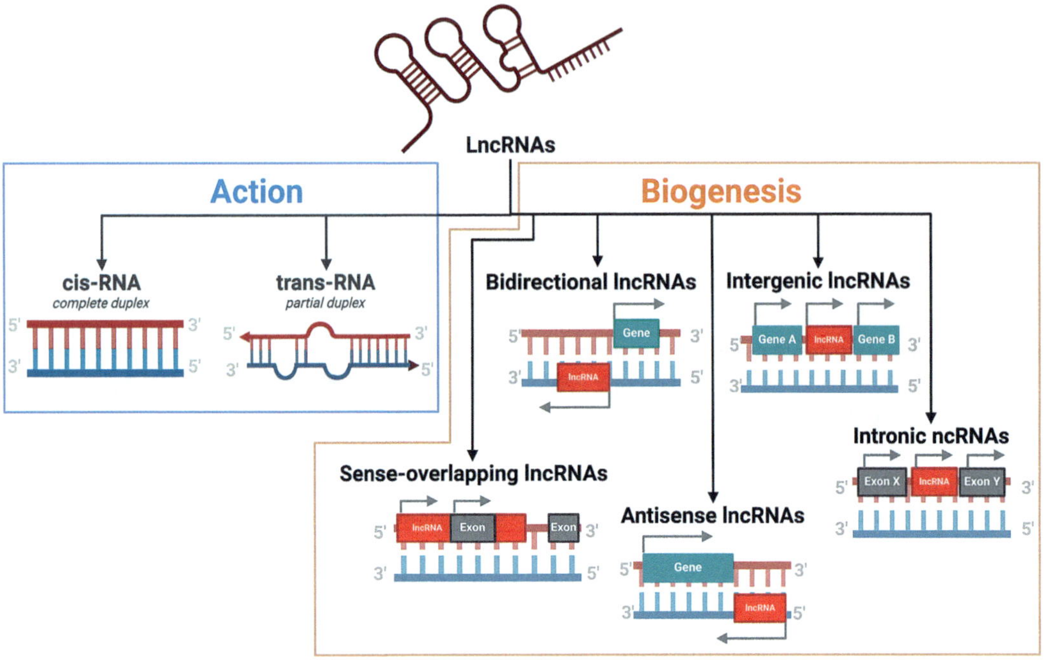

Fig. 7.3 Linear long non-coding RNA biogenesis and general function. LncRNAs are formed through transcription by RNA polymerase II; their biogenesis is similar to that of mRNAs. They can originate from intergenic regions, intronic coding regions, or regions antisense to other genes. Different types of lncRNAs include sense lncRNAs, which overlap with protein-coding genes on the same strand; antisense lncRNAs, which are transcribed from the opposite strand of protein-coding genes; and bidirectional lncRNAs, which are transcribed in the opposite direction to a nearby protein-coding gene. They can act in "cis" by regulating the expression of nearby genes on the same chromosome or in "trans" by regulating distant sites across the genome. This illustration was created with Biorender.com

7.3.1 PAE and Linear Long Non-coding RNAs

LncRNAs have emerged as critical regulators of gene expression and cellular processes, offering new avenues for understanding the molecular etiology of FASD. Although relatively less explored than miRNAs, lncRNA has been considered a critical regulator of gene expression, provided its key roles are implicated in neurodevelopmental disorders, including AUD and autism (Wilkinson and Campbell 2013; Fu et al. 2021; Liaci et al. 2022; van de Vondervoort et al. 2013).

Dysregulation of the long intergenic non-coding RNA *H19* is one of the first lines of evidence suggesting potential regulatory roles of long non-coding RNA in the context of PAE-associated deficits (Fig. 7.4). *H19* is a maternally expressed imprinted gene that produces a long non-coding RNA involved in various cellular processes, including development and growth (Monnier et al. 2013). The expression of this negative growth controller was significantly increased in the alcohol-exposed embryos compared to controls, potentially contributing to growth restrictions due to PAE (Marjonen et al. 2018). However, Bestry et al. found inconclusive evidence of hypomethylation at the *IGF2/H19* regions in somatic tissues due to potential differences in models and timing of ethanol exposure (Bestry et al. 2022).

More recently, Papageorgiou et al. explored how *H19* expression changes during developmental processes and as a consequence of PAE (Papageorgiou et al. 2023). While *H19* is predominantly found in the embryonic brain and decreases over time, PAE abolished or delayed the normal developmental changes in *H19* lev-

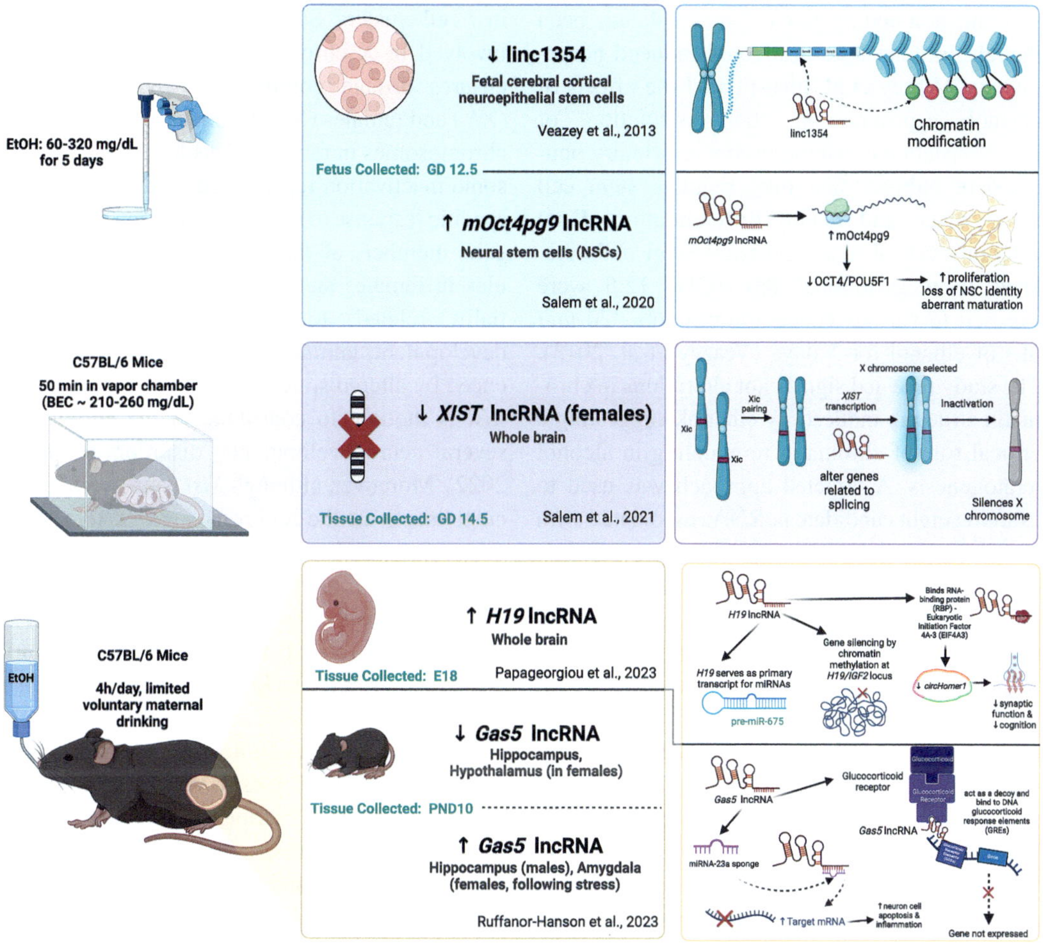

Fig. 7.4 Long non-coding RNAs of interest potentially regulating PAE-induced CNS dysfunction. Based on the current literature, this figure highlights the intricate relationship between PAE and lncRNA expression dynamics in fetal brain development. Panel 1 depicts the downregulation of linc1354 and upregulation of *Oct4pg9* in fetal cerebral cortical neuroepithelial stem cells following ethanol exposure. Panel 2 illustrates the downregulation of *XIST* in the whole brain of female fetuses. Panel 3 emphasizes two studies that employed the same moderate drinking paradigm in mice. The first half of panel 3 showcases PAE-induced increased *H19* in the whole brain at embryonic day (E)18, with proposed mechanisms including miRNA regulation and chromatin modification and also regulating the biogenesis of another non-coding circular RNA, *circHomer1* via regulating an RNA-binding protein, EIF4A3. The second half of panel 3 presents findings from Ruffanor-Hanson et al., demonstrating differential expression of *Gas5* lncRNA in the various brain regions, with potential implications in glucocorticoid signaling and regulations on cell apoptosis and inflammation via sponging miR-23a. This illustration was created with Biorender.com

els. A notable increase in the expression of *H19* was observed in the frontal cortex of adult male mice subjected to PAE. However, H19 levels were similar in female mice. Moreover, this study established an interaction between *H19* and the biogenesis of a circular RNA, *circHomer1*, that impacts cognition and learning (Fig. 7.4) (Hafez et al. 2022). Furthermore, *H19* harbors a miRNA-containing hairpin that serves as a template for miR-675. Also, *H19* acts as a molecular sponge for the let-7 family of miRNAs and *H19* dysregulation may exert regulatory effects via these miRNA molecules (Kallen et al. 2013). Therefore, *H19* regulation may involve a more complex mechanism that is yet to be appreciated.

Long non-coding RNA, Linc1354, has been thought to contribute to alcohol-induced neurotoxicity. Veazey et al. investigated the effects of ethanol exposure on the distribution of post-translational histone marks in primary neurosphere cultures that may regulate stem cell maintenance and neural differentiation. Fetal cerebral cortical neuroepithelial stem cells isolated from gestational day (GD) 12.5 were exposed to various concentrations (60–320 mg/dL) of ethanol for 5 days (Veazey et al. 2013). The study revealed significant alterations in chromatin structure induced by ethanol, suggesting a critical role of chromatin remodeling in alcohol teratogenesis. A targeted approach was used to examine eight candidate ncRNAs associated with neural stem cell differentiation. Data revealed an ethanol-mediated significant reduction of linc1354 levels across all ethanol concentrations tested. Although the exact role is unknown, dysregulation of linc1354 may contribute to aberrant gene expression patterns and disrupt neural development and chromatin remodeling, potentially predisposing individuals to FASD-related phenotypes. More recently, utilizing similar ex vivo neurosphere cultures, Salem et al. reported ethanol-induced increased expression of a lncRNA, *Oct4pg9*. Increased levels of *Oct4pg9* correlated with the reduction of the parent gene *Oct4/Pou5f1*, a transcription factor that plays a role in the pluripotency and self-renewing capacity of NSCs, leading to increased proliferation and expression of transcripts associated with neural maturation (Salem et al. 2021b).

Salem et al. also assessed the persistent effects of a single episode of binge-like maternal ethanol exposure at GD12.5 on the developmental trajectory of fetal cerebral cortical cells (Salem et al. 2021a). Salem et al. reported sex-specific alterations in developmental trajectory and cell cycle-related genes in fetal cerebral cortical cells at GD14.5 and identified a higher number of dysregulated genes in females than males (Salem et al. 2021a). Utilizing single-cell RNA sequencing and bioinformatics analysis, this study made an important observation that X-inactivation long non-coding RNA, *Xist*, was significantly downregulated due to PAE in 19 out of 33 cell clusters, and its antagonist, Tsix, was upregulated by PAE in 7 cell clusters. Studies have shown that *Xist* is involved in X chromosome inactivation, which ensures dosage compensation between males (XY) and females (XX) by silencing one of the X chromosomes in females. This loss of X chromosome inactivation represented a critical and sex-specific response to PAE. PAE also downregulated gene members of the *Xist* hub-containing modules in females and these genes were preferentially related to RNA splicing. Cortical development, lamination, and cell fate are influenced by altered spliceosome function, and thus, *Xist* is thought to contribute to the etiology of several neurodevelopmental disorders (Li et al. 2022). Moreover, although *Xist* is known to act in cis to inactivate the X chromosome, *Xist* can regulate autosomal genes to contribute to PAE-associated neurodevelopmental deficits. However, direct evidence linking *Xist* dysregulation to PAE-related neurodevelopmental outcomes is currently lacking and warrants further investigation.

Another lncRNA of interest is Growth arrest-specific transcript 5 (Gas5), which is known for its involvement in various cellular processes, including apoptosis, proliferation, stress response, anxiety, and memory (Banerjee et al. 2024). A recent study by Ruffanor-Hanson et al. highlighted a potential relationship between PAE, glucocorticoid signaling pathways, neuroimmune function and the expression of lncRNA, *Gas5*, during the early postnatal period (Ruffaner-Hanson et al. 2023). Employing a moderate PAE model in mice, the impact of PAE on *Gas5* expression across different brain regions was measured, with or without early-life maternal stress at postnatal day 10 (PND10). While no changes were observed in *Gas5* in the embryonic brain in this PAE paradigm (Papageorgiou et al. 2023), sex-specific alterations in *Gas5* lncRNA expression were observed. In males, PAE was associated with increased *Gas5* levels in the hippocampus at PND10, suggesting a potential dysregulation of glucocorticoid receptor signaling. Conversely, females exhibited decreased *Gas5* expression in response to PAE. These fascinating findings highlighted the importance of sex-specific considerations in FASD research. *Gas5* alterations were also brain region-specific; *Gas5* levels were upregulated in the amygdala in PAE-stressed mice, concurrent with augmented Toll-like Receptor (TLR4)-related proinflammatory immune factors in the amygdala. TLR4 has emerged as a critical immune pathway affected by PAE (Airapetov et al. 2021; Pascual et al. 2017). Interestingly, *Gas5* is known to sponge miR-23a, which targets *TLR4* (Gao and Huang 2021). Thus, *Gas5* may play a causal role in augmented TLR4 signaling in the context of PAE, which is unknown. These results emphasized the nuanced interplay between prenatal alcohol exposure, stress susceptibility, and immune regulation, particularly in brain regions crucial for stress regulation and neurodevelopment.

Together, these studies highlight the regulatory role of lncRNAs in modulating ethanol-related behaviors and underscore the importance of further research to elucidate the underlying biological mechanisms. Several lncRNAs hold great promise to be mechanistically linked in the context of PAE. The expression of lncRNAs *MALAT1* and brain-derived neurotrophic factor (*BDNF*) antisense are notably upregulated in various brain regions of human alcoholics and rats during alcohol withdrawal, suggesting its potential involvement in ethanol consumption and AUD pathophysiology (Kryger et al. 2012; Dulman et al. 2020). *NEAT1*, a lncRNA, displays dynamic responses to stress and is involved in lipid metabolism and inflammation (Kukharsky et al. 2020). An intriguing avenue for future research is the potential role of lncRNAs in accelerating aging (He et al. 2018). Several lncRNAs, such as *H19* and *MALAT*, are found to be involved in regulating growth and aging and may contribute to FASD. Furthermore, future research could gain insights into the role of these lncRNAs in aging or later-life health outcomes under PAE conditions.

7.4 Circular RNAs

CircRNAs are a novel category of long non-coding RNAs with high abundance, stability, diverse biological roles and conservation across species (Zhang et al. 2018). Circular RNAs are generated by pre-mRNA being processed through a back-splicing event leading to the circularization and covalent joining of back-spliced exons and/or introns of protein-coding genes (Barrett and Salzman 2016; Haque and Harries 2017) rather than canonical splicing that generates linear mRNA transcripts (Fig. 7.5a). CircRNAs were first described as early as the 1970s; however, circRNAs were thought to be splicing artifacts and were referred to as "junk" RNAs for about two decades until the recent applications of improved annotation tools following deep sequencing of ribosomal and linear RNA-depleted samples confirmed the existence of tens of thousands of circRNAs in multiple species (Ng et al. 2018b; Barrett and Salzman 2016; Haque and Harries 2017).

CircRNAs and their putative functions are an emerging area of research as they can regulate gene expression via multiple and diverse mechanisms (Fig. 7.5b) (Barrett and Salzman 2016; Chen 2020; Chen et al. 2021b). Among their varied roles, circRNAs function as molecular sponges, adeptly sequestering miRNAs and other regulatory RNA molecules (Jarlstad Olesen and Kristensen 2021). By doing so, they intricately modulate gene expression by indirectly influencing the availability and activity of miRNAs, thus exerting fine-tuned control over cellular processes (Ratti et al. 2020). Beyond their role as miRNA sponges, circRNAs engage in intricate partnerships with RNA-binding proteins (RBPs), orchestrating the fate of RNA molecules within the cellular milieu (Zang et al. 2020). CircRNAs may or may not influence the host gene expression by interacting with transcription and translation machinery that control parental gene expression or competing with linear mRNA splicing (Wang et al. 2020; Xu et al. 2020; Shao et al. 2021). In response to cellular cues and environmental stimuli, circRNAs undergo dynamic transformations, altering their expression profiles and subcellular localization. These adaptive responses allow circRNAs to participate in cellular signaling pathways and stress responses, showcasing their versatility and functional significance (Panda et al. 2017).

CircRNAs are highly enriched in the brain; however, the majority of circRNAs and their biological role during normal brain development and

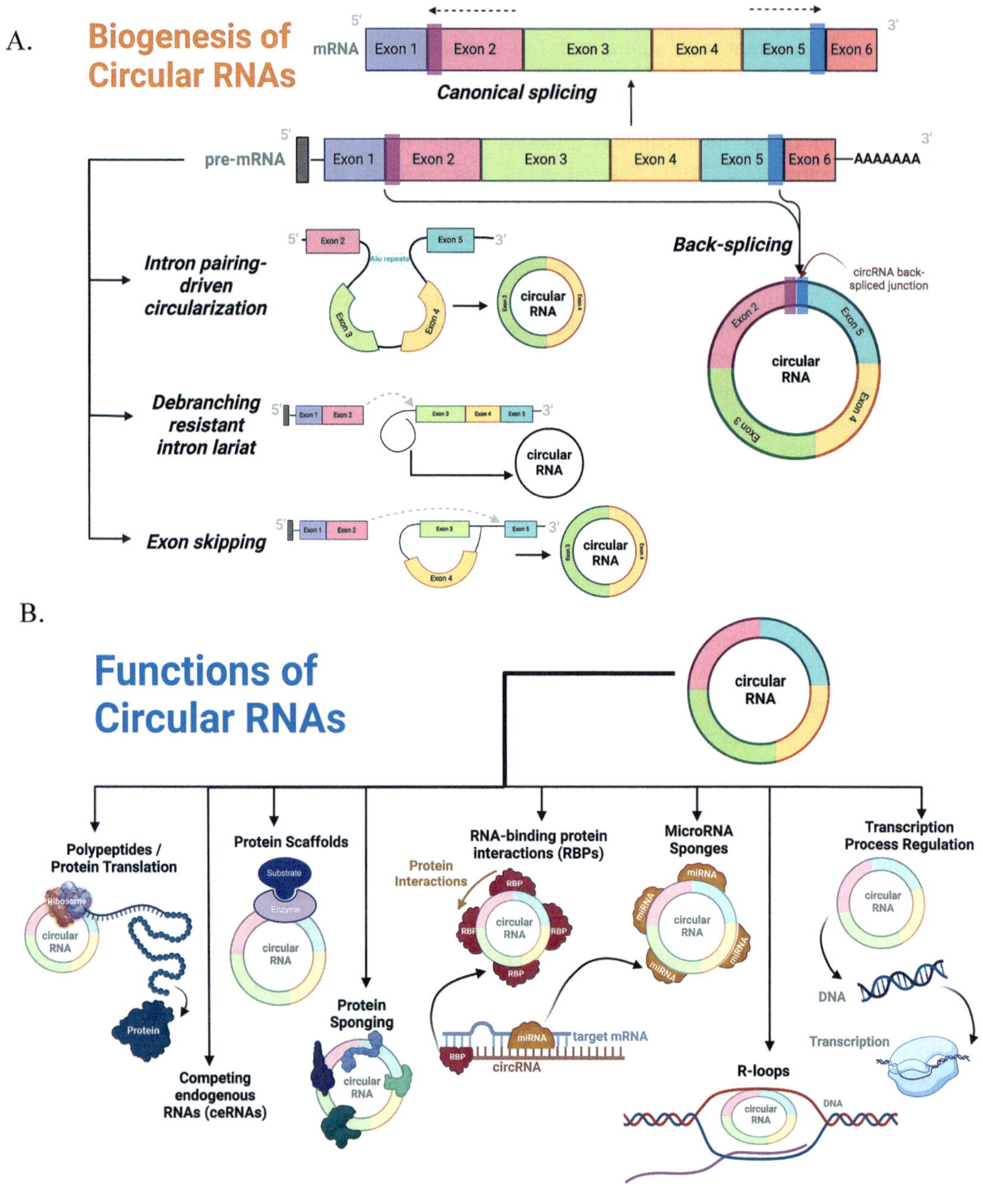

Fig. 7.5 Circular RNA biogenesis and their roles in regulating gene expression. (**a**) Circular RNA is formed when a downstream splice donor site is joined to an upstream splice acceptor site within a pre-mRNA transcript in a process known as back-splicing, forming a covalently closed loop structure, often excluding or including specific exons or introns. Part figure (**a**) shows how circular RNA biogenesis from pre-mRNA via the back-splicing process generates a unique transcript at the junction point compared to canonical splicing generating mRNA transcripts. Other methods of generating circRNAs include intron pairing-driven circularization, debranching resistant intron lariat, and exon skipping formation. The biogenesis of circRNAs is popularly studied through the back-splicing method, as the back-spliced junction can be analyzed through various primer designs using qRT-PCRs. (**b**) Circular RNAs can act in different ways to repress a target mRNA: miRNA sponging, binding to, and sequestering microRNAs. They can also interact with RNA-binding proteins to alter the stability of other RNA molecules. CircRNAs are involved in the regulation of transcription by interacting with transcriptional machinery or chromatin modifiers. They can endogenously compete with their "parent/host" gene. This illustration was created with Biorender.com

CNS pathology are still areas of active investigation. CircRNAs regulate synaptic and activity-dependent gene expression and alter neuronal function and behavior (Piwecka et al. 2017; Zimmerman et al. 2020). Given the tremendous implications for circRNAs in diagnosing and treating CNS disorders, this burgeoning field has multiplied in the number of studies conducted in recent years. CircRNAs play modulatory roles in psychiatric disorders (Zimmerman et al. 2020), addiction (Bu et al. 2019), alcohol use disorders (Vornholt et al. 2021) , and Alzheimer's disease (Dube et al. 2019).

CircRNAs display developmental stage-specific expression profiles (Reddy et al. 2017). CircRNAs play essential roles in embryonic and fetal development and neuronal maturation (Suenkel et al. 2020) and may respond to or regulate synaptic function (Rybak-Wolf et al. 2015; Paudel et al. 2020). CircRNAs have been described to be linked to Autism Spectrum Disorders (ASD) (Mai et al. 2022). ASD risk genes were found to be regulated by circRNA–miRNA interactions (Chen et al. 2020). Moreover, circRNA dysregulation has been documented in ischemic stroke and other neurovascular conditions that share common underlying mechanisms, including vascular dysfunction, cell death and apoptosis, inflammation, blood-brain barrier permeability, and peripheral-immune–CNS interactions (Liu et al. 2021a; Yang et al. 2021) that may be implicated in FASD-related dysfunction. However, very little is known about their roles in FASD.

7.4.1 PAE Induces circRNA Dysregulation in Embryonic and Adult Brain

A landmark study by Paudel et al. 2020 examined the effects of PAE on the differential expression of circRNAs in the developing mouse brain (Paudel et al. 2020). This study utilized a well-established model (Brady et al. 2012) of moderate levels of PAE in mice that previously demonstrated FASD-like neurodevelopmental deficits (Allan et al. 2014; Paudel et al. 2020). Utilizing an unbiased microarray platform with more than 14,000 probes to detect circRNA-specific splice junction sequences (Memczak 2013; Guo et al. 2014) in the whole brain tissue, samples collected at embryonic day 18 (E18) were examined in the non-PAE group vs. PAE offspring across both sexes. Results suggested that PAE alters circRNA expression in the brain in a unique sex-dimorphic manner. There was minimal overlap in PAE-induced changes in the E18 brain in male vs. female PAE offspring, only three circRNAs were dysregulated due to PAE, regardless of sex. Strikingly, considering sex as a variable, data revealed a unique pattern of circRNA expression in the developing brain, suggesting a robust negative correlation between female-to-male circRNA ratios in non-PAE and PAE mice. In male mice, 46 circRNAs were significantly upregulated, and nine circRNAs were downregulated in the PAE brain, compared to controls. Fourteen circRNAs were upregulated, and five circRNAs were downregulated in female PAE brains. These findings are intriguing and suggest that in addition to sex-specific circRNA expression in the normal developing brain, PAE alters circRNA expression in a sex-specific manner. This study further confirmed the importance of studying both sexes while considering PAE-induced molecular and neurobehavioral changes. Compellingly, these PAE-induced differentially expressed circRNAs in the prenatal brain preferentially originated from the genes crucial for neurogenesis and neuronal development. Bioinformatics exploration of these genes linked to differentially regulated circRNAs predicted the "GABA receptor signaling" and "dopamine-DARP32 feedback in cAMP signaling" are among the top pathways in female PAE mice. In contrast, synaptogenesis and Ephrin receptor signaling were the top canonical pathways related to the circRNA host genes in male PAE mice. In male PAE mice, these differentially regulated brain circRNAs were primarily associated with genes involved in the later stage of neuronal development (Paudel et al. 2020).

Notably, PAE preferentially reversed many sex-specific changes in circRNA expression that physiologically occur in the normal developing brain. Utilizing qualitative PCR, this study further validated two circRNA molecules of interest. *CircPtchd2* was upregulated in control E18 brains in females than in males, and such changes were not observed in PAE brains. *Ptchd2* is a sterol-sending domain-containing protein regulated by thyroid hormone signaling implicated in PAE-related behavioral dysfunction (Wilcoxon et al. 2005). Therefore, although unknown, thyroid signaling may impact the biogenesis of *circPtchd2*. The other molecule of interest was *CircSatb2*, which was upregulated by PAE in males but not in females. *Satb2* plays a critical role during CNS development in determining projection neuron identity. *Satb2* regulates hyperactivity, impulsivity, intellectual disability, spatial learning, and memory (Zhang et al. 2019b; Bissell et al. 2022). However, whether *circSatb2* has similar or additional effects as their linear mRNA molecules is unknown. Interestingly, *circSatb2* also showed a negative correlation with brain-derived neurotrophic factor, BDNF, and inhibitory neuron marker Glutamate decarboxylase 1 (*Gad1*) mRNA levels in the whole brain. However, their mRNA levels were not significantly different across sex or prenatal exposure. Although mechanisms are unknown, *circSatb2* can be involved in regulating these critical players of neurodevelopment and synaptic function.

7.4.2 *CircHomer1* Expression in the Adult Brain and Its Interaction with lncRNA *H19*

CircRNA expressions and their regulation can be region-specific, and their abundance can vary among various brain regions (Sekar and Liang 2019). It is unknown whether PAE-induced circRNA changes are developmental age-specific and whether these circRNA dysregulations due to PAE can be evident during adulthood. Using the same model of moderate PAE (Brady et al. 2012), circRNA expression was examined in the adult brain (PND 80–90) tissues. This study by Papageorgiou et al. focused on a particular circRNA of interest, *circHomer1*, and its expression levels across different brain regions. *CircHomer1* is an activity-dependent circRNA derived from Homer protein homolog 1 (*Homer1*). *CircHomer1* is particularly enriched in the adult brain and is known to regulate synaptic genes and cognitive function. Downregulation of *circHomer1* can impact the synaptic localization of numerous synaptic plasticity genes, including genes known to be implicated in FASD (Zimmerman et al. 2020; Hafez et al. 2022). *CircHomer1* was significantly downregulated due to PAE in the adult brain but not in the E18 whole brain (Papageorgiou et al. 2023). However, PAE-induced downregulation of *circHomer1* was sex-specific and was observed only in male PAE mice. Moreover, this downregulation was pronounced in the frontal cortex and hippocampus. This data demonstrated a brain region- and developmental age-specific expression pattern of this *circHomer1*.

Papageorgiou et al. further examined a potential interaction with a linear lncRNA *H19* and *circHomer1*. In contrast to *circHomer1* downregulation, *H19* levels were upregulated in these brain regions due to PAE. Moreover, *H19* knockdown in neuroblastoma cells upregulated *circHomer1* (Papageorgiou et al. 2023). Utilizing in silico analysis, authors proposed that H19 can bind and sequester eukaryotic initiation factor 4A-III (EIF4A3), an RNA-binding protein involved in *circHomer1* biogenesis (Fig. 7.4). This study introduced novel mechanistic insights into molecular networks comprising circRNA interactions with other non-coding RNAs in the context of PAE, which warrants future explorations. To summarize, these studies demonstrated circRNA alterations in a developmental age- and sex-specific manner resulting from PAE, potentially modulating the critical neuronal development and function underlying FASD-related disorders.

7.4.3 CircRNAs, a Potential Novel Modulator of Neuroimmune Dysfunction in PAE

The biological vulnerability to develop neurobehavioral and functional deficits in FASD populations is thought to be driven by neuroimmune dysfunction (Mukherjee et al. 2023, Noor and

Milligan 2018; Chap. 2 by Vella et al. in this volume). Our recent work, and that of others (Noor and Milligan 2018; Noor et al. 2020; Sanchez et al. 2017; Terasaki and Schwarz 2016; Wong et al. 2018), recognizes that PAE can re-program later-life immune challenges (Noor and Milligan 2018). Strikingly, a PAE-generated heightened susceptibility to develop touch hypersensitivity (allodynia) was unmasked only after a minor nerve injury, not under basal conditions (Sanchez et al. 2017), and PAE-induced heightened levels of TLR4-associated proinflammatory cytokines and transcription factors were observed in the spinal cord, where damaged peripheral axon terminals communicate with spinal pain projection neurons. Similar observations of heightened proinflammatory function were made with ours and other prior studies with peripheral immune stimulation with TLR4 agonist, lipopolysaccharide (LPS), in PAE rodents in vitro and in vivo studies (Noor et al. 2023b). However, the potential mechanism(s) underlying the persistent insidious effects of PAE, even at low to moderate levels, on immune function are unknown.

To date, several circRNAs have been identified to influence neuroimmune function, offering novel avenues for targeted interventions to ameliorate neuroinflammatory responses (Liu et al. 2022; Yang et al. 2021; Chen et al. 2021b). Recently, our group explored whether circRNAs are implicated in PAE-induced reprogramming of neuroimmune function. Specifically, we explored whether PAE-induced circRNA alterations during adulthood precipitate into exaggerated TLR4 signaling after the secondary immune challenge, leading to heightened CNS glial and peripheral proinflammatory immune responses. Utilizing a moderate PAE paradigm, we conducted an unbiased microarray-based circRNA profiling in peripheral blood immune cells and the spinal cord in adult PAE rats. Data suggested that PAE dysregulates circRNAs both in the peripheral blood (discussed in Sect. 7.5) and the spinal cord during adulthood (Noor et al. 2023a). Twenty-four circRNAs were found to be downregulated, and eight circRNAs were upregulated in the PAE spinal cord. Using Ingenuity Pathway Analysis (IPA), an additional unbiased approach was taken to determine the pathways and top networks of the genes associated with these differentially expressed circRNAs. Strikingly, this study generated compelling evidence suggesting these dysregulated circRNAs were preferentially generated from genes that are linked to a key transcription factor, nuclear factor-kappa (NF-κB) complex, which is one of the most studied downstream regulators of the TLR4 signaling pathway and a downstream effector of other proinflammatory cytokines such as IL-1β and TNF-α receptor signaling (Mitchell et al. 2016; Verstrepen et al. 2008). All these immune factors are known to impact CNS function under PAE conditions (Pascual et al. 2017; Terasaki and Schwarz 2016). We found that PAE downregulated circRNA derived from serine/threonine protein kinase, *Rps6Ka3*, and *Akt3* (Delaunoy et al. 2001; Ng et al. 2018a), genes involved in MAPK, Akt3, and P13/Akt/mTOR signaling. PAE upregulated circRNAs derived from *Gas7*, *Slc7a11*. A number of these genes and circRNAs derived from them have immune modulatory roles and are involved in neuropathology including Parkinson's disease, ASD, and cancer models (Bentea et al. 2021; Guo et al. 2023; Huang et al. 2021; Chen et al. 2021a).

A peripheral nerve injury model was utilized to examine whether PAE further modulates circRNA expression following adult-onset immune insults (Noor et al. 2023a). PAE not only dysregulated circRNA expression at basal levels but also influenced the circRNA profile in the spinal cord following nerve injury. More than 100 circRNAs were found to be dysregulated concurrent with allodynia in minor nerve-injured PAE rats, a modified model of inducing minimal nerve injury, where no allodynia was observed in control nerve-injured rats. This PAE-induced allodynic susceptibility was concurrent with 36 up- and 78 downregulated circRNAs in the spinal cord. Surprisingly, this secondary immune challenge (nerve injury) resulted in an overlapping yet distinct circRNA expression profile in PAE (allodynic) rats compared to control (non-allodynic) rats. PAE not only modulated the fold regulations of circRNAs that were dysregulated in common by nerve injury in control conditions,

PAE also generated unique spinal cord circRNA changes due to peripheral nerve injury. Notably, these circRNA host genes are related to various pain-relevant pathways, such as "glutamate receptor signaling," "circadian rhythm signaling," "G-protein receptor signaling," and "chemokine signaling," and yet again, the involvement of these parental genes with the NF-κB immune complex was confirmed. CircRNA biogenesis is often paired with active transcription of their parental genes. Thus, these data reinforced prior transcriptome studies in alcohol and PAE literature suggesting TLR4, and NF-κB-responsive genes are potential key modulators of PAE-related CNS dysfunction (Pascual et al. 2017, Terasaki and Schwarz 2016). These latest findings support the concept that dysregulated circRNAs may act as novel regulators in the context of PAE-induced neuroimmune sensitization during adulthood. However, this study was conducted using only female rats. Based on this extensive evidence of sex-specific differences in PAE and the neuroimmune mechanism underlying allodynia, future studies will identify potential sex dimorphism in circRNA expression due to PAE during adulthood.

Our knowledge of circRNA-mediated mechanisms regulating immune and neuroimmune systems is limited and is an active area of research. In recent years, circRNAs have been identified to modulate peripheral immune and glial-immune function (Zhou et al. 2024; Chen et al. 2021b). Among the differentially expressed spinal circRNAs in PAE rats under nerve injury conditions, some circRNAs are proven to be critical regulators of immune and neuroimmune function by recent studies. *CircItch* (Su et al. 2022; Liu et al. 2022) was downregulated in the PAE spinal cord following nerve injury. *CircItch* is involved in several biological functions, including cell proliferation, inflammation, and neovascularization, and acts as an anti-inflammatory factor. *CircItch* levels are downregulated in multiple disease models, and overexpression of *circItch* reduced proinflammatory cytokines (Kong et al. 2020). *CircItch* regulates proinflammatory cytokines such as IL-1β and TNF-α via regulating mRNA or via absorbing miR-33 (Su et al. 2022, Liu et al. 2022) , and miR-33 is known to positively regulate the NLR family pyrin domain containing 3 (NLRP3)/Caspase-1 pathway that is necessary for PAE-induced allodynia (Noor et al. 2023b). Therefore, while speculative, downregulated *circItch* during PAE may reduce miR-33 sponging effects and promote inflammation.

Another circRNA of interest was circRNA derived from the Ovarian Tumor Domain containing 7B (Otud7b), which displayed significant downregulation in the PAE spinal cord relative to controls. Otud7b is a deubiquitinase enzyme (Rothschild et al. 2018; Lei et al. 2019) that acts as a negative regulator that balances the strength and duration of canonical and non-canonical NF-κB signaling through feedback mechanisms (Rona and Pagano 2019). Its deficiency augments the non-canonical NF-κb pathway (Hu et al. 2013) and defective cell cycle progression (Komander et al. 2009). Therefore, it is possible that diminished *circOtud7b* may affect its linear transcripts and may create an inability to suppress the transcription of proinflammatory responses, a state of sensitized NF-κB signaling on glial-immune cells under PAE conditions. Additionally, our microarray data detected upregulated circRNAs originating from *Rims1* and *Tttc3* and downregulated circRNAs derived from *Stag1 and Parg*. Not only do these linear transcripts play a critical role in CNS function and immunity, but circRNAs that arose from these genes have been implicated in preclinical models of cancer, depression, ischemia and endogenous stress response, CNS injury, and inflammation (Ma et al. 2021; Yang et al. 2021; You et al. 2015). *CircTtc3* is known to sponge miR-372, which targets *TLR4* mRNA and inhibits glial immune responses and dysfunction (Yang et al. 2021). *CircStag1* downregulation increases the stability of FAAH (fatty acid amide hydrolase) mRNA, which induces astrocyte dysfunction, upregulates proinflammatory factors, and downregulates anti-inflammatory factors (Huang et al. 2020; Tanaka et al. 2019). While the exact downstream effects of these circRNAs and their expression pattern in the adult brain are unknown under PAE conditions, these circRNAs may play critical roles in PAE-induced susceptibility to proinflammatory

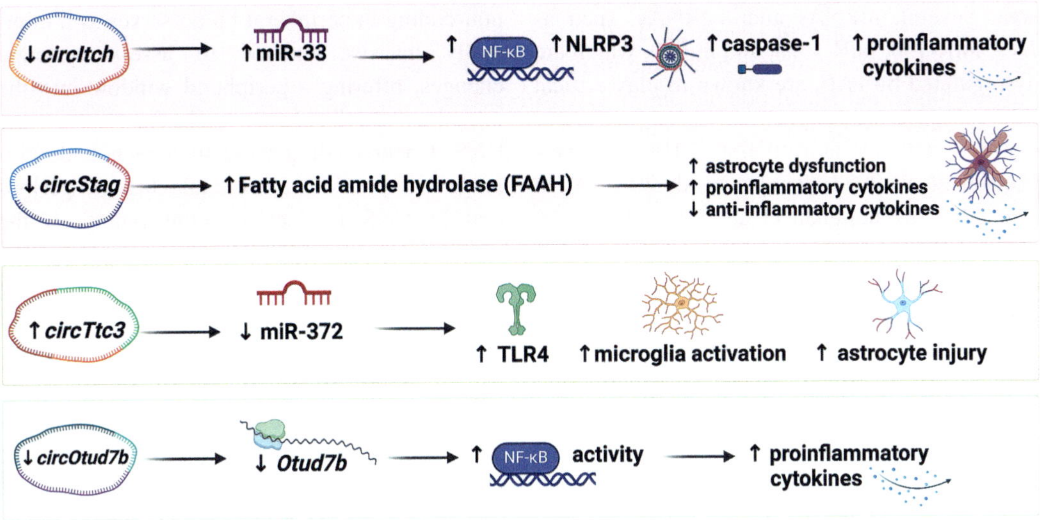

Fig. 7.6 Proposed mechanism(s) of how circular RNAs could modulate PAE-induced neuroimmune dysfunction. Based on the prior observations of PAE-induced susceptibility in developing proinflammatory immune bias leading to allodynia (Noor et al. 2020; Sanchez et al. 2017) and the concurrent changes in circRNA levels (adapted from Noor et al. 2023a), this diagram provides a few predicted mechanisms based on the current literature to show how these circular RNAs may promote neuroinflammation. Although these pathways are speculative, future research exploring the roles of these circular RNAs in the context of PAE may identify new targets contributing to FASD-related CNS deficits. Downregulation of *circItch* may result in increased miR-33 activity, promoting the activation of NF-κB and downstream factors such as NLRP3, producing proinflammatory cytokines. Downregulation of *circStag* results in increased stability of FAAH mRNA, which leads to increased inflammation in the CNS accompanied by astrocyte dysfunction. Upregulation of *circTtc3* sponges miR-372 leads to upregulation of TLR4 signaling and microglial and astrocytic dysfunction. PAE-induced downregulation of *circOtud7b* may downregulate its parental gene, *Otud7B* mRNA transcription, leading to increased NF-κB activity and the production of proinflammatory cytokines. This illustration was created with Biorender.com

neuroimmune responses. Based on the growing literature on these circRNAs of interest, a schematic of these hypothetical mechanisms of how these circRNA changes may promote PAE-induced neuroinflammation is depicted in Fig. 7.6, which is currently under investigation.

To summarize, studies described in this section demonstrated that PAE-induced dysregulation of circRNAs displays a distinct pattern dependent on tissue region, sex, and age and is further regulated by subsequent immune insult. This research identified several promising circRNA that may play critical roles in creating neuroimmune dysfunction observed under PAE conditions, which needs further confirmation. Future studies geared toward specific modulations of circRNA expression in various brain regions and examining neurobehavior to dissect the biological relevance of FASD would advance this research area. Moreover, these data also support the idea that targeting NF-κB signaling and downstream immune molecules or non-coding RNAs targeting these immune molecules may effectively mitigate FASD-related CNS pathologies (Noor et al. 2023b; Doremus-Fitzwater et al. 2020). As we navigate this intricate landscape of circRNAs and their interaction with other coding and non-coding RNAs, the convergence of clinical data and preclinical models holds the key to unlocking the full therapeutic potential of circRNAs in neuroinflammation.

7.5 Future Areas of Exploration in ncRNA Research

Although our knowledge of non-coding RNAs has expanded substantially in recent years, further research to elucidate the causal mechanisms is of prime importance for the FASD research area. Several miRNAs and lncRNAs, such as *H19*, miR-150, miR-135a, and miR-9, which are dysregulated by PAE, are known to play critical roles in regulating peripheral and glial proinflammatory function (Caldwell et al. 2018; Ruffaner-Hanson et al. 2023; Perales et al. 2022; Yuan et al. 2020; Balaraman et al. 2013). However, their role has yet to be explored in the context of PAE-induced aberrant neuroimmune signaling. Overall, expanding our focus on potential therapeutic avenues, ncRNA silencing/knockdown strategies could provide invaluable insights into the causal relationships between circRNA dysregulation and neurobehavioral alterations. Furthermore, delving into the intricate sponging interactions of circRNAs with miRNAs unveils a rich layer of regulatory mechanisms; further investigation may elucidate the causal relationships and functional consequences of circRNA dysregulation. Utilizing advanced technologies such as CRISPR-Cas9, RNA interference, or antisense oligonucleotides and computational predictions may enable a nuanced understanding of circRNA–microRNA interactions and may offer promising avenues for targeted interventions to prevent or reverse the long-lasting effects of PAE. Compellingly, specific loci may hold the long-term footprint of PAE-induced alterations of the ncRNAs (Laufer et al. 2013), suggesting PAE might alter the regulation of not only individual miRNAs but entire clusters of co-regulated ncRNAs which may help to identify potential circRNA–miRNA–mRNA interactions and drive future mechanistic studies and may pave the way for innovative treatment strategies.

7.5.1 Non-coding RNAs as Biomarkers of FASD

FASD is considered a whole-body disorder; the effects of prenatal alcohol exposure extend beyond the brain and impact various physiological systems. The dysregulation of non-coding RNAs in blood and other biofluid, such as saliva and cerebrospinal fluid, represents a promising avenue for potential biomarkers for diagnostic and prognostic purposes. The investigation of non-coding in peripheral blood serves as a minimally invasive method to assess systemic changes, offering a peripheral window into the intricate molecular landscape of the CNS. Understanding these molecular underpinnings may provide valuable insights into peripheral to CNS interactions contributing to the pathophysiology of FASD and may identify new therapeutics.

7.5.1.1 miRNAs as Blood Biomarkers of Alcohol Use and FASD

In recent years, significant progress has been made in generating clinical evidence supporting miRNAs as predictive biomarkers for adverse outcomes of FASD. Maternal and infant circulating miRNAs have provided novel insights and identified potential target miRNAs that may dictate FASD outcomes. A remarkable study by Sridevi Balaraman et al. reported elevated levels of 11 distinct miRNAs (Table 7.1) in maternal circulation during the second and third trimesters that distinguished infants who were affected by PAE (heavily exposed affected [HEa]) from those who were apparently unaffected at birth by PAE (heavily exposed unaffected [HEua]) or those who were unexposed (UE) (Balaraman et al. 2016). The maternal miRNA alterations in plasma predicted infant growth deficits following PAE and may be helpful in classifying difficult-to-diagnose FASD subpopulations. Also, a study by Gardiner et al. found maternal serum miRNA expression was associated with alcohol use during human pregnancy (Gardiner et al. 2016). This study identified a unique signature of serum miRNAs that can be used to predict maternal alcohol consumption in pregnant women and valuable as biomarkers of alcohol exposure.

Interestingly, sex plays a crucial variable in these predictive miRNA biomarkers. Salem et al. examined the association between infant sex and maternal plasma miRNA responses to PAE (Salem et al. 2020, 2021a). The study revealed that while most maternal miRNAs affected by PAE were not influenced by sex, some exhibited sex-specific responses. Disaggregating the data by fetal sex identified maternal miRNAs showing increased significance in response to PAE, indi-

Table 7.1 List of non-coding RNAs of interest as biomarkers

Biomarkers of interest	Alcohol exposure paradigm	Window of exposure	Time of sample collection	Sample type	Ref.
miR-663b, miR-320d, miR-30c-5p, miR-125a-5p, miR-21-5p	Self-reported	133 mL≈ 9 standard drinks/ occasion across pregnancy on an average of 1–2 days/week	Infant (males and females), 2 weeks of age	Circulating blood plasma, human	Mahnke et al. (2021)
miR-126-3p, miR-328-3p, miR-30c-5p, miR-24-3p, miR-146a-5p, miR-193a-5p, miR-103a-3p, miR-92a-3p, miR-143-3p, miR-140-5p, miR-23b-3p, miR-125a-5p, miR-194-5p,	Self-reported	133 mL≈ 9 standard drinks/ occasion across pregnancy on an average of 1–2 days/week	Infant (males and females), 6.5 months of age	Circulating blood plasma, human	Mahnke et al. (2021)
miR-222-5p, miR-187-5p, miR-299-3p, miR-491-3p, miR-885-3p, miR-518f-3p, miR-760, miR-671-5p, miR-449a, miR-204-5p, miR-519a-3p	Self-reported	Variable, throughout pregnancy	2nd and 3rd trimester (GD ~18 and ~32)	Maternal blood plasma, human	Balaraman et al. (2016)
miR-509-5p, miR-3119, miR-26a-2star, miR-1279, miR-4743, miR-4799-3p, miR-4657, miR-3942-3p, miR-3126-3p, miR-514b-5p	Self-reported	≥1 binge-drinking episode or ≥3 drinks/ week during pregnancy	At the admission time for delivery	Maternal blood serum, human	Gardiner et al. (2016)
circSmarcc1, circAgtpbp1, circGalnt9, circClasp2, circHlcs, circKdm2a, circArhgap10, circAsap2, circVopp1, circYwhaz, circDennd1b, circNsg2	Voluntary moderate drinking, 5% EtOH for 4 h/day	GD1 to GD21	7–8-month-old PAE females	Circulating blood leukocytes, PAE rat offspring	Noor et al. (2023a)

cating a potential link between fetal sex and maternal biomarker profiles. Moreover, the study demonstrated that PAE increased the number of significant miRNA cross-correlations, with variations observed based on both infant sex and birth outcome. These findings underscore the complexity of maternal–fetal interactions and the importance of considering fetal sex in predictive models based on extracellular miRNA profiling.

More recently, circulating microRNAs in infants have been explored as biomarkers of FASD. A study by Amanda Mahnke et al. examined whether exosomal miRNAs (exmiRNAs) can distinguish between alcohol-exposed and non-exposed infants and play critical roles in mediating PAE's effects on specific developmental outcomes, such as growth and cognition. By analyzing 148 miRNAs from both alcohol-exposed and control groups at two different time points (2 weeks and 6.5 months), the study revealed significant alterations in miRNA expression attributable to PAE. Specifically, PAE was found to significantly modify two miRNAs at 2 week-time points and 13 miRNAs at 6.5 month-time points (Mahnke et al. 2021). Notably, most PAE-responsive miRNAs were upregulated, indicating a potential mechanism through which alcohol exposure influences gene regulation during early development. Additionally, when the data were analyzed by sex, distinct patterns emerged, with miRNAs displaying significant elevation in female PAE infant plasma samples and others in male PAE samples. Correlation analysis further revealed that PAE was associated with increased coordinated expression of exmiRNAs across chromosomes, suggesting a broader impact on gene regulation pathways. In essence, this study established that exmiRNAs in infant plasma could serve as biomarkers for PAE and can predict adverse outcomes on growth and cognition. Expanding on prior investigations suggesting the predictive potential of maternal miRNAs for Fetal Alcohol Syndrome (FAS) and Prenatal Fetal Alcohol Spectrum (PFAS) disorders, this research establishes that exmiRNAs from infants can be indicative of fetal damage. These biomarkers could be crucial in identifying affected infants who may not exhibit obvious physical signs of PAE, allowing for early interventions during critical developmental periods. These studies align with recommendations from leading medical organizations emphasizing the need for new diagnostic tools for FASD.

7.5.1.2 Circular and Linear Long Non-coding RNAs as a Novel Group of Blood Biomarkers

CircRNAs are more stable than linear RNAs or miRNAs because of their closed secondary structure, constituting an enormous advantage as a novel class of clinical biomarkers (Meng et al. 2017). CircRNAs have been prevalent in current research due to their role as novel biomarkers that are detectable in blood and cerebrospinal fluid and saliva samples (Zhang et al. 2022a). Peripheral blood-identified circRNAs may serve as potential biomarkers for CNS pathologies (Piscopo et al. 2022; Liu et al. 2020; Lu and Xu 2016) and may reflect peripheral–CNS interactions and often an indicator of disease prognosis (Gaffo et al. 2019; Ravanidis et al. 2021). Numerous circRNAs are dysregulated in neurodegenerative conditions (e.g., Parkinson's disease and Alzheimer's disease) and ischemic and neurovascular conditions (Zhang and Bian 2021; Doxakis 2022). Blood-circulating circRNA dysregulation correlates to human physiological manifestations and can be used for early intervention or preventive strategies (Lu et al. 2022; Pan et al. 2020; He et al. 2021). Moving forward, integrating circRNA research into clinical practice could revolutionize diagnostic approaches and treatment paradigms.

Given substantial evidence that peripheral immune activation contributes to PAE-induced deficits (Bake et al. 2021; Sanchez et al. 2017; Noor et al. 2020; Terasaki and Schwarz 2016; Bodnar et al. 2016) during adulthood and its potential relevance to biomarker development, recently, our group examined circRNAs in the blood-circulating peripheral immune cells. A systematic profiling of circRNAs in peripheral blood immune cells suggested that PAE, even without further injury or immune insults, generates a unique circRNA expression profile during adulthood (Noor et al. 2023a). Although far fewer in numbers and with minimal overlap with circRNAs detected in the CNS tissues, PAE produced a distinctive signature with 18 differentially regulated circRNAs detected in the blood. Importantly, these differentially expressed circRNAs were derived from genes, including those associated with glucocorticoid receptor signaling (*Smarcc1*), RhoGTPase activating protein (*Arhgap10*), ATP/GTP binding Carboxypeptidase (*Agtpbp1*) and Vesicular, Overexpressed in Cancer, and Prosurvival Protein 1 (*Vopp1*). These genes are involved in cell–cell interactions, immune cell trafficking, inflammatory response, and genes linked to psychological and neurological diseases (Coleman et al. 2020; Sarowar and Grabrucker 2020). Although primarily driven by a different set of genes that harbor the differentially regulated circRNAs observed in blood, relative to the CNS tissues, the NF-κB pathway was identified as a top molecular network involved with PAE-induced changes. Interestingly, although in opposite directions, PAE-induced *circVopp1* dysregulation was observed in blood, as well as in the spinal cord, suggesting blood circRNAs can be novel predictive biomarkers of PAE and may reflect PAE-induced changes in the CNS. Future studies from clinical samples are needed to address whether these newly identified blood circRNAs are viable biomarkers for FASD-related adverse outcomes.

The FASD research area lacks research exploring linear lncRNAs as blood biomarkers, although lncRNAs hold great promise as effective biomarkers from preclinical models of autism and AUD (Honarmand Tamizkar et al. 2021). Future studies are needed to explore this potential avenue of research. Together, studies discussed in this section convincingly present the immense promise of integrating circRNAs and other non-coding RNAs into diagnostic and therapeutic strategies that may offer a paradigm shift toward precision medicine in FASD.

7.6 Summary and Conclusions

In essence, ncRNAs have emerged as pivotal regulators of the immediate and long-term effects of PAE, captivating researchers with their multifaceted functions and regulatory processes. The diverse roles of non-coding RNAs, as unveiled in this overview, signify a transformative era in FASD research. Although ncRNAs hold immense potential for new biomarkers and novel therapeutics, establishing standardized protocols for ncRNA detection and exploring their functional roles in-depth will be critical for harnessing their full potential. While preclinical models enable us to examine key mechanistic details of ncRNAs in PAE-related CNS dysfunction and offer new therapeutic targets, the field should also prioritize clinical studies to validate the identified circRNA and other ncRNA signatures and their associations with FASD-related deficits. Furthermore, employing sophisticated computational models to grasp the complex interplay among circRNAs and other non-coding RNAs and pinpointing the central "hub" molecules and critical pathways influenced by PAE holds significant promise for future research. Importantly, harnessing various categories of non-coding RNA molecules as a collective "molecular signature" could unveil novel avenues for drug discovery, enhance diagnostic capabilities, and introduce innovative therapeutic approaches in addressing the challenges posed by variable levels of prenatal alcohol exposure. Longitudinal studies could provide insights into the dynamic nature of ncRNA dysregulation, offering a temporal dimension crucial for clinical applicability. Additionally, exploring the potential of ncRNAs as therapeutic targets demands rigorous investigation into delivery mechanisms and safety profiles.

Acknowledgements This work is supported by NIH/NIAAA R01 AA029694, NIH/NIAAA P50 AA022534, and NIH 5T34GM145428 (U-RISE at UNM). While preparing this work, Grammarly was used to improve language and readability. After using this tool, the authors reviewed and edited the content as needed and take full responsibility for the publication's content.

References

Airapetov MI, Eresko SO, Bychkov ER, Lebedev AA, Shabanov PD (2021) Prenatal exposure to alcohol alters TLR4 signaling in the prefrontal cortex in rats. Biomed Khim 67:500–506

Aliperti V, Skonieczna J, Cerase A (2021) Long non-coding RNA (lncRNA) roles in cell biology, neurodevelopment and neurological disorders. Noncoding RNA 7

Allan AM, Goggin SL, Caldwell KK (2014) Prenatal alcohol exposure modifies glucocorticoid receptor subcellular distribution in the medial prefrontal cortex and impairs frontal cortex-dependent learning. PLoS One 9:e96200

Altounian M, Bellon A, Mann F (2023) Neuronal miR-17-5p contributes to interhemispheric cortical connectivity defects induced by prenatal alcohol exposure. Cell Rep 42:113020

Arzua T, Jiang C, Yan Y, Bai X (2021) The importance of non-coding RNAs in environmental stress-related developmental brain disorders: A systematic review of evidence associated with exposure to alcohol, anesthetic drugs, nicotine, and viral infections. Neurosci Biobehav Rev 128:633–647

Bake S, Pinson MR, Pandey S, Chambers JP, Mota R, Fairchild AE, Miranda RC, Sohrabji F (2021) Prenatal alcohol-induced sex differences in immune, metabolic and neurobehavioral outcomes in adult rats. Brain Behav Immun 98:86–100

Balaraman S, Tingling JD, Tsai PC, Miranda RC (2013) Dysregulation of microRNA expression and function contributes to the etiology of fetal alcohol spectrum disorders. Alcohol Res 35:18–24

Balaraman S, Schafer JJ, Tseng AM, Wertelecki W, Yevtushok L, Zymak-Zakutnya N, Chambers CD, Miranda RC (2016) Plasma miRNA profiles in pregnant women predict infant outcomes following prenatal alcohol exposure. PLoS One 11:e0165081

Banerjee D, Sultana S, Banerjee S (2024) Gas5 regulates early life stress-induced anxiety and spatial memory. J Neurochem 168(9):2999–3018

Bao A, Li F, Zhang M, Chen Y, Zhang P, Zhou X (2014) Impact of ozone exposure on the response to glucocorticoid in a mouse model of asthma: involvements of p38 MAPK and MKP-1. Respir Res 15:126

Barrett SP, Salzman J (2016) Circular RNAs: analysis, expression and potential functions. Development 143:1838–1847

Bentea E, Villers A, Moore C, Funk AJ, O'Donovan SM, Verbruggen L, Lara O, Janssen P, De Pauw L, Declerck NB, Depasquale EAK, Churchill MJ, Sato H, Hermans E, Arckens L, Meshul CK, Ris L, Mccullumsmith RE, Massie A (2021) Corticostriatal dysfunction and social interaction deficits in mice lacking the cystine/glutamate antiporter. Mol Psychiatry 26:4754–4769

Bestry M, Symons M, Larcombe A, Muggli E, Craig JM, Hutchinson D, Halliday J, Martino D (2022) Association of prenatal alcohol exposure with offspring DNA methylation in mammals: a systematic review of the evidence. Clin Epigenetics 14:12

Bissell S, Oliver C, Moss J, Heald M, Waite J, Crawford H, Kothari V, Rumbellow L, Walters G, Richards C (2022) The behavioural phenotype of SATB2-associated syndrome: a within-group and cross-syndrome analysis. J Neurodev Disord 14:25

Bodnar TS, Hill LA, Weinberg J (2016) Evidence for an immune signature of prenatal alcohol exposure in female rats. Brain Behav Immun 58:130–141

Brady ML, Allan AM, Caldwell KK (2012) A limited access mouse model of prenatal alcohol exposure that produces long-lasting deficits in hippocampal-dependent learning and memory. Alcohol Clin Exp Res 36:457–466

Bu Q, Long H, Shao X, Gu H, Kong J, Luo L, Liu B, Guo W, Wang H, Tian J, Zhao Y, Cen X (2019) Cocaine induces differential circular RNA expression in striatum. Transl Psychiatry 9:199

Caldwell KK, Hafez A, Solomon E, Cunningham M, Allan AM (2018) Arsenic exposure during embryonic development alters the expression of the long non-coding RNA growth arrest specific-5 (Gas5) in a sex-dependent manner. Neurotoxicol Teratol 66:102–112

Chen LL (2020) The expanding regulatory mechanisms and cellular functions of circular RNAs. Nat Rev Mol Cell Biol 21:475–490

Chen SY, Kannan M (2023) Neural crest cells and fetal alcohol spectrum disorders: Mechanisms and potential targets for prevention. Pharmacol Res 194:106855

Chen X, Liu J, Feng WK, Wu X, Chen SY (2015) MiR-125b protects against ethanol-induced apoptosis in neural crest cells and mouse embryos by targeting Bak 1 and PUMA. Exp Neurol 271:104–111

Chen YJ, Chen CY, Mai TL, Chuang CF, Chen YC, Gupta SK, Yen L, Wang YD, Chuang TJ (2020) Genome-wide, integrative analysis of circular RNA dysregulation and the corresponding circular RNA-microRNA-mRNA regulatory axes in autism. Genome Res 30:375–391

Chen HW, Zhang XX, Peng ZD, Xing ZM, Zhang YW, Li YL (2021a) The circular RNA circSlc7a11 promotes bone cancer pain pathogenesis in rats by modulating LLC-WRC 256 cell proliferation and apoptosis. Mol Cell Biochem 476:1751–1763

Chen M, Lai X, Wang X, Ying J, Zhang L, Zhou B, Liu X, Zhang J, Wei G, Hua F (2021b) Long non-coding RNAs and circular RNAs: insights into microglia and astrocyte mediated neurological diseases. Front Mol Neurosci 14:745066

Coleman SR, Smith ML, Spicer V, Lao Y, Mookherjee N, Hancock REW (2020) Overexpression of the small RNA PA0805.1 in Pseudomonas aeruginosa modulates the expression of a large set of genes and proteins, resulting in altered motility, cytotoxicity, and tobramycin resistance. mSystems 5

Delaunoy J, Abidi F, Zeniou M, Jacquot S, Merienne K, Pannetier S, Schmitt M, Schwartz C, Hanauer A (2001) Mutations in the X-linked RSK2 gene (RPS6KA3) in patients with Coffin-Lowry syndrome. Hum Mutat 17:103–116

Dong X, Chen K, Cuevas-Diaz Duran R, You Y, Sloan SA, Zhang Y, Zong S, Cao Q, Barres BA, Wu JQ (2015) Comprehensive identification of long non-coding RNAs in purified cell types from the brain reveals functional LncRNA in OPC fate determination. PLoS Genet 11:e1005669

Doremus-Fitzwater TL, Youngentob SL, Youngentob L, Gano A, Vore AS, Deak T (2020) Lingering effects of prenatal alcohol exposure on basal and ethanol-evoked expression of inflammatory-related genes in the CNS of adolescent and adult rats. Front Behav Neurosci 14:82

Doxakis E (2022) Insights into the multifaceted role of circular RNAs: implications for Parkinson's disease pathogenesis and diagnosis. NPJ Parkinsons Dis 8:7

Dube U, Del-Aguila JL, Li Z, Budde JP, Jiang S, Hsu S, Ibanez L, Fernandez MV, Farias F, Norton J, Gentsch J, Wang F, Salloway S, Masters CL, Lee JH, Graff-Radford NR, Chhatwal JP, Bateman RJ, Morris JC, Karch CM, Harari O, Cruchaga C, Dominantly Inherited Alzheimer Network (DIAN) (2019) An atlas of cortical circular RNA expression in Alzheimer disease brains demonstrates clinical and pathological associations. Nat Neurosci 22:1903–1912

Dulman RS, Wandling GM, Pandey SC (2020) Epigenetic mechanisms underlying pathobiology of alcohol use disorder. Curr Pathobiol Rep 8:61–73

Fan H, Li Y, Yuan F, Lu L, Liu J, Feng W, Zhang HG, Chen SY (2022) Up-regulation of microRNA-34a mediates ethanol-induced impairment of neural crest cell migration in vitro and in zebrafish embryos through modulating epithelial-mesenchymal transition by targeting Snail1. Toxicol Lett 358:17–26

Frankish A, Diekhans M, Ferreira AM, Johnson R, Jungreis I, Loveland J, Mudge JM, Sisu C, Wright J, Armstrong J, Barnes I, Berry A, Bignell A, Carbonell Sala S, Chrast J, Cunningham F, Di Domenico T, Donaldson S, Fiddes IT, García Girón C, Gonzalez JM, Grego T, Hardy M, Hourlier T, Hunt T, Izuogu OG, Lagarde J, Martin FJ, Martínez L, Mohanan S, Muir P, Navarro FCP, Parker A, Pei B, Pozo F, Ruffier M, Schmitt BM, Stapleton E, Suner MM, Sycheva I, Uszczynska-Ratajczak B, Xu J, Yates A, Zerbino D, Zhang Y, Aken B, Choudhary JS, Gerstein M, Guigó R, Hubbard TJP, Kellis M, Paten B, Reymond A, Tress ML, Flicek P (2019) GENCODE reference annotation for the human and mouse genomes. Nucleic Acids Res 47:D766–D773

Franklin L, Deitz J, Jirikowic T, Astley S (2008) Children with fetal alcohol spectrum disorders: problem behaviors and sensory processing. Am J Occup Ther 62:265–273

Fu GH, Chen W, Li HM, Wang YF, Liu L, Qian QJ (2021) A potential association of RNF219-AS1 with ADHD: Evidence from categorical analysis of clinical phenotypes and from quantitative exploration of executive function and white matter microstructure endophenotypes. CNS Neurosci Ther 27:603–616

Gaffo E, Boldrin E, Dal Molin A, Bresolin S, Bonizzato A, Trentin L, Frasson C, Debatin KM, Meyer LH, Te Kronnie G, Bortoluzzi S (2019) Circular RNA differential expression in blood cell populations and exploration of circRNA deregulation in pediatric acute lymphoblastic leukemia. Sci Rep 9:14670

Gangisetty O, Jabbar S, Wynne O, Sarkar DK (2017) MicroRNA-9 regulates fetal alcohol-induced changes in D2 receptor to promote prolactin production. J Endocrinol 235:1–14

Gao Z, Huang D (2021) lncRNA GAS5-mediated miR-23a-3p promotes inflammation and cell apoptosis by targeting TLR4 in a cell model of sepsis. Mol Med Rep 24

Gardiner AS, Gutierrez HL, Luo L, Davies S, Savage DD, Bakhireva LN, Perrone-Bizzozero NI (2016) Alcohol use during pregnancy is associated with specific alterations in microRNA levels in maternal serum. Alcohol Clin Exp Res 40:826–837

Giedd JN, Rapoport JL (2010) Structural MRI of pediatric brain development: what have we learned and where are we going? Neuron 67:728–734

Guo JU, Agarwal V, Guo H, Bartel DP (2014) Expanded identification and characterization of mammalian circular RNAs. Genome Biol 15:409

Guo Q, Xia L, Guo R, Xu W, Zhang Y, Zhao C, Zhang P, Bai T, Ni X, Hao C, Xia K, Li W (2023) Behavioural deficits of autism spectrum disorder and associations with different gene clusters: a study with the whole-genome transmission disequilibrium test. BMJ Paediatr Open 7

Hafez AK, Zimmerman AJ, Papageorgiou G, Chandrasekaran J, Amoah SK, Lin R, Lozano E, Pierotti C, Dell'Orco M, Hartley BJ, Alural B, Lalonde J, Esposito JM, Berretta S, Squassina A, Chillotti C, Voloudakis G, Shao Z, Fullard JF, Brennand KJ, Turecki G, Roussos P, Perlis RH, Haggarty SJ, Perrone-Bizzozero N, Brigman JL, Mellios N (2022) A bidirectional competitive interaction between circHomer1 and Homer1b within the orbitofrontal cortex regulates reversal learning. Cell Rep 38:110282

Haque S, Harries LW (2017) Circular RNAs (circRNAs) in health and disease. Genes (Basel) 8

He J, Tu C, Liu Y (2018) Role of lncRNAs in aging and age-related diseases. Aging Med (Milton) 1:158–175

He AT, Liu J, Li F, Yang BB (2021) Targeting circular RNAs as a therapeutic approach: current strategies and challenges. Signal Transduct Target Ther 6:185

Himmelrich M, Lutke CJ, Hargrave ET (2020) The lay of the land: Fetal alcohol spectrum disorder (FASD) as a whole body diagnosis. In: Begun AL, Murray MM (eds) The Routledge handbook of social work and addictive behaviors. Routledge, London

Honarmand Tamizkar K, Badrlou E, Aslani T, Brand S, Arsang-Jang S, Ghafouri-Fard S, Taheri M (2021) Dysregulation of NF-κB-associated LncRNAs in autism spectrum disorder. Front Mol Neurosci 14:747785

Hu H, Brittain GC, Chang JH, Puebla-Osorio N, Jin J, Zal A, Xiao Y, Cheng X, Chang M, Fu YX, Zal T, Zhu C, Sun SC (2013) OTUD7B controls non-canonical NF-kappaB activation through deubiquitination of TRAF3. Nature 494:371–374

Huang R, Zhang Y, Bai Y, Han B, Ju M, Chen B, Yang L, Wang Y, Zhang H, Xie C, Zhang Z, Yao H (2020) N 6-methyladenosine modification of fatty acid amide hydrolase messenger RNA in circular RNA STAG1-regulated astrocyte dysfunction and depressive-like behaviors. Biol Psychiatry 88:392–404

Huang Y, Ge W, Ding Y, Zhang L, Zhou J, Kong Y, Cui B, Gao B, Qian X, Wang W (2021) The circular RNA circSLC7A11 functions as a mir-330-3p sponge to accelerate hepatocellular carcinoma progression by regulating cyclin-dependent kinase 1 expression. Cancer Cell Int 21:636

Ignacio C, Mooney SM, Middleton FA (2014) Effects of acute prenatal exposure to ethanol on microRNA expression are ameliorated by social enrichment. Front Pediatr 2:103

Jabbar S, Reuhl K, Sarkar DK (2018) Prenatal alcohol exposure increases the susceptibility to develop aggressive prolactinomas in the pituitary gland. Sci Rep 8:7720

Jarlstad Olesen MT, Kristensen LS (2021) Circular RNAs as microRNA sponges: evidence and controversies. Essays Biochem 65:685–696

Kaikkonen MU, Lam MT, Glass CK (2011) Non-coding RNAs as regulators of gene expression and epigenetics. Cardiovasc Res 90:430–440

Kallen AN, Zhou XB, Xu J, Qiao C, Ma J, Yan L, Lu L, Liu C, Yi JS, Zhang H, Min W, Bennett AM, Gregory RI, Ding Y, Huang Y (2013) The imprinted H19 lncRNA antagonizes let-7 microRNAs. Mol Cell 52:101–112

Kleiber ML, Mantha K, Stringer RL, Singh SM (2013) Neurodevelopmental alcohol exposure elicits long-term changes to gene expression that alter distinct molecular pathways dependent on timing of exposure. J Neurodev Disord 5:6

Kleiber ML, Diehl EJ, Laufer BI, Mantha K, Chokroborty-Hoque A, Alberry B, Singh SM (2014) Long-term genomic and epigenomic dysregulation as a consequence of prenatal alcohol exposure: a model for fetal alcohol spectrum disorders. Front Genet 5:161

Komander D, Clague MJ, Urbe S (2009) Breaking the chains: structure and function of the deubiquitinases. Nat Rev Mol Cell Biol 10:550–563

Kong D, Gu R, Zhang C, Yin R (2020) Knockdown of hsa_circ_0059955 induces apoptosis and cell cycle arrest in nucleus pulposus cells via inhibiting. Drug Des Devel Ther 14:3951–3963

Kryger R, Fan L, Wilce PA, Jaquet V (2012) MALAT-1, a non protein-coding RNA is upregulated in the cerebellum, hippocampus and brain stem of human alcoholics. Alcohol 46:629–634

Kukharsky MS, Ninkina NN, An H, Telezhkin V, Wei W, Meritens CR, Cooper-Knock J, Nakagawa S, Hirose T, Buchman VL, Shelkovnikova TA (2020) Long noncoding RNA Neat1 regulates adaptive behavioural response to stress in mice. Transl Psychiatry 10:171

Laufer BI, Mantha K, Kleiber ML, Diehl EJ, Addison SM, Singh SM (2013) Long-lasting alterations to DNA methylation and ncRNAs could underlie the effects of fetal alcohol exposure in mice. Dis Model Mech 6:977–992

Lei S, He Z, Chen T, Guo X, Zeng Z, Shen Y, Jiang J (2019) Long noncoding RNA 00976 promotes pancreatic cancer progression through OTUD7B by sponging miR-137 involving EGFR/MAPK pathway. J Exp Clin Cancer Res 38:470

Li J, Ming Z, Yang L, Wang T, Liu G, Ma Q (2022) Long noncoding RNA. Genes Dis 9:1478–1492

Liaci C, Prandi L, Pavinato L, Brusco A, Maldotti M, Molineris I, Oliviero S, Merlo GR (2022) The emerging roles of long non-coding RNAs in intellectual disability and related neurodevelopmental disorders. Int J Mol Sci 23

Lim Y, Beane-Ebel JE, Tanaka Y, Ning B, Husted CR, Henderson DC, Xiang Y, Park IH, Farrer LA, Zhang H (2021) Exploration of alcohol use disorder-associated brain miRNA-mRNA regulatory networks. Transl Psychiatry 11:504

Liu L, Chen X, Chen YH, Zhang K (2020) Identification of circular RNA hsa_Circ_0003391 in peripheral blood is potentially associated with Alzheimer's disease. Front Aging Neurosci 12:601965

Liu X, Shen L, Han B, Yao H (2021a) Involvement of noncoding RNA in blood-brain barrier integrity in central nervous system disease. Noncoding RNA Res 6:130–138

Liu Y, Xu R, Gu H, Zhang E, Qu J, Cao W, Huang X, Yan H, He J, Cai Z (2021b) Metabolic reprogramming in macrophage responses. Biomark Res 9:1

Liu T, Huang T, Shang M, Han G (2022) CircRNA ITCH: insight into its role and clinical application prospect in tumor and non-tumor diseases. Front Genet 13:927541

Lu D, Xu AD (2016) Mini Review: circular RNAs as potential clinical biomarkers for disorders in the central nervous system. Front Genet 7:53

Lu H, Yang Y, Kuang D, Liu P, Yang J (2022) Expression profile of circRNA in peripheral blood mononuclear cells of patients with rheumatoid arthritis. BMC Med Genet 15:77

Lussier AA, Weinberg J, Kobor MS (2017) Epigenetics studies of fetal alcohol spectrum disorder: where are we now? Epigenomics 9:291–311

Ma X, Zhu G, Jiao T, Shao F (2021) Effects of circular RNA Ttc3/miR-148a/Rcan2 axis on inflammation and oxidative stress in rats with acute kidney injury induced by sepsis. Life Sci 272:119233

Mahnke AH, Salem NA, Tseng AM, Chung DD, Miranda RC (2018) Nonprotein-coding RNAs in fetal alcohol spectrum disorders. Prog Mol Biol Transl Sci 157:299–342

Mahnke AH, Sideridis GD, Salem NA, Tseng AM, Carter RC, Dodge NC, Rathod AB, Molteno CD, Meintjes EM, Jacobson SW, Miranda RC, Jacobson JL (2021) Infant circulating MicroRNAs as biomarkers of effect in fetal alcohol spectrum disorders. Sci Rep 11:1429

Mai TL, Chen CY, Chen YC, Chiang TW, Chuang TJ (2022) Trans-genetic effects of circular RNA expression quantitative trait loci and potential causal mechanisms in autism. Mol Psychiatry 27:4695–4706

Marjonen H, Toivonen M, Lahti L, Kaminen-Ahola N (2018) Early prenatal alcohol exposure alters imprinted gene expression in placenta and embryo in a mouse model. PLoS One 13:e0197461

Mattick JS, Amaral PP, Carninci P, Carpenter S, Chang HY, Chen LL, Chen R, Dean C, Dinger ME, Fitzgerald KA, Gingeras TR, Guttman M, Hirose T, Huarte M, Johnson R, Kanduri C, Kapranov P, Lawrence JB, Lee JT, Mendell JT, Mercer TR, Moore KJ, Nakagawa S, Rinn JL, Spector DL, Ulitsky I, Wan Y, Wilusz JE, Wu M (2023) Long non-coding RNAs: definitions, functions, challenges and recommendations. Nat Rev Mol Cell Biol 24:430–447

Mattson SN, Bernes GA, Doyle LR (2019) Fetal alcohol spectrum disorders: a review of the neurobehavioral deficits associated with prenatal alcohol exposure. Alcohol Clin Exp Res 43:1046–1062

Memczak S (2013) Circular RNAs are a large class of animal RNAs with regulatory potency. Nature 495:333–338

Meng S, Zhou H, Feng Z, Xu Z, Tang Y, Li P, Wu M (2017) CircRNA: functions and properties of a novel potential biomarker for cancer. Mol Cancer 16:94

Mercer TR, Mattick JS (2013) Structure and function of long noncoding RNAs in epigenetic regulation. Nat Struct Mol Biol 20:300–307

Mitchell S, Vargas J, Hoffmann A (2016) Signaling via the NFκB system. Wiley Interdiscip Rev Syst Biol Med 8:227–241

Monnier P, Martinet C, Pontis J, Stancheva I, Ait-Si-Ali S, Dandolo L (2013) H19 lncRNA controls gene expression of the Imprinted Gene Network by recruiting MBD1. Proc Natl Acad Sci USA 110:20693–20698

Mukherjee S, Tarale P, Sarkar DK (2023) Neuroimmune interactions in fetal alcohol spectrum disorders: potential therapeutic targets and intervention strategies. Cells 12

Ng GY, Lim YA, Sobey CG, Dheen T, Fann DY, Arumugam TV (2018a) Epigenetic regulation of inflammation in stroke. Ther Adv Neurol Disord 11:1756286418771815

Ng WL, Mohd Mohidin TB, Shukla K (2018b) Functional role of circular RNAs in cancer development and progression. RNA Biol 15:995–1005

Noor S, Milligan ED (2018) Lifelong impacts of moderate prenatal alcohol exposure on neuroimmune function. Front Immunol 9:1107

Noor S, Sanchez JJ, Sun MS, Pervin Z, Sanchez JE, Havard MA, Epler LT, Nysus MV, Norenberg JP, Wagner CR, Davies S, Wagner JL, Savage DD, Jantzie LL, Mellios N, Milligan ED (2020) The LFA-1 antagonist BIRT377 reverses neuropathic pain in prenatal alcohol-exposed female rats via actions on peripheral and central neuroimmune function in discrete pain-relevant tissue regions. Brain Behav Immun 87:339–358

Noor S, Pritha AN, Pasmay AA, Sanchez JE, Sanchez JJ, Fernandez-Oropeza AK, Sun MS, Dell'Orco M, Davies S, Savage DD, Mellios N, Milligan ED (2023a) Prenatal alcohol exposure dysregulates spinal and circulating immune cell circular RNA expression in adult female rats with chronic sciatic neuropathy. Front Neurosci 17:1180308

Noor S, Sun MS, Pasmay AA, Pritha AN, Ruffaner-Hanson CD, Nysus MV, Jimenez DC, Murphy M, Savage DD, Valenzuela CF, Milligan ED (2023b) Prenatal alcohol exposure promotes NLRP3 inflammasome-dependent immune actions following morphine treatment and paradoxically prolongs nerve injury-induced pathological pain in female mice. Alcohol Clin Exp Res (Hoboken) 47:2262–2277

Pan YH, Wu WP, Xiong XD (2020) Circular RNAs: promising biomarkers for age-related diseases. Aging Dis 11:1585–1593

Panczakiewicz AL, Glass L, Coles CD, Kable JA, Sowell ER, Wozniak JR, Jones KL, Riley EP, Mattson SN, CIFASD (2016) Neurobehavioral deficits consistent across age and sex in youth with prenatal alcohol exposure. Alcohol Clin Exp Res 40:1971–1981

Panda AC, Grammatikakis I, Munk R, Gorospe M, Abdelmohsen K (2017) Emerging roles and context of circular RNAs. Wiley Interdiscip Rev RNA 8

Papageorgiou G, Amoah SK, Pierotti C, Otero M, Eckel S, Coffey K, Allan AM, Caldwell KK, Mellios N (2023) Prenatal alcohol exposure results in brain region- and sex-specific changes in circHomer1 expression in adult mouse brain. Front Neurosci 17:1087950

Pappalardo-Carter DL, Balaraman S, Sathyan P, Carter ES, Chen WJ, Miranda RC (2013) Suppression and epigenetic regulation of MiR-9 contributes to ethanol teratology: evidence from zebrafish and murine fetal neural stem cell models. Alcohol Clin Exp Res 37:1657–1667

Pascual M, Montesinos J, Montagud-Romero S, Forteza J, Rodriguez-Arias M, Minarro J, Guerri C (2017) TLR4 response mediates ethanol-induced neurodevelopment alterations in a model of fetal alcohol spectrum disorders. J Neuroinflammation 14:145

Paudel P, Pierotti C, Lozano E, Amoah SK, Gardiner AS, Caldwell KK, Allan AM, Mellios N (2020) Prenatal alcohol exposure results in sex-specific alterations in circular RNA expression in the developing mouse brain. Front Neurosci 14:581895

Perales G, Westenskow M, Gutierrez R, Caldwell KK, Allan AM, Gardiner AS (2022) MicroRNA-150-5p is upregulated in the brain microvasculature during prenatal alcohol exposure and inhibits the angiogenic factor Vezf1. Alcohol Clin Exp Res 46:1953–1966

Piscopo P, Manzini V, Rivabene R, Crestini A, Le Pera L, Pizzi E, Veroni C, Talarico G, Peconi M, Castellano AE, D'Alessio C, Bruno G, Corbo M, Vanacore N, Lacorte E (2022) A plasma circular RNA profile differentiates subjects with Alzheimer's disease and mild cognitive impairment from healthy controls. Int J Mol Sci 23

Piwecka M, Glažar P, Hernandez-Miranda LR, Memczak S, Wolf SA, Rybak-Wolf A, Filipchyk A, Klironomos F, Cerda Jara CA, Fenske P, Trimbuch T, Zywitza V, Plass M, Schreyer L, Ayoub S, Kocks C, Kühn R, Rosenmund C, Birchmeier C, Rajewsky N (2017) Loss of a mammalian circular RNA locus causes miRNA deregulation and affects brain function. Science 357

Popova S, Lange S, Shield K, Mihic A, Chudley AE, Mukherjee RAS, Bekmuradov D, Rehm J (2016) Comorbidity of fetal alcohol spectrum disorder: a systematic review and meta-analysis. Lancet 387:978–987

Price A, Cook PA, Norgate S, Mukherjee R (2017) Prenatal alcohol exposure and traumatic childhood experiences: A systematic review. Neurosci Biobehav Rev 80:89–98

Ratti M, Lampis A, Ghidini M, Salati M, Mirchev MB, Valeri N, Hahne JC (2020) MicroRNAs (miRNAs) and long non-coding RNAs (lncRNAs) as new tools for cancer therapy: first steps from bench to bedside. Target Oncol 15:261–278

Ravanidis S, Bougea A, Karampatsi D, Papagiannakis N, Maniati M, Stefanis L, Doxakis E (2021) Differentially expressed circular RNAs in peripheral blood mononuclear cells of patients with Parkinson's disease. Mov Disord 36:1170–1179

Reddy AS, O'Brien D, Pisat N, Weichselbaum CT, Sakers K, Lisci M, Dalal JS, Dougherty JD (2017) A comprehensive analysis of cell type-specific nuclear RNA from neurons and glia of the brain. Biol Psychiatry 81:252–264

Reid N, Hayes N, Young SB, Akison LK, Moritz KM (2021) Caregiver-reported physical health status of children and young people with fetal alcohol spectrum disorder. J Dev Orig Health Dis 12:420–427

Rona G, Pagano M (2019) Mixed ubiquitin chains regulate DNA repair. Genes Dev 33:1615–1616

Rothschild DE, Mcdaniel DK, Ringel-Scaia VM, Allen IC (2018) Modulating inflammation through the negative regulation of NF-kappaB signaling. J Leukoc Biol. https://doi.org/10.1002/JLB.3MIR0817-346RRR. Online ahead of print, Published online

Ruffaner-Hanson CD, Fernandez-Oropeza AK, Sun MS, Caldwell KK, Allan AM, Savage DD, Valenzuela CF, Noor S, Milligan ED (2023) Prenatal alcohol exposure alters mRNA expression for stress peptides, glucocorticoid receptor function and immune factors in acutely stressed neonatal brain. Front Neurosci 17:1203557

Rybak-Wolf A, Stottmeister C, Glazar P, Jens M, Pino N, Giusti S, Hanan M, Behm M, Bartok O, Ashwal-Fluss R, Herzog M, Schreyer L, Papavasileiou P, Ivanov A, Ohman M, Refojo D, Kadener S, Rajewsky N (2015)

Circular RNAs in the mammalian brain are highly abundant, conserved, and dynamically expressed. Mol Cell 58:870–885

Salem NA, Mahnke AH, Wells AB, Tseng AM, Yevtushok L, Zymak-Zakutnya N, Wertlecki W, Chambers CD, Miranda RC, Collaborative Initiative on Fetal Alcohol Spectrum Disorders (CIFASD) (2020) Association between fetal sex and maternal plasma microRNA responses to prenatal alcohol exposure: evidence from a birth outcome-stratified cohort. Biol Sex Differ 11:51

Salem NA, Mahnke AH, Konganti K, Hillhouse AE, Miranda RC (2021a) Cell-type and fetal-sex-specific targets of prenatal alcohol exposure in developing mouse cerebral cortex. iScience 24:102439

Salem NA, Mahnke AH, Tseng AM, Garcia CR, Jahromi HK, Geoffroy CG, Miranda RC (2021b) A novel Oct4/Pou5f1-like non-coding RNA controls neural maturation and mediates developmental effects of ethanol. Neurotoxicol Teratol 83:106943

Sanchez JJ, Noor S, Davies S, Savage D, Milligan ED (2017) Prenatal alcohol exposure is a risk factor for adult neuropathic pain via aberrant neuroimmune function. J Neuroinflammation 14:254

Sarowar T, Grabrucker AM (2020) Rho GTPases in the amygdala-a switch for fears? Cells 9(9):1972

Sathyan P, Golden HB, Miranda RC (2007) Competing interactions between micro-RNAs determine neural progenitor survival and proliferation after ethanol exposure: evidence from an ex vivo model of the fetal cerebral cortical neuroepithelium. J Neurosci 27:8546–8557

Schaffner SL, Lussier AA, Baker JA, Goldowitz D, Hamre KM, Kobor MS (2020) Neonatal alcohol exposure in mice induces select differentiation- and apoptosis-related chromatin changes both independent of and dependent on sex. Front Genet 11:35

Sekar S, Liang WS (2019) Circular RNA expression and function in the brain. Noncoding RNA Res 4:23–29

Seyednejad SA, Sartor GC (2022) Noncoding RNA therapeutics for substance use disorder. Adv Drug Alcohol Res 2

Shao T, Pan YH, Xiong XD (2021) Circular RNA: an important player with multiple facets to regulate its parental gene expression. Mol Ther Nucleic Acids 23:369–376

Sibuh BZ, Quazi S, Panday H, Parashar R, Jha NK, Mathur R, Jha SK, Taneja P, Jha AK (2023) The emerging role of epigenetics in metabolism and endocrinology. Biology (Basel) 12

Soares AR, Pereira PM, Ferreira V, Reverendo M, Simões J, Bezerra AR, Moura GR, Santos MA (2012) Ethanol exposure induces upregulation of specific microRNAs in zebrafish embryos. Toxicol Sci 127:18–28

Statello L, Guo CJ, Chen LL, Huarte M (2021) Gene regulation by long non-coding RNAs and its biological functions. Nat Rev Mol Cell Biol 22:96–118

Su K, Yi Q, Dai X, Liu O (2022) Circular RNA ITCH: an emerging multifunctional regulator. Biomolecules 12

Suenkel C, Cavalli D, Massalini S, Calegari F, Rajewsky N (2020) A highly conserved circular RNA is required to keep neural cells in a progenitor state in the mammalian brain. Cell Rep 30:2170–2179.e5

Tal TL, Franzosa JA, Tilton SC, Philbrick KA, Iwaniec UT, Turner RT, Waters KM, Tanguay RL (2012) MicroRNAs control neurobehavioral development and function in zebrafish. FASEB J 26:1452–1461

Tanaka M, Yagyu K, Sackett S, Zhang Y (2019) Anti-inflammatory effects by pharmacological inhibition or knockdown of fatty acid amide hydrolase in BV2 microglial cells. Cells 8

Terasaki LS, Schwarz JM (2016) Effects of moderate prenatal alcohol exposure during early gestation in rats on inflammation across the maternal-fetal-immune interface and later-life immune function in the offspring. J NeuroImmune Pharmacol 11:680–692

Tominaga M, Caterina MJ (2004) Thermosensation and pain. J Neurobiol 61:3–12

Topiwala A, Allan CL, Valkanova V, Zsoldos E, Filippini N, Sexton C, Mahmood A, Fooks P, Singh-Manoux A, Mackay CE, Kivimaki M, Ebmeier KP (2017) Moderate alcohol consumption as risk factor for adverse brain outcomes and cognitive decline: longitudinal cohort study. BMJ 357:j2353

Treiber T, Treiber N, Meister G (2019) Regulation of microRNA biogenesis and its crosstalk with other cellular pathways. Nat Rev Mol Cell Biol 20:5–20

Tseng AM, Mahnke AH, Wells AB, Salem NA, Allan AM, Roberts VH, Newman N, Walter NA, Kroenke CD, Grant KA, Akison LK, Moritz KM, Chambers CD, Miranda RC, Collaborative Initiative on Fetal Alcohol Spectrum Disorders (2019) Maternal circulating miRNAs that predict infant FASD outcomes influence placental maturation. Life Sci Alliance 2

van de Vondervoort II, Gordebeke PM, Khoshab N, Tiesinga PH, Buitelaar JK, Kozicz T, Aschrafi A, Glennon JC (2013) Long non-coding RNAs in neurodevelopmental disorders. Front Mol Neurosci 6:53

Veazey KJ, Carnahan MN, Muller D, Miranda RC, Golding MC (2013) Alcohol-induced epigenetic alterations to developmentally crucial genes regulating neural stemness and differentiation. Alcohol Clin Exp Res 37:1111–1122

Verstrepen L, Bekaert T, Chau TL, Tavernier J, Chariot A, Beyaert R (2008) TLR-4, IL-1R and TNF-R signaling to NF-kappaB: variations on a common theme. Cell Mol Life Sci 65:2964–2978

Vijayakumar N, Op De Macks Z, Shirtcliff EA, Pfeifer JH (2018) Puberty and the human brain: Insights into adolescent development. Neurosci Biobehav Rev 92:417–436

Vornholt E, Drake J, Mamdani M, Mcmichael G, Taylor ZN, Bacanu SA, Miles MF, Vladimirov VI (2021) Identifying a novel biological mechanism for alcohol addiction associated with circRNA networks acting as potential miRNA sponges. Addict Biol 26:e13071

Wang LL, Zhang Z, Li Q, Yang R, Pei X, Xu Y, Wang J, Zhou SF, Li Y (2009) Ethanol exposure induces differential microRNA and target gene expression and teratogenic effects which can be suppressed by folic acid supplementation. Hum Reprod 24:562–579

Wang X, Li H, Lu Y, Cheng L (2020) Regulatory effects of circular RNAs on host genes in human cancer. Front Oncol 10:586163

Wilcoxon JS, Kuo AG, Disterhoft JF, Redei EE (2005) Behavioral deficits associated with fetal alcohol exposure are reversed by prenatal thyroid hormone treatment: a role for maternal thyroid hormone deficiency in FAE. Mol Psychiatry 10:961–971

Wilkinson B, Campbell DB (2013) Contribution of long noncoding RNAs to autism spectrum disorder risk. Int Rev Neurobiol 113:35–59

Wong EL, Lutz NM, Hogan VA, Lamantia CE, Mcmurray HR, Myers JR, Ashton JM, Majewska AK (2018) Developmental alcohol exposure impairs synaptic plasticity without overtly altering microglial function in mouse visual cortex. Brain Behav Immun 67:257–278

Xu X, Zhang J, Tian Y, Gao Y, Dong X, Chen W, Yuan X, Yin W, Xu J, Chen K, He C, Wei L (2020) CircRNA inhibits DNA damage repair by interacting with host gene. Mol Cancer 19:128

Yang B, Zang L, Cui J, Wei L (2021) Circular RNA TTC3 regulates cerebral ischemia-reperfusion injury and neural stem cells by miR-372-3p/TLR4 axis in cerebral infarction. Stem Cell Res Ther 12:125

You X, Vlatkovic I, Babic A, Will T, Epstein I, Tushev G, Akbalik G, Wang M, Glock C, Quedenau C, Wang X, Hou J, Liu H, Sun W, Sambandan S, Chen T, Schuman EM, Chen W (2015) Neural circular RNAs are derived from synaptic genes and regulated by development and plasticity. Nat Neurosci 18:603–610

Yuan F, Yun Y, Fan H, Li Y, Lu L, Liu J, Feng W, Chen SY (2020) MicroRNA-135a protects against ethanol-induced apoptosis in neural crest cells and craniofacial defects in zebrafish by modulating the Siah1/p38/p53 pathway. Front Cell Dev Biol 8:583959

Zang J, Lu D, Xu A (2020) The interaction of circRNAs and RNA binding proteins: An important part of circRNA maintenance and function. J Neurosci Res 98:87–97

Zhang M, Bian Z (2021) The emerging role of circular RNAs in Alzheimer's disease and Parkinson's disease. Front Aging Neurosci 13:691512

Zhang CR, Ho MF, Vega MC, Burne TH, Chong S (2015) Prenatal ethanol exposure alters adult hippocampal VGLUT2 expression with concomitant changes in promoter DNA methylation, H3K4 trimethylation and miR-467b-5p levels. Epigenetics Chromatin 8:40

Zhang Z, Yang T, Xiao J (2018) Circular RNAs: promising biomarkers for human diseases. EBioMedicine 34:267–274

Zhang P, Wu W, Chen Q, Chen M (2019a) Non-coding RNAs and their integrated networks. J Integr Bioinform 16

Zhang Q, Huang Y, Zhang L, Ding YQ, Song NN (2019b) Loss of Satb2 in the cortex and hippocampus leads to abnormal behaviors in mice. Front Mol Neurosci 12:33

Zhang X, Wang W, Zhu W, Dong J, Cheng Y, Yin Z, Shen F (2019c) Mechanisms and functions of long noncoding RNAs at multiple regulatory levels. Int J Mol Sci 20

Zhang D, Ji Y, Chen X, Chen R, Wei Y, Peng Q, Lin J, Yin J, Li H, Cui L, Lin Z, Cai Y (2022a) Peripheral blood circular RNAs as a biomarker for major depressive disorder and prediction of possible pathways. Front Neurosci 16:844422

Zhang Y, Zeng L, Lin D, Chang G, Zeng Y, Xia Y (2022b) Identification and characterization of nucleotide metabolism and neuroendocrine regulation-associated modification patterns in stomach adenocarcinoma with auxiliary prognostic assessment and immunotherapy response prediction. Front Endocrinol (Lausanne) 13:1076521

Zhang J, Li A, Gu R, Tong Y, Cheng J (2023a) Role and regulatory mechanism of microRNA mediated neuroinflammation in neuronal system diseases. Front Immunol 14:1238930

Zhang YF, Zhang HH, Su X, Wang Y, Gao GZ, Wang XD, Zhang T (2023b) Analysis of influencing factors and a predictive model of small airway dysfunction in adults. BMC Pulm Med 23

Zhou M, Li S, Huang C (2024) Physiological and pathological functions of circular RNAs in the nervous system. Neural Regen Res 19:342–349

Zhu S, Wu J, Hu J (2022) Non-coding RNA in alcohol use disorder by affecting synaptic plasticity. Exp Brain Res 240:365–379

Zimmerman AJ, Hafez AK, Amoah SK, Rodriguez BA, Dell'Orco M, Lozano E, Hartley BJ, Alural B, Lalonde J, Chander P, Webster MJ, Perlis RH, Brennand KJ, Haggarty SJ, Weick J, Perrone-Bizzozero N, Brigman JL, Mellios N (2020) A psychiatric disease-related circular RNA controls synaptic gene expression and cognition. Mol Psychiatry 25:2712–2727

Part II

Effects of Alcohol on the Adolescent Brain

Adolescent Alcohol Exposure Dysregulates Developing Cortical GABA Circuits

8

Avery R. Sicher and Nicole A. Crowley

Abstract

Adolescence is a critical developmental period during which physical, behavioral, and neurobiological maturation occurs. Within the brain, the prefrontal cortex is one of the last brain regions to undergo remodeling, often into adulthood. These relatively late developmental changes leave the prefrontal cortex uniquely vulnerable to insults beginning in adolescence—including alcohol exposure. Adolescents initiate alcohol consumption at a high rate, increasing the risk of lasting consequences through impairing the typical development of the prefrontal cortex. In this chapter, we discuss the development of prefrontal circuitry and the current literature investigating how alcohol influences prefrontal development. We primarily focus on preclinical studies in rodent models, which allow for the study of specific populations of neurons in the prefrontal cortex. We identify several future directions for adolescent alcohol research, including greater focus on neuropeptides and stronger understanding of sex differences in brain maturation and alcohol consumption.

Keywords

Ethanol · Adolescence · Development · Prefrontal · Cortex · GABA · Neuropeptide

A. R. Sicher
Department of Biology, The Pennsylvania State University, University Park, PA, USA

Neuroscience Graduate Program, Huck Institute of the Life Sciences, The Pennsylvania State University, University Park, PA, USA

N. A. Crowley (✉)
Department of Biology, The Pennsylvania State University, University Park, PA, USA

Neuroscience Graduate Program, Huck Institute of the Life Sciences, The Pennsylvania State University, University Park, PA, USA

Department of Pharmacology, Penn State College of Medicine, Hershey, PA, USA

Department of Biomedical Engineering, The Pennsylvania State University, University Park, PA, USA
e-mail: nzc27@psu.edu

8.1 Cortical Development During Adolescence

8.1.1 Introduction

Adolescence is the key transitional period between childhood and adulthood, spanning a time of physical, behavioral, and neurobiological development. Human adolescence constitutes roughly the second decade of life, although this definition has been updated based on additional neurobiological and social factors to encompass ages 10–24 years (Sawyer et al. 2018). Adolescents show a variety of characteristic

C. F. Valenzuela et al. (eds.), *Effects of Alcohol on the Brain across the Lifespan*, Advances in Experimental Medicine and Biology 1473, https://doi.org/10.1007/978-3-031-81908-7_8

behaviors, including heightened risk taking and sensation seeking, particularly in social situations (for review, see Spear 2000; Tomova et al. 2021). Risky substance use, particularly use of alcohol, is frequently initiated during adolescence. By grade 12 (age 17–18), about one-quarter of adolescents report alcohol consumption at least once in the previous 30 days (Miech et al. 2024). Although these numbers have been on the decline over the previous few decades, binge drinking, a problematic drinking pattern resulting in intoxication, remains prevalent in about 10% of high school seniors (Miech et al. 2024). Although the rates of adolescent alcohol consumption are declining, adolescent drinking increases the risk of many short- and long-term consequences. Alcohol consumption during adolescence, particularly binge or high-intensity drinking episodes, is associated with negative outcomes including risky driving decisions, violence, alcohol use disorders (AUD), and chronic organ damage (Vaca et al. 2020; Chung et al. 2018). Understanding the neurobiological mechanisms which may underlie the association between early alcohol consumption and these problematic outcomes could lead to improved therapeutic interventions.

Much of our mechanistic understanding of adolescent development comes from non-human primate and murine models. Rodents undergo a transitional period after weaning marked by behavioral, social, and neurobiological changes similar to that seen in human adolescents (Spear 2000). During this period, rodent models demonstrate several behaviors matching those seen in human adolescents (for review, see Spear 2014), suggesting that many of these behaviors and underlying neurobiology are evolutionarily conserved. The purpose of this chapter is to discuss the current understanding of adolescent alcohol's effects on prefrontal circuit development, primarily in studies using preclinical rodent models.

8.1.2 Adolescence in Humans and Rodent Models

The precise age range for what constitutes adolescence in both humans and rodents is debated. Adolescence covers but extends beyond the pubertal period (Spear 2000; Schneider 2013). It is now widely accepted that human adolescence continues into the early 20s, with the ages between 10 and 24 years comprising the neurobiological, behavioral, and social transition to adulthood (Sawyer et al. 2018).

Rodents show an accelerated developmental timeline compared to humans. In rodents, adolescence is now thought to span the fourth through eighth postnatal weeks, approximately postnatal day (PND) 28–56 (Schneider 2013; Spear 2015). Rodent adolescence begins after the age of weaning, which occurs between PND 21 and 26, and roughly coincides with pubertal onset in female mice and rats (Schneider 2013). A similar age range for adolescence is typically used for both mice and rats, though there are slight differences in the precise timings of puberty across species (Schneider 2013). The first 2 weeks of adolescence, PND 28–42, are defined as early adolescence and thought to represent ages 10–18 in humans (Spear 2015). The early adolescent window is primarily characterized by the hormonal changes associated with puberty, while the latter half of the rodent adolescent window is characterized by unique vulnerabilities to insults including alcohol use (Spear 2015). This later window is thought to represent emerging adulthood stages, roughly equivalent to ages 18–25 in humans (Spear 2015).

8.1.3 Overall Prefrontal Development During Adolescence

The adolescent brain of both humans and murine models is still undergoing critical development, particularly in cortical circuits (Crews et al. 2007). A major characteristic of adolescent brain development is the growth and remodeling of both gray and white matter, representing cell bodies and processes, respectively. Cortical gray matter volume changes according to an inverted U-shape over the lifespan, peaking during childhood (Gogtay et al. 2004), followed by a steady decline throughout adolescence and adulthood (Giedd et al. 1999). There are sex differences in the timing of cortical maturation, with frontal cortical gray matter volume peaking in adolescent girls 1 year earlier than in adolescent boys (Giedd et al. 1999). Synaptogenesis, the birth of new synaptic connections, occurs early in life, while synaptic pruning, the process by which neuronal communication is further refined through removal of excess connections between neurons, occurs in the frontal lobe during early adolescence in humans (Huttenlocher and Dabholkar 1997). This synaptic refinement continues well into early adulthood (Petanjek et al. 2011).

In both sexes, the prefrontal cortex (PFC) is one of the last regions of the brain to undergo major development during adolescence, with changes occurring across cellular, molecular, and circuit levels (Spear 2000). The broad functions of the PFC include executive control and goal-directed function (Fuster 2015). Because of its protracted developmental period, the PFC has been associated with various neuropsychiatric disorders which begin to present in adolescence (Drzewiecki and Juraska 2020). Compared to other earlier-developing regions, the PFC is also likely to be particularly vulnerable to insults during adolescence, including alcohol exposure. In humans, the PFC has traditionally been defined as regions innervated by the dorsomedial nucleus of the thalamus (Fuster 2015). These regions include Brodmann areas (BA) 9–12, 25, and 44–46 (Grossmann 2013). The PFC can be divided into subregions, including medial (medial portions of BA 9–12, BA 25) and lateral (lateral regions of BA 9–12, BA 44–46; Grossmann 2013). Although there are structural differences across species, rodents have a functionally analogous region of cortex which underlies similar cognitive functions including working memory that is termed the rodent PFC (Carlén 2017). The rodent PFC is often subdivided into the anterior cingulate, prelimbic, and infralimbic cortices (Laubach et al. 2018). As in humans, the rodent PFC undergoes a protracted development into adolescence, as discussed below. Despite these similarities, PFC growth and cytoarchitecture vary significantly across species and likely reflect the particular evolution and environments of that species (Kolk and Rakic 2022); therefore, interpretations across model organisms need to be considered in this context.

Along with gross structural developments in the PFC during adolescence, innervation of the PFC by several neurotransmitter systems also develops, including local gamma-aminobutyric acid (GABA) neurons and long-range dopaminergic (Kalsbeek et al. 1988) inputs. Integration of these synaptic contacts is a time-sensitive and fixed developmental process of critical importance, as proper development of the PFC GABAergic system in particular is necessary for excitatory/inhibitory balance and contributes to both the processing of afferent inputs and the orchestration of output responses, in the adult PFC. Excitatory and inhibitory balance is vital for overall PFC functioning and behavior, and disruptions to either can cause long-lasting and far-reaching consequences. In addition to its ongoing development in adolescence, the medial PFC has been implicated in alcohol-seeking behaviors and the transition toward alcohol dependence (Klenowski 2018). The PFC is therefore in a unique position to be especially vulnerable to insults during adolescence, including early alcohol use, setting the stage for long-term consequences.

8.1.4 Adolescent Development of GABAergic and Peptidergic Systems in the Prefrontal Cortex

During adolescence, the prefrontal inhibitory system goes through critical developmental remodeling and maturational changes, spanning cellular and molecular refinement to the development of overall broader cortical inhibition. There are developmental changes at the levels of GABAergic receptors, neurotransmitter function, and populations of GABAergic neurons. It is important to note that GABA signaling at each of these levels is influenced by both immediate and long-term hormonal shifts (for review, see Gilfarb and Leuner 2022).

First, GABAergic receptor composition undergoes developmental changes throughout postnatal development (for review, see Luján et al. 2005). Ionotropic GABAA receptors are composed of several subunits, and this composition changes throughout postnatal development.

Developmental changes in $GABA_a$ receptor subunit composition can be reflected in the kinetics of inhibitory postsynaptic currents, which has an impact on the time course of synaptic transmission. Across species, α_1 subunits gradually replace α_2 subunits in frontal cortex during early postnatal development (Duncan et al. 2010; Fritschy et al. 1994; Laurie et al. 1992). This corresponds with shorter-duration inhibitory postsynaptic currents, reflecting an overall increase in the temporal precision of cortical GABA signaling throughout postnatal development and adolescence (Bosman et al. 2005; Kilb 2012).

Maturation of $GABA_A$ receptors contributes to a tonic current which emerges in the prefrontal cortex throughout development (Centanni et al. 2017). This tonic current is mediated by extrasynaptic $GABA_A$ receptors containing the δ-subunit and is a critical regulator of pyramidal neuron excitability and firing properties in the mature PFC (Brickley and Mody 2012; Walker and Semyanov 2008). During adolescence, the number of pyramidal neurons in the prelimbic cortex receiving the tonic inhibitory current continues to increase, until nearly all pyramidal neurons express this current by adulthood (Centanni et al. 2017). Together, the electrophysiological changes due to developmental changes in $GABA_A$ receptor composition reflect the overall increase in inhibitory tone in the adolescent PFC.

GABA neurotransmitter signaling is initially excitatory in the developing brain due to higher intracellular chloride (Cl–) concentrations in neurons (Kilb 2012). During the first postnatal weeks, expression of a Cl– importer NKCC1 in the membrane is gradually replaced by the KCC2 Cl– exporter, reducing the intracellular Cl– concentration and leading to GABA exerting hyperpolarizing effects (Peerboom and Wierenga 2021). Importantly, the switch of GABA from excitatory to inhibitory has been found to occur at differential rates across discrete cell types elsewhere in the cortex (Ikeda et al. 2003), with similar cell-specific changes in the GABA transporter seen across development (Ikeda et al. 2003)—highlighting the need to understand the nuanced progression of individual populations of GABA neurons. Overall GABAergic transmission onto PFC glutamatergic neurons increases throughout adolescence (Cass et al. 2014; Kroon et al. 2019), though these experiments did not tie increases in spontaneous GABAergic signaling to specific populations. This increase in inhibitory drive is layer-specific in the PFC, with spontaneous inhibitory events increasing later in superficial layers as compared to deeper layers (Kroon et al. 2019). Increases in inhibitory signaling may be due to the changes in GABA neurotransmitter function or due to maturation of GABAergic neurons. Subpopulations of GABA neurons are traditionally classified based on the expression of co-expressed genetic molecular markers, including parvalbumin, somatostatin, corticotrophin-releasing factor, dynorphin, and vasoactive intestinal peptide (Gabbott et al. 1997; Xu et al. 2010; Brockway and Crowley 2020). These subpopulations show different developmental trajectories in their expression and in their emerging electrophysiological properties, and development of each cellular population and the overall network of GABAergic neurons contributes to the overall refinement of the excitatory/inhibitory balance in the PFC in adolescence.

8.1.4.1 Development of Somatostatin-Expressing GABAergic Neurons

Somatostatin (SST)-expressing neurons in the PFC are involved in a variety of affective behaviors including fear learning (Cummings and Clem 2020; Xu et al. 2019) and affective recognition and discrimination (Scheggia et al. 2020). Several preclinical studies have sought to characterize changes in SST throughout development to understand the potential role of this peptide and neuron population in affective disorders which often first present during adolescence. Derived from the medial ganglionic eminence, SST neurons are among the earliest born GABAergic neuron populations, and their innervation of the cortex increases during the first postnatal week in rodents (Forloni et al. 1990; Miyoshi et al. 2007; Tuncdemir et al. 2016). One study investigating adolescent development of GABAergic populations found that SST protein expression in PFC develops in an inverted-U pattern, increasing between postnatal weeks 3 and 6 before decreasing by adulthood in female mice, while male mice do not show appreciable changes across this same timespan (Du et al. 2018). Importantly, this study captures changes in the expression of SST peptide protein—suggesting developmental regulation of peptide expression, as there were no changes in SST-expressing cell density in the female PFC measured using immunohistochemical labeling of the SST protein. SST mRNA has been found elsewhere in the brain to be regulated by sex hormones (Argente et al. 1990) pointing toward a potential mechanism of sex-specific developmental growth. A "critical period" for cortical SST neuron electrophysiological development has been proposed to span the first two postnatal weeks in mice (Pan et al. 2019). During this "critical period" prior to PND 15, the intrinsic electrophysiological properties of cortical SST neurons, including membrane resistance and resting membrane potential, mature and stabilize (Pan et al. 2016, 2019). Despite early stabilization of these intrinsic properties, active properties including current-induced spiking continue to develop past PND 28 and into early adolescence (Pan et al. 2019; Kinnischtzke et al. 2012). Microcircuitry including SST neurons continues to evolve after weaning, as shown by increases in synaptic drive onto SST neurons (Pan et al. 2016). And interestingly, while SST neurons show stability in intrinsic properties throughout adolescence, they also show marked plasticity in response to learning events like fear conditioning (Koppensteiner et al. 2019).

8.1.4.2 Parvalbumin-Expressing GABAergic Neurons

Like SST-expressing neurons, parvalbumin (PV)-expressing neurons are derived from the medial ganglionic eminence (Miyoshi et al. 2007) and experience developmental regulation. Du et al. found developmental increases in expression of other GABAergic subpopulations such as PV-expressing neurons (Du et al. 2018)—an effect that replicates work across humans (Fung et al. 2010) and rats (Caballero et al. 2014). Both male and female mice show a steady increase in PV-expressing neurons throughout postnatal weeks 3–12 (Du et al. 2018). Others have found similar conclusions, with PV-expressing neurons increasing in rats around PND 45–55, at which point they appear to stabilize to their adulthood levels (Caballero et al. 2014). This increase in PV was driven by an increase in both cell number and PV protein expression and is correlated with increasing glutamatergic drive onto these neurons (Caballero et al. 2014). However, PFC PV neurons can be further classified as chandelier cells and basket cells, with only basket cells present in deeper cortical layers after about the second week of postnatal development, and developing excitatory drive onto these two subpopulations differs (Miyamae et al. 2017). This highlights the need to continuously re-evaluate our classification systems for GABAergic neurons and to consider them in new contexts. As with prefrontal SST neurons, prefrontal PV neurons show rapid development of intrinsic properties during the first two postnatal weeks, although synaptic communication between PV neurons and excitatory pyramidal neurons continues to increase past PND 28 and into early adolescence (Yang et al. 2014). During the later juvenile, preadolescent period into mid-adolescence (PND 14–50), PV neurons are integrated into local circuitry in an activity-dependent manner (Canetta et al. 2022). Together, studies in both prefrontal SST and PV neurons show that inhibitory neuron maturation does not conclude with stabilization of intrinsic properties, but that integration of these neurons into mature microcircuitry continues into adolescence.

8.1.4.3 Other Populations of GABAergic Neurons

Other GABAergic subpopulations have not been investigated as thoroughly as SST and PV. Calretinin-expressing neurons appear to decrease throughout adolescence, with lower expression of calretinin protein seen at PND 45–55 as compared to PND 25–35 (Caballero et al. 2014). In humans, neuropeptide-y expressing neurons appear to increase in density from approximately 4 to 7 years of age (Delalle et al. 1997). Interestingly, while this study found stable adulthood-like expression around years 8–10, they also noted strong variability across subjects.

And importantly, almost nothing is known about how these neurons developmentally regulate release and signaling through their peptidergic transmitters. While RNA expression levels hint more toward developmental regulation of peptide expression than true neuronal markers, the release of neuropeptides in general is still an emerging area of interest (Dao et al. 2019). Aberrant expression of these neurons during postnatal development is seen in neuropsychiatric conditions such as schizophrenia (Fung et al. 2010) highlighting the need to further understand the trajectory of these GABAergic neurons—and how to protect their natural growth trajectories. Also importantly, only a small percentage of GABA neurons need to be dysregulated during development to cause profound impacts on overall PFC excitatory inhibitory balance (Caballero et al. 2020)—setting the stage for disruptions such as adolescent alcohol consumption to have catastrophic ramifications on PFC circuit function.

8.1.5 Adolescent Development of Glutamatergic Circuits

Much like the GABAergic system, glutamatergic neurons both within and projecting to the developing PFC undergo profound synaptic pruning. Some rodent studies have suggested that pruning of synapses and neurons is greater in females than in males, with decreases in female medial PFC cell bodies between PND 35 and 45 (Willing and Juraska 2015). As no changes in GABAergic neuronal markers were seen, this implicates pruning of glutamatergic neurons. Ovarian hormones underlie this neuronal pruning, as gonadectomy prevents this neuron loss in female but not male rats (Koss et al. 2015). Similar to the pattern seen in SST and PV neurons, glutamatergic pyramidal neurons show stabilization of intrinsic properties primarily during the first two postnatal weeks (Kroon et al. 2019). This general trend coincides with morphological growth of these cells and is consistent across prefrontal layers. However, synaptic drive onto pyramidal neurons does develop in a layer-specific manner, reflecting continued integration of inhibitory neurons into pyramidal circuits during adolescence, especially in the superficial layers (Kroon et al. 2019).

Glutamatergic receptor signaling within the PFC undergoes developmental remodeling throughout adolescence as well (for review, see Luján et al. 2005). A developmental switch in *N*-methyl-*D*-aspartate (NMDA) receptor subunit expression has been reported, with NR2A gradually replacing NR2B (Ueda et al. 2015), although increases in NR2B-facilitated excitatory transmission from late adolescence into early adulthood have been shown (Flores-Barrera et al. 2014). These changes may depend on the cortical layer investigated or the PFC subregion targeted. Composition of kainite and metabotropic glutamate receptors is also developmentally regulated (for review, see Luján et al. 2005). Specific inputs onto pyramidal neurons develop as well, which regulates pyramidal neuron cell firing (Trantham-Davidson et al. 2017; Tseng and O'Donnell 2007). During adolescence, as expression of dopaminergic receptors increases in the prelimbic cortex, the number of pyramidal neurons responsive to D1 agonists increases (Trantham-Davidson et al. 2017). This coincides with an increase in dopaminergic innervation of the medial PFC throughout adolescence in male and female rodents (Willing et al. 2017; Naneix et al. 2012). Changes in receptor expression and inputs onto PFC glutamatergic neurons refine the overall excitatory activity during adolescence and may underlie behavioral maturation. However, there is much yet to be explored in regard to adolescent development of PFC glutamatergic systems and their interaction between these other inputs.

8.2 Adolescent Alcohol Consumption Can Lead to Sex-Specific, Long-Lasting Deleterious Consequences

8.2.1 Human and Animal Studies: Behavioral Implications

Any perturbation to prefrontal maturation is likely to produce lasting effects, and alcohol exposure is a particularly common insult to PFC development during adolescence. The late developmental maturation of the PFC may be causal to increased alcohol consumption in adolescence (Guerri and Pascual 2010), possibly reflecting an imbalance between executive control and the earlier-maturing reward-seeking systems (Steinberg 2008). Adolescents show some resilience to some of the acute negative consequences of alcohol drinking (e.g. hangover symptoms), supporting higher levels of alcohol drinking during a drinking event among adolescents (Spear 2014). Indeed, episodes of binge drinking, consuming at least 4 or 5 alcoholic drinks within a 2-h period in women and men, respectively, are reported by about 10% of adolescents under the legal drinking age (SAMHSA 2022). Within those who report binge drinking episodes, patterns of repeated binge drinking and occasions of high-intensity drinking, defined as consuming more than ten alcoholic drinks in one episode, are also prevalent (SAMHSA 2022). In addition to immediate risks of high levels of alcohol consumption like impaired driving and unsafe sexual intercourse, alcohol drinking in adolescents is associated with other long-term risk-taking behaviors, including fighting, use of other drugs of misuse, and unsafe driving (Chung et al. 2018; Vaca et al. 2020; DuRant et al. 1999). Earlier alcohol consumption, particularly in patterns of binge or high-intensity drinking, has been associated with higher risk of alcohol-related harm, including developing an alcohol use disorder (AUD) later in life (Grant and Dawson 1997; DeWit et al. 2000; Wells et al. 2004; Yuen et al. 2020). Importantly, this relationship between early alcohol use and risk of problematic alcohol consumption patterns is influenced by other factors, such as co-morbid mental health conditions and socioeconomic background (Wells et al. 2004). Taken together, these changes in outcomes surrounding risky decision making and related behaviors following adolescent alcohol consumption suggest long-lasting changes in the cortex are likely.

Indeed, problematic alcohol use in adolescents has been associated with lasting changes in the PFC. Human neuroimaging studies have found that adolescents with an AUD have reduced PFC volume (De Bellis et al. 2005), with a more pronounced reduction in adolescent women (Medina et al. 2008). Medina et al. demonstrated an interaction between sex and AUD condition on PFC volume—adolescent boys with an AUD show enlarged PFC volumes and adolescent girls with an AUD show reduced PFC volumes compared to same-sex counterparts without an AUD, respectively. Differences in the microstructure of frontostriatal connectivity have been identified even prior to alcohol use in adolescents who went on to engage in binge drinking (Jones and Nagel 2019). The preclinical animal literature has provided a breadth of information on the negative consequences of adolescent alcohol exposure, often demonstrating complementary behavioral changes to those seen in humans (Spear 2000). Conflicting results are seen in regard to whether adolescent alcohol consumption leads to greater consumption or preference for alcohol in adulthood. Adolescent mice that undergo a developmentally modified drinking in the dark (DID) binge-like model of alcohol consumption show significantly higher consumption of alcohol when compared to adults (Holstein et al. 2011; Lee et al. 2017; Younis et al. 2019; Van Hees et al. 2022). This effect can be driven by overall higher drinking in female rodents (Strong et al. 2010). However, not all rodent studies of adolescent drinking report an escalation of drinking during adulthood (Varlinskaya et al. 2017; Wooden et al. 2023; Nentwig et al. 2019; Chandler et al. 2022a, b; Sicher et al. 2024; for review, see Towner and Varlinskaya 2020). These conflicting results highlight a few key points. First, experimental differences in the precise developmental timing of access to alcohol may suggest that sex differences in alcohol consumption do not emerge until later developmental timepoints—potentially pointing toward hormone-driven mechanisms. Second, while consumption of alcohol is similar across males and females at early developmental timepoints, it may set the stage for exacerbated effects across the sexes if sex-dependent or sexually dimorphic brain circuits develop abnormally post-alcohol. This highlights the need for precise investigations of alcohol-induced changes in PFC circuits in both sexes. One factor potentially contributing to these conflicting reports is housing conditions, i.e. individual versus grouped housing. Many rodent alcohol exposure paradigms require subjects to be housed individually in order to track ethanol consumption (Crowley et al. 2019), though social isolation during adolescence can drive heightened alcohol consumption later in life (Chappell et al. 2013; Skelly et al. 2015; Lopez et al. 2011; for review, see Lodha et al. 2022). However, this effect is not universally observed (Pisu et al. 2011; Butler et al. 2014; Logue et al. 2014; Rivera-Irizarry et al. 2020). Other studies show that the combination of social isolation and ethanol exposure in adolescence, rather than exposure to one factor alone, produces the heightened drinking phenotype in adulthood (Chandler et al. 2022b). We recently reported no change in adulthood alcohol consumption following several paradigms of voluntary adolescent alcohol exposure regardless of housing condition (Sicher et al. 2024). Importantly, the variability in effect seen on adolescent drinking-induced escalation in adulthood drinking emphasizes the need to further understand social factors that underlie human adolescent alcohol consumption that are likely to be uncaptured in murine models.

While many studies investigating the effects of adolescent alcohol exposure do not explicitly test PFC neuronal function, persistent alterations in behaviors known to require the PFC can point toward its disruption. In both humans and murine models, this can include changes in risk taking, novelty seeking, cognitive flexibility, learning, and fear memory, and consumption of rewarding substances (for review see Sicher et al. 2022). In rodent models, behavioral changes induced by adolescent alcohol often indicate a lasting adolescent-like phenotype, i.e. alcohol-exposed adult mice show behaviors more comparable to adolescent mice than to control adult mice (Spear and Swartzwelder 2014). Intermittent ethanol exposure during both early (PND 24–45) and late (PND 45–65) adolescence has been shown to alter anxiety-like behavior in both male and female mice (Varlinskaya et al. 2020), as well as behavioral flexibility in both sexes (Van Hees et al. 2022). Exposure to vaporized ethanol during early adolescence (PND 28–44) can lead to altered threat-related behavior in adult rats, likely to be governed by PFC circuits (Landin and Chandler 2023). Importantly, Landin and Chandler demonstrated sex differences in the consequences of vaporized ethanol, with female rats showing enhanced freezing compared to male rats. Other groups have similarly shown sex differences in fear-related behaviors following a voluntary alcohol consumption model (Grizzell et al. 2023), suggesting changes in the PFC are likely. Exposure to ethanol via intragastric gavage during adolescence reduced social preference in male rats and altered the neuronal activation patterns underlying these social behaviors (Towner et al. 2022). In male rats, medial PFC neuronal activation was correlated with social behavior, an effect that was blocked by adolescent ethanol exposure. In female rats, there was a negative correlation between orbitofrontal cortex activity and social behavior that was not observed following adolescent alcohol exposure (Towner et al. 2022). This study captured broad changes in neuronal activation after social interaction test, although specific populations of neurons may be more vulnerable to adolescent alcohol. Together, these studies show lasting effects of adolescent alcohol exposure on various behaviors which can be linked to alterations in the PFC.

8.2.2 Adolescent Alcohol Effects on Prefrontal Glutamatergic, GABAergic, and Peptide Circuits

Because the prefrontal excitatory and inhibitory systems experience critical developmental changes during adolescence, these neurons and their connections are especially vulnerable to insults including alcohol use (Caballero et al. 2021). Additionally, any disruptions to prefrontal maturation are more likely to produce permanent dysfunction. Supporting this, many preclinical studies have demonstrated behavioral and neurobiological consequences persisting into adulthood following adolescent alcohol exposure (for review see Robinson et al. 2021). Studies thus far have largely centered on prefrontal pyramidal neurons, which represent the major excitatory population of neurons and which project to other cortical and subcortical regions. Adolescent alcohol exposure has been shown to cause altered intrinsic excitability of both pyramidal neurons and fast-spiking GABA neurons in the prelimbic (PL) cortex, a subdivision of the PFC in rodents (Sicher et al. 2023; Salling et al. 2018; Trantham-Davidson et al. 2017). These changes are driven—at least in part—by reductions in hyperpolarization-activated cation currents (Salling et al. 2018). Importantly, intermittent exposure to alcohol during adolescence also reduces PFC resting-state connectivity and functional connectivity between the PFC and structures in the striatum (Broadwater et al. 2018), suggesting that the PFC changes induced by adolescent alcohol exposure have profound downstream network implications. Previous work from our group has found that these changes in excitability can be long lasting—persisting at least 30 days after the cessation of adolescent alcohol consumption (Sicher et al. 2023). Others have shown that changes in pyramidal neuron excitability are layer-specific, with deeper-layer pyramidal neurons being more vulnerable to persistent changes following adolescent alcohol exposure than superficial pyramidal neurons (Galaj et al. 2020). Glutamatergic drive onto layer V pyramidal neurons is increased following a two-bottle choice access model (Klenowski et al. 2016). Differences in results may be due to differences in species used (mice versus rats) or exposure paradigms employed (Drinking in the Dark versus chronic intermittent vapor exposure, discussed further below).

Several groups have shown that adolescent alcohol produces unique effects on prefrontal neurons compared to alcohol exposure beginning later in life. Although our group has found hypoexcitability in layer II/III pyramidal neurons 24 h after cessation of our adolescent binge drinking model (Sicher et al. 2023), a previous study from our lab using an identical drinking paradigm beginning in adulthood revealed hyperexcitability of the same neuron population (Dao et al. 2021). Another study similarly found opposing effects of adolescent (PND 28–45) and early adulthood (PND 70–88) alcohol exposure on prefrontal pyramidal neurons using a chronic exposure paradigm (Galaj et al. 2020). These findings show the importance of considering developmental age when investigating the neurobiological consequences of alcohol exposure.

In addition to disruption of intrinsic properties, adolescent alcohol exposure interferes with dopaminergic modulation of glutamatergic neurotransmission in the PFC. Agonists for two dopaminergic receptor subtypes (D1 and D2) have opposing effects on pyramidal neuron activity—D1 agonists increase pyramidal neuron firing while D2 agonists reduce firing, likely driven by differing g-protein coupled actions of the receptor subtypes (Trantham-Davidson et al. 2017; Kebabian and Greengard 1971). In rats, exposure to alcohol from PND 28 to 42, during early adolescence, disrupts D1-modulated regulation of prelimbic pyramidal neuron firing without altering expression of D1 receptors themselves (Trantham-Davidson et al. 2017). This reduction of D1-mediated function is seen in male and female rats and in pyramidal neurons projecting to both the nucleus accumbens and basolateral amygdala (Obray et al. 2022). Although changes in D1 modulation of pyramidal neurons were seen across pyramidal neurons with different downstream targets, the effects of adolescent alcohol on synaptic input do depend on the output population of the pyramidal neurons. The balance of excitatory/inhibitory synaptic drive was shifted toward inhibition in accumbens-projecting pyramidal neurons after adolescent alcohol, but synaptic drive onto amygdala-projecting pyramidal neurons was not altered (Obray et al. 2022). Together, these findings reveal lasting effects of adolescent alcohol exposure for excitatory prefrontal circuitry, which could contribute to the behavioral consequences discussed previously.

The consequences of adolescent alcohol exposure at the cellular level for inhibitory neuron populations are still being uncovered, as there is a large diversity of subtypes to investigate. GABAergic neurons undergo profound developmental growth, with a greater change in overall inhibitory function compared to overall excitatory function (Caballero et al. 2021). Several studies have assessed the effects of exposure to alcohol during adolescence on some of the different subpopulations of GABAergic neurons in the

PFC and have reported cell type-specific changes following alcohol. Importantly, these effects are also dependent on the timing of alcohol exposure. As inhibitory neuron activity is critical for regulation of overall PFC activity and function, understanding the consequences of adolescent alcohol for these neurons is critical.

At a broad level, adolescent alcohol has been shown to interfere with the typical development of extrasynaptic GABA transmission which serves as an important regulator of pyramidal neuron excitability. Intermittent alcohol during early adolescence, from PND 28 to 42, blocked the development of a tonic GABA current onto layer V of the prelimbic cortex (Centanni et al. 2017). Loss of this current persisted at least 30 days after alcohol exposure concluded, suggesting that alcohol prevented this developmental milestone, rather than delaying it. Reduced expression of this current could lead to disinhibition and hyperexcitability of prefrontal pyramidal cells following adolescent alcohol, an effect that has been shown previously (Sicher et al. 2023; Salling et al. 2018).

Although differences in adolescent maturation have been observed across GABAergic cell populations, relatively few studies have investigated the effects of adolescent alcohol on individual inhibitory neuron populations. Fast-spiking interneurons in the prelimbic region of the PFC are vulnerable to intermittent ethanol exposure from PND 28 to 42, during early- to mid-adolescence, showing reduced evoked action potential firing (Trantham-Davidson et al. 2017). As seen in pyramidal neurons, dopaminergic modulation of fast-spiking interneuron activity, specifically mediated by the D1 receptor, was also lost following adolescent alcohol exposure (Trantham-Davidson et al. 2017). Binge alcohol exposure during early adolescence from PND 28 to 37 can also reduce the number of PV-expressing neurons in the PFC altogether (Rice et al. 2019), though it is unknown whether this loss was due to changes in the adolescent-typical increase of PV cells, reduced PV protein expression, or death of PV neurons. This loss in PV-expressing neurons also corresponded to a loss in myelination, indicating that synaptic transmission from PV-expressing cells may be impaired by alcohol (Rice et al. 2019). Complementary lines of evidence show increased excitability and medium afterhyperpolarization (mAHP) in PV interneurons in mice following an intermittent access paradigm beginning in mid-late adolescence at ~PND 42 (Joffe et al. 2020). These effects were more pronounced in female mice, as male mice showed altered mAHP without excitability changes, and the sexes differed in excitatory synaptic changes onto PV neurons. These results highlight the importance of considering developmental age of alcohol exposure, as alcohol exposure earlier in adolescence reduced the excitability of prefrontal PV-expressing neurons (Trantham-Davidson et al. 2017). Future studies should further determine the effects of adolescent alcohol on PV expression (i.e. cells vs protein loss). As a calcium binding protein, loss of PV within neurons could underlie the electrophysiological changes discussed above. Taken together, multiple lines of evidence suggest that not only are the number of PV-expressing neurons reduced, but the function of the remaining neurons is drastically compromised following adolescent alcohol consumption or exposure.

PFC SST neurons in the adult brain have been shown to be modified by events like adulthood alcohol exposure and stress exposure (Dao et al. 2021; Joffe et al. 2022; Fuchs et al. 2017), but little is known about developmental exposure to alcohol and this cell type. Recent work shows that a late adolescence model of intermittent access to alcohol, beginning between PND 42 and 49, reduces excitatory drive onto somatostatin neurons (Joffe et al. 2020). We have investigated adolescent alcohol-induced disruptions to the intrinsic properties of prefrontal SST neurons. Using the Drinking in the Dark paradigm to model binge-like drinking throughout adolescence, from PND 28 to 52 we found that prelimbic SST neurons show increases in current-induced action potential firing 24 h after the final binge session (Sicher et al. 2023). We found that this hyperexcitable phenotype remained even when electrophysiology experiments were performed 30 days after the cessation of alcohol consumption. These experiments did not find reductions in SST cell number, suggesting a potential to therapeutically restore the excitatory/inhibitory balance in the PFC. Interestingly, we have previously found contrasting results using identical methods to investigate the consequences of binge drinking in adult mice (Dao et al. 2021), further highlighting how adolescent consumption of alcohol uniquely dysregulates developing PFC neurons and circuits, as compared to when the adult brain is exposed to alcohol. Despite our previous finding that adult binge drinking increases SST-mediated inhibition onto other prelimbic inhibitory neurons (Dao et al. 2021), we did not see changes in SST connectivity with pyramidal or inhibitory neurons following adolescent alcohol (though these papers used slightly differing optogenetic and transgenic approaches). Importantly, little work thus far has assessed alterations in SST peptide function following adolescent alcohol exposure. SST hyperpolarizes neurons in the prefrontal cortex (Brockway et al. 2023), so changes in prefrontal SST expression or function following adolescent alcohol may have implications for prefrontal circuit function.

In addition to the studies above suggesting adolescent alcohol-induced dysregulation of pyramidal and inhibitory neurons, there have been important observations into adolescent alcohol's effects on other non-neuronal processes in the PFC that warrant further focused investigation. For example, adolescent alcohol exposure from PND 25 to 54 increases expression of perineuronal nets on PV-expressing cells without altering the expression of PV or choline acetyltransferase (CHAT) cells themselves in the PFC (Dannenhoffer et al. 2022). Perineuronal nets are components of the extracellular matrix that surround neurons and their processes and are an important regulator of PV neuron excitability and plasticity (Drzewiecki et al. 2020). The expression of perineuronal nets increases post-puberty in both sexes in rats (Drzewiecki et al. 2020), suggesting another indirect mechanism by which PV and overall PFC function can be disrupted by adolescent alcohol.

8.3 Many Variables Can Affect the Outcome of Adolescent Alcohol Exposure on Prefrontal Circuits

There are multiple variables that can influence the effect adolescent alcohol has on developing cortical circuits. Exposure paradigms can range from voluntary consumption (drinking in the dark, intermittent access) to forced consumption (intragastric gavage, intraperitoneal injections, and vaporized ethanol exposure), which can result in compounding stress effects (Crowley et al. 2019). All of these models can also be used for either acute or chronic exposure. The precise timing of exposure, and for how long, may capture differing developmental windows. Importantly, it is well known that alcohol and stress interact, and this is critical for the developing PFC as well (Shaw et al. 2020; Sicher et al. 2022). Stress itself can alter the developmental trajectory of the PFC (Page and Coutellier 2018) and adolescent alcohol work must be considered in the context of both general stressors and any stress introduced by the model, such as physical stress and housing conditions.

Further work is needed to better understand both the typical development of the PFC and the impacts of adolescent alcohol. A major limitation toward understanding how adolescent alcohol impacts nuanced development in the PFC is that the field lacks a comprehensive framework of the typical development of individual neurons, signaling molecules, and circuits in the PFC. Adolescent alcohol leads to complex changes elsewhere in the brain (Avegno et al. 2016)—and these projections can have profound implications for the PFC. Importantly, increasing amounts of evidence suggest the need to further understand neuropeptide populations in the PFC (Brockway and Crowley 2020). While these neuropeptide populations are often co-expressed with GABA, neuropeptides themselves exert a diverse array of actions in the PFC (Brockway et al. 2023)—all of which may be altered following adolescent alcohol. These peptidergic populations likely represent promising therapeutic targets (Crowley and Joffe 2022). In addition, greater attention should be paid to polysubstance misuse—including prescription medications, and other drugs taken both legally and illicitly throughout adolescence (Crowley et al. 2014; Ansell et al. 2023; Bedillion et al. 2021; Linden-Carmichael et al. 2021).

The field is also increasingly attuned to the need to further investigate sex differences in adolescent alcohol consumption and brain maturation. Drinking patterns in men and women are converging, with adolescent girls and young women representing a higher risk of problematic alcohol consumption (Slade et al. 2016; White et al. 2015; Cheng and Anthony 2017). Understanding how sex may influence behavioral and neurobiological outcomes of adolescent alcohol use is therefore becoming increasingly prudent. While some preclinical studies do not note sex differences in adolescent alcohol consumption (e.g. Sicher et al. 2023; Younis et al. 2019), this does not preclude more nuanced interactions with biological sex, as there are known interactions between sex hormones and drug misuse more broadly. Hormonal fluctuations of the menstrual cycle have been linked to probability of binge drinking behaviors in humans and macaques (Hayaki et al. 2020; Warren et al. 2020; Martel et al. 2017; Thomas and Czoty 2019; Mello et al. 1986; Hoffmann et al. 2024), and ovarian hormones can regulate brain reward circuitry in female mice (Hilderbrand and Lasek 2018a; Giacometti et al. 2022; Chen et al. 2022; Vandegrift et al. 2020; Tonn Eisinger et al. 2018; Satta et al. 2018). Additionally, adult female C57BL/6J mice consistently consume more alcohol (g/kg) than adult males in voluntary exposure model of binge drinking, an effect that appears to be dependent on estrogen (Satta et al. 2018; Hilderbrand and Lasek 2018b). Alcohol consumption has not been consistently shown to correlate with estrus cycle phase in mice (Satta et al. 2018), suggesting that the effects may be based on broader levels and not just daily fluctuations in estrogen and other gonadal hormones (Radke et al. 2021). Sex differences in adolescent alcohol consumption may not be consistently observed as the timing of alcohol exposure in relation to pubertal increases in gonadal hormone production may influence these results. Sex chromosome complement, assessed through the Four Core Genotypes model, has also been shown to influence alcohol intake independent of gonadal hormones, with XX mice consuming more than XY mice (Sneddon et al. 2022). Ongoing work using the Four Core Genotypes model in adolescent mice suggests that the role of sex chromosome complement in alcohol consumption may be age dependent (Aarde et al. 2022). How these hormone-driven escalations in binge drinking interact with brain development, particularly in the PFC, is unknown.

While considering biological sex when studying alcohol use and prefrontal development is important, gender identity is another important and emerging factor to consider. Studies of alcohol use in transgender and gender non-conforming populations are currently limited but growing. Hazardous alcohol use has been found repeatedly in transgender populations; however, much of our understanding of alcohol use thus far fails to separate the effects of biological sex and gender identity (Gilbert et al. 2018). Continuing to better understand how biological sex and gender identity separately relate to hazardous drinking patterns can improve interventions and screening processes for problematic alcohol use among transgender and gender non-conforming youth (Gilbert et al. 2018). Substance use, including episodes of problematic or binge drinking, is common in transgender youth (Day et al. 2017; Watson et al. 2019). Future studies including transgender and gender non-conforming youth are critical to improve treatment strategies and outcomes for these particularly vulnerable groups.

8.4 Conclusions

The PFC undergoes a variety of critical developmental changes during adolescence, a period in which alcohol use is commonly initiated. PFC maturation includes overall refinement of glutamatergic and inhibitory GABAergic transmission and circuitry which is necessary for proper function and behavior in adulthood. Alcohol consumption in adolescents is associated with a variety of long-term behavioral consequences and alterations in prefrontal neurobiology. Evidence from preclinical models shows that adolescent alcohol interferes with the development of prefrontal circuitry, though our understanding of adolescent alcohol's effects at the cellular level remains limited. Many factors can influence the consequences of adolescent alcohol exposure and should be considered in future studies, including biological sex and gender identity, sex hormone levels, paradigm and pattern of adolescent alcohol exposure (i.e., intermittent versus chronic exposure, binge and high-intensity drinking), and the developmental timing of alcohol exposure.

References

Aarde SM, Bagley JR, Jentsch JD (2022) Gonadal sex and sex-chromosome complement interact to affect ethanol consumption in adolescent four core genotypes mice. bioRxiv. https://doi.org/10.1101/2022.10.25.513748

Ansell EB, Bedillion MF, Farris SR, Gilbert JM, Koch MM, Thureen SE (2023) Cannabis use breaks in young adults: The highs and lows of tolerance breaks. Drug Alcohol Depend 249:109951. https://doi.org/10.1016/j.drugalcdep.2023.109951

Argente J, Chowen-Breed JA, Steiner RA, Clifton DK (1990) Somatostatin messenger RNA in hypothalamic neurons is increased by testosterone through activation of androgen receptors and not by aromatization to estradiol. Neuroendocrinology 52:342–349. https://doi.org/10.1159/000125618

Avegno EM, Salling MC, Borgkvist A, Mrejeru A, Whitebirch AC, Margolis EB, Sulzer D, Harrison NL (2016) Voluntary adolescent drinking enhances excitation by low levels of alcohol in a subset of dopaminergic neurons in the ventral tegmental area. Neuropharmacology 110:386–395

Bedillion MF, Blaine SK, Claus ED, Ansell EB (2021) The effects of alcohol and cannabis co-use on neurocognitive function, brain structure, and brain function. Curr Behav Neurosci Rep 8:134–149. https://doi.org/10.1007/s40473-021-00243-8

Bosman LWJ, Heinen K, Spijker S, Brussaard AB (2005) Mice lacking the major adult GABAA receptor subtype have normal number of synapses, but retain juvenile IPSC kinetics until adulthood. J Neurophysiol 94:338–346. https://doi.org/10.1152/jn.00084.2005

Brickley SG, Mody I (2012) Extrasynaptic GABAA receptors: their function in the CNS and implications for disease. Neuron 73:23–34. https://doi.org/10.1016/j.neuron.2011.12.012

Broadwater MA, Lee S, Yu Y, Zhu H, Crews FT, Robinson DL, Shih YI (2018) Adolescent alcohol exposure decreases frontostriatal resting-state functional connectivity in adulthood. Addict Biol 23(2):810–823. https://doi.org/10.1111/adb.12530

Brockway DF, Crowley NA (2020) Turning the 'tides on neuropsychiatric diseases: the role of peptides in the prefrontal cortex. Front Behav Neurosci 14

Brockway DF, Griffith KR, Aloimonos CM, Clarity TT, Moyer JB, Smith GC, Dao NC, Hossain MS, Drew PJ, Gordon JA, Kupferschmidt DA, Crowley NA (2023) Somatostatin peptide signaling dampens cortical circuits and promotes exploratory behavior. Cell Rep 42:112976. https://doi.org/10.1016/j.celrep.2023.112976

Butler TR, Carter E, Weiner JL (2014) Adolescent social isolation does not lead to persistent increases in anxiety-like behavior or ethanol intake in female Long-Evans rats. Alcohol Clin Exp Res 38:2199–2207. https://doi.org/10.1111/acer.12476

Caballero A, Flores-Barrera E, Cass DK, Tseng KY (2014) Differential regulation of parvalbumin and calretinin interneurons in the prefrontal cortex during adolescence. Brain Struct Funct 219:395–406. https://doi.org/10.1007/s00429-013-0508-8

Caballero A, Flores-Barrera E, Thomases DR, Tseng KY (2020) Downregulation of parvalbumin expression in the prefrontal cortex during adolescence causes enduring prefrontal disinhibition in adulthood. Neuropsychopharmacology 45:1527–1535. https://doi.org/10.1038/s41386-020-0709-9

Caballero A, Orozco A, Tseng KY (2021) Developmental regulation of excitatory-inhibitory synaptic balance in the prefrontal cortex during adolescence. Semin Cell Dev Biol 118:60–63. https://doi.org/10.1016/j.semcdb.2021.02.008

Canetta SE, Holt ES, Benoit LJ, Teboul E, Sahyoun GM, Ogden RT, Harris AZ, Kellendonk C (2022) Mature parvalbumin interneuron function in prefrontal cortex requires activity during a postnatal sensitive period. eLife 11:e80324. https://doi.org/10.7554/eLife.80324

Carlén M (2017) What constitutes the prefrontal cortex? Science 358:478–482. https://doi.org/10.1126/science.aan8868

Cass DK, Flores-Barrera E, Thomases DR, Vital WF, Caballero A, Tseng KY (2014) CB1 cannabinoid receptor stimulation during adolescence impairs the maturation of GABA function in the adult rat prefrontal cortex. Mol Psychiatry 19:536–543. https://doi.org/10.1038/mp.2014.14

Centanni SW, Burnett EJ, Trantham-Davidson H, Chandler LJ (2017) Loss of δ–GABAA receptor-mediated tonic currents in the adult prelimbic cortex following adolescent alcohol exposure. Addict Biol 22:616–628. https://doi.org/10.1111/adb.12353

Chandler CM, Hamid U, Maggio SE, Peng H, Pauly JR, Beckmann J, Nixon K, Bardo MT (2022a) Effects of adolescent alcohol exposure via oral gavage on adult alcohol drinking and co-use of alcohol and nicotine in Sprague Dawley rats. Drug Alcohol

Depend 232:109298. https://doi.org/10.1016/j.drugalcdep.2022.109298
Chandler CM, Shaykin JD, Peng H, Pauly JR, Nixon K, Bardo MT (2022b) Effects of voluntary adolescent intermittent alcohol exposure and social isolation on adult alcohol intake in male rats. Alcohol 104:13–21. https://doi.org/10.1016/j.alcohol.2022.07.008
Chappell AM, Carter E, McCool BA, Weiner JL (2013) Adolescent rearing conditions influence the relationship between initial anxiety-like behavior and ethanol drinking in male Long Evans rats. Alcohol Clin Exp Res 37:E394–E403. https://doi.org/10.1111/j.1530-0277.2012.01926.x
Chen H, Lu Y, Xiong R, Rosales CI, Coles C, Hamada K, Asad N, Thatcher GRJ, Lasek AW (2022) Effect of a brain-penetrant selective estrogen receptor degrader (SERD) on binge drinking in female mice. Alcohol Clin Exp Res 46:1313–1320. https://doi.org/10.1111/acer.14874
Cheng HG, Anthony JC (2017) A new era for drinking? Epidemiological evidence on adolescent male–female differences in drinking incidence in the United States and Europe. Soc Psychiatry Psychiatr Epidemiol 52:117–126. https://doi.org/10.1007/s00127-016-1318-0
Chung T, Creswell KG, Bachrach R, Clark DB, Martin CS (2018) Adolescent binge drinking. Alcohol Res 39:5–15. https://pmc.ncbi.nlm.nih.gov/articles/PMC6104966/
Crews F, He J, Hodge C (2007) Adolescent cortical development: A critical period of vulnerability for addiction. Pharmacol Biochem Behav 86:189–199. https://doi.org/10.1016/j.pbb.2006.12.001
Crowley NA, Joffe ME (2022) Developing breakthrough psychiatric treatments by modulating G protein-coupled receptors on prefrontal cortex somatostatin interneurons. Neuropsychopharmacology 47:389–390. https://doi.org/10.1038/s41386-021-01119-x
Crowley NA, Cody PA, Davis MI, Lovinger DM, Mateo Y (2014) Chronic methylphenidate exposure during adolescence reduces striatal synaptic responses to ethanol. Eur J Neurosci 39:548–556. https://doi.org/10.1111/ejn.12426
Crowley NA, Dao NC, Magee SN, Bourcier AJ, Lowery-Gionta EG (2019) Animal models of alcohol use disorder and the brain: From casual drinking to dependence. Transl Issues Psychol Sci 5:222–242. https://doi.org/10.1037/tps0000198
Cummings KA, Clem RL (2020) Prefrontal somatostatin interneurons encode fear memory. Nat Neurosci 23:61–74. https://doi.org/10.1038/s41593-019-0552-7
Dannenhoffer CA, Gómez-A A, Macht VA, Jawad R, Sutherland EB, Vetreno RP, Crews FT, Boettiger CA, Robinson DL (2022) Impact of adolescent intermittent ethanol exposure on interneurons and their surrounding perineuronal nets in adulthood. Alcohol Clin Exp Res 46:759–769. https://doi.org/10.1111/acer.14810
Dao NC, Brockway DF, Crowley NA (2019) In vitro optogenetic characterization of neuropeptide release from prefrontal cortical somatostatin neurons. Neuroscience 419:1–4. https://doi.org/10.1016/j.neuroscience.2019.08.014
Dao NC, Brockway DF, Suresh Nair M, Sicher AR, Crowley NA (2021) Somatostatin neurons control an alcohol binge drinking prelimbic microcircuit in mice. Neuropsychopharmacology. https://doi.org/10.1038/s41386-021-01050-1
Day JK, Fish JN, Perez-Brumer A, Hatzenbuehler ML, Russell ST (2017) Transgender youth substance use disparities: results from a population-based sample. J Adolesc Health 61:729–735. https://doi.org/10.1016/j.jadohealth.2017.06.024
De Bellis MD, Narasimhan A, Thatcher DL, Keshavan MS, Soloff P, Clark DB (2005) Prefrontal cortex, thalamus, and cerebellar volumes in adolescents and young adults with adolescent-onset alcohol use disorders and comorbid mental disorders. Alcohol Clin Exp Res 29:1590–1600. https://doi.org/10.1097/01.alc.0000179368.87886.76
Delalle I, Evers P, Kostović I, Uylings HBM (1997) Laminar distribution of neuropeptide Y-immunoreactive neurons in human prefrontal cortex during development. J Comp Neurol 379:515–522. https://doi.org/10.1002/(SICI)1096-9861(19970324)379:4<515::AID-CNE4>3.0.CO;2-5
DeWit DJ, Adlaf EM, Offord DR, Ogborne AC (2000) Age at first alcohol use: a risk factor for the development of alcohol disorders. Am J Psychiatry 157:745–750. https://doi.org/10.1176/appi.ajp.157.5.745
Drzewiecki CM, Juraska JM (2020) The structural reorganization of the prefrontal cortex during adolescence as a framework for vulnerability to the environment. Pharmacol Biochem Behav 199:173044. https://doi.org/10.1016/j.pbb.2020.173044
Drzewiecki CM, Willing J, Juraska JM (2020) Influences of age and pubertal status on number and intensity of perineuronal nets in the rat medial prefrontal cortex. Brain Struct Funct 225:2495–2507. https://doi.org/10.1007/s00429-020-02137-z
Du X, Serena K, Hwang WJ, Grech AM, Wu YWC, Schroeder A, Hill RA (2018) Prefrontal cortical parvalbumin and somatostatin expression and cell density increase during adolescence and are modified by BDNF and sex. Mol Cell Neurosci 88:177–188. https://doi.org/10.1016/j.mcn.2018.02.001
Duncan CE, Webster MJ, Rothmond DA, Bahn S, Elashoff M, Shannon Weickert C (2010) Prefrontal GABAA receptor α-subunit expression in normal postnatal human development and schizophrenia. J Psychiatr Res 44:673–681. https://doi.org/10.1016/j.jpsychires.2009.12.007
DuRant RH, Smith JA, Kreiter SR, Krowchuk DP (1999) The relationship between early age of onset of initial substance use and engaging in multiple health risk behaviors among young adolescents. Arch Pediatr Adolesc Med 153(3):286–291. https://doi.org/10.1001/archpedi.153.3.286
Flores-Barrera E, Thomases DR, Heng L-J, Cass DK, Caballero A, Tseng KY (2014) Late adolescent expression of GluN2B transmission in the prefrontal cortex

is input-specific and requires postsynaptic protein kinase A and D1 dopamine receptor signaling. Biol Psychiatry 75(6):508–516. https://doi.org/10.1016/j.biopsych.2013.07.033

Forloni G, Hohmann C, Coyle JT (1990) Developmental expression of somatostatin in mouse brain. I. Immunocytochemical studies. Dev Brain Res 53:6–25. https://doi.org/10.1016/0165-3806(90)90120-N

Fritschy JM, Paysan J, Enna A, Mohler H (1994) Switch in the expression of rat GABAA-receptor subtypes during postnatal development: an immunohistochemical study. J Neurosci 14:5302–5324. https://doi.org/10.1523/JNEUROSCI.14-09-05302.1994

Fuchs T, Jefferson SJ, Hooper A, Yee P-H, Maguire J, Luscher B (2017) Disinhibition of somatostatin-positive GABAergic interneurons results in an anxiolytic and antidepressant-like brain state. Mol Psychiatry 22:920–930. https://doi.org/10.1038/mp.2016.188

Fung SJ, Webster MJ, Sivagnanasundaram S, Duncan C, Elashoff M, Weickert CS (2010) Expression of interneuron markers in the dorsolateral prefrontal cortex of the developing human and in schizophrenia. AJP 167:1479–1488. https://doi.org/10.1176/appi.ajp.2010.09060784

Fuster J (2015) The prefrontal cortex. Academic, London

Gabbott PLA, Dickie BGM, Vaid RR, Headlam AJN, Bacon SJ (1997) Local-circuit neurones in the medial prefrontal cortex (areas 25, 32 and 24b) in the rat: Morphology and quantitative distribution. J Comp Neurol 377:465–499. https://doi.org/10.1002/(SICI)1096-9861(19970127)377:4<465::AID-CNE1>3.0.CO;2-0

Galaj E, Guo C, Huang D, Ranaldi R, Ma Y-Y (2020) Contrasting effects of adolescent and early-adult ethanol exposure on prelimbic cortical pyramidal neurons. Drug Alcohol Depend 216:108309. https://doi.org/10.1016/j.drugalcdep.2020.108309

Giacometti LL, Buck LA, Barker JM (2022) Estrous cycle and hormone regulation of stress-induced reinstatement of reward seeking in female mice. Addict Neurosci 4:100035. https://doi.org/10.1016/j.addicn.2022.100035

Giedd JN, Blumenthal J, Jeffries NO, Castellanos FX, Liu H, Zijdenbos A, Paus T, Evans AC, Rapoport JL (1999) Brain development during childhood and adolescence: a longitudinal MRI study. Nat Neurosci 2:861–863. https://doi.org/10.1038/13158

Gilbert PA, Pass LE, Keuroghlian AS, Greenfield TK, Reisner SL (2018) Alcohol research with transgender populations: A systematic review and recommendations to strengthen future studies. Drug Alcohol Depend 186:138–146. https://doi.org/10.1016/j.drugalcdep.2018.01.016

Gilfarb RA, Leuner B (2022) GABA system modifications during periods of hormonal flux across the female lifespan. Front Behav Neurosci 16. https://doi.org/10.3389/fnbeh.2022.802530

Gogtay N, Giedd JN, Lusk L, Hayashi KM, Greenstein D, Vaituzis AC, Nugent TF, Herman DH, Clasen LS, Toga AW, Rapoport JL, Thompson PM (2004) Dynamic mapping of human cortical development during childhood through early adulthood. Proc Natl Acad Sci USA 101:8174–8179. https://doi.org/10.1073/pnas.0402680101

Grizzell JA, Vanbaelinghem M, Westerman J, Saddoris MP (2023) Voluntary alcohol consumption during distinct phases of adolescence differentially alters adult fear acquisition, extinction and renewal in male and female rats. Stress 26:2278315. https://doi.org/10.1080/10253890.2023.2278315

Grant BF, Dawson DA (1997) Age at onset of alcohol use and its association with DSM-IV alcohol abuse and dependence: results from the national longitudinal alcohol epidemiologic survey. J Subst Abus 9:103–110. https://doi.org/10.1016/S0899-3289(97)90009-2

Grossmann T (2013) The role of medial prefrontal cortex in early social cognition. Front Hum Neurosci 7. https://doi.org/10.3389/fnhum.2013.00340

Guerri C, Pascual M (2010) Mechanisms involved in the neurotoxic, cognitive, and neurobehavioral effects of alcohol consumption during adolescence. Alcohol 44:15–26. https://doi.org/10.1016/j.alcohol.2009.10.003

Hayaki J, Holzhauer CG, Epstein EE, Cook S, Gaba A, Lorenzo AC, McCrady BS (2020) Menstrual cycle phase, alcohol consumption, alcohol cravings, and mood among women in outpatient treatment for alcohol use disorder. Psychol Addict Behav 34:680–689. https://doi.org/10.1037/adb0000576

Hilderbrand ER, Lasek AW (2018a) Estradiol enhances ethanol reward in female mice through activation of ERα and ERβ. Horm Behav 98:159–164. https://doi.org/10.1016/j.yhbeh.2018.01.001

Hilderbrand ER, Lasek AW (2018b) Studying sex differences in animal models of addiction: an emphasis on alcohol-related behaviors. ACS Chem Neurosci 9:1907–1916. https://doi.org/10.1021/acschemneuro.7b00449

Hoffmann S, Gerhardt S, Mühle C, Reinhard I, Reichert D, Bach P, Boroumand-Jazi R, Kuehner C, Aguilera A, Aslan A, Bahr NS, Belanger M, Deeken F, Ebrahimi C, Fischbach PC, Ganz M, Garbusow M, Großkopf CM, Heigert M, Hentschel A, Karl D, Liu S, Mazza M, Pelz P, Pinger M, Reichl M, Riemerschmid C, Rosenthal A, Steffen J, Strehle J, Wedemeyer F, Weiss F, Wenzel J, Wieder G, Wieland A, Zaiser J, Zech HG, Zimmermann S, Kornhuber J, Müller CP, Sommer WH, Spanagel R, Banaschewski T, Deserno L, Ebner-Priemer UW, Flor H, Kirsch P, Rietschel M, Vollstädt-Klein S, Walter H, Meyer-Lindenberg A, Rapp MA, Witt S, Smolka MN, Heinz A, Tost H, Kiefer F, Reichert M, Lenz B (2024) Associations of menstrual cycle and progesterone-to-estradiol ratio with alcohol consumption in alcohol use disorder: a sex-separated multicenter longitudinal study. AJP. https://doi.org/10.1176/appi.ajp.20230027

Holstein SE, Spanos M, Hodge CW (2011) Adolescent C57BL/6J mice show elevated alcohol intake, but reduced taste aversion, as compared to adult mice: a potential behavioral mechanism for binge drinking. Alcohol Clin Exp Res 35:1842–1851. https://doi.org/10.1111/j.1530-0277.2011.01528.x

Huttenlocher PR, Dabholkar AS (1997) Regional differences in synaptogenesis in human cerebral cortex. J Comp Neurol 387:167–178. https://doi.org/10.1002/(SICI)1096-9861(19971020)387:2<167::AID-CNE1>3.0.CO;2-Z

Ikeda M, Toyoda H, Yamada J, Okabe A, Sato K, Hotta Y, Fukuda A (2003) Differential development of cation-chloride cotransporters and Cl− homeostasis contributes to differential GABAergic actions between developing rat visual cortex and dorsal lateral geniculate nucleus. Brain Res 984:149–159. https://doi.org/10.1016/S0006-8993(03)03126-3

Joffe ME, Winder DG, Conn PJ (2020) Contrasting sex-dependent adaptations to synaptic physiology and membrane properties of prefrontal cortex interneuron subtypes in a mouse model of binge drinking. Neuropharmacology 178:108126. https://doi.org/10.1016/j.neuropharm.2020.108126

Joffe ME, Maksymetz J, Luschinger JR, Dogra S, Ferranti AS, Luessen DJ, Gallinger IM, Xiang Z, Branthwaite H, Melugin PR, Williford KM, Centanni SW, Shields BC, Lindsley CW, Calipari ES, Siciliano CA, Niswender CM, Tadross MR, Winder DG, Conn PJ (2022) Acute restraint stress redirects prefrontal cortex circuit function through mGlu5 receptor plasticity on somatostatin-expressing interneurons. Neuron. https://doi.org/10.1016/j.neuron.2021.12.027

Jones SA, Nagel BJ (2019) Altered frontostriatal white matter microstructure is associated with familial alcoholism and future binge drinking in adolescence. Neuropsychopharmacology 44:1076–1083. https://doi.org/10.1038/s41386-019-0315-x

Kalsbeek A, Voorn P, Buijs RM, Pool CW, Uylings HBM (1988) Development of the dopaminergic innervation in the prefrontal cortex of the rat. J Comp Neurol 269:58–72. https://doi.org/10.1002/cne.902690105

Kebabian JW, Greengard P (1971) Dopamine-sensitive adenyl cyclase: possible role in synaptic transmission. Science 174:1346–1349. https://doi.org/10.1126/science.174.4016.1346

Kilb W (2012) Development of the GABAergic system from birth to adolescence. Neuroscientist 18:613–630. https://doi.org/10.1177/1073858411422114

Kinnischtzke A, Sewall A, Berkepile J, Fanselow E (2012) Postnatal maturation of somatostatin-expressing inhibitory cells in the somatosensory cortex of GIN mice. Front Neural Circuits 6:33. https://doi.org/10.3389/fncir.2012.00033

Klenowski PM (2018) Emerging role for the medial prefrontal cortex in alcohol-seeking behaviors. Addict Behav 77:102–106. https://doi.org/10.1016/j.addbeh.2017.09.024

Klenowski PM, Fogarty MJ, Shariff M, Belmer A, Bellingham MC, Bartlett SE (2016) Increased synaptic excitation and abnormal dendritic structure of prefrontal cortex layer V pyramidal neurons following prolonged binge-like consumption of ethanol. eNeuro 3. https://doi.org/10.1523/ENEURO.0248-16.2016

Kolk SM, Rakic P (2022) Development of prefrontal cortex. Neuropsychopharmacology 47:41–57. https://doi.org/10.1038/s41386-021-01137-9

Koppensteiner P, Von Itter R, Melani R, Galvin C, Lee FS, Ninan I (2019) Diminished fear extinction in adolescents is associated with an altered somatostatin interneuron-mediated inhibition in the infralimbic cortex. Biol Psychiatry 86:682–692. https://doi.org/10.1016/j.biopsych.2019.04.035

Koss WA, Lloyd MM, Sadowski RN, Wise LM, Juraska JM (2015) Gonadectomy before puberty increases the number of neurons and glia in the medial prefrontal cortex of female, but not male, rats. Dev Psychobiol 57:305–312. https://doi.org/10.1002/dev.21290

Kroon T, van Hugte E, van Linge L, Mansvelder HD, Meredith RM (2019) Early postnatal development of pyramidal neurons across layers of the mouse medial prefrontal cortex. Sci Rep 9:5037. https://doi.org/10.1038/s41598-019-41661-9

Landin JD, Chandler LJ (2023) Adolescent alcohol exposure alters threat avoidance in adulthood. Front Behav Neurosci 16. https://doi.org/10.3389/fnbeh.2022.1098343

Laubach M, Amarante LM, Swanson K, White SR (2018) What, if anything, is rodent prefrontal cortex? eNeuro 5. https://doi.org/10.1523/ENEURO.0315-18.2018

Laurie D, Wisden W, Seeburg P (1992) The distribution of thirteen GABAA receptor subunit mRNAs in the rat brain. III. Embryonic and postnatal development. J Neurosci 12:4151–4172. https://doi.org/10.1523/JNEUROSCI.12-11-04151.1992

Lee KM, Coehlo MA, Solton NR, Szumlinski KK (2017) Negative affect and excessive alcohol intake incubate during protracted withdrawal from binge-drinking in adolescent, but not adult, mice. Front Psychol 8

Linden-Carmichael AN, Allen HK, Masters LD, Ansell EB, Lanza ST (2021) Age-varying trends in alcohol and cannabis co-occurring use: Implications for prescription drug misuse. Am J Drug Alcohol Abuse 47:209–219. https://doi.org/10.1080/00952990.2020.1831003

Lodha J, Brocato E, Wolstenholme JT (2022) Areas of convergence and divergence in adolescent social isolation and binge drinking: a review. Front Behav Neurosci 16

Logue S, Chein J, Gould T, Holliday E, Steinberg L (2014) Adolescent mice, unlike adults, consume more alcohol in the presence of peers than alone. Dev Sci 17:79–85. https://doi.org/10.1111/desc.12101

Lopez MF, Doremus-Fitzwater TL, Becker HC (2011) Chronic social isolation and chronic variable stress during early development induce later elevated ethanol intake in adult C57BL/6J mice. Alcohol 45:355–364. https://doi.org/10.1016/j.alcohol.2010.08.017

Luján R, Shigemoto R, López-Bendito G (2005) Glutamate and GABA receptor signalling in the developing brain. Neuroscience 130:567–580. https://doi.org/10.1016/j.neuroscience.2004.09.042

Martel MM, Eisenlohr-Moul T, Roberts B (2017) Interactive effects of ovarian steroid hormones on alcohol use and binge drinking across the menstrual cycle. J Abnorm Psychol 126:1104–1113. https://doi.org/10.1037/abn0000304

Medina KL, McQueeny T, Nagel BJ, Hanson KL, Schweinsburg AD, Tapert SF (2008) Prefrontal cortex volumes in adolescents with alcohol use disorders: unique gender effects. Alcohol Clin Exp Res 32:386–394. https://doi.org/10.1111/j.1530-0277.2007.00602.x

Mello NK, Bree MP, Mendelson JH (1986) Alcohol and food self-administration by female Macaque monkeys as a function of menstrual cycle phase. Physiol Behav 36:959–966. https://doi.org/10.1016/0031-9384(86)90460-9

Miech RA, Johnston LD, Patrick ME, O'Malley PM (2024) National survey results on drug use, 1975-2023: Overview and detailed results for secondary school students. Monitoring the future monograph series. Institute for Social Research, University of Michigan, Ann Arbor, MI. Available at https://monitoringthefuture.org/results/annual-reports/

Miyamae T, Chen K, Lewis DA, Gonzalez-Burgos G (2017) Distinct physiological maturation of parvalbumin-positive neuron subtypes in mouse prefrontal cortex. J Neurosci 37:4883–4902. https://doi.org/10.1523/JNEUROSCI.3325-16.2017

Miyoshi G, Butt SJB, Takebayashi H, Fishell G (2007) Physiologically distinct temporal cohorts of cortical interneurons arise from telencephalic Olig2-expressing precursors. J Neurosci 27:7786–7798. https://doi.org/10.1523/JNEUROSCI.1807-07.2007

Naneix F, Marchand AR, Scala GD, Pape J-R, Coutureau E (2012) Parallel maturation of goal-directed behavior and dopaminergic systems during adolescence. J Neurosci 32:16223–16232. https://doi.org/10.1523/JNEUROSCI.3080-12.2012

Nentwig TB, Starr EM, Chandler LJ, Glover EJ (2019) Absence of compulsive drinking phenotype in adult male rats exposed to ethanol in a binge-like pattern during adolescence. Alcohol 79:93–103. https://doi.org/10.1016/j.alcohol.2019.01.006

Obray JD, Landin JD, Vaughan DT, Scofield MD, Chandler LJ (2022) Adolescent alcohol exposure reduces dopamine 1 receptor modulation of prelimbic neurons projecting to the nucleus accumbens and basolateral amygdala. Addict Neurosci 4:100044. https://doi.org/10.1016/j.addicn.2022.100044

Page CE, Coutellier L (2018) Adolescent stress disrupts the maturation of anxiety-related behaviors and alters the developmental trajectory of the prefrontal cortex in a sex- and age-specific manner. Neuroscience 390:265–277. https://doi.org/10.1016/j.neuroscience.2018.08.030

Pan G, Yang J-M, Hu X-Y, Li X-M (2016) Postnatal development of the electrophysiological properties of somatostatin interneurons in the anterior cingulate cortex of mice. Sci Rep 6:28137. https://doi.org/10.1038/srep28137

Pan NC, Fang A, Shen C, Sun L, Wu Q, Wang X (2019) Early excitatory activity-dependent maturation of somatostatin interneurons in cortical layer 2/3 of mice. Cereb Cortex 29:4107–4118. https://doi.org/10.1093/cercor/bhy293

Peerboom C, Wierenga CJ (2021) The postnatal GABA shift: A developmental perspective. Neurosci Biobehav Rev 124:179–192. https://doi.org/10.1016/j.neubiorev.2021.01.024

Petanjek Z, Judaš M, Šimić G, Rašin MR, Uylings HBM, Rakic P, Kostović I (2011) Extraordinary neoteny of synaptic spines in the human prefrontal cortex. Proc Natl Acad Sci USA 108:13281–13286. https://doi.org/10.1073/pnas.1105108108

Pisu MG, Mostallino MC, Dore R, Maciocco E, Secci PP, Serra M (2011) Effects of voluntary ethanol consumption on emotional state and stress responsiveness in socially isolated rats. Eur Neuropsychopharmacol 21:414–425. https://doi.org/10.1016/j.euroneuro.2010.07.006

Radke AK, Sneddon EA, Frasier RM, Hopf FW (2021) Recent perspectives on sex differences in compulsion-like and binge alcohol drinking. Int J Mol Sci 22:3788. https://doi.org/10.3390/ijms22073788

Rice J, Coutellier L, Weiner JL, Gu C (2019) Region-specific interneuron demyelination and heightened anxiety-like behavior induced by adolescent binge alcohol treatment. Acta Neuropathol Commun. https://actaneurocomms.biomedcentral.com/articles/10.1186/s40478-019-0829-9. Accessed 20 Mar 2024

Rivera-Irizarry JK, Skelly MJ, Pleil KE (2020) Social isolation stress in adolescence, but not adulthood, produces hypersocial behavior in adult male and female C57BL/6J mice. Front Behav Neurosci 14. https://doi.org/10.3389/fnbeh.2020.00129

Robinson DL, Amodeo LR, Chandler LJ, Crews FT, Ehlers CL, Gómez-A A, Healey KL, Kuhn CM, Macht VA, Marshall SA, Swartzwelder HS, Varlinskaya EI, Werner DF (2021) The role of sex in the persistent effects of adolescent alcohol exposure on behavior and neurobiology in rodents. Int Rev Neurobiol 160:305–340. https://doi.org/10.1016/bs.irn.2021.07.007

Salling MC, Skelly MJ, Avegno E, Regan S, Zeric T, Nichols E, Harrison NL (2018) Alcohol consumption during adolescence in a mouse model of binge drinking alters the intrinsic excitability and function of the prefrontal cortex through a reduction in the hyperpolarization-activated cation current. J Neurosci 38:6207–6222. https://doi.org/10.1523/JNEUROSCI.0550-18.2018

Satta R, Hilderbrand ER, Lasek AW (2018) Ovarian hormones contribute to high levels of binge-like drinking by female mice. Alcohol Clin Exp Res 42:286–294. https://doi.org/10.1111/acer.13571

Sawyer SM, Azzopardi PS, Wickremarathne D, Patton GC (2018) The age of adolescence. Lancet Child

Adolesc Health 2:223–228. https://doi.org/10.1016/S2352-4642(18)30022-1
Scheggia D, Managò F, Maltese F, Bruni S, Nigro M, Dautan D, Latuske P, Contarini G, Gomez-Gonzalo M, Requie LM, Ferretti V, Castellani G, Mauro D, Bonavia A, Carmignoto G, Yizhar O, Papaleo F (2020) Somatostatin interneurons in the prefrontal cortex control affective state discrimination in mice. Nat Neurosci 23:47–60. https://doi.org/10.1038/s41593-019-0551-8
Schneider M (2013) Adolescence as a vulnerable period to alter rodent behavior. Cell Tissue Res 354:99–106. https://doi.org/10.1007/s00441-013-1581-2
Shaw GA, Dupree JL, Neigh GN (2020) Adolescent maturation of the prefrontal cortex: Role of stress and sex in shaping adult risk for compromise. Genes Brain Behav 19:e12626. https://doi.org/10.1111/gbb.12626
Sicher AR, Duerr A, Starnes WD, Crowley NA (2022) Adolescent alcohol and stress exposure rewires key cortical neurocircuitry. Front Neurosci 16
Sicher AR, Starnes WD, Griffith KR, Dao NC, Smith GC, Brockway DF, Crowley NA (2023) Adolescent binge drinking leads to long-lasting changes in cortical microcircuits in mice. Neuropharmacology 234:109561. https://doi.org/10.1016/j.neuropharm.2023.109561
Sicher AR, Liss A, Vozella V, Marsland P, Seemiller LR, Springer M, Starnes WD, Griffith KR, Smith GC, Astefanous A, Deak T, Roberto M, Varodayan FP, Crowley NA (2024) Voluntary adolescent alcohol exposure does not robustly increase adulthood consumption of alcohol in multiple mouse and rat models. Addict Neurosci. https://doi.org/10.1101/2024.04.30.591674
Skelly MJ, Chappl AE, Carter E, Weiner JL (2015) Adolescent social isolation increases anxiety-like behavior and ethanol intake and impairs fear extinction in adulthood: possible role of disrupted noradrenergic signaling. Neuropharmacology 97:149–159. https://doi.org/10.1016/j.neuropharm.2015.05.025
Slade T, Chapman C, Swift W, Keyes K, Tonks Z, Teesson M (2016) Birth cohort trends in the global epidemiology of alcohol use and alcohol-related harms in men and women: systematic review and metaregression. BMJ Open 6:e011827. https://doi.org/10.1136/bmjopen-2016-011827
Sneddon EA, Rasizer LN, Cavalco NG, Jaymes AH, Ostlie NJ, Minshall BL, Masters BM, Hughes MR, Hrncir H, Arnold AP, Radke AK (2022) Gonadal hormones and sex chromosome complement differentially contribute to ethanol intake preference and relapse-like behaviour in four core genotypes mice. Abstract Addiction Biology 27(5) 10.1111/adb.v27.5. https://doi.org/10.1111/adb.13222
Spear LP (2000) The adolescent brain and age-related behavioral manifestations. Neurosci Biobehav Rev 24:417–463. https://doi.org/10.1016/s0149-7634(00)00014-2
Spear LP (2014) Adolescents and alcohol: Acute sensitivities, enhanced intake, and later consequences. Neurotoxicol Teratol 41:51–59. https://doi.org/10.1016/j.ntt.2013.11.006
Spear LP (2015) Adolescent alcohol exposure: Are there separable vulnerable periods within adolescence? Physiol Behav 148:122–130. https://doi.org/10.1016/j.physbeh.2015.01.027
Spear LP, Swartzwelder HS (2014) Adolescent alcohol exposure and persistence of adolescent-typical phenotypes into adulthood: a mini-review. Neurosci Biobehav Rev 0:1–8. https://doi.org/10.1016/j.neubiorev.2014.04.012
Steinberg L (2008) A social neuroscience perspective on adolescent risk-taking. Dev Rev 28:78–106. https://doi.org/10.1016/j.dr.2007.08.002
Strong MN, Yoneyama N, Fretwell AM, Snelling C, Tanchuck MA, Finn DA (2010) "Binge" drinking experience in adolescent mice shows sex differences and elevated ethanol intake in adulthood. Horm and Behav 58:82–90. https://doi.org/10.1016/j.yhbeh.2009.10.008
Substance Abuse and Mental Health Services Administration (SAMHSA) (2022) Report to congress on the prevention and reduction of underage drinking. SAMHSA, U.S. Department of Health and Human Services, Washington, DC. https://www.stopalcoholabuse.gov/about-iccpud/data/national-reports/report-to-congress/default.aspx. Accessed 26 Mar 2024
Thomas SG, Czoty PW (2019) Effect of menstrual cycle on ethanol drinking in rhesus monkeys. Drug Alcohol Depend 194:205–209. https://doi.org/10.1016/j.drugalcdep.2018.11.001
Tomova L, Andrews JL, Blakemore S-J (2021) The importance of belonging and the avoidance of social risk taking in adolescence. Dev Rev 61:100981. https://doi.org/10.1016/j.dr.2021.100981
Tonn Eisinger KR, Larson EB, Boulware MI, Thomas MJ, Mermelstein PG (2018) Membrane estrogen receptor signaling impacts the reward circuitry of the female brain to influence motivated behaviors. Steroids 133:53–59. https://doi.org/10.1016/j.steroids.2017.11.013
Towner TT, Varlinskaya EI (2020) Adolescent ethanol exposure: anxiety-like behavioral alterations, ethanol intake, and sensitivity. Front Behav Neurosci 14
Towner TT, Applegate DT, Varlinskaya EI, Werner DF (2022) Impact of adolescent intermittent ethanol exposure on social investigation, social preference, and neuronal activation in cFos-LacZ male and female rats. Front Pharmacol 13:841657. https://doi.org/10.3389/fphar.2022.841657
Trantham-Davidson H, Centanni SW, Garr SC, New NN, Mulholland PJ, Gass JT, Glover EJ, Floresco SB, Crews FT, Krishnan HR, Pandey SC, Chandler LJ (2017) Binge-like alcohol exposure during adolescence disrupts dopaminergic neurotransmission in the adult prelimbic cortex. Neuropsychopharmacology 42:1024–1036. https://doi.org/10.1038/npp.2016.190
Tseng K-Y, O'Donnell P (2007) Dopamine modulation of prefrontal cortical interneurons changes during adolescence. Cereb Cortex 17:1235–1240. https://doi.org/10.1093/cercor/bhl034

Tuncdemir SN, Wamsley B, Stam FJ, Osakada F, Goulding M, Callaway EM, Rudy B, Fishell G (2016) Early somatostatin interneuron connectivity mediates the maturation of deep layer cortical circuits. Neuron 89(3):521–535. https://doi.org/10.1016/j.neuron.2015.11.020

Ueda S, Niwa M, Hioki H, Sohn J, Kaneko T, Sawa A, Sakurai T (2015) Sequence of molecular events during the maturation of the developing mouse prefrontal cortex. CXP 1:94–104. https://doi.org/10.1159/000430095

Younis RM, Wolstenholme JT, Bagdas D, Bettinger JC, Miles MF, Damas MI (2019) Adolescent but not adult ethanol binge drinking modulates ethanol behavioral effects in mice later in life. Pharmacol Biochem Behav 184:172740. https://doi.org/10.1016/j.pbb.2019.172740

Yuen WS, Chan G, Bruno R, Clare P, Mattick R, Aiken A, Boland V, McBride N, McCambridge J, Slade T, Kypri K, Horwood J, Hutchinson D, Najman J, De Torres C, Peacock A (2020) Adolescent alcohol use trajectories: risk factors and adult outcomes. Pediatrics 146(4):e20200440. https://doi.org/10.1542/peds.2020-0440

Vaca FE, Li K, Luk JW, Hingson RW, Haynie DL, Simons-Morton BG (2020) Longitudinal associations of 12th-grade binge drinking with risky driving and high-risk drinking. Pediatrics 145:e20184095. https://doi.org/10.1542/peds.2018-4095

Vandegrift BJ, Hilderbrand ER, Satta R, Tai R, He D, You C, Chen H, Xu P, Coles C, Brodie MS, Lasek AW (2020) Estrogen receptor α regulates ethanol excitation of ventral tegmental area neurons and binge drinking in female mice. J Neurosci 40:5196–5207. https://doi.org/10.1523/JNEUROSCI.2364-19.2020

Van Hees L, Didone V, Charlet-Briart M, Van Ingelgom T, Alexandre A, Quertemont E, Nguyen L, Laguesse S (2022) Voluntary alcohol binge-drinking in adolescent C57Bl6 mice induces delayed appearance of behavioural defects in both males and females. Addict Biol 27:e13102. https://doi.org/10.1111/adb.13102

Varlinskaya EI, Kim EU, Spear LP (2017) Chronic intermittent ethanol exposure during adolescence: Effects on stress-induced social alterations and social drinking in adulthood. Brain Res 1654:145–156. https://doi.org/10.1016/j.brainres.2016.03.050

Varlinskaya EI, Hosová D, Towner T, Werner DF, Spear LP (2020) Effects of chronic intermittent ethanol exposure during early and late adolescence on anxiety-like behaviors and behavioral flexibility in adulthood. Behav Brain Res 378:112292. https://doi.org/10.1016/j.bbr.2019.112292

Walker MC, Semyanov A (2008) Regulation of excitability by extrasynaptic GABAA receptors. In: Darlison MG (ed) Inhibitory regulation of excitatory neurotransmission. Springer, Berlin, pp 29–48. https://doi.org/10.1007/400_2007_030

Warren JG, Goodwin L, Gage SH, Rose AK (2020) The effects of menstrual cycle stage and hormonal contraception on alcohol consumption and craving: A pilot investigation. Compr Psychoneuroendocrinol 5:100022. https://doi.org/10.1016/j.cpnec.2020.100022

Watson RJ, Veale JF, Gordon AR, Clark BA, Saewyc EM (2019) Risk and protective factors for transgender youths' substance use. Prev Med Rep 15:100905. https://doi.org/10.1016/j.pmedr.2019.100905

Wells JE, Horwood LJ, Fergusson DM (2004) Drinking patterns in mid-adolescence and psychosocial outcomes in late adolescence and early adulthood. Addiction 99(12):1529–1541. https://doi.org/10.1111/j.1360-0443.2004.00918.x

White A, Castle I-JP, Chen CM, Shirley M, Roach D, Hingson R (2015) Converging patterns of alcohol use and related outcomes among females and males in the United States, 2002 to 2012. Alcohol Clin Exp Res 39:1712–1726. https://doi.org/10.1111/acer.12815

Willing J, Juraska JM (2015) The timing of neuronal loss across adolescence in the medial prefrontal cortex of male and female rats. Neuroscience 301:268–275. https://doi.org/10.1016/j.neuroscience.2015.05.073

Willing J, Cortes LR, Brodsky JM, Kim T, Juraska JM (2017) Innervation of the medial prefrontal cortex by tyrosine hydroxylase immunoreactive fibers during adolescence in male and female rats. Dev Psychobiol 59:583–589. https://doi.org/10.1002/dev.21525

Wooden JI, Peacoe LE, Anasooya Shaji C, Melbourne JK, Chandler CM, Bardo MT, Nixon K (2023) Adolescent intermittent ethanol drives modest neuroinflammation but does not escalate drinking in male rats. Cells 12:2572. https://doi.org/10.3390/cells12212572

Xu X, Roby KD, Callaway EM (2010) Immunochemical characterization of inhibitory mouse cortical neurons: Three chemically distinct classes of inhibitory cells. J Comp Neurol 518:389–404. https://doi.org/10.1002/cne.22229

Xu H, Liu L, Tian Y, Wang J, Li J, Zheng J, Zhao H, He M, Xu T-L, Duan S, Xu H (2019) A disinhibitory microcircuit mediates conditioned social fear in the prefrontal cortex. Neuron 102:668–682.e5. https://doi.org/10.1016/j.neuron.2019.02.026

Yang J-M, Zhang J, Yu Y-Q, Duan S, Li X-M (2014) Postnatal development of 2 microcircuits involving fast-spiking interneurons in the mouse prefrontal cortex. Cereb Cortex 24:98–109. https://doi.org/10.1093/cercor/bhs291

9 Impact of Neuroimmune System Activation by Adolescent Binge Alcohol Exposure on Adult Neurobiology

Victoria Macht, Sagan de Castro, and Ryan P. Vetreno

Abstract

Adolescence is a conserved neurodevelopmental period encompassing maturation of glia and the innate immune system that parallels refinement of brain structures, neurotransmitter systems, and neurocircuitry. Given the vast neurodevelopmental processes occurring during adolescence, spanning brain structural and neurocircuitry refinement to maturation of neurotransmitter systems, glia, and the innate immune system, insults incurred during this critical period of neurodevelopment, could have profound effects on brain function and behavior that persist into adulthood. Adolescent binge drinking is common and associated with many adverse outcomes that may underlie the lifelong increased risk of alcohol-related problems and development of an alcohol use disorder (AUD). In this chapter, we examined the impact of adolescent binge drinking models using the adolescent intermittent ethanol (AIE) model on adult neurobiology. These studies implicate proinflammatory neuroimmune signaling across glia and neurons in persistent AIE-induced neuropathology. Some of these changes are reversible, providing unique opportunities for the development of treatments to prevent many of the long-term consequences of adolescent alcohol misuse.

Keywords

Acetylcholine · Adolescent intermittent ethanol · Development · Epigenetic · HMGB1 · Innate immune system · Neurogenesis · TLRs

9.1 Introduction

9.1.1 Adolescence Development and Human Adolescent Binge Drinking

Adolescence is a conserved neurodevelopmental period encompassing cognitive, emotive, and social maturation that marks the transition from the prepubescent juvenile period to independence and adulthood. Across adolescence, frontal cortical and subcortical limbic regions undergo structural changes coincident with maturation of adult behaviors and executive function (Ernst et al.

V. Macht · S. de Castro
Bowles Center for Alcohol Studies, School of Medicine, University of North Carolina at Chapel Hill, Chapel Hill, NC, USA

R. P. Vetreno (✉)
Bowles Center for Alcohol Studies, School of Medicine, University of North Carolina at Chapel Hill, Chapel Hill, NC, USA

Department of Psychiatry, School of Medicine, University of North Carolina at Chapel Hill, Chapel Hill, NC, USA
e-mail: rvetreno@email.unc.edu

C. F. Valenzuela et al. (eds.), *Effects of Alcohol on the Brain across the Lifespan*, Advances in Experimental Medicine and Biology 1473, https://doi.org/10.1007/978-3-031-81908-7_9

2009; Tervo-Clemmens et al. 2023). Adolescent typical behaviors, including risk-taking, novelty- and sensation-seeking, environmental exploration, and social interaction, parallel brain maturation (Lenroot and Giedd 2006; Bava and Tapert 2010) and contribute to the transition from parental dependence to adult independence. While these adolescent typical behaviors are crucial for successful transition to independence, they also confer increased likelihood of experimentation with and initiation of drug use, including alcohol that could negatively affect neurodevelopment due to the heightened plasticity that characterizes the adolescent brain.

Unlike daily, heavy drinking patterns observed in adults with alcohol use disorder (**AUD**), adolescents generally consume alcohol intermittently (i.e., not every day) in social groups on weekends. Indeed, not only does alcohol use typically initiate during adolescence, but estimates suggest that up to 90% of all alcohol consumed by adolescents is consumed in binges (NSDUH 2020). Binge drinking produces high blood alcohol levels (i.e., ≥80 mg%) and roughly equates to the consumption of five or more alcohol beverages for males and four or more for females in a two-hour period. This may be due to an innate low responsivity to the sedative effects of alcohol, allowing adolescents to consume larger quantities of alcohol during a single drinking occasion (Vetreno et al. 2023). Further, adolescent binge drinking is quite prevalent, with approximately 20% of individuals ages 16–17 and 34% of individuals ages 18–25 engaging in binge drinking in the United States in the past month (NSDUH 2020), and approximately 10% of high school seniors endorsed high-intensity binge drinking, reporting 10+ drinks and 5.6% reporting 15+ drinks in a row (Patrick et al. 2013).

Adolescent binge drinking is associated with many adverse outcomes, including lasting neurological difficulties, school and legal problems, sexual violence, and increased risk for suicide and unintended injury (Dewit et al. 2000; Grant and Dawson 1997; Hingson et al. 2009; Sher and Gotham 1999; Grant et al. 2006). Further, evidence suggests that the developing adolescent cortex is more sensitive to alcohol-induced damage than the adult brain (Crews et al. 2000, 2007). There are also long-term consequences of adolescent drinking, including increased lifelong risk of alcohol-related problems and development of an AUD. Epidemiological evidence indicates that an early adolescent age of drinking onset is one of the strongest predictors for the development of AUD in adulthood (Grant et al. 2001; Grant and Dawson 1998). In fact, several large-scale human studies reported that individuals who initiate drinking at age 10 (Prescott and Kendler 1999) or 14 (Grant and Dawson 1997) have a 40–50% lifetime prevalence of AUD, which declines to an approximate 10% lifetime prevalence of AUD if drinking initiation started at or after age 20. An early adolescent age of drinking onset is also associated with increased risk for lifetime drug dependence, violence, and fights and injuries associated with alcohol use (Dewit et al. 2000; Grant and Dawson 1997; Hingson et al. 2009; Sher and Gotham 1999; Grant et al. 2006). These human studies indicate that adolescent initiation of alcohol drinking contributes to alcohol-related problems and increased risk for AUD development in adulthood. However, adult drinking confounds a clear delineation of the unique effects of an adolescent age of drinking onset on adult neurobiology in humans necessitating the use of animal models to delineate the discrete consequences of alcohol exposure during adolescence on adult neurobiology.

9.1.2 Adolescent Intermittent Ethanol Exposure Models Human Adolescent Binge Drinking

Adolescent development across mammalian species is highly conserved, allowing the use of animal models to investigate the impact of discreet adolescent binge drinking on neurodevelopment that persists into adulthood. The need to develop preclinical rodent models of underage binge drinking led to the establishment of the Neurobiology of Adolescent Drinking in Adulthood (**NADIA**) consortium and development of adolescent intermittent ethanol (**AIE**) rodent models of human adolescent binge drinking (Crews et al. 2019; Pascual et al. 2007; Coleman et al. 2011). Preclinical AIE models in rats and mice generally involve administering ethanol across adolescence in an intermittent (i.e., not consecutive days) binge-like fashion to mimic heavy weekend binge drinking endorsed by human adolescents, but not daily drinking associated with AUD. The age and intermittency of exposure in these models varies to some degree with some groups administering ethanol on a 2-day on/2-day off schedule throughout adolescence from postnatal day (**P**)25 to P54 (Macht et al. 2022a; Kipp et al. 2021). In contrast, other AIE models involve administration with different ethanol dosing intervals and shorter periods of exposure across adolescence (see, e.g., Pascual et al. 2007; Wooden et al. 2023; Swartzwelder et al. 2019). In addition, the route of ethanol administration varies across models with various groups employing intragastric (i.g.) administration, intraperitoneal (i.p.) administration, and vapor inhalation (Vetreno et al. 2020; Kyzar et al. 2017, 2019; Obray et al. 2022; Fernandez and Savage 2017). Importantly, regardless of the differences in route of administration or timing of exposure during adolescence, AIE treatment models produce blood ethanol concentrations >100 mg%, which models blood alcohol levels attained by human adolescents that engage in binge drinking (Lamminpaa et al. 1993; White et al. 2006; Kraus et al. 2005; Donovan 2009; Patrick et al. 2013; Orio et al. 2018). Assessments are then conducted in young adulthood through adulthood to determine the lasting consequences of adolescent alcohol exposure in adults. Further, adult characteristics are generally stable, allowing longitudinal studies and/or extended behavioral training without the confounds of adolescent age-related changes or additional adult ethanol exposure.

Using the preclinical AIE model of human adolescent binge drinking, multiple groups report that it promotes voluntary adult alcohol drinking (Alaux-Cantin et al. 2013; Broadwater et al. 2013; Lee et al. 2017; Pandey et al. 2015; Pascual et al. 2009; Rodd-Henricks et al. 2002; Toalston et al. 2015; Wille-Bille et al. 2017; Gass et al. 2014), with females drinking more alcohol than males (Amodeo et al. 2018). In addition, AIE also increases adult operant alcohol self-administration while reducing extinction (Amodeo et al. 2017). Interestingly, individual rats that consume the highest levels of alcohol during adolescence also show the largest increases in alcohol drinking in adulthood (Amodeo et al. 2017; Gass et al. 2014). The AIE-induced increases in adult alcohol drinking are accompanied by decreases in adult sensitivity to alcohol-induced motor impairments (White et al. 2002), reduced adult sensitivity to acute ethanol-induced alterations in resting state functional magnetic resonance imaging functional connectivity (Broadwater et al. 2018), and enhancement of the social facilitatory properties of alcohol in adult male rats (Varlinskaya et al. 2014). In addition to altering adult alcohol drinking and sensitivity, AIE increases impulsivity (Boutros et al. 2014; Miller et al. 2017), behavioral disinhibition (Ehlers et al. 2011; Gass et al. 2014), and anxiety-like behaviors in adulthood (Kyzar et al. 2017, Pandey et al. 2015, Sakharkar et al. 2016, Kokare et al. 2017, Lee et al. 2017, Vetreno et al. 2016, Coleman et al. 2014; Vetreno and Crews 2015). Assessment of cognitive functioning in adult AIE-treated rodents reveals persistent deficits associated with impairments in learning and behavioral flexibility (e.g., Morris water maze, Barnes maze) (Vetreno et al. 2019; Vetreno and Crews 2012; Coleman et al. 2011, 2014; Acheson et al. 2013; Swartzwelder et al. 2015; Vetreno and Crews 2015; Marco et al. 2017; Macht et al. 2020b, 2023). Specifically, adult AIE-treated male animals evidence reversal learning deficits on the Morris water maze while spatial learning is unimpaired whereas in adult AIE-treated female animals, both spatial and reversal learning are impaired (Macht et al. 2020b, 2023) suggesting sex differences in the presentation of behavioral deficits on measure of cognitive functioning. Collectively, these findings indicate that AIE treatment causes long-lasting adult increases in alcohol drinking, risky decision-making, reward seeking, and anxiety-like behavior as well as reduced executive function that all associated with increased risk for development of AUD. Although a detailed assessment of these findings is beyond the scope of this chapter, we encourage the reader to reference excellent reviews published by components of the NADIA consortium (see, e.g., Crews et al. 2019; Spear 2018; Robinson et al. 2021).

While persistent AIE-induced alterations in adult behavior are accompanied by many lasting brain structural and molecular changes, the innate immune system is emerging as a critical mediator of AIE-induced neurobiological changes across various brain regions that persist into adulthood. Therefore, this chapter will first introduce data describing adolescent maturation of brain glial cells and the innate immune system. We will then present preclinical findings on the effects of adolescent binge alcohol exposure on the innate immune system within the basal forebrain and hippocampus and neuroinflammatory contributions to neuropathology and behavioral dysfunction within these regions in adulthood. Together, the data presented in this chapter support a mechanistic role for persistent innate immune signaling in adolescent binge drinking induction of long-lasting adult pathology, contributing to increased risk for development of adult psychopathologies, including AUD.

9.2 The Innate Immune System in the Developing Brain

Our understanding of neuroinflammation in the CNS has grown rapidly over the last few decades, shifting from the perception of the brain as an immune privileged organ to inflammation contributing to most, if not all, neurological and psychological diseases of the brain (Zhang et al. 2023; Najjar et al. 2013; Cui et al. 2014). There are two main components of the immune system, namely the adaptive or acquired immune system, which utilizes immunological memory to mount a highly specific response to a pathogen it has previously encountered, and the innate immune system. While the role of adaptive immunity in cognitive processes is continuing to emerge across the field (Brynskikh et al. 2008), the vast majority of research on neuroimmunology with respect to AIE is on the innate immune system.

The innate immune system is evolutionarily conserved and mounts a generalized, non-specific immune response (as opposed to the more specific adaptive immune response), reacting broadly to cellular damage as well as foreign pathogens. This response includes identification of endogenous and exogenous pathogen-associated molecular patterns (**PAMPs**) and danger-associated molecular patterns (**DAMPs**), initiation of a general and appropriately scaled proinflammatory response, and ultimate resolution of that response in a temporally discrete fashion. The appropriate initiation and resolution of the innate immune response is a critical feature of a healthy nervous system, and dysfunction in the magnitude, specificity, and/or resolution of this response is frequently observed in a host of disease states, including AUD. Indeed, while transient, controlled neuroinflammation is considered beneficial to the host organism to fight off infection, chronic uncontrolled neuroinflammation is associated with increased and sustained production of proinflammatory cytokines, chemokines, reactive oxygen species, and other inflammatory mediators that can exert deleterious effects on the brain (Disabato et al. 2016). These effects become more complicated when viewed across development as the cells of the immune system are also undergoing critical maturational changes.

9.2.1 Maturation of Glia and the Innate Immune System Across Adolescence

Adolescence is characterized by functional maturation of glia and the innate immune system that parallels refinement of brain structures, neurotransmitter systems, and neurocircuitry that in humans continues into the mid-20s, predominantly occurring in regions associated with executive and emotive function (i.e., cortical and subcortical limbic regions). Glia are central players in many neurodevelopmental processes that occur during adolescence, such as myelination by oligodendrocytes, synaptogenesis by astrocytes, and microglial-mediated synaptic pruning (Paolicelli et al. 2011; Paolicelli and Gross 2011; Tremblay et al. 2011; Allen et al. 2012; Stevens et al. 2007). This suggests that not only do glia exhibit critical roles for typical development, but disruption of glial phenotypic maturation and modulation of glial-driven changes in neurocircuitry may underscore and contribute to long-term neuropathology. Beyond glia, innate immune signaling across all cells of the brain is emerging at the crossroads of typical neurodevelopment and pathogenesis of various disease states across the lifespan, including AUD (for a review, see Zengeler and Lukens 2021). Considering the divergent roles of innate immunity in typical development and disease pathogenesis, understanding maturational changes of innate immune cells is critical for understanding the long-term consequences of AIE.

Microglia were initially relegated as scavengers of the brain with the sole role of phagocytosing cellular debris. However, as research into their function has evolved, it has become increasingly clear that microglia play complex roles in neurodevelopment, from actively surveying the microenvironment to sculpting synapses and extracellular matrices (Crapser et al. 2021), suggesting that early-life insults on microglia may result in changes in neurocircuitry. Microglia in brain are classically categorized as exhibiting a homeostatic or surveying state under non-pathological conditions, although a more diverse picture of microglial phenotype heterogeneity under healthy and pathological conditions is continuing to emerge (Tan et al. 2020). Unlike other glial cells of the brain, microglia are of mesodermal origin and colonize the developing brain during early embryogenesis (embryonic day [**E**]8–E10) in rodents wherein they transition from macrophage-like immature cells to microglia (Schwarz et al. 2011), playing a critical role in brain maturation (Eyolfson et al. 2020; Melbourne et al. 2021). Microglia in the developing adolescent brain exhibit an activation-like morphology, characterized by an amoeboid shape (Schwarz et al. 2011, 2012), and undergo brain regional maturation toward a ramified profile that parallels neural maturation (Brenhouse and Schwarz 2016; Scheffel et al. 2012; Hanamsagar et al. 2018). This developmental phenotypic shift across adolescence of microglia could affect their responsivity to an immune challenge during adolescence. Indeed, evidence suggests that relative to adults, adolescent males have a greater anti-inflammatory cytokine response, reduced proinflammatory response, and attenuated sickness behavioral response following a lipopolysaccharide (**LPS**) innate immune challenge (Cai et al. 2016). During adolescence, microglia continue to function as key regulators of synaptic pruning through selective phagocytosis of synapses involving a C1q and C3 complement-mediated process. In fact, microglial complement-mediated phagocytosis of dopamine D1 receptors in the nucleus accumbens during development shapes development of male social play behavior (Kopec et al. 2018), whereas transgenic reductions of microglial number in mice impair synaptic pruning, resulting in increased levels of synapses and synaptic proteins (Wieghofer and Prinz 2016). Given that postnatal microglial populations are relatively permanent with little turnover (Male and Rezaie 2001), stimuli that negatively impact immature adolescent microglia could cause lasting changes in their function (Bilbo and Schwarz 2009), perhaps priming them toward a persistent proinflammatory phenotype (Melbourne et al. 2021; Qin and Crews 2012; Qin et al. 2007, 2008; Brenhouse and Schwarz 2016; Town et al. 2005).

Unlike microglia, astrocytes are derived from neural stem cells (**NSCs**) following neuronal differentiation early in embryogenesis, between approximately E18 and P7 in the rodent brain (Tien et al. 2012; Sauvageot and Stiles 2002). Although the majority of astrocytes are generated during early postnatal development, their fine peripheral processes, which constitute approximately 50% of the mature astrocyte volume, continue to grow and mature into adolescence, particularly in the medial prefrontal cortex (**PFC**) (Wolff 1970; Bandeira et al. 2009; Ge et al. 2012; Freeman 2010; Bushong et al. 2004; Testen et al. 2019). Astrocyte-derived factors, such as thrombospondins (**TSPs**), are crucial for the initiation of synaptogenesis wherein they promote formation and maturation of neuronal processes (Christopherson et al. 2005; Kucukdereli et al. 2011; Eroglu et al. 2009). While astrocytes play a crucial role in synaptogenesis and modulate synaptic physiology during adolescent development, generation and functional maturation of these cells during development remain to be fully elucidated. Similar to astrocytes, oligodendrocytes, which are the myelinating glial cells of the brain, are formed from NSCs during embryogenesis and early postnatal life from oligoprogenitor cells. Oligodendrocyte populations are relatively stable, persisting long into adulthood, with maturational profiles consistent with adolescent developmental maturation of white matter volume and structures involving maturation of myelin-forming oligodendrocytes during development to adulthood (Shaw et al. 2020; Xin et al. 2023).

Myelin-forming oligodendrocytes express specific proteins responsible for myelin formation, including myelin basic protein, proteolipid protein 1, myelin-associated glycoprotein, and myelin oligodendrocyte glycoprotein, all of which are thought to contribute to the structure of myelin formed by mature oligodendrocytes (Marques et al. 2016; Gonzalez-Perez and Alvarez-Buylla 2011). While oligodendrocytes were originally thought to be passive bystanders of microglia, a more active role of oligodendrocytes in innate immune signaling has begun to emerge. Accumulating evidence suggests that oligodendrocytes also express a variety of pattern recognition receptors (**PRRs**) as well as endogenously produce and release the chemokine monocyte chemoattractant protein-1 (**MCP1**) and the proinflammatory cytokine interleukin-1beta (**IL-1β**) under pathological conditions, including bacterial infection and demyelinating diseases (Ramesh et al. 2012; Vela et al. 2002).

Paralleling maturational changes in glia during development, the expression of many innate immune signaling molecules accompany adolescent brain maturation, including Toll-like receptors (**TLRs**). While much of our understanding of TLR involvement in neurodevelopment comes from developmental neuropathology rodent models, it is crucial to acknowledge their role in neurogenesis and processes associated with typical neurodevelopment (Barak et al. 2014). TLRs (i.e., TLR1-9) are expressed in the developing brain in temporally and spatially distinct patterns (Barak et al. 2014). Microglia express the entire repertoire of TLRs, whereas the expression of TLRs by other cells of the brain, including neurons, varies across neurodevelopment into adulthood (Okun et al. 2011; Olson and Miller 2004). In addition to responding to specific PAMPs, TLRs and other PRRs (e.g., receptor for advanced glycation end-products [**RAGE**]) also respond to endogenous ligands, such as high-mobility group box 1 (**HMGB1**). HMGB1 is a ubiquitously expressed nuclear protein involved in intracellular nuclear transcriptional regulation, chromatin stability, and extracellular neurogenesis and neurite outgrowth in the non-pathological brain through interactions with TLRs and RAGE. TLR4 and RAGE are the primary receptors for HMGB1 mediation of proinflammatory signaling induction through NF-κB (Tobon-Velasco et al. 2014; Chavakis et al. 2004; Liu et al. 2017; Park et al. 2004; Aucott et al. 2018; Andersson et al. 2018). HMGB1 contains three cysteine residues (i.e., C23, C45, and C106) that are modifiable by redox reactions to produce multiple HMGB1 redox isoforms with differing extracellular activities. The redox state of HMGB1 dictates receptor binding, as disulfide HMGB1, containing C23 and C45 disulfide linkage and C106 in a reduced form as a thiol, binds to TLR4 (Yang et al. 2015), whereas reduced HMGB1, in which all three cysteine residues are reduced, binds to RAGE (Tang et al. 2010; Janko et al. 2014). Expression levels of TLR4 and RAGE decrease in the rat cortex from P55 to P80, whereas expression levels of HMGB1 increase during this period (Vetreno et al. 2013; Vetreno and Crews 2012). Accumulating evidence also implicates cytokines and chemokines in mediating diverse neurodevelopmental processes, including neurogenesis, neurite outgrowth, and axonal pathfinding (Zengeler and Lukens 2021; Deverman and Patterson 2009). Thus, given the vast neurodevelopmental processes occurring during adolescence—spanning brain structural and neurocircuitry refinement to the maturation of neurotransmitter systems, glia, and the innate immune system—insults incurred during this critical period of neurodevelopment, such as adolescent binge drinking, could have profound effects on brain function and behavior that persist into adulthood.

9.2.2 Adolescent Intermittent Ethanol-Induced Reprogramming of the Innate Immune Response Across the Brain

As opposed to the peripheral immune response to insult, neuroinflammation in the brain can persist long after peripheral resolution of the initial insult. For instance, Qin et al. (2007) reported that a single systemic dose of the TLR4 ligand LPS to late adolescent mice rapidly increased mRNA and protein expression of the proinflammatory cytokine tumor necrosis factor alpha (**TNFα**) in the blood, liver, and brain. While the blood and liver TNFα response resolved within 12–24 h, upregulation of TNFα persisted in association with long-lasting microglial activation and gradual degeneration of dopaminergic neurons in the substantia nigra (Qin et al. 2007). These findings highlight that the resolution of the proinflammatory response in the brain is complex and can persist and contribute to neuropathology long after peripheral responses resolve. Although distinct from LPS, alcohol also produces complex innate immune signaling cascades in the brain and periphery that varies across intoxication and withdrawal phases.

During intoxication, alcohol suppresses innate immune signaling. For example, acute application of ethanol (50 mM) to primary microglial culture inhibits induction of microglial proinflammatory signaling pathways, assessed using deep proteome profiling, that parallels suppression of innate immune signaling molecules (Guergues et al. 2020). Acute alcohol exposure similarly alters innate immune responses to a second hit challenge as acute alcohol exposure blunts human blood-derived monocyte responsivity to an LPS challenge and attenuates proinflammatory cytokine induction to an LPS challenge in preclinical studies and culture models (Szabo et al. 1995; Mandrekar et al. 2002; Pruett et al. 2004; Szabo and Mandrekar 2009). Interestingly, this acute alcohol suppression of inflammatory response appears to be transient or dependent on the presence of alcohol as acute high-dose alcohol consumption in healthy male human volunteers transiently reduces serum levels of the proinflammatory chemokine MCP-1 two hr after alcohol intake that steadily increases through withdrawal by 12 h post-intake, paralleling the elimination of blood alcohol content (Neupane et al. 2016). Consistent with this observation, administration of ethanol to adult male mice in vivo transiently suppresses brain mRNA levels of proinflammatory signaling (TNFα and MCP-1) and microglial genes (CD68 and Iba1) that parallel elevations in blood ethanol levels. These findings are also true following AIE as in male rats with a history of AIE, an acute ethanol challenge in adulthood reduces innate immune signaling molecules in the hippocampus as assessed using microdialysis (Gano et al. 2019). However, during withdrawal when blood ethanol levels decrease, the expression of proinflammatory signaling and microglial genes increases (Walter and Crews 2017), consistent with ethanol acutely suppressing neuroinflammatory marker expression, whereas neuroinflammation increases during withdrawal. In contrast, in the absence of an acute ethanol or LPS stimulus, AIE models generally find persistently long-lasting induction of proinflammatory signaling molecules across the brain that do not resolve with abstinence. For example, Pascual et al. (2007) reported that AIE treatment of adolescent mice persistently upregulated cyclooxygenase-2 (**COX-2**) and iNOS in the adult neocortex, hippocampus, and cerebellum in addition to increasing expression of cell death markers. In rats, AIE treatment increases expression of HMGB1 and the HMGB1 receptors RAGE and various TLRs (i.e., TLR3 and TLR4) in the late adolescent PFC that persists into adulthood in association with upregulation of proinflammatory cytokines (e.g., TNFα), chemokines (e.g., MCP-1), and oxidases (e.g., NOX2) (Vetreno et al. 2013; Vetreno and Crews 2012). We have reported similar upregulation of proinflammatory signaling cascades in the post-mortem human orbitofrontal cortex of individuals with AUD and an adolescent age of drinking onset (Vetreno et al. 2013, 2021; Crews et al. 2013). Activation of TLR cascades led to activation of the nuclear transcription factor NF-κB and downstream induction of proinflammatory cytokines, chemokines, and other proinflammatory signaling mediators (Kawai and Akira 2007).

In addition to persistent induction of neuroinflammation, AIE induces a persistent microglial activation state throughout the adult brain (Vetreno et al. 2017a; Sanchez-Alavez et al. 2019; Walter et al. 2017; Mcclain et al. 2011) characterized by a hyper-ramified morphology that may contribute to the persistent, long-lasting upregulation of proinflammatory cytokines and chemokines. Similar evidence of microglial activation has been reported across the post-mortem human brain of individuals with AUD and an adolescent age of drinking onset (He and Crews 2008). Due to the critical role of microglia in adolescent brain development, persistent shifts of microglia to a proinflammatory phenotype could disrupt developmental synaptic pruning resulting in shifts in the trajectory of brain maturation through adolescence (Paolicelli and Ferretti 2017). Further, AIE appears to prime microglia for future insults as Walter et al. (2017) reported that AIE increased microglial expression of CD11b—a constitutively expressed microglial-specific marker upregulated in activated microglia (Kettenmann et al. 2013)—across the adult brain, sensitizing and exacerbating the microglial CD11b response to an adult stress challenge consistent with lasting alterations in microglial responsivity. This priming of microglia following AIE is highly translational and likely relevant to human pathology as microglial markers (e.g., Iba-1) are also increased in the post-mortem human adult brain of individuals with AUD and an adolescent age of drinking onset (He and Crews 2008).

While the majority of research to date has focused on the effect of AIE on neuroinflammation and microglia activation in males, accumulating evidence in more recent studies suggests sex differences in innate immune and microglial responsivity to ethanol. For instance, AIE treatment in mice increased serum levels of MCP-1, IL-17A, and MIP-1a, and PFC levels of IL-1β, MCP-1, IL-17A, MIP-1a, TLR4, and pNFκBp65 in late adolescent females but not males, an effect that was blocked in TLR4 KO mice (Pascual et al. 2017). In contrast, AIE treatment in rats increased IL-1β levels in the medial PFC of both late adolescent males and females, whereas expression of TLR4 was only increased in males. In the same study, AIE similarly shifted microglia to a more reactive proinflammatory phenotype in the late adolescent medial PFC of both male and female rats, whereas in the hippocampus, AIE treatment differentially shifts microglia to a proinflammatory phenotype with males showing greater shifts in the dentate gyrus, an effect that was not observed in females (Silva-Gotay et al. 2021). In young adult humans, acute alcohol exposure increased plasma protein levels of multiple proinflammatory signaling molecules in human females (e.g., IL-1β, IL-6, IL-4, MCP-1), an effect that was not observed in age-matched males. These findings suggest the adolescent alcohol induction of proinflammatory responses may activate microglial populations and/or induce neuroinflammation in a sex-specific manner suggesting that a more comprehensive examination of sex differences across the brain is an important future direction for the field.

While the mechanism underlying AIE-induced lasting priming of microglia is unknown, emerging studies suggest that perinatal perturbations cause lasting alterations in microglial function associated with epigenetic reprogramming (Dziabis and Bilbo 2022; Bilbo and Stevens 2017; Schwarz and Bilbo 2013; Bilbo and Schwarz 2012; Schwarz and Bilbo 2012; Schwarz et al. 2011; Williamson et al. 2011; Knuesel et al. 2014). Collectively, these findings suggest global changes in brain proinflammatory markers and microglial function that persist into adulthood that may be related to microglial epigenetic modifications. However, as the vast majority of work on innate immune signaling to date has focused on the basal forebrain and hippocampus, these next sections will provide an in-depth look to the emerging findings across these brain regions (Fig. 9.1). In the following sections, we will describe the consequences of AIE-induced neuroinflammation on cholinergic neurons of the basal forebrain and hippocampal neurogenesis.

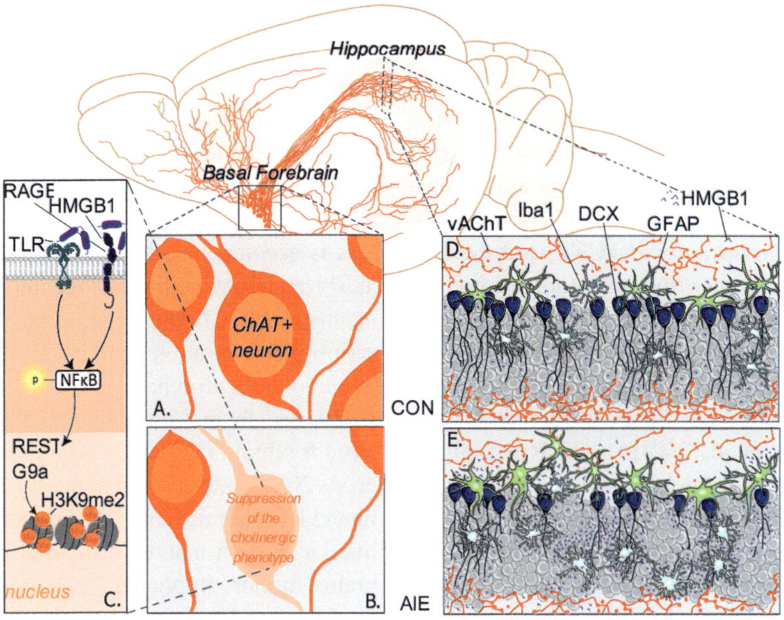

Fig. 9.1 Adolescent intermittent ethanol (AIE) treatment causes persistent induction of proinflammatory neuroimmune signaling across the adult brain, including the basal forebrain and hippocampal formation, relative to controls (CON). In the basal forebrain (**a**–**c**), AIE causes nuclear release of high-mobility group box 1 (HMGB1), which acts as an endogenous agonist at the proinflammatory pattern recognition receptors Toll-like receptor 4 (TLR4) and receptor for advanced glycation end-products (RAGE), leading to phosphorylation and nuclear translocation of NF-κB p65. Emerging studies report HMGB1 interactions with NFκB are linked to H3K9me2 gene silencing through recruitment of the methyltransferase G9a to repress gene transcription (**c**) (El Gazzar et al. 2009, 2010; Chang et al. 2011). Induction of HMGB1-TLR4/RAGE-pNF-κB p65 leads to epigenetic silencing of the cholinergic phenotype (e.g., TrkA, ChAT, vAChT) through REST/G9a-driven upregulation of H3K9me2 on cholinergic gene promoter (**c**). Conversely, in the hippocampus (**d**, **e**), AIE disrupts the milieu of the neurogenic niche involving increases in astrocytic (GFAP) and microglial (Iba1) proinflammatory markers as well as increases in HMGB1 (**e**), relative to controls (CON) (**d**). Loss of cholinergic terminals at hippocampal targets may be a mechanism driving induction of hippocampal neuroinflammation as acetylcholine provides important anti-inflammatory feedback. As a result of this disruption, AIE causes a persistent reduction in adult hippocampal neurogenesis (doublecortin; DCX) that does not resolve with abstinence (**e**)

9.3 Adolescent Intermittent Ethanol Exposure Causes a Persistent Loss of Basal Forebrain Cholinergic Neurons

9.3.1 Neuroinflammation Contributes to Persistent AIE-Induced Loss of the Basal Forebrain Cholinergic Neurons

Cholinergic neurons of the basal forebrain play a major regulatory role in learning and memory through their vast projections to the hippocampus, cortex, and other brain regions (Smith and Pang 2005; Mesulam et al. 1983). Generated early in embryonic development (Dinopoulos et al. 1992; Gould et al. 1989; Gould et al. 1991; Linke and Frotscher 1993), cholinergic neurons continue to undergo maturational refinement and consolidation of projections and receptors during adolescence (Matthews et al. 1974; Nadler et al. 1974; Zahalka et al. 1993). Multiple laboratories report AIE causes a loss of choline acetyltransferase (**ChAT**)-immunopositive neurons, which is the enzyme responsible for ACh synthesis, and somal shrinkage of the remaining cholinergic neurons in the basal forebrain of male and female mice and rats that persists well into adulthood despite prolonged abstinence from continued alcohol exposure (Vetreno et al. 2014; Coleman et al. 2011; Ehlers et al. 2011; Fernandez and Savage 2017; Swartzwelder et al. 2015; Vetreno and Crews 2018). Loss of the ACh-synthesizing enzyme ChAT in the basal forebrain is accompanied by decreased expression of the vesicular acetylcholine transporter (**vAChT**) and the cholinergic lineage transcription factor LIM homeobox protein 8 (**Lhx8**) as well as the high- and low-affinity nerve growth factor (**NGF**) receptors tropomyosin receptor kinase A (**TrkA**) and NGF receptor (Vetreno et al. 2019; Vetreno and Crews 2018; Crews et al. 2021a), which are highly expressed on basal forebrain cholinergic neurons (Vetreno et al. 2019; Vetreno and Crews 2018) and critical for cholinergic neuron survival and function (Isaev et al. 2017). The persistent adult loss of cholinergic neuron markers is accompanied by global reductions of cholinergic receptor gene expression (Coleman et al. 2011; Crews et al. 2023b) and diminished ACh efflux in basal forebrain target regions (Fernandez and Savage 2017) consistent with lasting degeneration of basal forebrain cholinergic neurons. Interestingly, this effect appears to be specific to adolescent binge alcohol exposure as identical intermittent ethanol treatment in adulthood does not affect basal forebrain cholinergic neuron populations (Vetreno et al. 2014). Further, assessment of postmortem human AUD basal forebrain tissue samples from individuals with an adolescent age of drinking onset reveals a similar loss of ChAT and vAChT expression (Vetreno et al. 2014) lending validity to the preclinical AIE model. The persistent dysregulation of the cholinergic neurotransmitter system likely manifests in cognitive, emotive, and behavioral deficits in adulthood. Indeed, emerging studies implicate the loss of basal forebrain cholinergic neurons in adult AIE-induced increases in impulsivity, behavioral disinhibition, and deficits in behavioral flexibility (Boutros et al. 2014; Ehlers et al. 2011; Macht et al. 2023; Vetreno et al. 2020). Further, ACh is a negative regulator of the innate immune response (Wang et al. 2004) suggesting that disruption of the central cholinergic system may not only directly impact cognitive function, but may contribute to the observed persistence of innate immune activation following AIE treatment.

Emerging studies implicate AIE-induced neuroimmune induction as a mechanism underlying the persistent loss of basal forebrain cholinergic neurons (see Fig. 9.1a, b). Cholinergic neuron loss is accompanied by increased expression of TLR4 and RAGE, the endogenous TLR/RAGE agonist HMGB1, the downstream nuclear transcriptional activation marker pNFκB p65, and induction of proinflammatory cytokines (e.g., IL-1β) and chemokines (e.g., MCP1) in the adult basal forebrain (Vetreno and Crews 2018; Crews et al. 2021a, 2023b). Treatment with the TLR4 ligand LPS mimics the AIE-induced loss of cholinergic neuron markers in the adult basal forebrain further implicating a proinflammatory mechanism. While these neuroimmune signaling molecules are expressed by glia, basal forebrain cholinergic neurons also express these neuroimmune markers suggesting concerted signaling across glia and neurons (Crews et al. 2021a). Interestingly, preventative interventions during AIE, including voluntary wheel running, the nonsteroidal anti-inflammatory drug indomethacin, and the cholinesterase inhibitor galantamine, all of which exert anti-inflammatory effects, block the AIE-induced increases of HMGB1-TLR4/RAGE-pNFκB p65 signaling and concomitant loss of cholinergic neurons in the adult basal forebrain (Vetreno et al. 2019, Vetreno and Crews 2018, Crews et al. 2021a). Further, blockade of HMGB1 using the selective HMGB1 antagonist glycyrrhizin prevents, whereas direct application of HMGB1 mimics, ethanol-induced loss of ChAT+ cholinergic neurons in an *ex vivo* basal forebrain slice culture model (Crews et al. 2023b), implicating HMGB1-mediated proinflammatory signaling in the AIE-induced loss of basal forebrain cholinergic neurons. Future studies are necessary to determine if HMGB1-mediated neuroinflammation underlies in vivo AIE-induced loss of basal forebrain cholinergic neurons. Together, these data implicate AIE-induced neuroimmune signaling in the persistent loss of basal forebrain cholinergic neurons.

9.3.2 Epigenetic Mechanisms Underlie AIE-induced Reversible Suppression of the Basal Forebrain Cholinergic Neuron Phenotype

Basal forebrain cholinergic neuron populations are decreased immediately following the conclusion of AIE and persist for at least 165 days following the conclusion of AIE treatment despite abstinence from continued ethanol exposure (Vetreno et al. 2014). The loss of cholinergic neuron markers was initially interpreted as a cell death mechanism. However, restorative post-AIE therapeutic interventions initiated after the onset of cholinergic neuron loss restore the adult AIE-induced loss of cholinergic neuron markers in male and female rats and reverse the increase of HMGB1-TLR4/RAGE-pNFκB p65 signals as well as increased pNFκB p65+ expression within adult ChAT+ neurons (Macht et al. 2023; Crews et al. 2021a; Vetreno et al. 2020). Restoration of the loss of basal forebrain cholinergic neurons is accompanied by reversal of the AIE-induced behavioral flexibility deficits in adult males and spatial learning deficits in female rats on the Morris water maze (Vetreno et al. 2020; Macht et al. 2023). While NF-κB signaling cascades have largely been ascribed to glial cells with a particular emphasis on microglia, NF-κB p65 is also constitutively expressed in neurons of the hippocampus, cortex, and basal forebrain (Kaltschmidt et al. 1994; Crews and Vetreno 2022; Crews et al. 2021a). The observed loss and subsequent restoration of cholinergic neurons post-AIE suggest that either (1) cholinergic neurons die and are restored via genesis of new cholinergic neurons or (2) these neurons undergo a reversible loss of the cholinergic neuron phenotype. Indeed, it has been reported that under certain pathological conditions, neurogenesis occurs in non-neurogenic brain regions (Emsley et al. 2005). However, expression of the pan-neuronal marker NeuN is unchanged in the adult AIE-treated basal forebrain and assessment of cell proliferation and neurogenesis using the proliferation marker BrdU did not reveal colocalization with ChAT+ neurons following restorative therapeutics (Vetreno et al. 2019; Crews et al. 2021a). Thus, post-AIE restoration of cholinergic neurons in the absence of cholinergic neuron generation suggests that AIE-induced neuroinflammation does not cause cell death, but rather the loss of the cholinergic phenotype that is reversible.

Emerging studies suggest that alcohol exposure and neuroimmune induction can elicit long-lasting epigenetic changes in cellular programming through chromatin remodeling (Montesinos et al. 2016; Pandey et al. 2017; Wolstenholme et al. 2017; Vetreno et al. 2019; Crews et al. 2021a). Epigenetic modifications involve histone acetylation and histone and DNA methylation, which can enhance or repress gene transcription without changing the underlying DNA sequence (Kouzarides 2007; Berger et al. 2009). Acetylation and methylation at histone 3 lysine 9 (**H3K9**) can activate and repress gene transcription, respectively (Wang et al. 2008), and the mammalian *Chat* gene contains transcriptional enhancers and repressors that are involved in maturation of basal forebrain cholinergic neurons (Pu et al. 1993; Hersh and Shimojo 2003; Hersh et al. 1993). The AIE-induced loss of basal forebrain cholinergic neuron markers is accompanied by increases of H3K9 dimethylation (**H3K9me2**) occupancy at *Chat* and *Trka* gene promoters as well as DNA methylation at the CpG island of the *Chat* promoter (Crews et al. 2021a, Vetreno et al. 2020). Increased H3K9me2 and DNA methylation are processes generally associated with stable transcriptional gene repression (Fernandes et al. 2017; Goll and Bestor 2005) that may contribute to the persistent, long-lasting AIE-induced loss of cholinergic neuron markers. In support of an epigenetic repressive mechanism underlying the loss of cholinergic neuron phenotype, post-AIE voluntary exercise exposure and galantamine treatment restores the aberrant AIE-induced increase of chromatin and DNA methylation at cholinergic gene promoters to baseline control levels that parallels restoration of cholinergic neuron markers in the adult male basal forebrain (Crews et al. 2021a, Vetreno et al. 2020). The *Chat* and *Lhx8* gene promoters contain the consensus 21-base-pair DNA RE1 binding sequence that binds the transcriptional repressor RE1-silencing transcription factor (**REST**), and REST is known to regulate expression of *Chat* and other cholinergic genes (Shimojo and Hersh 2004). The histone methyltransferase G9a (EHMT2), which is recruited by the repressive transcription factor REST, reversibly suppresses gene transcription through dimethylation of H3K9 (Ballas et al. 2005; Roopra et al. 2004). Interestingly, AIE treatment increases REST and G9a expression in the adult basal forebrain paralleling to increased occupancy of H3K9me2 at cholinergic gene promoters, an effect that is also observed in the postmortem basal forebrain of individuals with AUD and an adolescent age of drinking onset (Crews et al. 2023b). While the contributions of REST and G9a to epigenetic repression of the cholinergic neuron phenotype in vivo following AIE treatment remain to be fully investigated, *ex vivo* basal forebrain slice culture mechanistic studies reveal that direct stimulation of TLR4 with LPS and application of ethanol mimics the AIE-induced epigenetic silencing of the cholinergic neuron phenotype. Further, these *ex vivo* slice culture studies revealed increased occupancy of REST and H3K9me2 at the *Chat* and *Lhx8* gene promoters, and targeted pharmacological and siRNA blockade of G9a and REST reversed the loss of ChAT+ cholinergic neurons (Crews et al. 2023b; Crews and Vetreno 2022). Lhx8 regulates multiple cholinergic genes (Cho et al. 2014), including the high-affinity NGF receptor TrkA (Tomioka et al. 2014) that is known to promote cholinergic differentiation as well as the survival and maintenance of basal forebrain cholinergic neurons (Lucidi-Phillipi et al. 1996; Fagan et al. 1997). An Lhx8-TrkA-NGF positive feedback loop is hypothesized to maintain cholinergic neurons in their highly differentiated state (Tomioka et al. 2014). REST repression of Lhx8 is consistent with suppression of the cholinergic transcriptome and reduced neuronal excitability and, perhaps, diminished cerebral excitability. The somal shrinkage and loss of ChAT phenotype in BFCNs may represent a protective mechanism or an initial phase of degeneration.

REST repression of mature neuronal genes during development has been suggested to protect immature neurons, allowing for growth and development of axons and dendrites before REST is removed allowing expression of mature neuronal and synaptic genes (Zhao et al. 2017). In the adult brain, neuronal REST expression is low although it increases with age and in response to insults, suggesting it may be neuroprotective (Thiel et al. 2015). Indeed, reduced nuclear REST has been reported in several dementia disorders (e.g., Alzheimer's disease) and is consistent with mouse conditional knockout studies reporting that REST deletion leads to age-related neurodegeneration (Lu et al. 2014). REST is hypothesized to reduce excitability, protecting neurons since high levels of REST expression are found in the brain of humans cognitively intact at 90–100 years of age (Aron et al. 2022; Zullo et al. 2019). Thus, while adolescence is a period of heightened vulnerability to the deleterious effects of adolescent binge drinking, it may also represent a unique period of resiliency wherein insults are reversible through epigenetic mechanisms. Taken together, reversal of AIE-induced increases of HMGB1-TLR4/RAGE-pNFκB p65 signaling and cholinergic gene promoter REST-G9a-H3K9me2 methylation that accompanies the restoration of adult basal forebrain cholinergic neurons supports a role for proinflammatory neuroimmune signaling in reversible epigenetic silencing of the cholinergic phenotype.

9.4 Adolescent Intermittent Ethanol Exposure Disrupts Innate Immune Signaling in the Microenvironment of the Hippocampal Neurogenic Niche

9.4.1 Reduced Cell Proliferation and Increased Cell Death Contribute to Persistent AIE-Induced Loss of Hippocampal Neurogenesis

The hippocampus, which is known for its critical role in the execution of learning and memory-related cognitive functions, is also highly susceptible to alcohol-related neuroimmune-driven structural damage (Crews and Vetreno 2014), resulting in a total loss of volume after AIE

(Vetreno et al. 2016, 2017b; Gass et al. 2014; Ehlers et al. 2013b), an effect that is also observed in human adolescents with AUD (De Bellis et al. 2000). These AIE-induced changes in hippocampal structure are associated with reduced function on hippocampal-dependent learning and memory tasks, including novel object recognition memory (Macht et al. 2020b; Vetreno and Crews 2015) and spatial and non-spatial reversal learning (Coleman et al. 2014; Sey et al. 2019) as well as anxiety-like behaviors in adulthood (Vetreno and Crews 2015). Loss of function on hippocampal-dependent memory-related tasks has been linked to alterations in cellular and signaling integrity within the hippocampal formation, suggesting that AIE changes the hippocampal enviornmental milieu. This is due, in part, to the unique cellular structure of the hippocampal formation as it is one of only two regions in the adult mammalian brain wherein neurons are continuously generated and functionally integrated into the existing hippocampal neurocircuitry (He and Crews 2007; Toni and Schinder 2015). Survival of these newborn neurons into adulthood is highly dependent on and sensitive to their microenvironment, making adult hippocampal neurogenesis an interesting marker of overall brain health and plasticity. Emerging evidence further suggests a direct role of newborn neurons in a variety of hippocampal-related behaviors that are impaired after AIE (Cameron and Glover 2015; Anacker and Hen 2017; Webler et al. 2019; Martinez-Canabal et al. 2023). Conversely, loss of adult hippocampal neurogenesis is characteristic of most neurodegenerative and neuropsychiatric disorders, which also exhibit increased chronic neuroinflammatory signaling (Abdipranoto et al. 2008). This emphasizes that reductions in adult neurogenesis are often indicative of a dysfunctional proinflammatory hippocampal microenvironment (Macht et al. 2020a; Chintamen et al. 2020), which is a complex and delicate construction of vascularization, glial-driven trophic support, and low levels of innate immune signaling.

The neurogenic consequences of binge ethanol exposure across adolescence is far more severe than the consequences of similar ethanol exposure across adulthood. For example, while ethanol treatment in adulthood transiently reduces hippocampal neurogenesis in rats, subsequent abstinence following adult ethanol exposure restores neurogenesis to control levels (Crews and Nixon 2009). In contrast, AIE causes reductions of hippocampal neurogenesis that persists well into adulthood despite prolonged abstinence (Broadwater et al. 2014; Vetreno and Crews 2015; Sakharkar et al. 2016; Swartzwelder et al. 2019). This persistent reduction is evident in both male and female rats (Macht et al. 2023; Nwachukwu et al. 2022a) and has been replicated across various models of adolescent ethanol exposure in both rodents (for review, see Macht et al. 2020a) and primates (Taffe et al. 2010). Interestingly, identical intermittent ethanol treatment in adulthood does not persistently affect hippocampal neurogenesis, emphasizing the heightened vulnerabiltiy of the adolescent brain to neurogenic insults (Broadwater et al. 2014).

The AIE-induced loss of hippocampal neurogenesis involves both reduced proliferation of neuroprogenitor cells and increased cellular apoptosis of these differentiating newborn neurons. Indeed, AIE treatment decreases expression of the proliferating cell marker Ki-67 (Ehlers et al. 2013a; Broadwater et al. 2014; Sakharkar et al. 2016; Vetreno and Crews 2015) while increasing expression of cell death markers, including cleaved caspase-3 and Fluoro Jade (Swartzwelder et al. 2019; Broadwater et al. 2014; Vetreno and Crews 2015; Ehlers et al. 2013a) consistent with the AIE-induced loss of neurogenesis involving decreased progenitor proliferation and increased cell death. Indeed, AIE treatment induces hippocampal Death Receptor 3 (**DR3**), the DR3 ligand TL1A, and downstream induction of the caspase cell death signaling cascade in adult AIE-treated rats (Liu et al. 2020). Of note, these findings are also evident in post-mortem human AUD brain and positively correlated with expresison of the cell death marker cleaved caspase 3 (Liu et al. 2020). The increased hippocampal DR3-TL1A caspase cascade is associated with induction of proinflammatory neuroimmune signaling and accumulating evidence implicates neuroimmune system activation in the AIE-induced disruption of hippocampal neurogenesis (Vetreno et al. 2018; Swartzwelder et al. 2019; Liu et al. 2020). Further work elucidated that cleaved caspase 3 is specifically upregulated in cells expressing the microtubule-associated protein doublecortin that labels immature newborn neurons: doublecortin (Macht et al. 2021). This suggests that in addition to general mobilizaiton of cell death machinery across the dentate gyrus, AIE specifically increases activation of cell death machinery in adult neuroborn neurons.

9.4.2 Persistent AIE Induction of Innate Immune Signaling in the Hippocampal Neurogenic Niche

While glia are often the focus when discussing the immune system in the brain, neurons are also important effectors of innate immune function as AIE induces persistent expression of many proinflammatory mediators in neurons of the hippocampus. For example, AIE increases expression of HMGB1, COX-2, RAGE, and activated phospho-NFκB p65 in neurons in the granule cell layer of the hippocampus (Macht et al. 2021; Vetreno et al. 2018; Swartzwelder et al. 2019). Similarly, MCP1 is expressed in neurons and upregulated after AIE in the polymorphic layer of the dorsal hippocampal dentate gyrus (Macht et al. 2021). These increases in hippocampal innate immune signaling cascades have been directly linked to the loss of hippocampal neurogenesis across a multitude of studies (for review, see Macht et al. 2020a). For example, Vetreno et al. (2018) reported that AIE causes an upregulation of hippocampal TLRs as well as phosphorylation of NF-κB p65 in doublecortin-positive newborn neurons. This innate immune induction is coupled with increases in proinflammatory gene expression of NF-κB p65 target genes, including TNFα and IκBα. Excitingly, voluntary exercise and treatment with the nonsteroidal anti-inflammatory compound indomethacin not only block AIE-induced induction of these proinflammatory markers in the neurogenic niche, but this blockade also prevents AIE-induced loss of doublecortin-positive neurons and concurrent increases of the cell death enzyme cleaved caspase-3. Indomethacin treatment after AIE similarly restores survivabiltiy of doublecortin-positive neurons while decreasing AIE induction in neuronal expression of COX-2 and HMGB1 (Macht et al. 2023). This restoration in the balance of innate immune signaling in the neurogenic niche also restores reversal learning deficits, further emphasizing the critical role of hippocampal neurogenesis and innate immune signalig in reversal learning.

Emerging studies reveal that changes in the basal forebrain cholinergic system may play an important role in both AIE-induced loss of hippocampal neurogenesis and induction of hippocampal innate immune signaling as the hippocampus is densely innervated by basal forebrain cholinergic neurons (see Fig. 9.1c). Indeed, lesion of basal forebrain cholinergic neurons with 192IgG-Saporin reduces hippocampal neurogenesis and causes microglial activation implicating cholinergic signaling in regulation of hippocampal neurogenesis and neuroinflammation (Cooper-Kuhn et al. 2004; Dobryakova et al. 2017; Field et al. 2012). Cholinergic involvement in AIE-induced loss of huppocampal neurogenesis was first elucidated by findings that interventions, such as exercise, which restore hippocampal neurogenesis also restore the AIE-induced loss of basal forebrain cholinergic neurons (Vetreno et al. 2018; Vetreno and Crews 2018). More recent work found that the hippocampus does exhibit decreased innervation by cholinergic neurons after AIE as evidenced by reductions in hippocampal expression of vAChT (see Fig. 9.1d; Macht et al. 2023). Interestingly, increasing synaptic acetylcholine levels with the repeated administration of the cholinesterase inhibitor galantamine was able to reverse induction of cleaved caspase-3 in immature neurons, further emphasizing the critical role of cholinergic signaling in mediating restoration of hippocampal neurogenesis after AIE and linking basal forebrain cholinergic dysfunction to hippocampal cellular survival (Macht et al. 2021). Collectively, this suggests that one of the mechanisms by which acetylcholine may improve newborn neuron survival is through inhibtion of proinflammatory signaling cascades in the neurogenic niche.

9.4.3 Persistent AIE-Induced Alterations in Microglial and Astrocytic Function in the Adult Hippocampus

Several studies find that microglia, important modulators of hippocampal neurogenesis and effectors of innate immune function, are dysregulated after adolescent ethanol exposure. For example, Marshall et al. (2020) reported that microglia after AIE are more likely to exhibit a dystrophic phenotype in the hippocampus (see Fig. 9.1d)—that is, exhibiting morphological aberrations including fragmentation of the cytoplasm (cytorrhexis) as well as processes that are fragmented, beaded, or bare of fine arborizations. This morphological phenotype suggests severe cellular distress and is often evident in neurodegenerative diseases and aging (Streit et al. 2014). In addition to microglial dystrophy, AIE induces microglial proinflammatory signaling cascades as indicated by increased pNFκB p65 colocalization with Iba-1+IR microglia within the adult hippocamopal dentate gyrus (Vetreno et al. 2018). There is some heterogeneity of microglial phenotype across subregions of the hippocampus, and further elucidation of region-specific AIE-effects on microglia remains an important area of investigation. For example, AIE increases Iba-1+IR cells in the dentate gyrus of both males and females (see Fig. 9.1d), but in CA1 only females exhibit increases in Iba-1+IR microglial number (Nwachukwu et al. 2022b). Other studies have similarly found that AIE reduces overall microglial cell number with surviving microglia exhibiting decreased length of processes, coupled with an increased sensitivity to later acute ethanol-induced microglial proliferation, activation (evidenced by increased *Iba1* mRNA), and microglial apoptosis further emphasizing that AIE impairs microglial ability to effectively respond to later insults. Hyper-responsivity of microglia to later psychological stressors following AIE has been evidenced in other studies as well. For example, the proinflammatory microglial marker CD11b is increased in the dentate gyrus, CA3, and CA1 regions of the hippocampus in response to an acute stress challenge following AIE (Walter et al. 2017). Collectively, these findings indicate that AIE causes distress to hippocampal microglia, resulting in dystrophic morphology, increased proinflammatory signaling, impaired regulation of trophic support, and a reduced ability to respond effectively to later environmental insults.

Emerging evidence suggests that astrocytes are similarly disrupted by AIE in the hippocampus (see Fig. 9.1c, d). Increased glial fibrillary acidic protein (**GFAP**) is often used as an index of astrocyte reactivity, and GFAP+IR is increased after AIE in both males and females across the hippocampal dorsal dentate gyrus, CA2/3, and CA1 regions (Nwachukwu et al. 2022b). Somewhat incongruently, Gomez et al. (2018) find that hippocampal astrocytic branch complexity is only transiently increased after AIE, suggesting AIE causes dramatic but short-lived structural alterations in astrocytic structure, whereas Healey et al. (2020) reported long-lasting changes in astrocytic surface area in the adult mPFC following AIE treatment. Nevertheless, the functional consequences of AIE on astrocyte functionality are more severe. For example, AIE increases astrocyte connexin 43 hemichannels and pannexin-1 porosity, indicating an increased rate of diffusion between astrocytes and the extracellular environment that is positively correlated with increased hippocampal levels of IL-1β, TNFα, and IL6, and reversed with anti-inflammatory blockade (Gomez et al. 2018) further emphasizing the critical and emerging role of astrocytes in adolescent ethanol-induced persistent neuroinflammatory signaling in the hippocampal neurogenic niche. Astrocytes also exhibit reduced proximity to glutamatergic synapses in the hippocampus after AIE (Healey et al. 2020); as astrocytes are key regulators of glutamate update, this loss of proximity to glutamatergic synapses could indicate an increase in sensitivity to glutamate-induced cellular damage. Astrocytes also form glial scars and are critical barriers in severe neural damage, such as with traumatic brain injury. Emerging evidence suggests that AIE alters the progression to later hippocampal traumatic brain injury pathology (Mira et al. 2020), but the specific role of astrocytes in this process remains an important area of investigation.

9.4.4 Persistent AIE-Induced Proinflammatory Signaling Disrupts Hippocampal Networks

Astrocytes, microglia, and neurons exhibit robust cross-communication to mediate cellular networks, much of which is dependent upon intercellular innate immune signaling (Macht 2016). This suggests that AIE-driven effects on these individual cell populations within the hippocampus likely has compounding consequences when considering their collective, interrelated function, much of which has yet to be explored. For example, the persistent AIE-induced alterations in hippocampal neurogenesis may also contribute to adult stress-induced depression (Eisch and Petrik 2012), impaired cognitive flexibility (Anacker and Hen 2017), and cognitive decline in adulthood (Vetreno and Crews 2015). However, as with the basal forebrain, while the effects of AIE appear persistent in the absence of interventions, appropriate interventions have revealed that these effects are not necessarily appropriate. At the center of both the persistence and reversibilty of AIE-induced neuropathy is innate immune signaling, and in fact many broad anti-inflammatory interventions may be effective in rescuing AIE-induced hippocampal pathology, giving hope for a diverse set of effective therapeutic options.

9.5 Reversibility of Adolescent Intermittent Ethanol-Induced Neuropathology: Future Anti-inflammatory Therapeutics Targeting Persistent Neuroinflammation to Treat Addiction

9.5.1 Anti-inflammatory Anti-cholinesterases and Acetylcholine

While the immunomodulatory actions of ACh have been well established (Wang et al. 2004; Cox et al. 2020; Zorbaz et al. 2022), loss of cholinergic neurons could contribute to the observed microglial activation state and persistent neuroinflammation observed in the preclinical AIE model and post-mortem human AUD brain (Field et al. 2012; Vetreno et al. 2014; Walter et al. 2017; He and Crews 2008). Acetylcholine suppresses LPS-induced release of HMGB1, and nicotinic ACh receptor (nAChR) activation prevents activity of the NF-κB signaling pathway (Wang et al. 2004). The cholinesterase inhibitors galantamine and donepezil, which increase ACh bioavailability, have been shown to block AIE-induced HMGB1-RAGE/TLR4-NF-κB p65 neuroimmune signaling and neurodegeneration (Crews et al. 2021a; Macht et al. 2021; Swartzwelder et al. 2019). Galantamine, which is an alkaloid isolated from the Caucasian snowdrop (*Galanthus woronowii*) with FDA approval for the treatment of Alzheimer's disease (Lilienfeld 2002; Haake et al. 2020; Hampel et al. 2018), is a selective and reversible acetylcholinesterase inhibitor. In humans, galantamine has been shown to decrease circulating levels of the proinflammatory cytokine TNFα in individuals with metabolic syndrome associated with diabetes (Consolim-Colombo et al. 2017). Galantamine is not only a cholinesterase inhibitor, but also purported to be positive allosteric ligand at nAChRs, including α7 nAChRs (Wazea et al. 2018) and potentiates cholinergic transmission by positively modulating the response of nAChR to ACh and their agonists (Samochocki et al. 2003; Dajas-Bailador et al. 2003). α7 nAChRs are present on microglia and neurons (Ulloa 2005; Azam et al. 2003) and play a role in modulating neuroinflammation (Ren et al. 2017; Wazea et al. 2018; Sitapara et al. 2014). In addition to its benefit as an anti-inflammatory drug, galantamine also shows therapeutic promise in preclinical and clinical substance use disorder studies (Carroll et al. 2019; Maclean et al. 2018; Sofuoglu and Carroll 2011; Mann et al. 2006). In rodents, galantamine decreases alcohol intake in alcohol-preferring rats (Doetkotte et al. 2005), reduces cue-induced heroin seeking in rats following prolonged opioid withdrawal (Liu et al. 2012), and suppresses reinstatement of methamphetamine-seeking in rats (Koseki et al. 2014). In human clinical studies, galantamine is reported to decrease the number of alcohol drinks consumed during relapse in individuals with AUD (Mann et al. 2006). Galantamine is also reported to decrease opioid use (Carroll et al. 2019), cocaine use (Sofuoglu and Carroll 2011; Carroll et al. 2018), and nicotine use (Maclean et al. 2018; Mann et al. 2006), relative to placebo controls. Additional studies are needed to investigate the therapeutic potential of targeting the cholinergic system to block neuroinflammation and treat AUD-associated neuropathology.

9.5.2 G9a/Ehmt2 Inhibitors

Numerous studies have reported that EHMT2 inhibition across neuropathology models blocks induction of proinflammatory cytokines (e.g., TNFα, IL-1β, Il6), microglial morphology changes (i.e., increased Iba-1), and upregulation of proinflammatory nuclear transcription factor NFκB p65 while increasing microglial "M2-like" anti-inflammatory signaling molecules (Yang et al. 2023; Grinan-Ferre et al. 2019; Bellver-Sanchis et al. 2024; Li et al. 2022). EHMT2 inhibition decreases early-life stress-induced hippocampal microglial activation (Wang et al. 2017). Preclinical in vivo and in vitro microglial models report pathological increases of EHMT2, and EHMT2 inhibition reduces expression of microglial-enriched proinflammatory signaling molecules (e.g., TNFα, IL6, IL-1β) and transcription factors (e.g., NFκB p65) while increasing anti-inflammatory microglial-associated homeostatic genes (e.g., IL-4, IL-10, Arg1) (Yang et al. 2023; Zheng et al. 2019; Grinan-Ferre et al. 2019; Zou et al. 2022; Bellver-Sanchis et al. 2024; Li et al. 2022), suggesting that EHMT2 mediates, in part, reprogramming of microglia to a proinflammatory phenotype and persistent neuroinflammation.

9.5.3 Targeting Microglia

Preclinical AIE and post-mortem human AUD studies implicate upregulation of innate immune signaling and shifts of microglia to a proinflammatory phenotype in alcohol-induced neuropathology. Minocycline, an FDA-approved tetracycline antibiotic with anti-inflammatory and microglial activation inhibitor properties (Plane et al. 2010; Kobayashi et al. 2013; Elewa et al. 2006), prevents chronic adult EtOH-induced microglial priming in the mouse cortex (Qin and Crews 2012). In an early-life stress rat model, minocycline blocked increased microglial Iba-1 expression and mRNA/protein expression of proinflammatory TNFα, IL-1β, and IL-6 as well as expression of the repressive epigenetic marker H3K9me2 (Wang et al. 2017). Minocycline has also been reported to reduce alcohol withdrawal-induced anxiety-like behavior and alcohol reinstatement (Wu et al. 2011; Gajbhiye et al. 2018). Thus, blockade of microglial activation and proinflammatory signaling may represent a therapeutic target for the treatment of alcohol-induced neuropathology.

9.5.4 Inhibitors of HMGB1

Glycyrrhizin, which is a constituent of *Glycyrrhiza glabra* (licorice) classified by the FDA as Generally Recognized As Safe (GRAS) as a food additive, is a potent, selective HMGB1 inhibitor that binds to and inactivates released HMGB1 (Mollica et al. 2007). Crystallography analysis reveals that glycyrrhizin binds a pocket in HMGB1 blocking the proinflammatory activities of HMGB1 (Mollica et al. 2007). Intraperitoneal administration of glycyrrhizin decreases ethanol withdrawal-induced upregulation of HMGB1 and induction of proinflammatory signaling genes in the adult rat brain (Whitman et al. 2013). Both *ex vivo* and in vivo studies find remarkable anti-inflammatory activity of glycyrrhizin in alcohol and proinflammatory models (Crews and Vetreno 2022; Crews et al. 2021b; Zou and Crews 2014; Whitman et al. 2013) suggesting that glycyrrhizin may be useful in targeting neuroinflammation in AUD.

9.6 Conclusions and Future Directions

Collectively, these findings highlight the sensitive nature of the adolescent developmental window, where alcohol exposure can disrupt maturation of glia and innate immune signaling cascades in parallel with changes in the refinement of neurocircuitry. The molecular consequences of this developmental disruption induced by adolescent intermittent binge alcohol can tip the balance toward the development of an AUD later in life by increasing risk factors, including impaired behavioral flexibility and escalated voluntary drinking. Mechanistically, the intersection of the adolescent developing innate immune and cholinergic systems is emerging as a key hub at the crossroads of alcohol-induced adult neuropathology as well as emerging therapeutic interventions. However, there are several key areas that remain underexplored. Specifically, four areas stand out as important directions for future investigations for the field: (1) sex differences, (2) neuronal and innate immune responsivity to subsequent challenges, (3) multi-hits of developmental alcohol exposure, and (4) persistence of therapeutic interventions.

While several studies have begun to elucidate findings across males and females, this remains an understudied area of investigation—especially as adolescent females exhibit higher rates of binge alcohol drinking than males for the first time in recorded history (SAMHSA 2022). Emerging findings suggest that many of the molecular cascades—notably cholinergic, neurogenic, and HMGB1-TLR4-driven innate immune—are similarly disrupted across males and females in a variety of brain regions (Macht et al. 2023). However, the vast majority of investigated innate immune pathways are not linked to the X-chromosome, despite the fact that the X-chromosome contains more immune-related genes than any other chromosome in the human genome (for reviews, see Bianchi et al. 2012, Libert et al. 2010). These X-linked innate immune genes are thought to underlie many of the sex differences observed across disorders, including some cancers, autoimmune disorders, schizophrenia, vaccine responsivity, and more (Syrett and Anguera 2019). For example, while several human and preclinical studies have linked TLR7 to alcohol drinking (Lovelock et al. 2022), alcohol-related cellular toxicity (Coleman et al. 2017), and AUD pathology (Crews et al. 2023a), no studies have examined this in AIE-treated females. While *Tlr7* mRNA is upregulated in the rat hippocampus after AIE treatment in adult males (Vetreno et al. 2018), one would predict that females would be more sensitive to this ethanol-induced innate immune response due to their higher levels of TLR7 expression, which has been linked to this gene escaping X-linked inactivation (Souyris et al. 2019; Berghofer et al. 2006). Furthermore, many studies of innate immune signaling focus on mRNA, largely due to methodological limitations and complications associated with measuring proteins. However, protein and mRNA can and frequently do diverge (Jiang et al. 2023). Therefore, more thorough analyses of many of these neuroimmunological endpoints using protein will advance the field significantly.

The innate immune system is extraordinarily complex, and its efficacy requires a multitude of time-dependent and intersecting signaling cascades to both initiate and resolve the immune response in an efficient and stimulus-specific manner. Findings that AIE produces persistent, low-level induction of proinflammatory signaling cascades in the adult basal forebrain and hippocampus that do not resolve with abstinence suggest a state of chronic low-grade proinflammatory signaling that is often reported in the aging literature. One of the characteristics in the aged brain is the inefficiency in induction and resolution of innate immune responsivity paralleled by a lack of response specificity. However, few studies have examined how the brain and body respond to an additional innate immune challenge after AIE, and this remains an important area of future investigation. For example, AIE sensitizes the TLR3-driven fever response in adulthood in males but attenuated the fever response in females, according to core body temperature (Gano et al. 2024). Despite these divergent physiological changes across sex following

AIE exposure, both males and females exhibit similar poly I:C-associated gene induction of TLR3, IκBα, and IL-1β across liver, spleen, and the paraventricular nucleus of the hypothalamus relative to ethanol-naïve controls suggesting alternative and yet unknown innate immune pathways or neuronal circuits may be driving these effects. In contrast, LPS-driven TLR4 activation blunts the *Il1* and *IκBα* mRNA responsivity in peripheral blood samples of AIE male rats relative to ethanol-naïve controls. While a history AIE exposure did not affect peripheral mRNA responsivity of IL-6, IL-1, TNF-α, or IκBα in females, brain regional differences in innate immune responsivity remain to be evaluated following LPS. Acute ethanol (2.5 g/kg ethanol) challenge in adulthood revealed that there may be some subtle sex differences in IL-6, TNF-α, and IκBα ethanol-mRNA responsivity in adulthood (Vore et al. 2021). However, the subtlety of some of these findings may require increased power to resolve sex differences. In addition, while this study examined mRNA changes during intoxication, ethanol produces different changes during withdrawal, and changes during withdrawal across brain regions also remain an important future area of investigation.

To complicate matters, alcohol produces an evolving pathology that varies by developmental age of exposure as well as age of assessment. These effects become potentially confounded when one considers that individuals with a history of prenatal or adolescent exposure are more likely to drink or develop an AUD in adolescence and/or adulthood. For example, individuals prenatally exposed to alcohol often begin drinking at an age younger than their peers, are more likely to be hospitalized as a result of alcohol use, and are more likely to develop AUD (Moore and Riley 2015; Maya-Enero et al. 2021). Moreover, common mechanistic links may underlie alcohol pathology across the lifespan—particularly innate immune signaling cascades and cholinergic dysfunction (for review, see Macht et al. 2022b). This suggests that double or even triple-hit studies of alcohol exposure across the lifespan remain an important understudied area of investigation for the field.

Finally, the discovery that anti-inflammatory therapeutics can reverse AIE-induced changes at the molecular and behavioral level provides promise for novel pharmacotherapies. Excitingly, mounting evidence suggests that while many of the AIE-induced neuropathological changes are persistent, they are not permanent. In fact, many of these changes appear to be epigenetically regulated, opening the opportunity for assessment of new therapeutic targets. However, as the efficacy of most interventions to date is examined shortly after the conclusion of the treatment, it is unclear whether this recovery is permanent once treatment is suspended. Or, if treatment remains effective for a period, do yet unrealized age-related deficits become unmasked in senescence? These questions are critical to resolve when contemplating the translational nature of therapeutic interventions in future clinical populations.

The accumulating body of work on the molecular consequences of AIE on adult brain pathology and its potential recovery has widespread implications for our understanding of the developmental mechanisms across brain disorders. In humans, alcohol increases the vulnerability to or accelerates the pathogenesis of a multitude of disorders, including diseases that emerge in senescence such as Alzheimer's disease and Parkinson's disease (Seemiller et al. 2024). Early-life exposure to alcohol may have long-term consequences that set the stage for accelerated progression of these diseases that are often overlooked. Therefore, these same molecular insights into the persistent pathogenesis of alcohol-related brain and innate immune dysfunction may also yield novel innovations for treatment strategies across a wide array of brain disorders—not only in terms of mechanisms but also in terms of critical developmental windows for intervention timing.

References

Abdipranoto A, Wu S, Stayte S, Vissel B (2008) The role of neurogenesis in neurodegenerative diseases and its implications for therapeutic development. CNS Neurol Disord Drug Targets 7:187–210

Acheson SK, Bearison C, Risher ML, Abdelwahab SH, Wilson WA, Swartzwelder HS (2013) Effects of acute or chronic ethanol exposure during adolescence on behavioral inhibition and efficiency in a modified water maze task. PLoS One 8:e77768

Alaux-Cantin S, Warnault V, Legastelois R, Botia B, Pierrefiche O, Vilpoux C, Naassila M (2013) Alcohol intoxications during adolescence increase motivation for alcohol in adult rats and induce neuroadaptations in the nucleus accumbens. Neuropharmacology 67:521–531

Allen NJ, Bennett ML, Foo LC, Wang GX, Chakraborty C, Smith SJ, Barres BA (2012) Astrocyte glypicans 4 and 6 promote formation of excitatory synapses via GluA1 AMPA receptors. Nature 486:410–414

Amodeo LR, Kneiber D, Wills DN, Ehlers CL (2017) Alcohol drinking during adolescence increases consumptive responses to alcohol in adulthood in Wistar rats. Alcohol 59:43–51

Amodeo LR, Wills DN, Sanchez-Alavez M, Nguyen W, Conti B, Ehlers CL (2018) Intermittent voluntary ethanol consumption combined with ethanol vapor exposure during adolescence increases drinking and alters other behaviors in adulthood in female and male rats. Alcohol 73:57–66

Anacker C, Hen R (2017) Adult hippocampal neurogenesis and cognitive flexibility—linking memory and mood. Nat Rev Neurosci 18:335–346

Andersson U, Yang H, Harris H (2018) High-mobility group box 1 protein (HMGB1) operates as an alarmin outside as well as inside cells. Semin Immunol 38:40–48

Aron L, Zullo J, Yankner BA (2022) The adaptive aging brain. Curr Opin Neurobiol 72:91–100

Aucott H, Lundberg J, Salo H, Klevenvall L, Damberg P, Ottosson L, Andersson U, Holmin S, Erlandsson Harris H (2018) Neuroinflammation in response to intracerebral injections of different HMGB1 redox isoforms. J Innate Immun 10:215–227

Azam L, Winzer-Serhan U, Leslie FM (2003) Co-expression of alpha7 and beta2 nicotinic acetylcholine receptor subunit mRNAs within rat brain cholinergic neurons. Neuroscience 119:965–977

Ballas N, Grunseich C, Lu DD, Speh JC, Mandel G (2005) REST and its corepressors mediate plasticity of neuronal gene chromatin throughout neurogenesis. Cell 121:645–657

Bandeira F, Lent R, Herculano-Houzel S (2009) Changing numbers of neuronal and non-neuronal cells underlie postnatal brain growth in the rat. Proc Natl Acad Sci USA 106:14108–14113

Barak B, Feldman N, Okun E (2014) Toll-like receptors as developmental tools that regulate neurogenesis during development: an update. Front Neurosci 8:272

Bava S, Tapert SF (2010) Adolescent brain development and the risk for alcohol and other drug problems. Neuropsychol Rev 20:398–413

Bellver-Sanchis A, Geng Q, Navarro G, Avila-Lopez PA, Companys-Alemany J, Marsal-Garcia L, Larramona-Arcas R, Miro L, Perez-Bosque A, Ortuno-Sahagun D, Banerjee DR, Choudhary BS, Soriano FX, Poulard C, Pallas M, Du HN, Grinan-Ferre C (2024) G9a inhibition promotes neuroprotection through GMFB regulation in Alzheimer's disease. Aging Dis 15(1):311–337

Berger SL, Kouzarides T, Shiekhattar R, Shilatifard A (2009) An operational definition of epigenetics. Genes Dev 23:781–783

Berghofer B, Frommer T, Haley G, Fink L, Bein G, Hackstein H (2006) TLR7 ligands induce higher IFN-alpha production in females. J Immunol 177:2088–2096

Bianchi I, Lleo A, Gershwin ME, Invernizzi P (2012) The X chromosome and immune associated genes. J Autoimmun 38:J187–J192

Bilbo SD, Schwarz JM (2009) Early-life programming of later-life brain and behavior: a critical role for the immune system. Front Behav Neurosci 3:14

Bilbo SD, Schwarz JM (2012) The immune system and developmental programming of brain and behavior. Front Neuroendocrinol 33:267–286

Bilbo S, Stevens B (2017) Microglia: the brain's first responders. Cerebrum 2017:cer-14-17

Boutros N, Semenova S, Liu W, Crews FT, Markou A (2014) Adolescent intermittent ethanol exposure is associated with increased risky choice and decreased dopaminergic and cholinergic neuron markers in adult rats. Int J Neuropsychopharmacol 18

Brenhouse HC, Schwarz JM (2016) Immunoadolescence: Neuroimmune development and adolescent behavior. Neurosci Biobehav Rev 70:288–299

Broadwater M, Varlinskaya EI, Spear LP (2013) Effects of voluntary access to sweetened ethanol during adolescence on intake in adulthood. Alcohol Clin Exp Res 37:1048–1055

Broadwater MA, Liu W, Crews FT, Spear LP (2014) Persistent loss of hippocampal neurogenesis and increased cell death following adolescent, but not adult, chronic ethanol exposure. Dev Neurosci 36:297–305

Broadwater MA, Lee SH, Yu Y, Zhu H, Crews FT, Robinson DL, Shih YI (2018) Adolescent alcohol exposure decreases frontostriatal resting-state functional connectivity in adulthood. Addict Biol 23:810–823

Brynskikh A, Warren T, Zhu J, Kipnis J (2008) Adaptive immunity affects learning behavior in mice. Brain Behav Immun 22:861–869

Bushong EA, Martone ME, Ellisman MH (2004) Maturation of astrocyte morphology and the establishment of astrocyte domains during postnatal hippocampal development. Int J Dev Neurosci 22:73–86

Cai KC, Van Mil S, Murray E, Mallet JF, Matar C, Ismail N (2016) Age and sex differences in immune response following LPS treatment in mice. Brain Behav Immun 58:327–337

Cameron HA, Glover LR (2015) Adult neurogenesis: beyond learning and memory. Annu Rev Psychol 66:53–81

Carroll KM, Nich C, Devito EE, Shi JM, Sofuoglu M (2018) Galantamine and computerized cognitive

behavioral therapy for cocaine dependence: a randomized clinical trial. J Clin Psychiatry 79
Carroll KM, Devito EE, Yip SW, Nich C, Sofuoglu M (2019) Double-blind placebo-controlled trial of galantamine for methadone-maintained individuals with cocaine use disorder: secondary analysis of effects on illicit opioid use. Am J Addict 28:238–245
Chang Y, Levy D, Horton JR, Peng J, Zhang X, Gozani O, Cheng X (2011) Structural basis of SETD6-mediated regulation of the NF-kB network via methyl-lysine signaling. Nucleic Acids Res 39:6380–6389
Chavakis T, Bierhaus A, Nawroth PP (2004) RAGE (receptor for advanced glycation end products): a central player in the inflammatory response. Microbes Infect 6:1219–1225
Chintamen S, Imessadouene F, Kernie SG (2020) Immune regulation of adult neurogenic niches in health and disease. Front Cell Neurosci 14:571071
Cho HH, Cargnin F, Kim Y, Lee B, Kwon RJ, Nam H, Shen R, Barnes AP, Lee JW, Lee S, Lee SK (2014) Isl1 directly controls a cholinergic neuronal identity in the developing forebrain and spinal cord by forming cell type-specific complexes. PLoS Genet 10:e1004280
Christopherson KS, Ullian EM, Stokes CC, Mullowney CE, Hell JW, Agah A, Lawler J, Mosher DF, Bornstein P, Barres BA (2005) Thrombospondins are astrocyte-secreted proteins that promote CNS synaptogenesis. Cell 120:421–433
Coleman LG Jr, He J, Lee J, Styner M, Crews FT (2011) Adolescent binge drinking alters adult brain neurotransmitter gene expression, behavior, brain regional volumes, and neurochemistry in mice. Alcohol Clin Exp Res 35:671–688
Coleman LG Jr, Liu W, Oguz I, Styner M, Crews FT (2014) Adolescent binge ethanol treatment alters adult brain regional volumes, cortical extracellular matrix protein and behavioral flexibility. Pharmacol Biochem Behav 116:142–151
Coleman LG Jr, Zou J, Crews FT (2017) Microglial-derived miRNA let-7 and HMGB1 contribute to ethanol-induced neurotoxicity via TLR7. J Neuroinflammation 14:22
Consolim-Colombo FM, Sangaleti CT, Costa FO, Morais TL, Lopes HF, Motta JM, Irigoyen MC, Bortoloto LA, Rochitte CE, Harris YT, Satapathy SK, Olofsson PS, Akerman M, Chavan SS, Mackay M, Barnaby DP, Lesser ML, Roth J, Tracey KJ, Pavlov VA (2017) Galantamine alleviates inflammation and insulin resistance in patients with metabolic syndrome in a randomized trial. JCI Insight 2
Cooper-Kuhn CM, Winkler J, Kuhn HG (2004) Decreased neurogenesis after cholinergic forebrain lesion in the adult rat. J Neurosci Res 77:155–165
Cox MA, Bassi C, Saunders ME, Nechanitzky R, Morgado-Palacin I, Zheng C, Mak TW (2020) Beyond neurotransmission: acetylcholine in immunity and inflammation. J Intern Med 287:120–133
Crapser JD, Arreola MA, Tsourmas KI, Green KN (2021) Microglia as hackers of the matrix: sculpting synapses and the extracellular space. Cell Mol Immunol 18:2472–2488
Crews FT, Nixon K (2009) Mechanisms of neurodegeneration and regeneration in alcoholism. Alcohol Alcohol 44:115–127
Crews FT, Vetreno RP (2014) Neuroimmune basis of alcoholic brain damage. Int Rev Neurobiol 118:315–357
Crews FT, Vetreno RP (2022) Cholinergic REST-G9a gene repression through HMGB1-TLR4 neuroimmune signaling regulates basal forebrain cholinergic neuron phenotype. Front Mol Neurosci 15:992627
Crews FT, Braun CJ, Hoplight B, Switzer RC 3rd, Knapp DJ (2000) Binge ethanol consumption causes differential brain damage in young adolescent rats compared with adult rats. Alcohol Clin Exp Res 24:1712–1723
Crews F, He J, Hodge C (2007) Adolescent cortical development: a critical period of vulnerability for addiction. Pharmacol Biochem Behav 86:189–199
Crews FT, Qin L, Sheedy D, Vetreno RP, Zou J (2013) High mobility group box 1/Toll-like receptor danger signaling increases brain neuroimmune activation in alcohol dependence. Biol Psychiatry 73:602–612
Crews FT, Robinson DL, Chandler LJ, Ehlers CL, Mulholland PJ, Pandey SC, Rodd ZA, Spear LP, Swartzwelder HS, Vetreno RP (2019) Mechanisms of persistent neurobiological changes following adolescent alcohol exposure: NADIA Consortium findings. Alcohol Clin Exp Res 43:1806–1822
Crews FT, Fisher R, Deason C, Vetreno RP (2021a) Loss of basal forebrain cholinergic neurons following adolescent binge ethanol exposure: recovery with the cholinesterase inhibitor galantamine. Front Behav Neurosci 15:652494
Crews FT, Zou J, Coleman LG, JR. (2021b) Extracellular microvesicles promote microglia-mediated pro-inflammatory responses to ethanol. J Neurosci Res 99(8):1940–1956
Crews FT, Coleman LG Jr, Macht VA, Vetreno RP (2023a) Targeting persistent changes in neuroimmune and epigenetic signaling in adolescent drinking to treat alcohol use disorder in adulthood. Pharmacol Rev 75:380–396
Crews FT, Fisher RP, Qin L, Vetreno RP (2023b) HMGB1 neuroimmune signaling and REST-G9a gene repression contribute to ethanol-induced reversible suppression of the cholinergic neuron phenotype. Mol Psychiatry 28(12):5159–5172
Cui C, Shurtleff D, Harris RA (2014) Neuroimmune mechanisms of alcohol and drug addiction. Int Rev Neurobiol 118:1–12
Dajas-Bailador FA, Heimala K, Wonnacott S (2003) The allosteric potentiation of nicotinic acetylcholine receptors by galantamine is transduced into cellular responses in neurons: Ca2+ signals and neurotransmitter release. Mol Pharmacol 64:1217–1226
De Bellis MD, Clark DB, Beers SR, Soloff PH, Boring AM, Hall J, Kersh A, Keshavan MS (2000) Hippocampal volume in adolescent-onset alcohol use disorders. Am J Psychiatry 157:737–744

Deverman BE, Patterson PH (2009) Cytokines and CNS development. Neuron 64:61–78
Dewit DJ, Adlaf EM, Offord DR, Ogborne AC (2000) Age at first alcohol use: a risk factor for the development of alcohol disorders. Am J Psychiatry 157:745–750
Dinopoulos A, Uylings HB, Parnavelas JG (1992) The development of neurons in the nuclei of the horizontal and vertical limb of the diagonal band of Broca of the rat: a qualitative and quantitative analysis of Golgi preparations. Brain Res Dev Brain Res 65:65–74
Disabato DJ, Quan N, Godbout JP (2016) Neuroinflammation: the devil is in the details. J Neurochem 139(Suppl 2):136–153
Dobryakova YV, Kasianov A, Zaichenko MI, Stepanichev MY, Chesnokova EA, Kolosov PM, Markevich VA, Bolshakov AP (2017) Intracerebroventricular administration of (192)IgG-saporin alters expression of microglia-associated genes in the dorsal but not ventral hippocampus. Front Mol Neurosci 10:429
Doetkotte R, Opitz K, Kiianmaa K, Winterhoff H (2005) Reduction of voluntary ethanol consumption in alcohol-preferring Alko alcohol (AA) rats by desoxypeganine and galanthamine. Eur J Pharmacol 522:72–77
Donovan JE (2009) Estimated blood alcohol concentrations for child and adolescent drinking and their implications for screening instruments. Pediatrics 123:e975–e981
Dziabis JE, Bilbo SD (2022) Microglia and Sensitive Periods in Brain Development. Curr Top Behav Neurosci 53:55–78
Ehlers CL, Criado JR, Wills DN, Liu W, Crews FT (2011) Periadolescent ethanol exposure reduces adult forebrain ChAT+IR neurons: correlation with behavioral pathology. Neuroscience 199:333–345
Ehlers CL, Liu W, Wills DN, Crews FT (2013a) Periadolescent ethanol vapor exposure persistently reduces measures of hippocampal neurogenesis that are associated with behavioral outcomes in adulthood. Neuroscience 244:1–15
Ehlers CL, Oguz I, Budin F, Wills DN, Crews FT (2013b) Peri-adolescent ethanol vapor exposure produces reductions in hippocampal volume that are correlated with deficits in prepulse inhibition of the startle. Alcohol Clin Exp Res 37:1466–1475
Eisch AJ, Petrik D (2012) Depression and hippocampal neurogenesis: a road to remission? Science 338:72–75
Elewa HF, Hilali H, Hess DC, Machado LS, Fagan SC (2006) Minocycline for short-term neuroprotection. Pharmacotherapy 26:515–521
El Gazzar M, Yoza BK, Chen X, Garcia BA, Young NL, Mccall CE (2009) Chromatin-specific remodeling by HMGB1 and linker histone H1 silences proinflammatory genes during endotoxin tolerance. Mol Cell Biol 29:1959–1971
El Gazzar M, Liu T, Yoza BK, Mccall CE (2010) Dynamic and selective nucleosome repositioning during endotoxin tolerance. J Biol Chem 285:1259–1271
Emsley JG, Mitchell BD, Kempermann G, Macklis JD (2005) Adult neurogenesis and repair of the adult CNS with neural progenitors, precursors, and stem cells. Prog Neurobiol 75:321–341
Ernst M, Romeo RD, Andersen SL (2009) Neurobiology of the development of motivated behaviors in adolescence: a window into a neural systems model. Pharmacol Biochem Behav 93:199–211
Eroglu C, Allen NJ, Susman MW, O'Rourke NA, Park CY, Ozkan E, Chakraborty C, Mulinyawe SB, Annis DS, Huberman AD, Green EM, Lawler J, Dolmetsch R, Garcia KC, Smith SJ, Luo ZD, Rosenthal A, Mosher DF, Barres BA (2009) Gabapentin receptor alpha2delta-1 is a neuronal thrombospondin receptor responsible for excitatory CNS synaptogenesis. Cell 139:380–392
Eyolfson E, Khan A, Mychasiuk R, Lohman AW (2020) Microglia dynamics in adolescent traumatic brain injury. J Neuroinflammation 17:326
Fagan AM, Garber M, Barbacid M, Silos-Santiago I, Holtzman DM (1997) A role for TrkA during maturation of striatal and basal forebrain cholinergic neurons in vivo. J Neurosci 17:7644–7654
Fernandes J, Arida RM, Gomez-Pinilla F (2017) Physical exercise as an epigenetic modulator of brain plasticity and cognition. Neurosci Biobehav Rev 80:443–456
Fernandez GM, Savage LM (2017) Adolescent binge ethanol exposure alters specific forebrain cholinergic cell populations and leads to selective functional deficits in the prefrontal cortex. Neuroscience 361:129–143
Field RH, Gossen A, Cunningham C (2012) Prior pathology in the basal forebrain cholinergic system predisposes to inflammation-induced working memory deficits: reconciling inflammatory and cholinergic hypotheses of delirium. J Neurosci 32:6288–6294
Freeman MR (2010) Specification and morphogenesis of astrocytes. Science 330:774–778
Gajbhiye SV, Tripathi RK, Petare A, Potey AV, Shankar A (2018) Minocycline in alcohol withdrawal induced anxiety and alcohol relapse in rats. Curr Clin Pharmacol 13:65–72
Gano A, Vore AS, Sammakia MN, Deak T (2019) Assessment of extracellular cytokines in the hippocampus of the awake behaving rat using large-molecule microdialysis combined with multiplex arrays after acute and chronic ethanol exposure. Alcohol Clin Exp Res 43:640–654
Gano A, Wojcik H, Danseglio NC, Kelliher K, Varlinskaya EI, Deak T (2024) Adolescent intermittent ethanol (AIE) sensitized fever in male Sprague Dawley rats exposed to poly I:C in adulthood. Brain Behav Immun 120:82–97
Gass JT, Glen WB Jr, Mcgonigal JT, Trantham-Davidson H, Lopez MF, Randall PK, Yaxley R, Floresco SB, Chandler LJ (2014) Adolescent alcohol exposure reduces behavioral flexibility, promotes disinhibition, and increases resistance to extinction of ethanol self-administration in adulthood. Neuropsychopharmacology 39:2570–2583
Ge WP, Miyawaki A, Gage FH, Jan YN, Jan LY (2012) Local generation of glia is a major astrocyte source in postnatal cortex. Nature 484:376–380

Goll MG, Bestor TH (2005) Eukaryotic cytosine methyltransferases. Annu Rev Biochem 74:481–514

Gomez GI, Falcon RV, Maturana CJ, Labra VC, Salgado N, Rojas CA, Oyarzun JE, Cerpa W, Quintanilla RA, Orellana JA (2018) Heavy alcohol exposure activates astroglial hemichannels and pannexons in the hippocampus of adolescent rats: effects on neuroinflammation and astrocyte arborization. Front Cell Neurosci 12:472

Gonzalez-Perez O, Alvarez-Buylla A (2011) Oligodendrogenesis in the subventricular zone and the role of epidermal growth factor. Brain Res Rev 67:147–156

Gould E, Farris TW, Butcher LL (1989) Basal forebrain neurons undergo somatal and dendritic remodeling during postnatal development: a single-section Golgi and choline acetyltransferase analysis. Brain Res Dev Brain Res 46:297–302

Gould E, Woolf NJ, Butcher LL (1991) Postnatal development of cholinergic neurons in the rat: I. Forebrain. Brain Res Bull 27:767–789

Grant BF, Dawson DA (1997) Age at onset of alcohol use and its association with DSM-IV alcohol abuse and dependence: results from the National Longitudinal Alcohol Epidemiologic Survey. J Subst Abus 9:103–110

Grant BF, Dawson DA (1998) Age of onset of drug use and its association with DSM-IV drug abuse and dependence: results from the National Longitudinal Alcohol Epidemiologic Survey. J Subst Abus 10:163–173

Grant BF, Stinson FS, Harford TC (2001) Age at onset of alcohol use and DSM-IV alcohol abuse and dependence: a 12-year follow-up. J Subst Abus 13:493–504

Grant JD, Scherrer JF, Lynskey MT, Lyons MJ, Eisen SA, Tsuang MT, True WR, Bucholz KK (2006) Adolescent alcohol use is a risk factor for adult alcohol and drug dependence: evidence from a twin design. Psychol Med 36:109–118

Grinan-Ferre C, Marsal-Garcia L, Bellver-Sanchis A, Kondengaden SM, Turga RC, Vazquez S, Pallas M (2019) Pharmacological inhibition of G9a/GLP restores cognition and reduces oxidative stress, neuroinflammation and beta-Amyloid plaques in an early-onset Alzheimer's disease mouse model. Aging (Albany NY) 11:11591–11608

Guergues J, Wohlfahrt J, Zhang P, Liu B, Stevens SM, JR. (2020) Deep proteome profiling reveals novel pathways associated with pro-inflammatory and alcohol-induced microglial activation phenotypes. J Proteome 220:103753

Haake A, Nguyen K, Friedman L, Chakkamparambil B, Grossberg GT (2020) An update on the utility and safety of cholinesterase inhibitors for the treatment of Alzheimer's disease. Expert Opin Drug Saf 19:147–157

Hampel H, Mesulam MM, Cuello AC, Farlow MR, Giacobini E, Grossberg GT, Khachaturian AS, Vergallo A, Cavedo E, Snyder PJ, Khachaturian ZS (2018) The cholinergic system in the pathophysiology and treatment of Alzheimer's disease. Brain 141:1917–1933

Hanamsagar R, Alter MD, Block CS, Sullivan H, Bolton JL, Bilbo SD (2018) Generation of a microglial developmental index in mice and in humans reveals a sex difference in maturation and immune reactivity. Glia 66:460

He J, Crews FT (2007) Neurogenesis decreases during brain maturation from adolescence to adulthood. Pharmacol Biochem Behav 86:327–333

He J, Crews FT (2008) Increased MCP-1 and microglia in various regions of the human alcoholic brain. Exp Neurol 210:349–358

Healey KL, Kibble S, Hodges S, Reissner KJ, Testen A, Wills TA, Acheson SK, Siemsen BM, Mcfaddin JA, Scofield MD, Swartzwelder HS (2020) Enduring alterations in hippocampal astrocytesynaptic proximity following adolescent alcohol exposure: reversal by gabapentin. Neural Regen Res 15:1496–1501

Hersh LB, Shimojo M (2003) Regulation of cholinergic gene expression by the neuron restrictive silencer factor/repressor element-1 silencing transcription factor. Life Sci 72:2021–2028

Hersh LB, Kong CF, Sampson C, Mues G, Li YP, Fisher A, Hilt D, Baetge EE (1993) Comparison of the promoter region of the human and porcine choline acetyltransferase genes: localization of an important enhancer region. J Neurochem 61:306–314

Hingson RW, Edwards EM, Heeren T, Rosenbloom D (2009) Age of drinking onset and injuries, motor vehicle crashes, and physical fights after drinking and when not drinking. Alcohol Clin Exp Res 33:783–790

Isaev NK, Stelmashook EV, Genrikhs EE (2017) Role of nerve growth factor in plasticity of forebrain cholinergic neurons. Biochemistry (Mosc) 82:291–300

Janko C, Filipovic M, Munoz LE, Schorn C, Schett G, Ivanovic-Burmazovic I, Herrmann M (2014) Redox modulation of HMGB1-related signaling. Antioxid Redox Signal 20:1075–1085

Jiang D, Cope AL, Zhang J, Pennell M (2023) On the decoupling of evolutionary changes in mRNA and protein levels. Mol Biol Evol 40

Kaltschmidt C, Kaltschmidt B, Neumann H, Wekerle H, Baeuerle PA (1994) Constitutive NF-kappa B activity in neurons. Mol Cell Biol 14:3981–3992

Kawai T, Akira S (2007) Signaling to NF-kappaB by Toll-like receptors. Trends Mol Med 13:460–469

Kettenmann H, Kirchhoff F, Verkhratsky A (2013) Microglia: new roles for the synaptic stripper. Neuron 77:10–18

Kipp BT, Nunes PT, Galaj E, Hitchcock B, Nasra T, Poynor KR, Heide SK, Reitz NL, Savage LM (2021) Adolescent ethanol exposure alters cholinergic function and apical dendritic branching within the orbital frontal cortex. Neuroscience 473:52–65

Knuesel I, Chicha L, Britschgi M, Schobel SA, Bodmer M, Hellings JA, Toovey S, Prinssen EP (2014) Maternal immune activation and abnormal brain development across CNS disorders. Nat Rev Neurol 10:643–660

Kobayashi K, Imagama S, Ohgomori T, Hirano K, Uchimura K, Sakamoto K, Hirakawa A, Takeuchi H, Suzumura A, Ishiguro N, Kadomatsu K (2013)

Minocycline selectively inhibits M1 polarization of microglia. Cell Death Dis 4:e525
Kokare DM, Kyzar EJ, Zhang H, Sakharkar AJ, Pandey SC (2017) Adolescent alcohol exposure-induced changes in alpha-melanocyte stimulating hormone and neuropeptide Y pathways via histone acetylation in the brain during adulthood. Int J Neuropsychopharmacol 20:758–768
Kopec AM, Smith CJ, Ayre NR, Sweat SC, Bilbo SD (2018) Microglial dopamine receptor elimination defines sex-specific nucleus accumbens development and social behavior in adolescent rats. Nat Commun 9:3769
Koseki T, Mouri A, Suzuki S, Nakajima A, Mamiya T, Yan Y, Nabeshima T (2014) Galantamine attenuates reinstatement of cue-induced methamphetamine-seeking behavior in mice. Addict Biol 19:1–4
Kouzarides T (2007) Chromatin modifications and their function. Cell 128:693–705
Kraus CL, Salazar NC, Mitchell JR, Florin WD, Guenther B, Brady D, Swartzwelder SH, White AM (2005) Inconsistencies between actual and estimated blood alcohol concentrations in a field study of college students: do students really know how much they drink? Alcohol Clin Exp Res 29:1672–1676
Kucukdereli H, Allen NJ, Lee AT, Feng A, Ozlu MI, Conatser LM, Chakraborty C, Workman G, Weaver M, Sage EH, Barres BA, Eroglu C (2011) Control of excitatory CNS synaptogenesis by astrocyte-secreted proteins Hevin and SPARC. Proc Natl Acad Sci USA 108:E440–E449
Kyzar EJ, Zhang H, Sakharkar AJ, Pandey SC (2017) Adolescent alcohol exposure alters lysine demethylase 1 (LSD1) expression and histone methylation in the amygdala during adulthood. Addict Biol 22:1191–1204
Kyzar EJ, Zhang H, Pandey SC (2019) Adolescent alcohol exposure epigenetically suppresses amygdala arc enhancer RNA expression to confer adult anxiety susceptibility. Biol Psychiatry 85(11):904–914
Lamminpaa A, Vilska J, Korri UM, Riihimaki V (1993) Alcohol intoxication in hospitalized young teenagers. Acta Paediatr 82:783–788
Lee KM, Coehlo MA, Solton NR, Szumlinski KK (2017) Negative affect and excessive alcohol intake incubate during protracted withdrawal from binge-drinking in adolescent, but not adult, mice. Front Psychol 8:1128
Lenroot RK, Giedd JN (2006) Brain development in children and adolescents: insights from anatomical magnetic resonance imaging. Neurosci Biobehav Rev 30:718–729
Li Y, Li G, Zhang L, Li Y, Zhao Z (2022) G9a promotes inflammation in Streptococcus pneumoniae induced pneumonia mice by stimulating M1 macrophage polarization and H3K9me2 methylation in FOXP1 promoter region. Ann Transl Med 10:583
Libert C, Dejager L, Pinheiro I (2010) The X chromosome in immune functions: when a chromosome makes the difference. Nat Rev Immunol 10:594–604
Lilienfeld S (2002) Galantamine—a novel cholinergic drug with a unique dual mode of action for the treatment of patients with Alzheimer's disease. CNS Drug Rev 8:159–176
Linke R, Frotscher M (1993) Development of the rat septohippocampal projection: tracing with DiI and electron microscopy of identified growth cones. J Comp Neurol 332:69–88
Liu H, Lai M, Zhou X, Zhu H, Liu Y, Sun A, Ma B, Zhang F, Zhou W (2012) Galantamine attenuates the heroin seeking behaviors induced by cues after prolonged withdrawal in rats. Neuropharmacology 62:2515–2521
Liu T, Zhang L, Joo D, Sun SC (2017) NF-kappaB signaling in inflammation. Signal Transduct Target Ther 2
Liu W, Vetreno RP, Crews FT (2020) Hippocampal TNF-death receptors, caspase cell death cascades, and IL-8 in alcohol use disorder. Mol Psychiatry 26(6):2254–2262
Lovelock DF, Liu W, Langston SE, Liu J, Van Voorhies K, Giffin KA, Vetreno RP, Crews FT, Besheer J (2022) The Toll-like receptor 7 agonist imiquimod increases ethanol self-administration and induces expression of Toll-like receptor related genes. Addict Biol 27:e13176
Lu T, Aron L, Zullo J, Pan Y, Kim H, Chen Y, Yang TH, Kim HM, Drake D, Liu XS, Bennett DA, Colaiacovo MP, Yankner BA (2014) REST and stress resistance in ageing and Alzheimer's disease. Nature 507:448–454
Lucidi-Phillipi CA, Clary DO, Reichardt LF, Gage FH (1996) TrkA activation is sufficient to rescue axotomized cholinergic neurons. Neuron 16:653–663
Macht VA (2016) Neuro-immune interactions across development: A look at glutamate in the prefrontal cortex. Neurosci Biobehav Rev 71:267–280
Macht V, Crews FT, Vetreno RP (2020a) Neuroimmune and epigenetic mechanisms underlying persistent loss of hippocampal neurogenesis following adolescent intermittent ethanol exposure. Curr Opin Pharmacol 50:9–16
Macht V, Elchert N, Crews F (2020b) Adolescent alcohol exposure produces protracted cognitive-behavioral impairments in adult male and female rats. Brain Sci 10
Macht V, Vetreno R, Elchert N, Crews F (2021) Galantamine prevents and reverses neuroimmune induction and loss of adult hippocampal neurogenesis following adolescent alcohol exposure. J Neuroinflammation 18:212
Macht V, Vetreno RP, Elchert N, Fisher RP, Crews FT (2022a) Indomethacin restores adolescent ethanol persistent innate immune gene induction, loss of hippocampal cholinergic innervation and neurogenesis, and reversal learning deficits in adult male and female rats. ACER
Macht VA, Vetreno RP, Crews FT (2022b) Cholinergic and neuroimmune signaling interact to impact adult hippocampal neurogenesis and alcohol pathology across development. Front Pharmacol 13:849997
Macht V, Vetreno R, Elchert N, Fisher R, Crews F (2023) Indomethacin restores loss of hippocampal neurogen-

esis and cholinergic innervation and reduces innate immune expression and reversal learning deficits in adult male and female rats following adolescent ethanol exposure. Alcohol Clin Exp Res 47(3):470–485
Maclean RR, Waters AJ, Brede E, Sofuoglu M (2018) Effects of galantamine on smoking behavior and cognitive performance in treatment-seeking smokers prior to a quit attempt. Hum Psychopharmacol 33:e2665
Male D, Rezaie P (2001) Colonisation of the human central nervous system by microglia: the roles of chemokines and vascular adhesion molecules. Prog Brain Res 132:81–93
Mandrekar P, Dolganiuc A, Bellerose G, Kodys K, Romics L, Nizamani R, Szabo G (2002) Acute alcohol inhibits the induction of nuclear regulatory factor kappa B activation through CD14/toll-like receptor 4, interleukin-1, and tumor necrosis factor receptors: a common mechanism independent of inhibitory kappa B alpha degradation? Alcohol Clin Exp Res 26:1609–1614
Mann K, Ackermann K, Diehl A, Ebert D, Mundle G, Nakovics H, Reker T, Richter G, Schmidt LG, Driessen M, Rettig K, Opitz K, Croissant B (2006) Galantamine: a cholinergic patch in the treatment of alcoholism: a randomized, placebo-controlled trial. Psychopharmacology 184:115–121
Marco EM, Penasco S, Hernandez MD, Gil A, Borcel E, Moya M, Gine E, Lopez-Moreno JA, Guerri C, Lopez-Gallardo M, Rodriguez De Fonseca F (2017) Long-term effects of intermittent adolescent alcohol exposure in male and female rats. Front Behav Neurosci 11:233
Marques S, Zeisel A, Codeluppi S, Van Bruggen D, Mendanha Falcao A, Xiao L, Li H, Haring M, Hochgerner H, Romanov RA, Gyllborg D, Munoz Manchado A, LA Manno G, Lonnerberg P, Floriddia EM, Rezayee F, Ernfors P, Arenas E, Hjerling-Leffler J, Harkany T, Richardson WD, Linnarsson S, Castelo-Branco G (2016) Oligodendrocyte heterogeneity in the mouse juvenile and adult central nervous system. Science 352:1326–1329
Marshall SA, Mcclain JA, Wooden JI, Nixon K (2020) Microglia dystrophy following binge-like alcohol exposure in adolescent and adult male rats. Front Neuroanat 14:52
Martinez-Canabal A, Lopez-Oropeza G, Sotres-Bayon F (2023) Hippocampal neurogenesis facilitates cognitive flexibility in a fear discrimination task. Front Behav Neurosci 17:1331928
Matthews DA, Nadler JV, Lynch GS, Cotman CW (1974) Development of cholinergic innervation in the hippocampal formation of the rat. I. Histochemical demonstration of acetylcholinesterase activity. Dev Biol 36:130–141
Maya-Enero S, Ramis-Fernandez SM, Astals-Vizcaino M, Garcia-Algar O (2021) Neurocognitive and behavioral profile of fetal alcohol spectrum disorder. An Pediatr (Engl Ed) 95:208 e1–208 e9
Mcclain JA, Morris SA, Deeny MA, Marshall SA, Hayes DM, Kiser ZM, Nixon K (2011) Adolescent binge alcohol exposure induces long-lasting partial activation of microglia. Brain Behav Immun 25(Suppl 1):S120–S128
Melbourne JK, Chandler CM, Van Doorn CE, Bardo MT, Pauly JR, Peng H, Nixon K (2021) Primed for addiction: A critical review of the role of microglia in the neurodevelopmental consequences of adolescent alcohol drinking. Alcohol Clin Exp Res 45:1908–1926
Mesulam MM, Mufson EJ, Wainer BH, Levey AI (1983) Central cholinergic pathways in the rat: an overview based on an alternative nomenclature (Ch1-Ch6). Neuroscience 10:1185–1201
Miller KM, Risher ML, Acheson SK, Darlow M, Sexton HG, Schramm-Sapyta N, Swartzwelder HS (2017) Behavioral inefficiency on a risky decision-making task in adulthood after adolescent intermittent ethanol exposure in rats. Sci Rep 7:4680
Mira RG, Lira M, Quintanilla RA, Cerpa W (2020) Alcohol consumption during adolescence alters the hippocampal response to traumatic brain injury. Biochem Biophys Res Commun 528:514–519
Mollica L, De Marchis F, Spitaleri A, Dallacosta C, Pennacchini D, Zamai M, Agresti A, Trisciuoglio L, Musco G, Bianchi ME (2007) Glycyrrhizin binds to high-mobility group box 1 protein and inhibits its cytokine activities. Chem Biol 14:431–441
Montesinos J, Pascual M, Rodriguez-Arias M, Minarro J, Guerri C (2016) Involvement of TLR4 in the long-term epigenetic changes, rewarding and anxiety effects induced by intermittent ethanol treatment in adolescence. Brain Behav Immun 53:159–171
Moore EM, Riley EP (2015) What happens when children with fetal alcohol spectrum disorders become adults? Curr Dev Disord Rep 2:219–227
Nadler JV, Matthews DA, Cotman CW, Lynch GS (1974) Development of cholinergic innervation in the hippocampal formation of the rat. II. Quantitative changes in choline acetyltransferase and acetylcholinesterase activities. Dev Biol 36:142–154
Najjar S, Pearlman DM, Alper K, Najjar A, Devinsky O (2013) Neuroinflammation and psychiatric illness. J Neuroinflammation 10:43
Neupane SP, Skulberg A, Skulberg KR, Aass HC, Bramness JG (2016) Cytokine changes following acute ethanol intoxication in healthy men: a crossover study. Mediat Inflamm 2016:3758590
NSDUH (2020) Center for Behavioral Health Statistics and Quality. 2019 National Survey on Drug Use and Health Public Use File Codebook. Substance Abuse and Mental Health Services Administration, Rockville, MD
Nwachukwu KN, Healey KL, Swartzwelder HS, Marshall SA (2022a) The influence of sex on hippocampal neurogenesis and neurotrophic responses on the persistent effects of adolescent intermittent ethanol exposure into adulthood. Neuroscience 506:68–79
Nwachukwu KN, King DM, Healey KL, Swartzwelder HS, Marshall SA (2022b) Sex-specific effects of adolescent intermittent ethanol exposure-induced dysregulation of hippocampal glial cells in adulthood. Alcohol 100:31–39

Obray JD, Landin JD, Vaughan DT, Scofield MD, Chandler LJ (2022) Adolescent alcohol exposure reduces dopamine 1 receptor modulation of prelimbic neurons projecting to the nucleus accumbens and basolateral amygdala. Addict Neurosci 4

Okun E, Griffioen KJ, Mattson MP (2011) Toll-like receptor signaling in neural plasticity and disease. Trends Neurosci 34:269–281

Olson JK, Miller SD (2004) Microglia initiate central nervous system innate and adaptive immune responses through multiple TLRs. J Immunol 173:3916–3924

Orio L, Anton M, Rodriguez-Rojo IC, Correas A, Garcia-Bueno B, Corral M, De Fonseca FR, Garcia-Moreno LM, Maestu F, Cadaveira F (2018) Young alcohol binge drinkers have elevated blood endotoxin, peripheral inflammation and low cortisol levels: neuropsychological correlations in women. Addict Biol 23:1130–1144

Pandey SC, Sakharkar AJ, Tang L, Zhang H (2015) Potential role of adolescent alcohol exposure-induced amygdaloid histone modifications in anxiety and alcohol intake during adulthood. Neurobiol Dis 82:607–619

Pandey SC, Kyzar EJ, Zhang H (2017) Epigenetic basis of the dark side of alcohol addiction. Neuropharmacology 122:74–84

Paolicelli RC, Ferretti MT (2017) Function and dysfunction of microglia during brain development: consequences for synapses and neural circuits. Front Synaptic Neurosci 9:9

Paolicelli RC, Gross CT (2011) Microglia in development: linking brain wiring to brain environment. Neuron Glia Biol 7:77–83

Paolicelli RC, Bolasco G, Pagani F, Maggi L, Scianni M, Panzanelli P, Giustetto M, Ferreira TA, Guiducci E, Dumas L, Ragozzino D, Gross CT (2011) Synaptic pruning by microglia is necessary for normal brain development. Science 333:1456–1458

Park JS, Svetkauskaite D, He Q, Kim JY, Strassheim D, Ishizaka A, Abraham E (2004) Involvement of toll-like receptors 2 and 4 in cellular activation by high mobility group box 1 protein. J Biol Chem 279:7370–7377

Pascual M, Blanco AM, Cauli O, Minarro J, Guerri C (2007) Intermittent ethanol exposure induces inflammatory brain damage and causes long-term behavioural alterations in adolescent rats. Eur J Neurosci 25:541–550

Pascual M, Boix J, Felipo V, Guerri C (2009) Repeated alcohol administration during adolescence causes changes in the mesolimbic dopaminergic and glutamatergic systems and promotes alcohol intake in the adult rat. J Neurochem 108:920–931

Pascual M, Montesinos J, Marcos M, Torres JL, Costa-Alba P, Garcia-Garcia F, Laso FJ, Guerri C (2017) Gender differences in the inflammatory cytokine and chemokine profiles induced by binge ethanol drinking in adolescence. Addict Biol 22(6):1829–1841

Patrick ME, Schulenberg JE, Martz ME, Maggs JL, O'Malley PM, Johnston LD (2013) Extreme binge drinking among 12th-grade students in the United States: prevalence and predictors. JAMA Pediatr 167:1019–1025

Plane JM, Shen Y, Pleasure DE, Deng W (2010) Prospects for minocycline neuroprotection. Arch Neurol 67:1442–1448

Prescott CA, Kendler KS (1999) Genetic and environmental contributions to alcohol abuse and dependence in a population-based sample of male twins. Am J Psychiatry 156:34–40

Pruett SB, Zheng Q, Fan R, Matthews K, Schwab C (2004) Ethanol suppresses cytokine responses induced through Toll-like receptors as well as innate resistance to Escherichia coli in a mouse model for binge drinking. Alcohol 33:147–155

Pu H, Zhai P, Gurney M (1993) Enhancer, silencer, and growth factor responsive regulatory sequences in the promoter for the mouse choline acetyltransferase gene. Mol Cell Neurosci 4:131–142

Qin L, Crews FT (2012) Chronic ethanol increases systemic TLR3 agonist-induced neuroinflammation and neurodegeneration. J Neuroinflammation 9:130

Qin L, Wu X, Block ML, Liu Y, Breese GR, Hong JS, Knapp DJ, Crews FT (2007) Systemic LPS causes chronic neuroinflammation and progressive neurodegeneration. Glia 55:453–462

Qin L, He J, Hanes RN, Pluzarev O, Hong JS, Crews FT (2008) Increased systemic and brain cytokine production and neuroinflammation by endotoxin following ethanol treatment. J Neuroinflammation 5:10

Ramesh G, Benge S, Pahar B, Philipp MT (2012) A possible role for inflammation in mediating apoptosis of oligodendrocytes as induced by the Lyme disease spirochete Borrelia burgdorferi. J Neuroinflammation 9:72

Ren C, Tong YL, Li JC, Lu ZQ, Yao YM (2017) The protective effect of alpha 7 nicotinic acetylcholine receptor activation on critical illness and its mechanism. Int J Biol Sci 13:46–56

Robinson DL, Amodeo LR, Chandler LJ, Crews FT, Ehlers CL, Gomez AA, Healey KL, Kuhn CM, Macht VA, Marshall SA, Swartzwelder HS, Varlinskaya EI, Werner DF (2021) The role of sex in the persistent effects of adolescent alcohol exposure on behavior and neurobiology in rodents. Int Rev Neurobiol 160:305–340

Rodd-Henricks ZA, Bell RL, Kuc KA, Murphy JM, Mcbride WJ, Lumeng L, LI, T. K. (2002) Effects of ethanol exposure on subsequent acquisition and extinction of ethanol self-administration and expression of alcohol-seeking behavior in adult alcohol-preferring (P) rats: I. Periadolescent exposure. Alcohol Clin Exp Res 26:1632–1641

Roopra A, Qazi R, Schoenike B, Daley TJ, Morrison JF (2004) Localized domains of G9a-mediated histone methylation are required for silencing of neuronal genes. Mol Cell 14:727–738

Sakharkar AJ, Vetreno RP, Zhang H, Kokare DM, Crews FT, Pandey SC (2016) A role for histone acetylation mechanisms in adolescent alcohol exposure-induced deficits in hippocampal brain-derived neurotrophic factor expression and neurogenesis markers in adulthood. Brain Struct Funct 221:4691–4703

SAMHSA (2022) Center for Behavioral Health Statistics and Quality. 2022 National Survey on Drug Use and Health

Samochocki M, Hoffle A, Fehrenbacher A, Jostock R, Ludwig J, Christner C, Radina M, Zerlin M, Ullmer C, Pereira EF, Lubbert H, Albuquerque EX, Maelicke A (2003) Galantamine is an allosterically potentiating ligand of neuronal nicotinic but not of muscarinic acetylcholine receptors. J Pharmacol Exp Ther 305:1024–1036

Sanchez-Alavez M, Nguyen W, Mori S, Wills DN, Otero D, Ehlers CL, Conti B (2019) Time course of microglia activation and brain and blood cytokine/chemokine levels following chronic ethanol exposure and protracted withdrawal in rats. Alcohol 76:37–45

Sauvageot CM, Stiles CD (2002) Molecular mechanisms controlling cortical gliogenesis. Curr Opin Neurobiol 12:244–249

Scheffel J, Regen T, Van Rossum D, Seifert S, Ribes S, Nau R, Parsa R, Harris RA, Boddeke HW, Chuang HN, Pukrop T, Wessels JT, Jurgens T, Merkler D, Bruck W, Schnaars M, Simons M, Kettenmann H, Hanisch UK (2012) Toll-like receptor activation reveals developmental reorganization and unmasks responder subsets of microglia. Glia 60:1930–1943

Schwarz JM, Bilbo SD (2012) Sex, glia, and development: interactions in health and disease. Horm Behav 62:243–253

Schwarz JM, Bilbo SD (2013) Adolescent morphine exposure affects long-term microglial function and later-life relapse liability in a model of addiction. J Neurosci 33:961–971

Schwarz JM, Hutchinson MR, Bilbo SD (2011) Early-life experience decreases drug-induced reinstatement of morphine CPP in adulthood via microglial-specific epigenetic programming of anti-inflammatory IL-10 expression. J Neurosci 31:17835–17847

Schwarz JM, Sholar PW, Bilbo SD (2012) Sex differences in microglial colonization of the developing rat brain. J Neurochem 120:948–963

Seemiller LR, Flores-Cuadra J, Griffith KR, Smith GC, Crowley NA (2024) Alcohol and stress exposure across the lifespan are key risk factors for Alzheimer's disease and cognitive decline. Neurobiol Stress 29:100605

Sey NYA, Gomez AA, Madayag AC, Boettiger CA, Robinson DL (2019) Adolescent intermittent ethanol impairs behavioral flexibility in a rat foraging task in adulthood. Behav Brain Res 373:112085

Shaw GA, Dupree JL, Neigh GN (2020) Adolescent maturation of the prefrontal cortex: Role of stress and sex in shaping adult risk for compromise. Genes Brain Behav 19:e12626

Sher KJ, Gotham HJ (1999) Pathological alcohol involvement: a developmental disorder of young adulthood. Dev Psychopathol 11:933–956

Shimojo M, Hersh LB (2004) Regulation of the cholinergic gene locus by the repressor element-1 silencing transcription factor/neuron restrictive silencer factor (REST/NRSF). Life Sci 74:2213–2225

Silva-Gotay A, Davis J, Tavares ER, Richardson HN (2021) Alcohol drinking during early adolescence activates microglial cells and increases frontolimbic Interleukin-1 beta and Toll-like receptor 4 gene expression, with heightened sensitivity in male rats compared to females. Neuropharmacology 197:108698

Sitapara RA, Antoine DJ, Sharma L, Patel VS, Ashby CR Jr, Gorasiya S, Yang H, Zur M, Mantell LL (2014) The alpha7 nicotinic acetylcholine receptor agonist GTS-21 improves bacterial clearance in mice by restoring hyperoxia-compromised macrophage function. Mol Med 20:238–247

Smith HR, Pang KC (2005) Orexin-saporin lesions of the medial septum impair spatial memory. Neuroscience 132:261–271

Sofuoglu M, Carroll KM (2011) Effects of galantamine on cocaine use in chronic cocaine users. Am J Addict 20:302–303

Souyris M, Mejia JE, Chaumeil J, Guery JC (2019) Female predisposition to TLR7-driven autoimmunity: gene dosage and the escape from X chromosome inactivation. Semin Immunopathol 41:153–164

Spear LP (2018) Effects of adolescent alcohol consumption on the brain and behaviour. Nat Rev Neurosci 19:197–214

Stevens B, Allen NJ, Vazquez LE, Howell GR, Christopherson KS, Nouri N, Micheva KD, Mehalow AK, Huberman AD, Stafford B, Sher A, Litke AM, Lambris JD, Smith SJ, John SW, Barres BA (2007) The classical complement cascade mediates CNS synapse elimination. Cell 131:1164–1178

Streit WJ, Xue QS, Tischer J, Bechmann I (2014) Microglial pathology. Acta Neuropathol Commun 2:142

Swartzwelder HS, Acheson SK, Miller KM, Sexton HG, Liu W, Crews FT, Risher ML (2015) Adolescent intermittent alcohol exposure: deficits in object recognition memory and forebrain cholinergic markers. PLoS One 10:e0140042

Swartzwelder HS, Healey KL, Liu W, Dubester K, Miller KM, Crews FT (2019) Changes in neuroimmune and neuronal death markers after adolescent alcohol exposure in rats are reversed by donepezil. Sci Rep 9:12110

Syrett CM, Anguera MC (2019) When the balance is broken: X-linked gene dosage from two X chromosomes and female-biased autoimmunity. J Leukoc Biol 106:919–932

Szabo G, Mandrekar P (2009) A recent perspective on alcohol, immunity, and host defense. Alcohol Clin Exp Res 33:220–232

Szabo G, Mandrekar P, Catalano D (1995) Inhibition of superantigen-induced T cell proliferation and monocyte IL-1 beta, TNF-alpha, and IL-6 production by acute ethanol treatment. J Leukoc Biol 58:342–350

Taffe MA, Kotzebue RW, Crean RD, Crawford EF, Edwards S, Mandyam CD (2010) Long-lasting reduction in hippocampal neurogenesis by alcohol con-

sumption in adolescent nonhuman primates. Proc Natl Acad Sci USA 107:11104–11109
Tan YL, Yuan Y, Tian L (2020) Microglial regional heterogeneity and its role in the brain. Mol Psychiatry 25:351–367
Tang D, Loze MT, Zeh HJ, Kang R (2010) The redox protein HMGB1 regulates cell death and survival in cancer treatment. Autophagy 6:1181–1183
Tervo-Clemmens B, Calabro FJ, Parr AC, Fedor J, Foran W, Luna B (2023) A canonical trajectory of executive function maturation from adolescence to adulthood. Nat Commun 14:6922
Testen A, Ali M, Sexton HG, Hodges S, Dubester K, Reissner KJ, Swartzwelder HS, Risher ML (2019) Region-specific differences in morphometric features and synaptic colocalization of astrocytes during development. Neuroscience 400:98–109
Thiel G, Ekici M, Rossler OG (2015) RE-1 silencing transcription factor (REST): a regulator of neuronal development and neuronal/endocrine function. Cell Tissue Res 359:99–109
Tien AC, Tsai HH, Molofsky AV, Mcmahon M, Foo LC, Kaul A, Dougherty JD, Heintz N, Gutmann DH, Barres BA, Rowitch DH (2012) Regulated temporal-spatial astrocyte precursor cell proliferation involves BRAF signalling in mammalian spinal cord. Development 139:2477–2487
Toalston JE, Deehan GA Jr, Hauser SR, Engleman EA, Bell RL, Murphy JM, Mcbride WJ, Rodd ZA (2015) The reinforcing properties of ethanol are quantitatively enhanced in adulthood by peri-adolescent ethanol, but not saccharin, consumption in female alcohol-preferring (P) rats. Alcohol 49:513–518
Tobon-Velasco JC, Cuevas E, Torres-Ramos MA (2014) Receptor for AGEs (RAGE) as mediator of NF-kB pathway activation in neuroinflammation and oxidative stress. CNS Neurol Disord Drug Targets 13:1615–1626
Tomioka T, Shimazaki T, Yamauchi T, Oki T, Ohgoh M, Okano H (2014) LIM homeobox 8 (Lhx8) is a key regulator of the cholinergic neuronal function via a tropomyosin receptor kinase A (TrkA)-mediated positive feedback loop. J Biol Chem 289:1000–1010
Toni N, Schinder AF (2015) Maturation and Functional Integration of New Granule Cells into the Adult Hippocampus. Cold Spring Harb Perspect Biol 8:a018903
Town T, Nikolic V, Tan J (2005) The microglial "activation" continuum: from innate to adaptive responses. J Neuroinflammation 2:24
Tremblay ME, Stevens B, Sierra A, Wake H, Bessis A, Nimmerjahn A (2011) The role of microglia in the healthy brain. J Neurosci 31:16064–16069
Ulloa L (2005) The vagus nerve and the nicotinic anti-inflammatory pathway. Nat Rev Drug Discov 4:673–684
Varlinskaya EI, Truxell E, Spear LP (2014) Chronic intermittent ethanol exposure during adolescence: effects on social behavior and ethanol sensitivity in adulthood. Alcohol 48:433–444
Vela JM, Molina-Holgado E, Arevalo-Martin A, Almazan G, Guaza C (2002) Interleukin-1 regulates proliferation and differentiation of oligodendrocyte progenitor cells. Mol Cell Neurosci 20:489–502
Vetreno RP, Crews FT (2012) Adolescent binge drinking increases expression of the danger signal receptor agonist HMGB1 and Toll-like receptors in the adult prefrontal cortex. Neuroscience 226:475–488
Vetreno RP, Crews FT (2015) Binge ethanol exposure during adolescence leads to a persistent loss of neurogenesis in the dorsal and ventral hippocampus that is associated with impaired adult cognitive functioning. Front Neurosci 9:35
Vetreno RP, Crews FT (2018) Adolescent binge ethanol-induced loss of basal forebrain cholinergic neurons and neuroimmune activation are prevented by exercise and indomethacin. PLoS One 13:e0204500
Vetreno RP, Qin L, Crews FT (2013) Increased receptor for advanced glycation end product expression in the human alcoholic prefrontal cortex is linked to adolescent drinking. Neurobiol Dis 59:52–62
Vetreno RP, Broadwater M, Liu W, Spear LP, Crews FT (2014) Adolescent, but not adult, binge ethanol exposure leads to persistent global reductions of choline acetyltransferase expressing neurons in brain. PLoS One 9:e113421
Vetreno RP, Yaxley R, Paniagua B, Crews FT (2016) Diffusion tensor imaging reveals adolescent binge ethanol-induced brain structural integrity alterations in adult rats that correlate with behavioral dysfunction. Addict Biol 21:939–953
Vetreno RP, Patel Y, Patel U, Walter TJ, Crews FT (2017a) Adolescent intermittent ethanol reduces serotonin expression in the adult raphe nucleus and upregulates innate immune expression that is prevented by exercise. Brain Behav Immun 60:333–345
Vetreno RP, Yaxley R, Paniagua B, Johnson GA, Crews FT (2017b) Adult rat cortical thickness changes across age and following adolescent intermittent ethanol treatment. Addict Biol 22:712–723
Vetreno RP, Lawrimore CJ, Rowsey PJ, Crews FT (2018) Persistent adult neuroimmune activation and loss of hippocampal neurogenesis following adolescent ethanol exposure: blockade by exercise and the anti-inflammatory drug indomethacin. Front Neurosci 12:200
Vetreno RP, Bohnsack JP, Kusumo H, Liu W, Pandey SC, Crews FT (2019) Neuroimmune and epigenetic involvement in adolescent binge ethanol-induced loss of basal forebrain cholinergic neurons: Restoration with voluntary exercise. Addict Biol 25(2):e12731
Vetreno RP, Bohnsack JP, Kusumo H, Liu W, Pandey SC, Crews FT (2020) Neuroimmune and epigenetic involvement in adolescent binge ethanol-induced loss of basal forebrain cholinergic neurons: Restoration with voluntary exercise. Addict Biol 25:e12731
Vetreno RP, Qin L, Coleman LG Jr, Crews FT (2021) Increased toll-like receptor-MyD88-NFkappaB-proinflammatory neuroimmune signaling in the

orbitofrontal cortex of human alcohol use disorder. Alcohol Clin Exp Res 45(9):1747–1761
Vetreno RP, Campbell J, Crews FT (2023) A multicomponent ethanol response battery across a cumulative dose ethanol challenge reveals diminished adolescent rat ethanol responsivity relative to adults. Adv Drug Alcohol Res 3
Vore AS, Barney TM, Gano A, Varlinskaya EI, Deak T (2021) Adolescent intermittent ethanol (AIE) produces sex specific alterations in adult neuroimmune gene expression and ethanol sensitivity that are independent of ethanol metabolism. Neuropharmacology 195:108635
Walter TJ, Crews FT (2017) Microglial depletion alters the brain neuroimmune response to acute binge ethanol withdrawal. J Neuroinflammation 14:86
Walter TJ, Vetreno RP, Crews FT (2017) Alcohol and stress activation of microglia and neurons: brain regional effects. Alcohol Clin Exp Res 41:2066–2081
Wang H, Liao H, Ochani M, Justiniani M, Lin X, Yang L, Al-Abed Y, Wang H, Metz C, Miller EJ, Tracey KJ, Ulloa L (2004) Cholinergic agonists inhibit HMGB1 release and improve survival in experimental sepsis. Nat Med 10:1216–1221
Wang Z, Zang C, Rosenfeld JA, Schones DE, Barski A, Cuddapah S, Cui K, Roh TY, Peng W, Zhang MQ, Zhao K (2008) Combinatorial patterns of histone acetylations and methylations in the human genome. Nat Genet 40:897–903
Wang HT, Huang FL, Hu ZL, Zhang WJ, Qiao XQ, Huang YQ, Dai RP, Li F, Li CQ (2017) Early-life social isolation-induced depressive-like behavior in rats results in microglial activation and neuronal histone methylation that are mitigated by minocycline. Neurotox Res 31:505–520
Wazea SA, Wadie W, Bahgat AK, El-Abhar HS (2018) Galantamine anti-colitic effect: Role of alpha-7 nicotinic acetylcholine receptor in modulating Jak/STAT3, NF-kappaB/HMGB1/RAGE and p-AKT/Bcl-2 pathways. Sci Rep 8:5110
Webler RD, Fulton S, Perera TD, Coplan JD (2019) Maturational phase of hippocampal neurogenesis and cognitive flexibility. Neurosci Lett 711:134414
White AM, Bae JG, Truesdale MC, Ahmad S, Wilson WA, Swartzwelder HS (2002) Chronic-intermittent ethanol exposure during adolescence prevents normal developmental changes in sensitivity to ethanol-induced motor impairments. Alcohol Clin Exp Res 26:960–968
White AM, Kraus CL, Swartzwelder H (2006) Many college freshmen drink at levels far beyond the binge threshold. Alcohol Clin Exp Res 30:1006–1010
Whitman BA, Knapp DJ, Werner DF, Crews FT, Breese GR (2013) The cytokine mRNA increase induced by withdrawal from chronic ethanol in the sterile environment of brain is mediated by CRF and HMGB1 release. Alcohol Clin Exp Res 37:2086–2097
Wieghofer P, Prinz M (2016) Genetic manipulation of microglia during brain development and disease. Biochim Biophys Acta 1862:299–309
Wille-Bille A, De Olmos S, Marengo L, Chiner F, Pautassi RM (2017) Long-term ethanol self-administration induces DeltaFosB in male and female adolescent, but not in adult, Wistar rats. Prog Neuro-Psychopharmacol Biol Psychiatry 74:15–30
Williamson LL, Sholar PW, Mistry RS, Smith SH, Bilbo SD (2011) Microglia and memory: modulation by early-life infection. J Neurosci 31:15511–15521
Wolff JR (1970) The astrocyte as link between capillary and nerve cell. Triangle 9:153–164
Wolstenholme JT, Mahmood T, Harris GM, Abbas S, Miles MF (2017) Intermittent ethanol during adolescence leads to lasting behavioral changes in adulthood and alters gene expression and histone methylation in the PFC. Front Mol Neurosci 10:307
Wooden JI, Peacoe LE, Anasooya Shaji C, Melbourne JK, Chandler CM, Bardo MT, Nixon K (2023) Adolescent intermittent ethanol drives modest neuroinflammation but does not escalate drinking in male rats. Cells 12
Wu Y, Lousberg EL, Moldenhauer LM, Hayball JD, Robertson SA, Coller JK, Watkins LR, Somogyi AA, Hutchinson MR (2011) Attenuation of microglial and IL-1 signaling protects mice from acute alcohol-induced sedation and/or motor impairment. Brain Behav Immun 25(Suppl 1):S155–S164
Xin W, Kaneko M, Roth RH, Zhang A, Nocera S, Ding JB, Stryker MP, Chan JR (2023) Adolescent oligodendrogenesis and myelination restrict experience-dependent neuronal plasticity in adult visual cortex. bioRxiv. https://doi.org/10.1101/2023.09.29.560231
Yang H, Wang H, Ju Z, Ragab AA, Lundback P, Long W, Valdes-Ferrer SI, HE M, Pribis JP, Li J, Lu B, Gero D, Szabo C, Antoine DJ, Harris HE, Golenbock DT, Meng J, Roth J, Chavan SS, Andersson U, Billiar TR, Tracey KJ, Al-Abed Y (2015) MD-2 is required for disulfide HMGB1-dependent TLR4 signaling. J Exp Med 212:5–14
Yang H, Chen Z, Gao W (2023) EHMT2 affects microglia polarization and aggravates neuronal damage and inflammatory response via regulating HMOX1. Transl Neurosci 14:20220276
Zahalka EA, Seidler FJ, Lappi SE, Yanai J, Slotkin TA (1993) Differential development of cholinergic nerve terminal markers in rat brain regions: implications for nerve terminal density, impulse activity and specific gene expression. Brain Res 601:221–229
Zengeler KE, Lukens JR (2021) Innate immunity at the crossroads of healthy brain maturation and neurodevelopmental disorders. Nat Rev Immunol 21:454–468
Zhang W, Xiao D, Mao Q, Xia H (2023) Role of neuroinflammation in neurodegeneration development. Signal Transduct Target Ther 8:267
Zhao Y, Zhu M, Yu Y, Qiu L, Zhang Y, He L, Zhang J (2017) Brain REST/NRSF is not only a silent repressor but also an active protector. Mol Neurobiol 54:541–550
Zheng Y, Liu A, Wang ZJ, Cao Q, Wang W, Lin L, Ma K, Zhang F, Wei J, Matas E, Cheng J, Chen GJ, Wang X, Yan Z (2019) Inhibition of EHMT1/2 rescues synaptic

and cognitive functions for Alzheimer's disease. Brain 142:787–807

Zorbaz T, Madrer N, Soreq H (2022) Cholinergic blockade of neuroinflammation: from tissue to RNA regulators. Neuronal Signal 6:NS20210035

Zou JY, Crews FT (2014) Release of neuronal HMGB1 by ethanol through decreased HDAC activity activates brain neuroimmune signaling. PLoS One 9:e87915

Zou J, Walter TJ, Barnett A, Rohlman A, Crews FT, Coleman LG, JR. (2022) Ethanol induces secretion of proinflammatory extracellular vesicles that inhibit adult hippocampal neurogenesis through G9a/GLP-epigenetic signaling. Front Immunol 13:866073

Zullo JM, Drake D, Aron L, O'Hern P, Dhamne SC, Davidsohn N, Mao CA, Klein WH, Rotenberg A, Bennett DA, Church GM, Colaiacovo MP, Yankner BA (2019) Regulation of lifespan by neural excitation and REST. Nature 574:359–364

Part III

Effects of Alcohol on Aging

10 Excessive Alcohol Use as a Risk Factor for Alzheimer's Disease: Epidemiological and Preclinical Evidence

Paige E. Anton, Nicole M. Maphis, David N. Linsenbardt, and Leon G. Coleman Jr.

Abstract

Alcohol use has recently emerged as a modifiable risk factor for Alzheimer's disease (AD). However, the neurobiological mechanisms by which alcohol interacts with AD pathogenesis remain poorly understood. In this chapter, we review the epidemiological and preclinical support for the interaction between alcohol use and AD. We hypothesize that alcohol use increases the rate of accumulation of specific AD-relevant pathologies during the prodromal phase and exacerbates dementia onset and progression. We find that alcohol consumption rates are increasing in adolescence, middle age, and aging populations. In tandem, rates of AD are also on the rise, potentially as a result of this increased alcohol use throughout the lifespan. We then review the biological processes in common between alcohol use disorder and AD as a means to uncover potential mechanisms by which they interact; these include oxidative stress, neuroimmune function, metabolism, pathogenic tauopathy development and spread, and neuronal excitatory/inhibitory balance (EIB). Finally, we provide some forward-thinking suggestions we believe this field should consider. In particular, the inclusion of alcohol use assessments in longitudinal studies of AD and more preclinical studies on alcohol's impacts using better animal models of late-onset Alzheimer's disease (LOAD).

Keywords

Ethanol · Alzheimer's disease · Dementia · Oxidative stress · Neuroimmune · Metabolism · Excitation · Inhibition

P. E. Anton · L. G. Coleman Jr. (✉)
Bowles Center for Alcohol Studies, University of North Carolina at Chapel Hill School of Medicine, Chapel Hill, NC, USA

Department of Pharmacology, University of North Carolina at Chapel Hill School of Medicine, Chapel Hill, NC, USA
e-mail: leon_coleman@med.unc.edu

N. M. Maphis · D. N. Linsenbardt (✉)
Department of Neurosciences and New Mexico Alcohol Research Center, School of Medicine, University of New Mexico Health Sciences Center, Albuquerque, NM, USA
e-mail: Dlinsenbardt@salud.unm.edu

10.1 Introduction

10.1.1 Alzheimer's Disease

Alzheimer's disease (AD), the most common type of dementia, is a chronic neurodegenerative disease predominantly defined by the accumulation of amyloid beta (Aβ) as extracellular plaques and pathologically modified protein tau as neuro-

C. F. Valenzuela et al. (eds.), *Effects of Alcohol on the Brain across the Lifespan*, Advances in Experimental Medicine and Biology 1473, https://doi.org/10.1007/978-3-031-81908-7_10

fibrillary tangles (NFTs; Knopman et al. 2021; Korczyn and Grinberg 2024). Clinically, AD is characterized by progressive memory impairments, difficulties in thinking, and neuropsychiatric symptoms (Apostolova et al. 2014; Li et al. 2014), including depression (Benoit et al. 2012; der Mussele et al. 2012), apathy (Ota et al. 2011), aggression (Guadagna et al. 2012), and psychosis (Gilley et al. 2004). AD is currently the seventh leading cause of death in the United States, but with global AD cases expected to reach 153 million by 2050 (GBD 2019 Dementia Forecasting Collaborators 2022; The Alzheimer's Association 2024), it is beginning to outpace other causes of death and therefore remains a pressing public health concern.

Despite the high disease burden, preventative treatments for AD do not exist, partially due to an incomplete understanding of the complex etiology of the disease (Scheltens et al. 2021). Approximately 5% of AD cases are hereditary (familial AD) and primarily arise from mutations in genes leading to abnormal amyloid beta (Aβ) processing such as *amyloid precursor protein (App)*, *presenilin 1 (Ps1)*, and *presenilin 2 (Ps2)* (Lanoiselée et al. 2017). However, *most* AD diagnoses (>95%) occur later in life and are referred to as late-onset Alzheimer's disease (LOAD) or sporadic AD. LOAD is characterized by an estimated 20-year preclinical phase of pathological aggregation (Vermunt et al. 2019), during which time changes in brain structure and function occur without any outward clinical symptoms (Fig. 10.1). Following the preclinical phase, the prodromal phase is characterized by progressive memory deficits in which mild cognitive impairment (MCI) is first diagnosed likely due to the now extensive pathology that has accumulated during the preclinical stage (Hampel and Lista 2012). Earlier AD prevalence studies used clinical symptoms to classify patients as having AD, but we have since learned that this approach led to a ~30% misdiagnosis rate compared to current blood/CSF and neuroimaging approaches (Hansson 2021; Hansson et al. 2022). Further advancements in these technologies have even

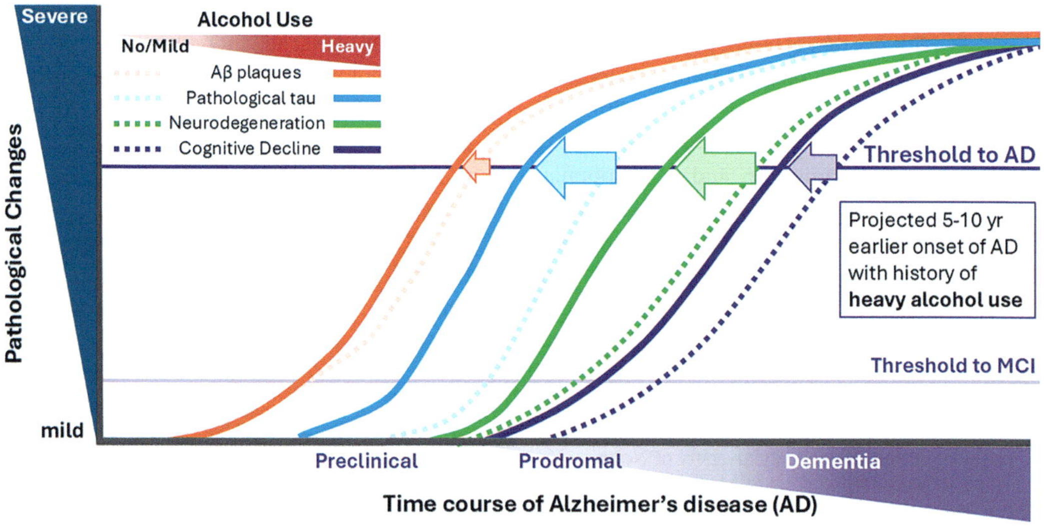

Fig. 10.1 Pathogenesis of late-onset Alzheimer's disease (LOAD, AD) and the proposed impact of alcohol misuse. Alcohol misuse throughout key developmental periods could both accelerate onset and exacerbate progression of AD measured by the biomarkers in the ATN framework: the amyloid beta plaques (A), pathological tau (T), and neurodegeneration (N). Development of these pathologic factors often develops before the outward manifestations of clinically diagnosed AD dementia (i.e., cognitive decline). There is an approximate 10–20-year period of preclinical changes, where cognition is not impacted. The Prodromal phase (Mild Cognitive Impairment, MCI) is characterized by the first appearance of cognitive decline. Dementia of the Alzheimer's type can exist as mild, moderate, and severe. Our hypothesis is that heavy alcohol use (solid line) can exacerbate and accelerate onset of AD (dotted line indicates "normal" AD pathogenesis), as indicated by the arrows

allowed us to begin to mechanistically characterize each of the phases of AD: preclinical, prodromal (MCI), and AD (Jack et al. 2018; Janelidze et al. 2020; Moscoso et al. 2021; Leuzy et al. 2022), as well as to determine the clinical efficacy of AD therapeutic candidates (Angioni et al. 2022; Hansson et al. 2023). This work continues, but has thus far identified a host of pathological changes involved in LOAD progression, including disruptions in amyloid processing leading to amyloid beta (Aβ) plaque accumulation (Hampel et al. 2021), changes in cerebral vascularization (Klohs 2020; Fisher et al. 2022; Cai et al. 2023), neuroinflammation (Franceschi and Campisi 2014; Pinti et al. 2016), metabolism (Maghsudi et al. 2020), autophagy (Aman et al. 2021; Caponio et al. 2022), pathological tau aggregation (Guillozet et al. 2003; Binder et al. 2005; Chiu et al. 2017) and spread (Walsh and Selkoe 2016; Wegmann et al. 2019), and alterations in neuronal excitatory/inhibitory balance (EIB) (Kang et al. 2020; Radulescu et al. 2023). Thus, there are many potential neurobiological mechanisms regulating LOAD, each of which has the potential to influence LOAD progression.

One approach used to identify mechanisms mediating LOAD is via genome-wide association studies (GWAS). These population genetic studies identify single-nucleotide polymorphisms (SNPs) statistically associated with certain diseases, like LOAD, to ferret out specific gene products (proteins) highly likely to be involved in the disease of interest (ADGC et al. 2019; Bellenguez et al. 2022). These studies have identified alleles associated with LOAD that involve neuroimmune function (Jansen et al. 2019; Andrews et al. 2020; Jorfi et al. 2023), particularly microglia (Thorlakur et al. 2013; ARUK Consortium et al. 2017; Efthymiou and Goate 2017), autophagy dysfunction (Barrachina et al. 2006; Ginsberg et al. 2010; Nixon 2013; Wolfe et al. 2013; Peric and Annaert 2015; Almeida et al. 2020; Lee et al. 2022), and lipid metabolism (Katzov et al. 2004; Mace et al. 2005; Liu et al. 2013; Huang and Mahley 2014), as well as *MAPT* (microtubule-associated protein tau) gene expression and circulating tau levels (ADGC et al. 2019; Sarnowski et al. 2022) and excitatory/inhibitory neuron populations (Gazestani et al. 2023). One GWAS for AD/dementia identified over 90 independent variants across 75 susceptibility loci, including genes enriched in amyloid plaque and NFT formation, cholesterol metabolism, endocytosis/phagocytosis, and innate immune function (Andrews et al. 2023). While these GWAS studies corroborate mechanisms discussed in this chapter, they also collectively emphasize the complexity of AD pathogenesis and progression.

In parallel to exploring the molecular genetic underpinnings of LOAD using GWAS, the field has leveraged advances in assessing brain function to explore the diseased brain directly. Positron emission tomography (PET), functional magnetic resonance imaging (MRI), and electroencephalogram (EEG) have been the tools of choice. Work done in this space has revealed that tau-PET alone, not Aβ-PET or anatomical MRI, accurately predicts the conversion from MCI to all-cause dementia (Groot et al. 2024). Additional evidence suggests that microglial-mediated neuroinflammation may play a role in the development and progression of AD. A translocator protein (TSPO) PET ligand associated with microglia was found to potentially act as a connective factor between the formation of Aβ plaques and the spread of pathogenic tau (Rossano et al. 2024). Furthermore, TSPO expression was linked to the severity of AD, brain volume, cognitive decline, and the accumulation of both Aβ plaques and pTau (Rossano et al. 2024). Long-term EEG monitoring studies have also observed an increase in the frequency of subclinical epileptiform discharge (SED) in patients compared with non-demented controls (Vossel et al. 2013, 2016, 2017; Beagle et al. 2017; Lam et al. 2020; Horvath et al. 2021). Interestingly, SEDs and other epileptiform activity are associated with earlier disease onset (Amatniek et al. 2006; Scarmeas et al. 2009; Irizarry et al. 2012; Vossel et al. 2013) as well as accelerated disease progression (Horvath et al. 2021; Yeh et al. 2022; Kamondi et al. 2024). Thus, the emergence of neuronal hyperexcitability observed in LOAD may be driven by the accumulation of Aβ plaques and/or pathological tau, or more insidiously, the instigator of pathological processes that facilitate the production, spread, and accumulation of these pathological proteins (Harris et al. 2020).

Using the revised ATN (Amyloid/Tau/Neurodegeneration) classification system (Jack et al. 2018; van der Flier and Scheltens 2022) (Fig. 10.1) and armed with new molecular tools (single molecule array, SiMoA) (Li and Mielke 2019; Chen et al. 2020; Emeršič et al. 2020; Hanes et al. 2020; Karikari et al. 2020; Kvartsberg et al. 2020; Meyer et al. 2020a; Thijssen et al. 2020; Ashton et al. 2021; Chatterjee et al. 2021; Fowler et al. 2022; Thomas et al. 2022), researchers have begun to evaluate the accuracy of blood-based biomarkers which may outperform traditional cerebrospinal fluid (CSF) metrics for tracking both disease progression and success of clinical therapeutics (Barthélemy et al. 2024). Notably, these sensitive clinical blood assessments have substantiated the traditional ATN framework by being able to robustly measure Aβ plaques (**A**) (Meyer et al. 2020b), multiple pathogenic species of tau (pT181 (Mielke et al. 2018; Qin et al. 2022; Tzartos et al. 2022), pT217 (Barthélemy et al. 2020, 2023), pT231 (Ashton et al. 2021; Wisch et al. 2024), **T**), as well as, total tau (Dage et al. 2016), and neurofilament light (NfL) (Bacioglu et al. 2016) to measure neurodegeneration (**N**). Notably, they have continued to single out pathologically modified tau (pTau) as the strongest predictor of conversion from MCI to AD (Barthélemy et al. 2020, 2023; Karikari et al. 2020; Ashton et al. 2021; Moscoso et al. 2021). Importantly, these blood-based biomarkers are beginning to work their way into clinical practice (Olsson et al. 2016) and clinical trials (Angioni et al. 2022)—a trend that will hopefully continue to shed light on LOAD progression.

Despite our limited current understanding of LOAD progression, there is an approximate 20-year preclinical phase prior to outward neurological symptom presentation, i.e., cognitive decline (Jack et al. 2018; Long and Holtzman 2019) (Fig. 10.1) where interventions and/or environmental and lifestyle factors could greatly influence AD risk (Eid et al. 2019; Frigerio et al. 2019; Yu et al. 2020; Wieckowska-Gacek et al. 2021). For example, smoking (Peters et al. 2008), heavy alcohol use (Schwarzinger et al. 2018; Rehm et al. 2019), diabetes (Athanasaki et al. 2022), hypertension (Tang et al. 2023), and obesity (Pedditzi et al. 2016) have all been demonstrated to potentially increase the risk of developing AD or accelerating the onset of AD (Fig. 10.1). On the other hand, engaging in a combination of protective activities, maintaining an active social life (Shafighi et al. 2023), and chronic disease maintenance have been shown to collectively reduce the risk of AD by up to 40–60% (Livingston et al. 2020). Unfortunately, by the time individuals are diagnosed with mild cognitive impairment (MCI) due to AD (Bradfield and Ames 2020), which can be defined as individuals performing at least 1.5 standard deviations below normal on memory tasks, they have most likely already accumulated significant AD-relevant pathology (Fig. 10.1). Therefore, developing a better understanding of how specific common lifestyle factors, for example alcohol use, contribute to AD pathogenesis and progression is critical to understanding disease etiology and uncovering novel therapeutic targets.

10.1.2 Alcohol Use Throughout the Lifespan

Alcohol is widely accessible and frequently used, with over 84% of adults aged 18 and older reporting lifetime use (NSDUH 2023). Alcohol misuse is defined as any drinking behavior that jeopardizes one's well-being or the well-being of others and includes two primary patterns of excessive alcohol consumption: heavy alcohol use (>4 drinks on any day or >14 drinks per week for men, and >3 drinks on any day or >7 drinks per week for women) and binge drinking (4–5 drinks consumed within 2 hours) (NIAAA 2024). Alcohol use disorder (AUD), a serious medical condition resulting from patterns of excessive use, is characterized by an impaired ability to reduce or cease alcohol use despite negative social, occupational, or health consequences. In 2021, nearly 28.6 million adults aged 18 and

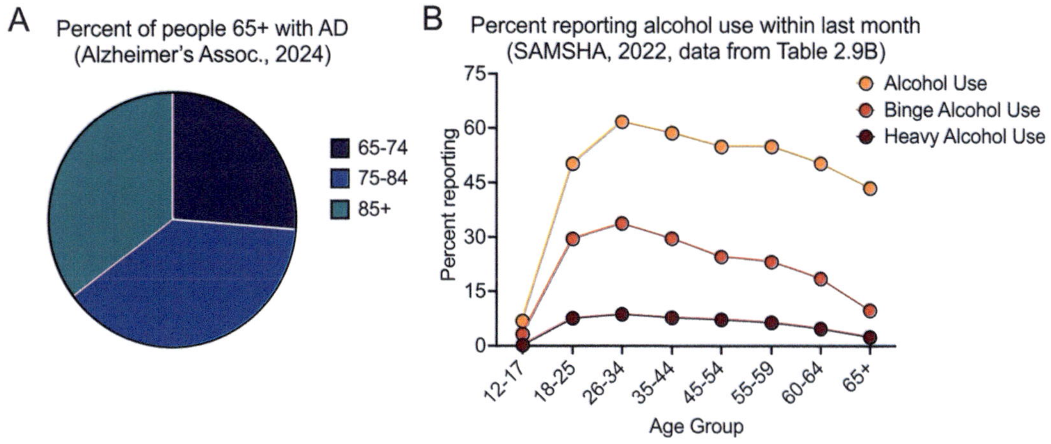

Fig. 10.2 Epidemiology of AD and alcohol use. (**a**) The percentage of people 65+ living with late-onset AD (LOAD) broken down by age, data from 2024. (**b**) Percentage of individuals reporting alcohol use throughout the lifespan, SAMHSA data from Table 2.9B, 2022

older in the United States had an AUD (NSDUH 2021), yet only 1 in 10 individuals seek and receive treatment (Mintz et al. 2023). Further, harmful drinking behaviors are not exclusive to people suffering from AUD, as described in detail below.

While alcohol misuse is typically associated with younger adults (18–25 years old), as nearly one-third of individuals in this age group report past month heavy or binge drinking (Fig. 10.2b), current data suggest that heavy alcohol use and binge drinking is increasing among older demographics including middle-aged adults (35–50 years) (Fig. 10.2b). Similar trends are being observed in older populations. A meta-analysis that analyzed surveys from 2000 to 2016 revealed significant increases in alcohol use among adults aged 50 and above, with a notable rise in binge drinking among those 30 and older, particularly in the oldest age group (65+ years) (Grucza et al. 2018). The most recent NSDUH 2022 data also found that 43.4% of people 65+ consumed alcohol within the past month, with 9.7% reporting binge drinking, and 2.4% reporting heavy alcohol use (Fig. 10.2b) (NSDUH 2022/2012). Over the last decade, the percentage of individuals 65+ consuming alcohol in the past month increased from 41.2% in 2012 (Fig. 10.3a) to 43.4% in 2022 (Fig. 10.3b), with the most substantial rise occurring within the binge alcohol use category from 6.2% in 2012 to 7.3% in 2022 (Fig. 10.3a, b). Interestingly, although men have historically consumed more alcohol than women, this gender gap in alcohol misuse is narrowing across all age groups (Keyes et al. 2019). As but one example, from 1997 to 2014 women over 60 experienced the largest increase in binge drinking behavior (Breslow et al. 2017). To be clear, adults who engage in excessive drinking are increasing in age, and excessive drinking rates continue to increase among older adults. Coupled with the fact that fewer than 15% of individuals with an AUD seek and receive treatment (Degenhardt et al. 2017; Glantz et al. 2020; GBD 2021 Diseases and Injuries Collaborators et al. 2024), it is imperative that we develop a better understanding of the risks alcohol consumption poses to brain health, particularly neurodegenerative diseases such as LOAD, hereafter referred to as Alzheimer's disease (AD).

10.2 Alcohol and Alzheimer's Disease: Epidemiological Evidence for Alcohol Misuse as an AD Risk Factor

Unraveling the influence of alcohol on the development and progression of AD is challenging; in part because heavy alcohol use was previously

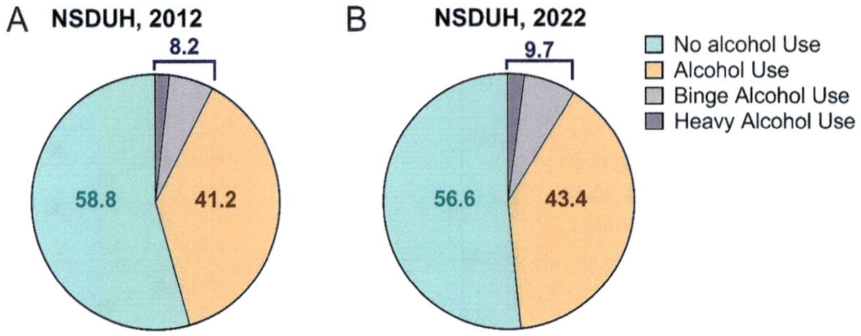

Fig. 10.3 The percentage of 65+ years old and type of alcohol use throughout the last month demonstrates a substantial increase from 2012 to 2022 (NSDUH 2022/2012). (**a**) Percentage of individuals 65+ in 2012 shows that 58.8% did not consume alcohol within the last month, while 41.2% of people engaged in alcohol use. 8.2% of those 65 and older engaged in binge drinking or heavy alcohol use. (**b**) Percentage of individuals 65+ consuming alcohol (43.4%); 9.7% reported engaging in binge alcohol use or heavy alcohol use

used as exclusion criterion for AD diagnosis (Tyas 2001). Although chronic heavy drinking can cause unique forms of neurodegeneration, like alcohol-related dementia (Moriyama et al. 2006) or Wernicke's-Korsakoff syndrome (Oudman et al. 2022), evidence from the past two decades supports that alcohol misuse throughout the lifespan is a modifiable risk factor for AD (Schwarzinger et al. 2018; Rehm et al. 2019). Adolescence is a key developmental period wherein high levels of alcohol consumption are often first experienced (Spear 2013). It is also a time of rapid brain development (Spear 2000, 2018). Yet, there is no data on how adolescent alcohol exposure impacts the risk of AD development later in life in clinical populations. However, teenagers who participate in alcohol misuse exhibit decreased brain volume in areas typically affected in AD, including the frontal and temporal lobes (Phillips et al. 2021), as well as impaired attention span and difficulties in memory (Lees et al. 2020). Because impaired cognitive performance in adolescence is associated with AD-related dementia later in life (Moceri et al. 2000; Huang et al. 2018), alcohol-related brain exposure and/or damage during adolescence may possibly promote AD pathogenesis.

The field has made more headway evaluating the impact of alcohol use on adults, in large part because many more adults drink alcohol, but also because there are fewer ethical considerations than working with minors. To this end, a landmark retrospective cohort study of adult men and women 20 years or older in France found that an alcohol use disorder (AUD) in early adulthood or middle age was the *strongest* modifiable risk factor for the development of dementia (hazard ratio/HR: ~3.3) (Schwarzinger et al. 2018). In a separate study, the presence of alcohol-related brain damage was associated with an earlier dementia onset (Zhao et al. 2024), with AUD being associated with an increased risk of both vascular (HR: 2.3) and other dementias, including AD (HR: 2.36). Other work suggests heavy alcohol use is associated with risk for AD independent of AUD, although AUD diagnosis was not defined in the study (Jeon et al. 2023). In this large retrospective cohort study assessing the baseline and change in alcohol consumption patterns over time, adults 40 years and older who engaged in sustained heavy drinking (i.e., >30 g/day) had an 8% increased risk for the development of subsequent AD 6 years earlier compared to participants who drank moderately or did not drink at all (Jeon et al. 2023). Furthermore, participants who increased their drinking from study-defined mild/moderate levels to heavy drinking also saw an increase in all-cause dementia (HR: 1.37) (Jeon et al. 2023). In studies with a longer follow-up period, such as the HUNT study, participants who reported drinking 5 or more times 2 weeks prior to the start of the study (during young adulthood-midlife) had an increased risk of AD diagnosis later in life compared to individuals who drank infrequently (Langballe et al. 2015). Together, these data sug-

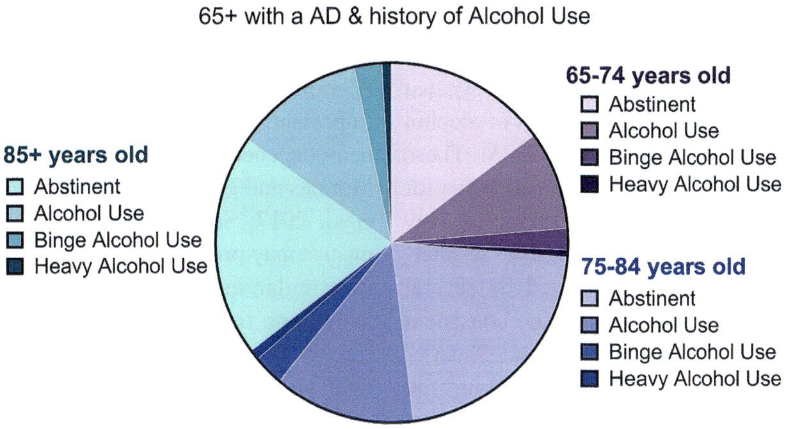

Fig. 10.4 Overlap of alcohol use history within the population of patients with an Alzheimer's disease diagnosis. Using the NSDUH 2022 percentages (Figs. 10.2a and 10.3b) we can speculate that within the percentages of US adults 65+ with AD, there are approximately 10% that are 65–74 years old with a history of alcohol use, 2% with a history of binge drinking, and more than 0.6% with a history of heavy alcohol use. In patients with an Alzheimer's diagnosis that are 75–84 years old, that percentage is troublingly higher where 13% report alcohol consumption, nearly 3% report consuming binge alcohol, and nearly 1% engage in heavy alcohol use. 85+ years and older with a diagnosis also have astonishingly high rates of alcohol use (11% consume alcohol, 2.5% consume binge alcohol, and nearly 1% engage in heavy alcohol use)

gest that heavy alcohol use, perhaps from late adolescence through midlife, promotes the development of AD and other related dementias (Fig. 10.4).

A growing body of literature also indicates that heavy alcohol use in advanced age increases the risk for AD (Mukamal et al. 2003; Koch et al. 2019). Older adults aged 65 years and older who drank more than 14 alcoholic drinks per week were found to have increased odds for AD and all-cause dementia compared to older adults who had low or moderate drinking behaviors (Mukamal et al. 2003). These data were corroborated by findings in participants with MCI at baseline, which found that those reporting drinking 14 or more drinks/week had an increased hazard ratio (HR) for dementia (HR: 1.72) than those drinking less than 1.0/week (Koch et al. 2019). However, in addition to heavy drinking increasing the risk for disease development, it may also accelerate disease onset. A retrospective study in patients with an AD diagnosis suggests individuals who had 2 or more drinks a day exhibited disease onset 4 years earlier than those compared to individuals who had less than 2 drinks a day (Harwood et al. 2010). These data from adolescence to advanced age together support alcohol use at all developmental time periods as having the potential to impact AD diagnosis later in life.

Since age is the largest non-modifiable risk factor for AD, and by 2030 all members of the baby-boom generation will be at least 65+ (Guerreiro and Bras 2015), there is a high potential for a wave of new AD diagnosis. The current age breakdown of those 65 and older with an AD diagnosis is as follows: 1.83 million are 65–74 years old, 2.67 million are 75–84, and 2.42 are 85+ (Rajan et al. 2021) (Fig. 10.2a). However, the population of Americans aged 65 and older is estimated to rise from 58 million in 2022 to 83 million by 2050 (The Alzheimer's Association 2024). Thus, taking the percentages of individuals reporting alcohol use within the last month (Fig. 10.3b data), and extrapolating to the percentage of individuals within each age group who have a diagnosis of AD (Fig. 10.2a), may provide some insight into our epidemiological future. To this end, it is likely that a growing number of individuals with an AD diagnosis will have a history of alcohol use that includes binge drinking and/or heavy alcohol use (Fig. 10.4).

Despite the negative outcomes alcohol has on brain health and likely AD outcomes in particular, a small set of studies have suggested decreased rates of AD with low levels of alcohol use (Sabia et al. 2018; Jeon et al. 2023). These results may be mediated by a J-shaped curve for alcohol as a risk factor for AD, with all but very low exposure increasing AD risk. However, interpretation of these studies is limited by several confounding factors. First, there is no consistent definition of "light," "moderate," and "heavy" across many studies, and the rate of drinking and time since last drink among participants are not regularly reported. Second, these studies do not account for the cause of abstinence, which may be in direct response to an AUD diagnosis, or other illness or confounding factors known to influence both AD development and drinking behavior, such as education, social activity, mental health, and diet. That even relatively low levels of alcohol consumption have recently been found to be associated with reduced brain volume is difficult to reconcile with the studies listed above (Topiwala et al. 2022). Regardless, the epidemiological observations that low alcohol-consuming individuals may be less prone to developing AD, while high alcohol-consuming individuals have an increased risk of developing AD, directly support the hypothesis that AD risk scales as a function of relative alcohol exposure.

10.3 Alcohol and Alzheimer's Disease: Shared Disruption of Biological Processes

Alcohol-related brain damage and AD pathogenesis lead to similar pathological changes in oxidative stress (Gella and Durany 2009; Qin and Crews 2012a), neuroimmune function (Eikelenboom et al. 2010; Pascual et al. 2021), metabolism (Volkow et al. 2015; Butterfield and Halliwell 2019; Tomasi et al. 2019; Popova et al. 2024), pathogenic tauopathy development and spread (Liu et al. 2012; Mohamed et al. 2013; DeVos et al. 2018; Gibbons et al. 2018; Bell et al. 2020; Gu and Liu 2020; Annadurai et al. 2021; Downs et al. 2022; Tucker et al. 2022), and neuronal excitatory/inhibitory balance (EIB) (Born 2015; Anastacio et al. 2022; Huang et al. 2022; Alberto et al. 2023; Barbour et al. 2023). Importantly, all of these pathologies promote neurodegeneration (Andersen 2004; Pasantes-Morales and Tuz 2006; González et al. 2014; Jha et al. 2017; Strang et al. 2019). Thus, alcohol misuse may promote AD pathogenesis by initiating similar insults and/or by exacerbating and facilitating those caused directly by AD. Given recent evidence that the brain is more vulnerable to alcohol-related damage from adolescence and into aging than we had previously appreciated (Coleman et al. 2011; Vetreno and Crews 2012; Coleman et al. 2014; Vetreno et al. 2014; Salling et al. 2018; Tapia-Rojas et al. 2018; Barnett et al. 2022; Khan et al. 2023; Anton et al. 2024), exposure to alcohol at virtually ***all stages of life*** has the potential to impact these molecular mechanisms to influence AD outcomes. In the next section, we will discuss the shared mechanisms of pathogenesis in AD and alcohol misuse throughout the lifespan.

10.3.1 Oxidative Stress

Alcohol (i.e., ethanol) freely passes through the blood–brain barrier (BBB) and is toxic to neurons at high concentrations (Zimatkin and Deitrich 1997; Wilson and Matschinsky 2020). Although the liver is the primary site of ethanol metabolism, ethanol is also locally metabolized in the brain via catalase and CYP2E1 leading to the creation of acetaldehyde, a toxic intermediate (Aragon et al. 1992). Regardless of where it is produced, acetaldehydes are damaging to the neural microenvironment through the formation of protein and DNA adducts, and the production of oxidative species during its metabolism into acetate by acetaldehyde dehydrogenase (Nakamura et al. 2003). Oxidizing agents such as hydrogen peroxide (H_2O_2) and free radicals (superoxide and hydroxyl radical) are also generated as by-products of CYP2E1 and catalase activity (Zimatkin and Deitrich 1997). Free radicals are harmful to the neurons through disrupting the cell membranes and consequential lipid peroxidation (Hernández et al. 2016). Ethanol also induces permeability of mitochondrial membranes and leakage of superoxide, which further perpetuates oxidative stress and can initiate apoptosis (González et al. 2007). Thus, ethanol, which is mainly metabolized in the liver, can also be locally metabolized in the brain and quite toxic to neurons, as well as other cell types.

Although all CNS cell types are vulnerable to ethanol-induced oxidative stress, glial cells play an important role in the production of oxidative factors because of their specialized functions (Montoliu et al. 1995; Qin and Crews 2012a). Repeated binge ethanol exposure in mice activates NADPH oxidase (NOX) in microglia, the resident macrophages of the CNS, leading to superoxide production and subsequent neurodegeneration (Qin and Crews 2012a). Astrocytes are an important site of ethanol metabolism in the brain and thus are a potent source of superoxide, further elevating oxidative-induced damage (Montoliu et al. 1995). Further, the impact of ethanol-related oxidative stress is exaggerated by depletion of the brain's natural antioxidant defense system by ethanol exposure (Reddy et al. 1999). Given the numerous pathways through which ethanol-induced oxidative stress can be amplified, its presence is unsurprising in rodent models of excessive ethanol exposure and in post-mortem human brain tissue from individuals with AUDs (Montoliu et al. 1995; Reddy et al. 1999; Qin and Crews 2012a).

Oxidative stress is also believed to be an early participant in AD pathogenesis (Rapoport 2003). Markers of lipid peroxidation, protein oxidation, and DNA oxidation are increased in the brains of individuals with AD compared to healthy controls (Gella and Durany 2009). Recent data have shown that this may be due in part to the dysregulation of a few key mitochondria proteins known to regulate oxidative stress, so much so that the degree of expression is able to predict LOAD pathogenesis (Yan et al. 2024). A mechanistic role for oxidative stress has also been demonstrated in vitro, where neuroblastoma cell lines treated with Aß under oxidative conditions were found to promote Aß aggregation (Zheng et al. 2006). On the other hand, antioxidant intervention in vivo mitigates neuropathology and cognitive dysfunction in aged 3x-Tg mice (Young and Franklin 2019). And finally, to make matters worse, the presence of Aß promotes oxidative stress itself directly through inhibition of mitochondria superoxide dismutase, creating a damaging positive feedback loop of oxidative stress in AD (Tamagno et al. 2021). Additional work in both a mouse model of tauopathy and antecedent AD brain tissue found that soluble tau pathology could also exacerbate oxidative stress through mitochondrial disruption (Kopeikina et al. 2011). Importantly, oxidative stress has been observed in many primary tauopathies, like Pick's disease, corticobasal degeneration (Castellani et al. 1995), and frontal-temporal dementia (Martínez et al. 2008), as well as secondary tauopathies, like AD (Stamer et al. 2002; David et al. 2005). These data suggest that the presence of both pathologic species relevant to AD, namely Aß and pathological tau, can exacerbate and are mediated by oxidative stress.

Thus, both excessive alcohol use and AD prominently feature damage from oxidative stress. As oxidative stress is a central mechanism of ethanol-related neuronal injury in adolescence (Vetreno et al. 2014; Pelicao et al. 2016; Barnett et al. 2022), and adolescence is the time excessive alcohol consumption is most initiated, it remains possible that cumulative oxidative stress beginning in adolescence exhausts neural resources that would otherwise be used to mitigate AD-induced oxidative stress. Likewise, because impaired mitochondria function and antioxidant depletion are a hallmark of brain aging (Ionescu-Tucker and Cotman 2021), any avenue by which ethanol increases oxidative load would likely be exacerbated with increasing age. Despite the overlap in oxidative stress, to date clinical trials using antioxidants for AD have not been successful (Polidori and Nelles 2014). While more research in this area is needed, to the extent that AD outcomes are impacted by oxidative stress, ethanol-induced increases in oxidative stress remain an important potential contributor to AD risk.

10.3.2 Neuroimmune Dysregulation

Heightened neuroinflammation is a critical factor that facilitates pathogenesis of both AD and heavy alcohol use (Cribbs et al. 2012; Qin et al. 2021; Barnett et al. 2022). In response to alcohol, damaged neurons release danger-associated molecular patterns (DAMPs), such as high mobility group box 1 (HMGB1) (Crews et al. 2013; Coleman et al. 2017). DAMPs, like HMGB1, bind to pattern recognition receptors (PRRs) including toll like receptors (TLRs) and receptors for advanced glycation end products (RAGE) on neurons and glial cells. TLR and RAGE activation can initiate pro-inflammatory signaling cascades including nuclear factor kappa-light-chain-enhancer of activated B cells (NF-κb) and inflammasomes (Montesinos et al. 2016). These pathways culminate in the production of several pro-inflammatory cytokines (TNF-α, IL-1β, IL-6) (Alfonso-Loeches et al. 2010; Coleman et al. 2017). Prolonged production of pro-inflammatory cytokines from repeated binge or chronic ethanol exposure also upregulates cytokine receptors (TNF-R, IL-1R) and TLRs to sustain these pro-inflammatory signaling loops (Qin and Crews 2012b; Vetreno and Crews 2012). Ultimately, the sustained production of pro-inflammatory cytokines can lead to neuronal death via apoptosis (Liu et al. 2021; Qin et al. 2021). The pro-inflammatory effects of ethanol are evident in mouse and rat models of chronic ethanol exposure and binge ethanol exposure (Crews and Vetreno 2014; Erickson et al. 2018; Pascual et al. 2021), as well as in post-mortem brain tissue of individuals with AUD, which specifically found increases in HMGB1, TLR, cytokines, and neurodegeneration (Qin and Crews 2012a; Crews et al. 2013; Coleman et al. 2017).

Similar innate immune pathways are implicated in AD. RAGE receptor expression levels in the hippocampus of AD patients correlates with more severe disease pathology (Lue et al. 2001), and microarray analysis of human brain tissue reveals an increase in innate immune signaling genes in the brains of people with AD compared to healthy controls (Cribbs et al. 2012). Increased expression of DAMPs, TLRs, and cytokines is also elevated in the brains of aged people compared to young controls and may suggest dysfunctional innate immune signaling precedes overt AD pathology (Cribbs et al. 2012). Heightened indicators of neuroinflammation are present in mouse models of AD as well, and suppression of pro-inflammatory cytokines slows Aβ accumulation and neurotoxicity (McAlpine et al. 2009; Heneka et al. 2013). Notably, the presence of pathologic tau is known to accelerate neuroinflammation through NLRP3 inflammasome signaling (Heneka et al. 2013; Ising et al. 2019; Jiang et al. 2021), and neuroinflammation, in turn, is known to influence and increase pathologic tau species (Bhaskar et al. 2010; Weston et al. 2021). HMGB1 promotes neurodegeneration in transgenic mouse models of AD without affecting plaque deposition, suggesting these immune pathways might also worsen neurodegeneration parallel to Aβ (Fujita et al. 2016). However, the release of HMGB1 promotes oligomeric aggregation of pathogenic tau which can, in turn, induce cellular senescence, tauopathy progression, and cognitive deficits (Gaikwad et al. 2021). Mitigating neuroinflammation through anti-inflammatory therapies like minocycline reduces neuropathology in mouse models of AD, further supporting a mechanistic role of the innate immune system in disease progression (Stirling et al. 2005; Seabrook et al. 2006; Choi et al. 2007; Noble et al. 2009). A few epidemiological studies (Delanty and Vaughan 1998; in't Veld et al. 2001; Zandi et al. 2002; Kotilinek et al. 2008; Vlad et al. 2008; Imbimbo et al. 2010), but not all (ADAPT Research Group et al. 2007; Szekely et al. 2008), have demonstrated a reduced prevalence of AD among users of immunosuppressants like non-steroidal anti-inflammatory drugs (NSAIDs), with the strongest effect seeming to have been in patients on the longest duration (Delanty and Vaughan 1998; Etminan et al. 2003; Vlad et al. 2008). To that end, a 2-year double-masked pharmacoprevention trial, which enrolled 195 AD family history-positive elderly with a mean age of 63 years, administered naproxen 2x daily or placebo to their clinical trial participants. Unfortunately, naproxen-treated individuals had more adverse events, and no reduction in the rate of AD progression compared to placebo controls and no notable treatment effects on any underlying neuroimaging or CSF biomarkers present (Meyer et al. 2019). Despite these null findings, interest remains high in using NSAIDs as potential AD treatments (Hershey and Lipton 2019; Rivers-Auty et al. 2020).

Microglia, the brain's resident macrophages, play a critical role in physiology and disease (Wolf et al. 2016). They are particularly important orchestrators of neuroinflammation in several neurodegenerative disorders, like AD (Hickman et al. 2018; Gao et al. 2023). Single-cell RNAseq revealed the remarkable complexity of microglial responses that can occur in a wide variety of neurodegenerative disease states (Provenzano et al. 2021). Ethanol acts as a DAMP, leading microglia to acquire complex phenotypes that can either facilitate recovery (Marshall et al. 2013) or promote injury following ethanol exposure (Chastain and Sarkar 2014). The presence of ethanol and DAMPs induces pro-inflammatory polarization of microglia, characterized by the upregulation of Iba-1 expression, increased cell surface receptors like CD68, and the production of pro-inflammatory cytokines (Chastain and Sarkar 2014). Altered microglia reactivity to ethanol is believed to promote neurogenesis during recovery from ethanol exposure (Nixon et al. 2008). Additionally, ethanol has been shown to induce microglial phagocytosis, which may aid in debris clearance and recovery following ethanol-induced neurotoxicity (Fernandez-Lizarbe et al. 2009). However, pro-inflammatory microglia are also associated with neuronal death in models of chronic or repeated binge ethanol exposure in mice and rats (Chastain and Sarkar 2014; Yang et al. 2014). Microglia contribute to neuronal death in these models through potent production of pro-inflammatory cytokines. For instance, depletion of microglia in vivo in mice and ex vivo in rat brain slice cultures mitigates both pro-inflammatory cytokine expression and ethanol-related injury (Walter et al. 2017; Coleman et al. 2020). Furthermore, in vivo and ex vivo ethanol exposure decreases microglia phagocytosis of Aβ in mice and rats (Marsland et al. 2022). These data suggest dysregulated microglia inflammatory responses to ethanol promote neurotoxicity.

Microglia also play critical roles in both promoting and mitigating AD pathology. Similar to ethanol exposure, microglia acquire diverse phenotypes throughout both aging (Provenzano et al. 2021) and AD progression (Hamelin et al. 2016; Mathys et al. 2017; Miao et al. 2023). In AD, damage-associated microglia (DAMs) upregulate genes associated with phagocytic activity aimed at reducing Aβ (Keren-Shaul et al. 2017). Soluble and oligomeric forms of Aβ as well as pathologic tau can induce pro-inflammatory polarization of microglia to promote disease progression (Brelstaff et al. 2021; Bo et al. 1995; Walter et al. 2007; Jin et al. 2008; Song et al. 2011; Solito and Sastre 2012; Dalgediene et al. 2013; Majerova et al. 2014; Koss et al. 2016; Yang et al. 2017; Luciunaite et al. 2020; Zhang et al. 2024). However, in the proximity of Aβ plaques, microglia can enhance phagocytic activity (Hamelin et al. 2016; Wang et al. 2016; Keren-Shaul et al. 2017; Ennerfelt et al. 2022). Microglia may also play a central role in spreading pathogenic tau, since their depletion suppresses tau propagation (Asai et al. 2015). Notably, both Aβ and tau may act together to induce unique microglia subtypes that could define both early- and late-stage AD (Kim et al. 2022). As AD advances, chronic stimulation of microglia by Aβ can ultimately impair phagocytosis and cause release of pro-inflammatory cytokines that promote tau pathology and neurodegeneration (Miao et al. 2023). Pathological tau can also prime neuroinflammation through NF-κb and IL-1β signaling (Jiang et al. 2021). Although these data suggest microglia dysfunction is driven by AD neuropathology, additional evidence shows microglia dysregulation *precedes* neuropathology. For example, genes associated with impaired microglia phagocytosis are identified as strong risk factors for LOAD (Efthymiou and Goate 2017; Novikova et al. 2019). Additionally, microglia lose their dynamic functions with advanced aging and respond to stimuli with exaggerated pro-inflammatory responses but reduced phagocytic capacity (i.e., "priming") (Norden and Godbout

2013). Therefore, additional activation by environmental exposures, such as heavy alcohol use, may exaggerate microglia priming and leave these cells in a chronically stimulated state that could accelerate or promote AD pathology, increasing risk for the disease.

Emerging preclinical evidence suggests that alcohol exposure in adolescence or advanced age has profound impacts on neuroinflammation and associated AD-like neuropathology. Intermittent ethanol exposure in a 3xTg genetic mouse model of AD elevates cytokine expression, markers of microglia reactivity, and promotes Aβ accumulation and neuronal death in adulthood compared to vehicle-exposed mice (Barnett et al. 2022). Treatment with the anti-inflammatory compound minocycline prevented this enhanced pathology. These data suggest adolescent exposure has lasting impacts on AD progression into adulthood that are neuroimmune-dependent. A persistent increase in innate immune genes also occurs in wild-type mice exposed to binge ethanol (Vetreno and Crews 2012), which can lead to deficits in cholinergic neuronal populations and cognitive inflexibility later in adulthood (Coleman et al. 2011, 2014). This is consistent with studies in aged wild-type mice and rats, which consistently find that ethanol increases markers of neuroinflammation and neuronal death more so in aged brains compared to young brains (Kane et al. 2013; Marsland et al. 2022; Anton et al. 2024). These studies suggest a heightened risk of neurodegeneration due to ethanol-induced increases in neuroinflammation in advanced age. Together these data find differences in the effects of ethanol exposure on neuroimmune function positioned to exacerbate AD outcomes.

10.3.3 Metabolic Dysfunction in Brain with AUD and AD: A Possible Link with Neuroinflammation

Both AUD and AD feature metabolic dysfunction. Glucose is the essential energy source of the brain (Mergenthaler et al. 2013), and in the adult brain, neurons have the biggest appetite (Howarth et al. 2012). Fortunately, fluorodeoxyglucose positron emission tomography (FDG-PET) has evolved to be a sensitive neuroimaging biomarker. Studies using FDG-PET have found hypometabolism in key brain regions over the course of both AUD (Thanos et al. 2008; Tomasi et al. 2019) and AD (Mosconi et al. 2008; Fouquet et al. 2009; Kobylecki et al. 2015). In human AUD brain, the loss in FDG-PET correlates strongly with alcohol-related brain damage across brain regions (Tomasi et al. 2019). In AD, reductions in FDG uptake predict the conversion from mild cognitive impairment to AD (Fouquet et al. 2009), although recent work suggests this may not be the case (Smailagic et al. 2018). There is also evidence that Aβ and tau drive cognitive decline early in AD, while glucose hypometabolism drives decline at later stages of the disease (Hammond et al. 2020). These reductions in glucose uptake and/or utilization suggest a shift in the primary metabolic pathways used by neurons in both diseases.

In the setting of AUD, there is evidence to suggest a shift toward acetate metabolism (Volkow et al. 2015), likely due to the oxidative metabolism of alcohol. The oxidative metabolism of alcohol results in the production of acetate (Wilson and Matschinsky 2020), which is either converted to acetyl-CoA for use in the tricarboxylic acid cycle (TCA) or converted to malonyl-CoA by acetyl-CoA carboxylase to promote lipogenesis. Interestingly, although it is well known that alcohol promotes lipidosis in the peripheral tissues such as the liver, we recently found that ethanol-induced neuronal lipidosis promotes AD pathology, with pro-inflammatory microglia driving neuronal lipidosis (Barnett et al. 2024). Thus, alcohol-induced alterations to metabolic processes may directly exacerbate AD pathology.

The underlying causes of the metabolic shifts observed in AUD and AD warrant further investigation. Altered glucose utilization or a shift to an alternative energy source such as acetate or lipids can lead to neuronal dysfunction with impairments in neurotransmitter production, synaptic function, and neuronal hyperexcitability (Rorbach-Dolata and Piwowar 2019) that can result in neuron death and cognitive abnormalities (Suh et al. 2003; McDonald et al. 2023). We believe studying the impact of pro-inflammatory microglia on neuronal function is of particular importance following recent evidence that neurons can perform glycolysis independently (Li et al. 2023) and that microglia may take up more glucose than astrocytes from the periphery, which can also be detected using FDG-PET (Xiang et al. 2021; Gnorich et al. 2023). Further, microglia take up glucose at high levels in vivo (Xiang et al. 2021; Gnorich et al. 2023), and pro-inflammatory microglia engage in high levels of glycolysis (Orihuela et al. 2016). Changes in glial cell activation state may therefore directly alter neuronal metabolism and activity during disease progression, and as such may represent a viable approach to mitigating neurodegeneration arising from ethanol exposure and/or AD. This is complicated by known differences in neuronal metabolism at different stages of life (Dienel 2019), which have yet to be fully described. Investigation into the interactions between age, ethanol exposure, and metabolic outcomes may reveal how ethanol exposure throughout the lifespan impacts AD development and subsequently AD-related pathologies.

10.3.4 Pathological Tau Formation and Spread

In AD, although Aβ plaques are likely the first pathology to form, pTau levels have been shown to be more predictive of AD onset (Malpetti et al. 2020; Binette et al. 2022) and to correlate better with the progression of AD-associated neurodegeneration (brain atrophy) and cognitive decline (Thijssen et al. 2020, 2021; Teunissen et al. 2021). In fact, many consider AD to be an amyloid-induced tauopathy. Microtubules and their accessory proteins, referred to as microtubule-associated proteins (MAPs), exist throughout the entire organism and play key roles in mitotic and meiotic spindle formation, neuronal development, and polarization, among many other diverse regulatory cellular processes (Goodson and Jonasson 2018). Microtubule-associated protein tau (encoded by the gene *Mapt*) is primarily associated with stabilizing these microtubules and regulating axonal transport in the brain (Wang and Mandelkow 2016). This interaction between microtubules and tau is tightly regulated by post-translational modifications (PTMs), such as phosphorylation throughout development (Yu et al. 2009). However, in AD, as well as other neurodegenerative tauopathies, tau undergoes a variety of PTMs, like hyperphosphorylation (Šimić et al. 2016), throughout the preclinical, prodromal, and symptomatic phases of AD. Recent work has suggested that these PTMs can be heterogeneous, but that they undergo a characteristic pattern of pathogenesis (Wesseling et al. 2020). Importantly, tau can undergo a variety of other PTMs that include acetylation, ubiquitination, methylation, oxidation, and nitration (Alquezar et al. 2021; Carroll et al. 2021), among others. Tau is a highly modifiable protein with more than 80 potential phosphorylation sites (Noble et al. 2013). However, in AD, tau becomes abnormally hyperphosphorylated at a ratio 3 times greater than physiological tau (Wang et al. 2013), which then aggregate into neurofibrillary tangles (NFTs) (Dujardin et al. 2020; Wesseling et al. 2020). The accumulation of PTMs by tau is dependent on its length, which varies due to alternative splicing. For instance, the generated tau protein can differ in the number N-terminal inserts (0N, 1N, or 2N), and the microtubule binding repeats (MTBRs, 3R or 4R), with 0N3R tau only present during fetal development (Andreadis 2005, 2006; Liu and Gong 2008; McMillan et al. 2008). All six of these isoforms can be present throughout the lifespan, but 3R and 4R isoforms are typically present in a 1:1 ratio (Goedert et al. 1989). However, the ratio of 3R to 4R tau plays a crucial role in maintaining tau homeostasis (Ginsberg et al. 2006; Conrad et al. 2007). In AD this balance is disrupted resulting in a 2:1 ratio of 4R to 3R, and this imbalance is thought to facilitate PTM of tau (Alquezar et al. 2021). Pathological tau (pTau) is thought to sequentially develop and spread throughout the human brain, originating from within the entorhinal cortex (EC), hippocampus (HP), and limbic areas before spreading to neocortical areas (Braak and Braak 1991a, b; Mufson et al. 2016). This spread can occur in various ways, such as through extracellular synaptic vesicle release and subsequent endocytosis (Vogels et al. 2019, 2020; Robert et al. 2021). Regardless, relative neural activity is a *key* component of this process (Pooler et al. 2013; Holmes et al. 2014; Yamada et al. 2014; Wu et al. 2016). Thus, pTau production and spread, which are exacerbated by neural activity, are two pathological hallmarks of AD pathogenesis.

Even though levels of CSF tau, indicative of neurodegeneration, are transiently elevated in Wernicke's encephalopathy (WE), this neuronal damage is different than AD (Matsushita et al. 2008). Emerging clinical reports using CSF-based biomarkers have found that older, cognitively intact participants in a frequent drinking group (≥ 1 time/week) had higher CSF p-Tau/Aβ42, and higher abnormalities in pTau and total-Tau levels, suggesting that AD biomarkers like pTau could be a strong indicator of future AD risk in this alcohol-consuming population (Wang et al. 2021). While other studies suggest that these cognitive deficits and abnormal CSF profiles from older demented alcohol-dependent patients could be masking an AD diagnosis (Azuar et al. 2021). Recent preclinical evidence agrees that excessive alcohol use can exacerbate intraneuronal Aβ and pTau in areas like the EC potentially due to lysosomal dysfunction (Tucker et al. 2022) and/or neuroinflammation (Barnett et al. 2022; Anton et al. 2024). Regardless, elucidating alcohol's role in pathogenic tau development and spread as an AD risk factor will be crucial moving forward.

10.3.5 Excitatory Inhibitory Balance and Tauopathy, Linked Pathologic Processes

An observation common to both AUD (Correas et al. 2021) and AD (Lauterborn et al. 2021; Javed et al. 2022; Scaduto et al. 2023; Soula et al. 2023) is neural hyperexcitability, driven by disruptions in excitatory inhibitory balance (EIB). Glutamatergic and GABAergic neurons are the main excitatory and inhibitory neurotransmitters in the central nervous system (CNS), and they play a crucial role in regulating EIB (Sears and Hewett 2021). Proper inhibitory signaling is necessary for regulating the global spatiotemporal network activity and proper neural processing (Buzsáki et al. 2007). GABAergic neurons, which have significant diversity (Markram et al. 2004; Huang and Paul 2019), primarily modulate inhibitory activity in the brain by interacting with excitatory neurons (Fritschy and Brünig 2003; Sears and Hewett 2021). Dysfunction in GABAergic interneurons, particularly parvalbumin-expressing (PV) interneurons (Bartos et al. 2007), contribute to hyperexcitability in neuronal networks associated with AD (Verret et al. 2012; Cattaud et al. 2018; Petrache et al. 2019; Chung et al. 2020; Mattson 2020; Xu et al. 2020; Aouci et al. 2022).

Excessive alcohol use and the development of alcohol dependence can be characterized by excessive neural activity and somatic withdrawal (Heilig et al. 2010; Koob and Volkow 2016). This dysfunctional EIB is thought to be a result of alcohol's direct facilitation of $GABA_A$ receptors (Valenzuela and Jotty 2015; Olsen and Liang 2017) and inhibition of glutamatergic receptors (Möykkynen and Korpi 2012) acutely leading to hypoactivity (Dharavath et al. 2023). In a study of young adult binge drinkers, not only were cortical GABA levels reduced (Marinkovic et al. 2022), but there were also alterations in theta power and synchrony (Huang et al. 2022), suggesting that the GABA system is particularly vulnerable with continued high alcohol use. Alcohol-induced hyperexcitability, which (1) occurs as a compensatory adaptation to its acute CNS depressing actions (Valenzuela 1997; Correas et al. 2021), (2) is exacerbated by multiple withdrawal periods (Becker 1998), and (3) is observed in individuals with a history of excessive alcohol use (Gimenez-Gomez et al. 2023), could be a potential mechanism of alcohol-induced increases in pTau production and spread. Hence, GABAergic neurons, pivotal for maintaining inhibitory tone, may undergo selective targeting and modification during repetitive alcohol exposure and ensuing withdrawal phases, potentially rendering them more susceptible to pTau aggregation—this may represent a key pathogenic process linking alcohol misuse throughout the lifespan to an increased risk of AD.

Excessive neuronal activity is a well-known driver of pTau spread (Brunello et al. 2019). In mouse models of tauopathy, inhibitory cells are thought to be the first to undergo tauopathy aggregation, while excitatory synapses stayed relatively intact leading to increased hyperexcitability (Shimojo et al. 2020; Kudo et al. 2023). Moreover, when extracellular vesicles containing pTau were injected into the brains of C57BL6/J mice, pTau preferentially accumulated in GABAergic interneurons (Ruan et al. 2020), suggesting a potential vulnerability within inhibitory neurons that drives hyperexcitability in AD. There are also reports of reductions in glutamate clearance from the synaptic cleft which would contribute to further increases in hyperexcitability (Hunsberger et al. 2015). Emerging work building on these findings is focusing on using EIB as a biomarker for MCI (Cope et al. 2022; Javed et al. 2022). Elevated neural hyperexcitability in the hippocampus (HP) and frontal cortex (FC) of adults with a first-degree AD-diagnosed relative is consistent with these findings (Bassett et al. 2006). Interestingly, in patients with mild cognitive impairment (MCI), hyperactivity shifts into hypoactivity during the transition from MCI to AD (Dickerson et al. 2005; Celone et al. 2006) as a direct result of neurodegeneration. While the cell-type specificity of pTau vulnerability remains an active area of investigation, alterations in EIB may be among the first observable markers associated with AD pathology.

Pathogenic tau (pTau) in AD is observed early in AD in layer II of the entorhinal cortex (EC) (Braak and Braak 1991a, b; Kaufman et al. 2017, 2018) and then propagates throughout interconnected networks like the hippocampus (van Groen et al. 2003; Nilssen et al. 2019; Ohara et al. 2023) at rates thought to be dictated by relative neuronal activity (Pooler et al. 2013; Yamada et al. 2014; Wu et al. 2016). This relationship between neural activity and tau is likely what mediates the ability of CSF total-tau levels to predict seizure activity in AD cases (Tábuas-Pereira et al. 2019), as well as the recent observations of seizures and seizure-like activity in AD and rodent models of AD (Scharfman 2012). For example, recent studies in the 5xFAD mouse model of AD found that decreasing hyperexcitability was effective at reducing the development of molecular pathology (Barbour et al. 2024) and that this may be due directly to decreases in pTau propagation known to be exacerbated by seizures in the same mouse model (Barbour et al. 2023). In fact, there is now emerging evidence from multiple groups that seizure activity is a common feature of mouse models of AD (Palop et al. 2007; Roberson et al. 2011; Bezzina et al. 2015; Lisgaras and Scharfman 2022, 2023; Anna et al. 2023; Hole et al. 2024). Taken together these data support neuronal network hyperexcitability as a key underlying mechanism behind tau seeding and spreading in the context of AD and associated tauopathies.

Increases in neural EIB (i.e., hyperexcitability) have been observed in many animal models of alcohol use and are ascribed to compensatory alterations in neurotransmission resulting from repeated alcohol exposures (Mihic and Harris 1995; Tabakoff and Hoffman 1996; Valenzuela and Harris 1997; Valenzuela 1997). These alterations to EIB appear to occur throughout the brain (Pati et al. 2020, 2022; Downs et al. 2022), and as discussed above in the context of seizures, this relative increase in neural activity may be directly responsible for driving increased tau spread. We are actively exploring this possibility and to date have found some initial evidence that at least alcohol consumption in the P301S model of AD may indeed increase EIB (Maphis et al. 2024). There is a clear need for further preclinical research on the role of alcohol-induced alterations to EIB dysfunction and AD pathology, particularly focused on brain regions known to be involved in the development of both AUD and AD. In addition to exploring alcohol/AD relationships with preclinical models, advanced neuroimaging technologies and new sensitive blood-based biomarkers should be combined with accurate measures of alcohol use history in the clinic. Unfortunately, time is likely running out for millions of individuals—there is an urgent need to combine preclinical and clinical efforts toward uncovering the mechanisms responsible for alcohol-induced increases in AD risk.

10.4 Alcohol and Alzheimer's Disease: Future Directions

There is now mounting evidence from the past two decades which strongly supports alcohol use as a major risk factor for AD. Studies examining the effects of alcohol misuse in adolescence (Pascale et al. 2022), young adulthood (de Goede et al. 2021; Kekkonen et al. 2021), midlife (Anttila et al. 2004; Sabia et al. 2018; Kivimäki et al. 2020; Chosy et al. 2022; Zhao et al. 2024), and advanced age (Heymann et al. 2016; Kivimäki et al. 2020) all indicate that alcohol exacerbates risk for AD. However, to date there are many gaps in our knowledge that need to be addressed. First, epidemiological studies often fail to track patterns of alcohol use throughout the lifespan. Second, these studies typically only assess AD-related outcomes at one or two points in time. Third, many large studies on alcohol use critical to our understanding of AD progression have not assessed alcohol use past the age of 65, when LOAD is most commonly diagnosed. Fourth, while there are active human longitudinal studies on the impact of alcohol use during adolescence and young adulthood when drinking rates are highest and the brain is still developing (Lisdahl et al. 2018), we are unaware of any that measure AD-related outcomes. All the above issues would be addressed with comprehensive lifespan studies that use neuroimaging, blood-based biomarkers, and alcohol use assessments throughout the preclinical and prodromal phases of AD. We argue that additional characterization of oxidative stress, neuroinflammation, metabolic dysregulation, pTau pathogenesis, and EIB is particularly beneficial to include.

There is a lack of preclinical studies examining the impact of alcohol on AD so more research is needed to gain even fundamental knowledge on how alcohol affects the development and progression of AD. Many alcohol/AD studies utilize the 3xTg animal model (Castano-Prat et al. 2019; Hoffman et al. 2019; Frausto et al. 2022; Tucker et al. 2022; Walter et al. 2022; Sanna et al. 2023), characterized by mutations in both APP and tau, while others have used tauopathy models featuring tau mutations exclusively (P301S, PS19) (Catavero et al. 2022; Downs et al. 2022; Maphis et al. 2024). While these mutations are clearly useful in driving pathogenesis, they do so at very young ages, which is inconsistent with the majority of AD cases (i.e., LOAD) (Drummond and Wisniewski 2017; Tai et al. 2020). These studies in transgenic mice are currently and will continue to be informative to this emerging field, but we should be prepared to quickly contextualize them using alternative approaches. For example, conducting a study to assess alcohol's influence on survival rates in a range of AD models could provide invaluable knowledge on which specific pathogenic pathways in AD are most perturbed in the context of alcohol. These data would also aid in prioritizing mouse models and molecular systems that better recapitulate LOAD. Moreover, there is a need for improved mouse models of LOAD. Fortunately, the national consortium, MODEL-AD (Wilcock and Lamb 2024), is actively working on creating and validating novel models for LOAD using humanized apolipoprotein epsilon 4 varient (APOEε4, strongest genetic predictor of LOAD, Sienski et al. 2021), as well as incorporation of non-mutant humanized Aβ and tau. These models will be particularly instrumental in evaluating the effects of alcohol exposure on AD at a variety of developmental stages and can do so in a timeframe that is greatly accelerated compared to human studies.

In summation, both epidemiological and preclinical studies implicate alcohol use as a risk factor for AD and AD-related dementia. These studies identify key molecular pathways that include oxidative stress, neuroimmune dysfunction, metabolic dysregulation, pTau formation/spread, and excitatory/inhibitory balance. We have outlined some potential future work that we believe will be essential to fully define the nature of alcohol-related risk in AD, which we are hopeful will identify key pathological mechanisms setting the stage for the development of therapeutic targets.

Acknowledgments This work was supported in part by National Institute on Alcohol Abuse and Alcoholism grants AA025120 (DNL), AA0251-05S1(DNL), AA022534 (DNL), AA015614 (DNL), AA028924 (LGC), U54AA030463 (LGC), AA007573 (PEA), Loan Repayment Program-Research on Emerging Areas Critical to Human Health (LRP-REACH; NMM), and an Institutional Research and Career Development (IRACDA, NIGMS, K12, GM088021; NMM), AA018108 (NM).

References

ADAPT Research Group, Lyketsos CG, Breitner JCS et al (2007) Naproxen and celecoxib do not prevent AD in early results from a randomized controlled trial. Neurology 68:1800–1808. https://doi.org/10.1212/01.wnl.0000260269.93245.d2

ADGC, TEADI, and CHARGE et al (2019) Genetic meta-analysis of diagnosed Alzheimer's disease identifies new risk loci and implicates Aβ, tau, immunity and lipid processing. Nat Genet 51:414–430. https://doi.org/10.1038/s41588-019-0358-2

Alberto GE, Klorig DC, Goldstein AT, Godwin DW (2023) Alcohol withdrawal produces changes in excitability, population discharge probability, and seizure threshold. Alcohol Clin Exp Res 47:211–218. https://doi.org/10.1111/acer.15004

Alfonso-Loeches S, Pascual-Lucas M, Blanco AM et al (2010) Pivotal role of TLR4 receptors in alcohol-induced neuroinflammation and brain damage. J Neurosci 30:8285–8295. https://doi.org/10.1523/jneurosci.0976-10.2010

Almeida MF, Bahr BA, Kinsey ST (2020) Endosomal-lysosomal dysfunction in metabolic diseases and Alzheimer's disease. Int Rev Neurobiol 154:303–324. https://doi.org/10.1016/bs.irn.2020.02.012

Alquezar C, Arya S, Kao AW (2021) Tau post-translational modifications: dynamic transformers of tau function, degradation, and aggregation. Front Neurol 11:595532. https://doi.org/10.3389/fneur.2020.595532

Aman Y, Schmauck-Medina T, Hansen M et al (2021) Autophagy in healthy aging and disease. Nat Aging 1:634–650. https://doi.org/10.1038/s43587-021-00098-4

Amatniek JC, Hauser WA, DelCastillo-Castaneda C et al (2006) Incidence and predictors of seizures in patients with Alzheimer's disease. Epilepsia 47:867–872. https://doi.org/10.1111/j.1528-1167.2006.00554.x

Anastacio HTD, Matosin N, Ooi L (2022) Neuronal hyperexcitability in Alzheimer's disease: what are the drivers behind this aberrant phenotype? Transl Psychiatry 12:257. https://doi.org/10.1038/s41398-022-02024-7

Andersen JK (2004) Oxidative stress in neurodegeneration: cause or consequence? Nat Med 10:S18–S25. https://doi.org/10.1038/nrn1434

Andreadis A (2005) Tau gene alternative splicing: expression patterns, regulation and modulation of function in normal brain and neurodegenerative diseases. Biochim Biophys Acta (BBA) Mol Basis Dis 1739:91–103. https://doi.org/10.1016/j.bbadis.2004.08.010

Andreadis A (2006) Alternative splicing and disease. Prog Mol Subcell Biol 44:89–107. https://doi.org/10.1007/978-3-540-34449-0_5

Andrews SJ, Fulton-Howard B, Goate A (2020) Interpretation of risk loci from genome-wide association studies of Alzheimer's disease. Lancet Neurol 19:326–335. https://doi.org/10.1016/s1474-4422(19)30435-1

Andrews SJ, Renton AE, Fulton-Howard B et al (2023) The complex genetic architecture of Alzheimer's disease: novel insights and future directions. eBioMedicine 90:104511. https://doi.org/10.1016/j.ebiom.2023.104511

Angioni D, Delrieu J, Hansson O et al (2022) Blood biomarkers from research use to clinical practice: what must be done? A report from the EU/US CTAD Task Force. J Prev Alzheimers Dis 9:569–579. https://doi.org/10.14283/jpad.2022.85

Anna BS, Cattaud V, Bezzina C et al (2023) Neuronal hyperexcitability in the Tg2576 mouse model of Alzheimer's disease—the influence of sleep and noradrenergic transmission. Neurobiol Aging 123:35–48. https://doi.org/10.1016/j.neurobiolaging.2022.11.017

Annadurai N, Sanctis JBD, Hajdúch M, Das V (2021) Tau secretion and propagation: Perspectives for potential preventive interventions in Alzheimer's disease and other tauopathies. Exp Neurol 113756. https://doi.org/10.1016/j.expneurol.2021.113756

Anton PE, Rutt LN, Kaufman ML et al (2024) Binge ethanol exposure in advanced age elevates neuroinflammation and early indicators of neurodegeneration and cognitive impairment in female mice. Brain Behav Immun 116:303–316. https://doi.org/10.1016/j.bbi.2023.12.034

Anttila T, Helkala E-L, Viitanen M et al (2004) Alcohol drinking in middle age and subsequent risk of mild cognitive impairment and dementia in old age: a prospective population based study. BMJ 329:539

Aouci R, Soudany ME, Maakoul Z et al (2022) Dlx5/6 expression levels in mouse GABAergic neurons regulate adult parvalbumin neuronal density and anxiety/compulsive behaviours. Cells 11:1739. https://doi.org/10.3390/cells11111739

Apostolova LG, Di LJ, Duffy EL et al (2014) Risk factors for behavioral abnormalities in mild cognitive impairment and mild Alzheimer's disease. Dement Geriatr Cogn 37:315–326. https://doi.org/10.1159/000351009

Aragon CMG, Rogan F, Amit Z (1992) Ethanol metabolism in rat brain homogenates by a catalase-H2O2

system. Biochem Pharmacol 44:93–98. https://doi.org/10.1016/0006-2952(92)90042-h
ARUK Consortium, EADI, GERAD/PERADES, CHARGE, ADGC, Sims R et al (2017) Rare coding variants in PLCG2, ABI3, and TREM2 implicate microglial-mediated innate immunity in Alzheimer's disease. Nat Genet 49:1373–1384. https://doi.org/10.1038/ng.3916
Asai H, Ikezu S, Tsunoda S et al (2015) Depletion of microglia and inhibition of exosome synthesis halt tau propagation. Nat Neurosci 18:1584–1593. https://doi.org/10.1038/nn.4132
Ashton NJ, Pascoal TA, Karikari TK et al (2021) Plasma p-tau231: a new biomarker for incipient Alzheimer's disease pathology. Acta Neuropathol 1–16. https://doi.org/10.1007/s00401-021-02275-6
Athanasaki A, Melanis K, Tsantzali I et al (2022) Type 2 diabetes mellitus as a risk factor for Alzheimer's disease: review and meta-analysis. Biomedicines 10:778. https://doi.org/10.3390/biomedicines10040778
Azuar J, Bouaziz-Amar E, Cognat E et al (2021) Cerebrospinal fluid biomarkers in patients with alcohol use disorder and persistent cognitive impairment. Alcohol Clin Exp Res 45:561–565. https://doi.org/10.1111/acer.14554
Bacioglu M, Maia LF, Preische O et al (2016) Neurofilament light chain in blood and CSF as marker of disease progression in mouse models and in neurodegenerative diseases. Neuron 91:56–66. https://doi.org/10.1016/j.neuron.2016.05.018
Barbour AJ, Gourmaud S, Stewart DA et al (2023) Seizures exacerbate excitatory-inhibitory imbalance and tau seeding effects in 5XFAD mice. Alzheimers Dement 19. https://doi.org/10.1002/alz.067907
Barbour AJ, Gourmaud S, Lancaster E et al (2024) Seizures exacerbate excitatory: inhibitory imbalance in Alzheimer's disease and 5XFAD mice. Brain 147:2169–2184. https://doi.org/10.1093/brain/awae126
Barnett AM, David E, Rohlman AR et al (2022) Adolescent binge alcohol enhances early Alzheimer's disease pathology in adulthood through proinflammatory neuroimmune activation. Front Pharmacol 13:884170. https://doi.org/10.3389/fphar.2022.884170
Barnett AG, Dawkins L, Zou J et al (2024) Loss of neuronal lysosomal acid lipase drives amyloid pathology in Alzheimer's disease. bioRxiv. https://doi.org/10.1101/2024.06.09.596693
Barrachina M, Maes T, Buesa C, Ferrer I (2006) Lysosome-associated membrane protein 1 (LAMP-1) in Alzheimer's disease. Neuropathol Appl Neurobiol 32:505–516. https://doi.org/10.1111/j.1365-2990.2006.00756.x
Barthélemy NR, Bateman RJ, Hirtz C et al (2020) Cerebrospinal fluid phospho-tau T217 outperforms T181 as a biomarker for the differential diagnosis of Alzheimer's disease and PET amyloid-positive patient identification. Alzheimers Res Ther 12:26. https://doi.org/10.1186/s13195-020-00596-4
Barthélemy NR, Saef B, Li Y et al (2023) CSF tau phosphorylation occupancies at T217 and T205 represent improved biomarkers of amyloid and tau pathology in Alzheimer's disease. Nat Aging 3:391–401. https://doi.org/10.1038/s43587-023-00380-7
Barthélemy NR, Salvadó G, Schindler SE et al (2024) Highly accurate blood test for Alzheimer's disease is similar or superior to clinical cerebrospinal fluid tests. Nat Med 30:1085–1095. https://doi.org/10.1038/s41591-024-02869-z
Bartos M, Vida I, Jonas P (2007) Synaptic mechanisms of synchronized gamma oscillations in inhibitory interneuron networks. Nat Rev Neurosci 8:45–56. https://doi.org/10.1038/nrn2044
Bassett SS, Yousem DM, Cristinzio C et al (2006) Familial risk for Alzheimer's disease alters fMRI activation patterns. Brain 129:1229–1239. https://doi.org/10.1093/brain/awl089
Beagle AJ, Darwish SM, Ranasinghe KG et al (2017) Relative incidence of seizures and myoclonus in Alzheimer's disease, dementia with Lewy bodies, and frontotemporal dementia. J Alzheimers Dis 60(1):211–223. https://doi.org/10.3233/jad-170031
Becker HC (1998) Kindling in alcohol withdrawal. Alcohol Health Res World 22:25–33
Bell BJ, Malvankar MM, Tallon C, Slusher BS (2020) Sowing the seeds of discovery: tau-propagation models of Alzheimer's disease. ACS Chem Neurosci 11:3499–3509. https://doi.org/10.1021/acschemneuro.0c00531
Bellenguez C, Küçükali F, Jansen IE et al (2022) New insights into the genetic etiology of Alzheimer's disease and related dementias. Nat Genet 54:412–436. https://doi.org/10.1038/s41588-022-01024-z
Benoit M, Berrut G, Doussaint J et al (2012) Apathy and depression in mild Alzheimer's disease: a cross-sectional study using diagnostic criteria. J Alzheimers Dis 31:325–334. https://doi.org/10.3233/jad-2012-112003
Bezzina C, Verret L, Juan C et al (2015) Early onset of hypersynchronous network activity and expression of a marker of chronic seizures in the Tg2576 mouse model of Alzheimer's disease. PLoS One 10:e0119910. https://doi.org/10.1371/journal.pone.0119910
Bhaskar K, Konerth M, Kokiko-Cochran ON et al (2010) Regulation of tau pathology by the microglial fractalkine receptor. Neuron 68:19–31. https://doi.org/10.1016/j.neuron.2010.08.023
Binder LI, Guillozet-Bongaarts AL, Garcia-Sierra F, Berry RW (2005) Tau, tangles, and Alzheimer's disease. Biochim Biophys Acta (BBA) Mol Basis Dis 1739:216–223. https://doi.org/10.1016/j.bbadis.2004.08.014
Binette AP, Palmqvist S, Bali D et al (2022) Combining plasma phospho-tau and accessible measures to evaluate progression to Alzheimer's dementia in mild cognitive impairment patients. Alzheimers Res Ther 14:46. https://doi.org/10.1186/s13195-022-00990-0
Bo RD, Angeretti N, Lucca E et al (1995) Reciprocal control of inflammatory cytokines, IL-1 and IL-6, and beta-amyloid production in cultures. Neurosci Lett 188:70–74. https://doi.org/10.1016/0304-3940(95)11384-9

Born HA (2015) Seizures in Alzheimer's disease. Neuroscience 286:251–263. https://doi.org/10.1016/j.neuroscience.2014.11.051

Braak H, Braak E (1991a) Neuropathological stageing of Alzheimer-related changes. Acta Neuropathol 82:239–259. https://doi.org/10.1007/bf00308809

Braak H, Braak E (1991b) Demonstration of amyloid deposits and neurofibrillary changes in whole brain sections. Brain Pathol 1

Bradfield NI, Ames D (2020) Mild cognitive impairment: narrative review of taxonomies and systematic review of their prediction of incident Alzheimer's disease dementia. BJPsych Bull 44:67–74. https://doi.org/10.1192/bjb.2019.77

Brelstaff JH, Mason M, Katsinelos T et al (2021) Microglia become hypofunctional and release metalloproteases and tau seeds when phagocytosing live neurons with P301S tau aggregates. Sci Adv 7:eabg4980. https://doi.org/10.1126/sciadv.abg4980

Breslow RA, Castle IP, Chen CM, Graubard BI (2017) Trends in alcohol consumption among older Americans: National Health Interview Surveys, 1997 to 2014. Alcohol Clin Exp Res 41:976–986. https://doi.org/10.1111/acer.13365

Brunello CA, Merezhko M, Uronen R-L, Huttunen HJ (2019) Mechanisms of secretion and spreading of pathological tau protein. Cell Mol Life Sci 1–24. https://doi.org/10.1007/s00018-019-03349-1

Butterfield DA, Halliwell B (2019) Oxidative stress, dysfunctional glucose metabolism and Alzheimer disease. Nat Rev Neurosci 20:148–160. https://doi.org/10.1038/s41583-019-0132-6

Buzsáki G, Kaila K, Raichle M (2007) Inhibition and brain work. Neuron 56:771–783. https://doi.org/10.1016/j.neuron.2007.11.008

Cai C, Zambach SA, Grubb S et al (2023) Impaired dynamics of precapillary sphincters and pericytes at first-order capillaries predict reduced neurovascular function in the aging mouse brain. Nat Aging 3:173–184. https://doi.org/10.1038/s43587-022-00354-1

Caponio D, Veverová K, Zhang S et al (2022) Compromised autophagy and mitophagy in brain ageing and Alzheimer's diseases. Aging Brain 2:100056. https://doi.org/10.1016/j.nbas.2022.100056

Carroll T, Guha S, Nehrke K, Johnson GVW (2021) Tau post-translational modifications: potentiators of selective vulnerability in sporadic Alzheimer's disease. Biology 10:1047. https://doi.org/10.3390/biology10101047

Castano-Prat P, Perez-Mendez L, Perez-Zabalza M et al (2019) Altered slow (<1 Hz) and fast (beta and gamma) neocortical oscillations in the 3xTg-AD mouse model of Alzheimer's disease under anesthesia. Neurobiol Aging 79:142–151. https://doi.org/10.1016/j.neurobiolaging.2019.02.009

Castellani R, Smith MA, Richey PL et al (1995) Evidence for oxidative stress in Pick disease and corticobasal degeneration. Brain Res 696:268–271. https://doi.org/10.1016/0006-8993(95)00535-x

Catavero CM, Marsh AE, Downs AM et al (2022) Effects of long-term alcohol consumption on behavior in the P301S (Line PS19) tauopathy mouse model. Biorxiv 2022(07):12.499737. https://doi.org/10.1101/2022.07.12.499737

Cattaud V, Bezzina C, Rey CC et al (2018) Early disruption of parvalbumin expression and perineuronal nets in the hippocampus of the Tg2576 mouse model of Alzheimer's disease can be rescued by enriched environment. Neurobiol Aging 72:147–158. https://doi.org/10.1016/j.neurobiolaging.2018.08.024

Celone KA, Calhoun VD, Dickerson BC et al (2006) Alterations in memory networks in mild cognitive impairment and Alzheimer's disease: an independent component analysis. J Neurosci 26:10222–10231. https://doi.org/10.1523/jneurosci.2250-06.2006

Chastain LG, Sarkar DK (2014) Chapter four. Role of microglia in regulation of ethanol neurotoxic action. Int Rev Neurobiol 118:81–103. https://doi.org/10.1016/b978-0-12-801284-0.00004-x

Chatterjee P, Pedrini S, Stoops E et al (2021) Plasma glial fibrillary acidic protein is elevated in cognitively normal older adults at risk of Alzheimer's disease. Transl Psychiatry 11:27. https://doi.org/10.1038/s41398-020-01137-1

Chen Y, Chang L, Blennow K et al (2020) An ultrasensitive immunoassay for detection of p-tau181 in blood. Alzheimers Dement 16. https://doi.org/10.1002/alz.041238

Chiu M-J, Fan L-Y, Chen T-F et al (2017) Plasma tau levels in cognitively normal middle-aged and older adults. Front Aging Neurosci 9:51. https://doi.org/10.3389/fnagi.2017.00051

Choi Y, Kim H-S, Shin KY et al (2007) Minocycline attenuates neuronal cell death and improves cognitive impairment in Alzheimer's disease models. Neuropsychopharmacology 32:2393–2404. https://doi.org/10.1038/sj.npp.1301377

Chosy EJ, Edland S, Launer L, White LR (2022) Midlife alcohol consumption and later life cognitive impairment: Light drinking is not protective and APOE genotype does not change this relationship. PLoS One 17:e0264575. https://doi.org/10.1371/journal.pone.0264575

Chung H, Park K, Jang HJ et al (2020) Dissociation of somatostatin and parvalbumin interneurons circuit dysfunctions underlying hippocampal theta and gamma oscillations impaired by amyloid β oligomers in vivo. Brain Struct Funct 225:935–954. https://doi.org/10.1007/s00429-020-02044-3

Coleman L, He J, Lee J et al (2011) Adolescent binge drinking alters adult brain neurotransmitter gene expression, behavior, brain regional volumes, and neurochemistry in mice. Alcohol Clin Exp Res 35:671–688. https://doi.org/10.1111/j.1530-0277.2010.01385.x

Coleman L, Liu W, Oguz I et al (2014) Adolescent binge ethanol treatment alters adult brain regional volumes, cortical extracellular matrix protein and behavioral flexibility. Pharmacol Biochem Behav 116:142–151. https://doi.org/10.1016/j.pbb.2013.11.021

Coleman L, Zou J, Qin L, Crews FT (2017) HMGB1/IL-1beta complexes regulate neuroimmune responses

in alcoholism. Brain Behav Immun 72:61–77. https://doi.org/10.1016/j.bbi.2017.10.027
Coleman LG, Zou J, Crews FT (2020) Microglial depletion and repopulation in brain slice culture normalizes sensitized proinflammatory signaling. J Neuroinflammation 17:27. https://doi.org/10.1186/s12974-019-1678-y
Conrad C, Zhu J, Conrad C et al (2007) Single molecule profiling of tau gene expression in Alzheimer's disease. J Neurochem 103:1228–1236. https://doi.org/10.1111/j.1471-4159.2007.04857.x
Cope ZA, Murai T, Rizzo SJS (2022) Emerging electroencephalographic biomarkers to improve preclinical to clinical translation in Alzheimer's disease. Front Aging Neurosci 14:805063. https://doi.org/10.3389/fnagi.2022.805063
Correas A, Cuesta P, Rosen BQ et al (2021) Compensatory neuroadaptation to binge drinking: Human evidence for allostasis. Addict Biol 26:e12960. https://doi.org/10.1111/adb.12960
Crews FT, Vetreno RP (2014) Neuroimmune basis of alcoholic brain damage. Int Rev Neurobiol 118:315–357. https://doi.org/10.1016/b978-0-12-801284-0.00010-5
Crews FT, Qin L, Sheedy D et al (2013) High mobility group box 1/Toll-like receptor danger signaling increases brain neuroimmune activation in alcohol dependence. Biol Psychiatry 73:602–612. https://doi.org/10.1016/j.biopsych.2012.09.030
Cribbs DH, Berchtold NC, Perreau V et al (2012) Extensive innate immune gene activation accompanies brain aging, increasing vulnerability to cognitive decline and neurodegeneration: a microarray study. J Neuroinflammation 9:1–18. https://doi.org/10.1186/1742-2094-9-179
Dage JL, Wennberg AMV, Airey DC et al (2016) Levels of tau protein in plasma are associated with neurodegeneration and cognitive function in a population-based elderly cohort. Alzheimers Dement 12:1226–1234. https://doi.org/10.1016/j.jalz.2016.06.001
Dalgediene I, Lasickiene R, Budvytyte R et al (2013) Immunogenic properties of amyloid beta oligomers. J Biomed Sci 20:10. https://doi.org/10.1186/1423-0127-20-10
David DC, Hauptmann S, Scherping I et al (2005) Proteomic and functional analyses reveal a mitochondrial dysfunction in P301L tau transgenic mice. J Biol Chem 280:23802–23814. https://doi.org/10.1074/jbc.m500356200
Degenhardt L, Glantz M, Evans-Lacko S et al (2017) Estimating treatment coverage for people with substance use disorders: an analysis of data from the World Mental Health Surveys. World Psychiatry 16:299–307. https://doi.org/10.1002/wps.20457
de Goede J, van der Mark-Reeuwijk KG, Braun KP et al (2021) Alcohol and brain development in adolescents and young adults: a systematic review of the literature and advisory report of the Health Council of the Netherlands. Adv Nutr 12:1379–1410. https://doi.org/10.1093/advances/nmaa170
Delanty N, Vaughan C (1998) Risk of Alzheimer's disease and duration of NSAID use. Neurology 51:652. https://doi.org/10.1212/wnl.51.2.652
der Mussele SV, Bekelaar K, Bastard NL et al (2012) Prevalence and associated behavioral symptoms of depression in mild cognitive impairment and dementia due to Alzheimer's disease: Depressive symptoms in MCI and Alzheimer's disease. Int J Geriatr Psych 28:947–958. https://doi.org/10.1002/gps.3909
DeVos SL, Corjuc BT, Oakley DH et al (2018) Synaptic tau seeding precedes tau pathology in human Alzheimer's disease brain. Front Neurosci (Switz) 12:267. https://doi.org/10.3389/fnins.2018.00267
Dharavath RN, Pina-Leblanc C, Tang VM et al (2023) GABAergic signaling in alcohol use disorder and withdrawal: pathological involvement and therapeutic potential. Front Neural Circuits 17:1218737. https://doi.org/10.3389/fncir.2023.1218737
Dickerson BC, Salat DH, Greve DN et al (2005) Increased hippocampal activation in mild cognitive impairment compared to normal aging and AD. Neurology 65:404–411. https://doi.org/10.1212/01.wnl.0000171450.97464.49
Dienel GA (2019) Brain glucose metabolism: integration of energetics with function. Physiol Rev 99:949–1045. https://doi.org/10.1152/physrev.00062.2017
Downs AM, Catavero CM, Kasten MR, McElligott ZA (2022) Tauopathy and alcohol consumption interact to alter locus coeruleus excitatory transmission and excitability in male and female mice. Alcohol. https://doi.org/10.1016/j.alcohol.2022.08.008
Drummond E, Wisniewski T (2017) Alzheimer's disease: experimental models and reality. Acta Neuropathol 133:155–175. https://doi.org/10.1007/s00401-016-1662-x
Dujardin S, Commins C, Lathuiliere A et al (2020) Tau molecular diversity contributes to clinical heterogeneity in Alzheimer's disease. Nat Med 1–8. https://doi.org/10.1038/s41591-020-0938-9
Efthymiou AG, Goate AM (2017) Late onset Alzheimer's disease genetics implicates microglial pathways in disease risk. Mol Neurodegener 12:1–12. https://doi.org/10.1186/s13024-017-0184-x
Eid A, Mhatre I, Richardson JR (2019) Gene-environment interactions in Alzheimer's disease: A potential path to precision medicine. Pharmacol Ther 199:173–187. https://doi.org/10.1016/j.pharmthera.2019.03.005
Eikelenboom P, van Exel E, Hoozemans JJ et al (2010) Neuroinflammation—an early event in both the history and pathogenesis of Alzheimer's disease. Neurodegener Dis 7:38–41. https://doi.org/10.1159/000283480
Emeršič A, Karikari TK, Rodríguez-Lantero J et al (2020) CSF phosphorylated tau-217 is increased in Alzheimer's and Creutzfeldt-Jakob diseases and correlates with amyloid pathology. Alzheimers Dement 16. https://doi.org/10.1002/alz.045296
Ennerfelt H, Frost EL, Shapiro DA et al (2022) SYK coordinates neuroprotective microglial responses in

neurodegenerative disease. Cell 185:4135–4152 e22. https://doi.org/10.1016/j.cell.2022.09.030

Erickson EK, Grantham EK, Warden AS, Harris RA (2018) Neuroimmune signaling in alcohol use disorder. Pharmacol Biochem Be 177:34–60. https://doi.org/10.1016/j.pbb.2018.12.007

Etminan M, Gill S, Samii A (2003) Effect of non-steroidal anti-inflammatory drugs on risk of Alzheimer's disease: systematic review and meta-analysis of observational studies. BMJ 327:128. https://doi.org/10.1136/bmj.327.7407.128

Fernandez-Lizarbe S, Pascual M, Guerri C (2009) Critical role of TLR4 response in the activation of microglia induced by ethanol. J Immunol 183:4733–4744. https://doi.org/10.4049/jimmunol.0803590

Fisher RA, Miners JS, Love S (2022) Pathological changes within the cerebral vasculature in Alzheimer's disease: New perspectives. Brain Pathol 32:e13061. https://doi.org/10.1111/bpa.13061

Fouquet M, Desgranges B, Landeau B et al (2009) Longitudinal brain metabolic changes from amnestic mild cognitive impairment to Alzheimer's disease. Brain 132:2058–2067. https://doi.org/10.1093/brain/awp132

Fowler CJ, Stoops E, Rainey-Smith SR et al (2022) Plasma p-tau181/Aβ1-42 ratio predicts Aβ-PET status and correlates with CSF-p-tau181/Aβ1-42 and future cognitive decline. Alzheimers Dement Diagn Assess Dis Monit 14:e12375. https://doi.org/10.1002/dad2.12375

Franceschi C, Campisi J (2014) Chronic inflammation (inflammaging) and its potential contribution to age-associated diseases. J Gerontol Ser 69:S4–S9. https://doi.org/10.1093/gerona/glu057

Frausto DM, Engen PA, Naqib A et al (2022) Impact of alcohol-induced intestinal microbiota dysbiosis in a rodent model of Alzheimer's disease. Front Aging 3:916336. https://doi.org/10.3389/fragi.2022.916336

Frigerio CS, Wolfs L, Fattorelli N et al (2019) The major risk factors for Alzheimer's disease: age, sex, and genes modulate the microglia response to Abeta plaques. Cell Rep 27(1293–1306):e6. https://doi.org/10.1016/j.celrep.2019.03.099

Fritschy J-M, Brünig I (2003) Formation and plasticity of GABAergic synapses: physiological mechanisms and pathophysiological implications. Pharmacol Ther 98:299–323. https://doi.org/10.1016/s0163-7258(03)00037-8

Fujita K, Motoki K, Tagawa K et al (2016) HMGB1, a pathogenic molecule that induces neurite degeneration via TLR4-MARCKS, is a potential therapeutic target for Alzheimer's disease. Sci Rep 6:31895. https://doi.org/10.1038/srep31895

Gaikwad S, Puangmalai N, Bittar A et al (2021) Tau oligomer induced HMGB1 release contributes to cellular senescence and neuropathology linked to Alzheimer's disease and frontotemporal dementia. Cell Rep 36:109419. https://doi.org/10.1016/j.celrep.2021.109419

Gao C, Jiang J, Tan Y, Chen S (2023) Microglia in neurodegenerative diseases: mechanism and potential therapeutic targets. Signal Transduct Target Ther 8:359. https://doi.org/10.1038/s41392-023-01588-0

Gazestani V, Kamath T, Nadaf NM et al (2023) Early Alzheimer's disease pathology in human cortex involves transient cell states. Cell 186:4438–4453.e23. https://doi.org/10.1016/j.cell.2023.08.005

GBD 2019 Dementia Forecasting Collaborators (2022) Estimation of the global prevalence of dementia in 2019 and forecasted prevalence in 2050: an analysis for the Global Burden of Disease Study 2019. Lancet Public Health 7:e105–e125. https://doi.org/10.1016/s2468-2667(21)00249-8

GBD 2021 Diseases and Injuries Collaborators, Ferrari AJ, Santomauro DF et al (2024) Global incidence, prevalence, years lived with disability (YLDs), disability-adjusted life-years (DALYs), and healthy life expectancy (HALE) for 371 diseases and injuries in 204 countries and territories and 811 subnational locations, 1990–2021: a systematic analysis for the Global Burden of Disease Study 2021. Lancet. https://doi.org/10.1016/s0140-6736(24)00757-8

Gella A, Durany N (2009) Oxidative stress in Alzheimer disease. Cell Adhes Migr 3:88–93. https://doi.org/10.4161/cam.3.1.7402

Gibbons GS, Lee VMY, Trojanowski JQ (2018) Mechanisms of cell-to-cell transmission of pathological tau: a review. JAMA Neurol 76:101. https://doi.org/10.1001/jamaneurol.2018.2505

Gilley DW, Bienias JL, Wilson RS et al (2004) Influence of behavioral symptoms on rates of institutionalization for persons with Alzheimer's disease. Psychol Med 34:1129–1135. https://doi.org/10.1017/s0033291703001831

Gimenez-Gomez P, Le T, Martin GE (2023) Modulation of neuronal excitability by binge alcohol drinking. Front Mol Neurosci 16:1098211. https://doi.org/10.3389/fnmol.2023.1098211

Ginsberg SD, Che S, Counts SE, Mufson EJ (2006) Shift in the ratio of three-repeat tau and four-repeat tau mRNAs in individual cholinergic basal forebrain neurons in mild cognitive impairment and Alzheimer's disease. J Neurochem 96:1401–1408. https://doi.org/10.1111/j.1471-4159.2005.03641.x

Ginsberg SD, Alldred MJ, Counts SE et al (2010) Microarray analysis of hippocampal CA1 neurons implicates early endosomal dysfunction during Alzheimer's disease progression. Biol Psychiatry 68:885–893. https://doi.org/10.1016/j.biopsych.2010.05.030

Glantz MD, Bharat C, Degenhardt L et al (2020) The epidemiology of alcohol use disorders cross-nationally: Findings from the World Mental Health Surveys. Addict Behav 102:106128. https://doi.org/10.1016/j.addbeh.2019.106128

Gnorich J, Reifschneider A, Wind K et al (2023) Depletion and activation of microglia impact metabolic connectivity of the mouse brain. J Neuroinflammation 20:47. https://doi.org/10.1186/s12974-023-02735-8

Goedert M, Spillantini MG, Potier MC et al (1989) Cloning and sequencing of the cDNA encoding an isoform of microtubule-associated protein tau contain-

ing four tandem repeats: differential expression of tau protein mRNAs in human brain. EMBO J 8:393–399. https://doi.org/10.1002/j.1460-2075.1989.tb03390.x
González A, Pariente JA, Salido GM (2007) Ethanol stimulates ROS generation by mitochondria through Ca2+ mobilization and increases GFAP content in rat hippocampal astrocytes. Brain Res 1178:28–37. https://doi.org/10.1016/j.brainres.2007.08.040
González H, Elgueta D, Montoya A, Pacheco R (2014) Neuroimmune regulation of microglial activity involved in neuroinflammation and neurodegenerative diseases. J Neuroimmunol 274:1–13. https://doi.org/10.1016/j.jneuroim.2014.07.012
Goodson HV, Jonasson EM (2018) Microtubules and microtubule-associated proteins. Cold Spring Harb Perspect Biol 10:a022608. https://doi.org/10.1101/cshperspect.a022608
Groot C, Smith R, Collij LE et al (2024) Tau positron emission tomography for predicting dementia in individuals with mild cognitive impairment. JAMA Neurol 81. https://doi.org/10.1001/jamaneurol.2024.1612
Grucza RA, Sher KJ, Kerr WC et al (2018) Trends in adult alcohol use and binge drinking in the early 21st-century United States: a meta-analysis of 6 national survey series. Alcohol Clin Exp Res 42:1939–1950. https://doi.org/10.1111/acer.13859
Gu J, Liu F (2020) Tau in Alzheimer's disease: pathological alterations and an attractive therapeutic target. Curr Méd Sci 40:1009–1021. https://doi.org/10.1007/s11596-020-2282-1
Guadagna S, Esiri MM, Williams RJ, Francis PT (2012) Tau phosphorylation in human brain: relationship to behavioral disturbance in dementia. Neurobiol Aging 33:2798–2806. https://doi.org/10.1016/j.neurobiolaging.2012.01.015
Guerreiro R, Bras J (2015) The age factor in Alzheimer's disease. Genome Med 7:106. https://doi.org/10.1186/s13073-015-0232-5
Guillozet AL, Weintraub S, Mash DC, Mesulam MM (2003) Neurofibrillary tangles, amyloid, and memory in aging and mild cognitive impairment. Arch Neurol 60:729–736. https://doi.org/10.1001/archneur.60.5.729
Hamelin L, Lagarde J, Dorothée G et al (2016) Early and protective microglial activation in Alzheimer's disease: a prospective study using 18 F-DPA-714 PET imaging. Brain 139:1252–1264. https://doi.org/10.1093/brain/aww017
Hammond TC, Xing X, Wang C et al (2020) β-amyloid and tau drive early Alzheimer's disease decline while glucose hypometabolism drives late decline. Commun Biol 3:352. https://doi.org/10.1038/s42003-020-1079-x
Hampel H, Lista S (2012) From inherited to sporadic AD—crossing the biomarker bridge. Nat Rev Neurol 8:598–600. https://doi.org/10.1038/nrneurol.2012.202
Hampel H, Hardy J, Blennow K et al (2021) The amyloid-β pathway in Alzheimer's disease. Mol Psychiatry 26:5481–5503. https://doi.org/10.1038/s41380-021-01249-0
Hanes J, Kovac A, Kvartsberg H et al (2020) Evaluation of a novel immunoassay to detect p-Tau Thr127 in the CSF to distinguish Alzheimer disease from other dementias. Neurology 95. https://doi.org/10.1212/wnl.0000000000010814
Hansson O (2021) Biomarkers for neurodegenerative diseases. Nat Med 27:954–963. https://doi.org/10.1038/s41591-021-01382-x
Hansson O, Edelmayer RM, Boxer AL et al (2022) The Alzheimer's Association appropriate use recommendations for blood biomarkers in Alzheimer's disease. Alzheimers Dement 18:2669–2686. https://doi.org/10.1002/alz.12756
Hansson O, Blennow K, Zetterberg H, Dage J (2023) Blood biomarkers for Alzheimer's disease in clinical practice and trials. Nat Aging 3:506–519. https://doi.org/10.1038/s43587-023-00403-3
Harris SS, Wolf F, Strooper BD, Busche MA (2020) Tipping the scales: peptide-dependent dysregulation of neural circuit dynamics in Alzheimer's disease. Neuron 107:417–435. https://doi.org/10.1016/j.neuron.2020.06.005
Harwood DG, Kalechstein A, Barker WW et al (2010) The effect of alcohol and tobacco consumption, and apolipoprotein E genotype, on the age of onset in Alzheimer's disease. Int J Geriatr Psychiatry 25:511–518. https://doi.org/10.1002/gps.2372
Heilig M, Egli M, Crabbe JC, Becker HC (2010) REVIEW: Acute withdrawal, protracted abstinence and negative affect in alcoholism: are they linked? Addict Biol 15:169–184. https://doi.org/10.1111/j.1369-1600.2009.00194.x
Heneka MT, Kummer MP, Stutz A et al (2013) NLRP3 is activated in Alzheimer's disease and contributes to pathology in APP/PS1 mice. Nature 493:674–678. https://doi.org/10.1038/nature11729
Hernández JA, López-Sánchez RC, Rendón-Ramírez A (2016) Lipids and oxidative stress associated with ethanol-induced neurological damage. Oxidative Med Cell Longev 2016:1543809. https://doi.org/10.1155/2016/1543809
Hershey LA, Lipton RB (2019) Naproxen for presymptomatic Alzheimer disease: Is this the end, or shall we try again? Neurology 92:829–830. https://doi.org/10.1212/wnl.0000000000007233
Heymann D, Stern Y, Cosentino S et al (2016) The association between alcohol use and the progression of Alzheimer's disease. Curr Alzheimer Res 13:1356–1362. https://doi.org/10.2174/1567205013666160603005035
Hickman S, Izzy S, Sen P et al (2018) Microglia in neurodegeneration. Nat Neurosci 21:1359–1369. https://doi.org/10.1038/s41593-018-0242-x
Hoffman JL, Faccidomo S, Kim M et al (2019) Alcohol drinking exacerbates neural and behavioral pathology in the 3xTg-AD mouse model of Alzheimer's disease. Int Rev Neurobiol 148:169–230. https://doi.org/10.1016/bs.irn.2019.10.017
Hole KL, Zhu B, Huggon L et al (2024) TauP301L disengages from the proteosome core complex and

neurogranin coincident with enhanced neuronal network excitability. Cell Death Dis 15:429. https://doi.org/10.1038/s41419-024-06815-2

Holmes BB, Furman JL, Mahan TE et al (2014) Proteopathic tau seeding predicts tauopathy in vivo. Proc Natl Acad Sci USA 111:E4376–E4385. https://doi.org/10.1073/pnas.1411649111

Horvath AA, Papp A, Zsuffa J et al (2021) Subclinical epileptiform activity accelerates the progression of Alzheimer's disease: A long-term EEG study. Clin Neurophysiol 132:1982–1989. https://doi.org/10.1016/j.clinph.2021.03.050

Howarth C, Gleeson P, Attwell D (2012) Updated energy budgets for neural computation in the neocortex and cerebellum. J Cereb Blood Flow Metab 32:1222–1232. https://doi.org/10.1038/jcbfm.2012.35

Huang Y, Mahley RW (2014) Apolipoprotein E: structure and function in lipid metabolism, neurobiology, and Alzheimer's diseases. Neurobiol Dis 72(Pt A):3–12. https://doi.org/10.1016/j.nbd.2014.08.025

Huang ZJ, Paul A (2019) The diversity of GABAergic neurons and neural communication elements. Nat Rev Neurosci 20:563–572. https://doi.org/10.1038/s41583-019-0195-4

Huang AR, Strombotne KL, Horner EM, Lapham SJ (2018) Adolescent cognitive aptitudes and later-in-life Alzheimer disease and related disorders. JAMA Netw Open 1:e181726. https://doi.org/10.1001/jamanetworkopen.2018.1726

Huang S, White DR, Marinkovic K (2022) Alterations of theta power and synchrony during encoding in young adult binge drinkers: Subsequent memory effects associated with retrieval after 48 h and 6 months. Front Psychol 13:1061016. https://doi.org/10.3389/fpsyg.2022.1061016

Hunsberger HC, Rudy CC, Batten SR et al (2015) P301L tau expression affects glutamate release and clearance in the hippocampal trisynaptic pathway. JNC 132:169–182. https://doi.org/10.1111/jnc.12967

Imbimbo BP, Solfrizzi V, Panza F (2010) Are NSAIDs useful to treat Alzheimer's disease or mild cognitive impairment? Front Aging Neurosci 2:19. https://doi.org/10.3389/fnagi.2010.00019

in't Veld BA, Ruitenberg A, Hofman A et al (2001) Nonsteroidal antiinflammatory drugs and the risk of alzheimer's disease. N Engl J Med 345:1515–1521. https://doi.org/10.1056/nejmoa010178

Ionescu-Tucker A, Cotman CW (2021) Emerging roles of oxidative stress in brain aging and Alzheimer's disease. Neurobiol Aging 107:86–95. https://doi.org/10.1016/j.neurobiolaging.2021.07.014

Irizarry MC, Jin S, He F et al (2012) Incidence of new-onset seizures in mild to moderate Alzheimer disease. Arch Neurol 69:368–372. https://doi.org/10.1001/archneurol.2011.830

Ising C, Venegas C, Zhang S et al (2019) NLRP3 inflammasome activation drives tau pathology. Nature 575:669–673. https://doi.org/10.1038/s41586-019-1769-z

Jack CR, Bennett DA, Blennow K et al (2018) NIA-AA Research Framework: Toward a biological definition of Alzheimer's disease. Alzheimers Dement 14:535–562. https://doi.org/10.1016/j.jalz.2018.02.018

Janelidze S, Mattsson N, Palmqvist S et al (2020) Plasma P-tau181 in Alzheimer's disease: relationship to other biomarkers, differential diagnosis, neuropathology and longitudinal progression to Alzheimer's dementia. Nat Med 26:379–386. https://doi.org/10.1038/s41591-020-0755-1

Jansen IE, Savage JE, Watanabe K et al (2019) Genome-wide meta-analysis identifies new loci and functional pathways influencing Alzheimer's disease risk. Nat Genet 51:404–413. https://doi.org/10.1038/s41588-018-0311-9

Javed E, Suárez-Méndez I, Susi G et al (2022) E/I unbalance and aberrant oscillation dynamics predict preclinical Alzheimer's disease. Biorxiv 2022(12):22.521549. https://doi.org/10.1101/2022.12.22.521549

Jeon KH, Han K, Jeong S-M et al (2023) Changes in alcohol consumption and risk of dementia in a nationwide cohort in South Korea. JAMA Netw Open 6:e2254771. https://doi.org/10.1001/jamanetworkopen.2022.54771

Jha SK, Jha NK, Kumar D et al (2017) Linking mitochondrial dysfunction, metabolic syndrome and stress signaling in Neurodegeneration. Biochim Biophys Acta (BBA) Mol Basis Dis 1863:1132–1146. https://doi.org/10.1016/j.bbadis.2016.06.015

Jiang S, Maphis NM, Binder J et al (2021) Proteopathic tau primes and activates interleukin-1β via myeloid-cell-specific MyD88- and NLRP3-ASC-inflammasome pathway. Cell Rep 36:109720. https://doi.org/10.1016/j.celrep.2021.109720

Jin JJ, Kim HD, Maxwell JA et al (2008) Toll-like receptor 4-dependent upregulation of cytokines in a transgenic mouse model of Alzheimer's disease. J Neuroinflammation 5:23. https://doi.org/10.1186/1742-2094-5-23

Jorfi M, Maaser-Hecker A, Tanzi RE (2023) The neuroimmune axis of Alzheimer's disease. Genome Med 15:6. https://doi.org/10.1186/s13073-023-01155-w

Kamondi A, Grigg-Damberger M, Löscher W et al (2024) Epilepsy and epileptiform activity in late-onset Alzheimer disease: clinical and pathophysiological advances, gaps and conundrums. Nat Rev Neurol 20:162–182. https://doi.org/10.1038/s41582-024-00932-4

Kane CJ, Phelan KD, Douglas JC et al (2013) Effects of ethanol on immune response in the brain: region-specific changes in aged mice. J Neuroinflammation 10:1–4. https://doi.org/10.1186/1742-2094-10-66

Kang S, Hayashi Y, Bruyns-Haylett M et al (2020) Model-predicted balance between neural excitation and inhibition was maintained despite of age-related decline in sensory evoked local field potential in rat barrel cortex. Front Syst Neurosci 14:24. https://doi.org/10.3389/fnsys.2020.00024

Karikari TK, Pascoal TA, Ashton NJ et al (2020) Blood phosphorylated tau 181 as a biomarker for Alzheimer's disease: a diagnostic performance and prediction modelling study using data from four prospective cohorts. Lancet Neurol 19:422–433. https://doi.org/10.1016/s1474-4422(20)30071-5

Katzov H, Chalmers K, Palmgren J et al (2004) Genetic variants of ABCA1 modify Alzheimer disease risk and quantitative traits related to beta-amyloid metabolism. Hum Mutat 23:358–367. https://doi.org/10.1002/humu.20012

Kaufman SK, Thomas TL, Tredici KD et al (2017) Characterization of tau prion seeding activity and strains from formaldehyde-fixed tissue. Acta Neuropathologica Commun 5:41. https://doi.org/10.1186/s40478-017-0442-8

Kaufman SK, Tredici KD, Thomas TL et al (2018) Tau seeding activity begins in the transentorhinal/entorhinal regions and anticipates phospho-tau pathology in Alzheimer's disease and PART. Acta Neuropathol 136:57–67. https://doi.org/10.1007/s00401-018-1855-6

Kekkonen V, Kallioniemi E, Kaarre O et al (2021) Heavy drinking from adolescence to young adulthood is associated with an altered cerebellum. Alcohol 92:35–40. https://doi.org/10.1016/j.alcohol.2021.02.002

Keren-Shaul H, Spinrad A, Weiner A et al (2017) A unique microglia type associated with restricting development of Alzheimer's disease. Cell 169:1276–1290 e17. https://doi.org/10.1016/j.cell.2017.05.018

Keyes KM, Jager J, Mal-Sarkar T et al (2019) Is there a recent epidemic of women's drinking? A critical review of national studies. Alcohol Clin Exp Res 43:1344–1359. https://doi.org/10.1111/acer.14082

Khan KM, Rosa GB-DL, Biggerstaff N et al (2023) Adolescent ethanol drinking promotes hyperalgesia, neuroinflammation and serotonergic deficits in mice that persist into adulthood. Brain Behav Immun 107:419–431. https://doi.org/10.1016/j.bbi.2022.07.160

Kim DW, Tu KJ, Wei A et al (2022) Amyloid-beta and tau pathologies act synergistically to induce novel disease stage-specific microglia subtypes. Mol Neurodegener 17:83. https://doi.org/10.1186/s13024-022-00589-x

Kivimäki M, Singh-Manoux A, Batty GD et al (2020) Association of alcohol-induced loss of consciousness and overall alcohol consumption with risk for dementia. JAMA Netw Open 3:e2016084. https://doi.org/10.1001/jamanetworkopen.2020.16084

Klohs J (2020) An integrated view on vascular dysfunction in Alzheimer's disease. Neurodegener Dis 19:109–127. https://doi.org/10.1159/000505625

Knopman DS, Amieva H, Petersen RC et al (2021) Alzheimer disease. Nat Rev Dis Primers 7:33. https://doi.org/10.1038/s41572-021-00269-y

Kobylecki C, Langheinrich T, Hinz R et al (2015) 18F-florbetapir PET in patients with frontotemporal dementia and Alzheimer disease. J Nucl Med 56:386–391. https://doi.org/10.2967/jnumed.114.147454

Koch M, Fitzpatrick AL, Rapp SR et al (2019) Alcohol consumption and risk of dementia and cognitive decline among older adults with or without mild cognitive impairment. JAMA Netw Open 2:e1910319. https://doi.org/10.1001/jamanetworkopen.2019.10319

Koob GF, Volkow ND (2016) Neurobiology of addiction: a neurocircuitry analysis. Lancet Psychiatry 3:760–773. https://doi.org/10.1016/s2215-0366(16)00104-8

Kopeikina KJ, Carlson GA, Pitstick R et al (2011) Tau accumulation causes mitochondrial distribution deficits in neurons in a mouse model of tauopathy and in human Alzheimer's disease brain. Am J Pathol 179:2071–2082. https://doi.org/10.1016/j.ajpath.2011.07.004

Korczyn AD, Grinberg LT (2024) Is Alzheimer disease a disease? Nat Rev Neurol 20:245–251. https://doi.org/10.1038/s41582-024-00940-4

Koss DJ, Jones G, Cranston A et al (2016) Soluble pre-fibrillar tau and β-amyloid species emerge in early human Alzheimer's disease and track disease progression and cognitive decline. Acta Neuropathol 132:875–895. https://doi.org/10.1007/s00401-016-1632-3

Kotilinek LA, Westerman MA, Wang Q et al (2008) Cyclooxygenase-2 inhibition improves amyloid-β-mediated suppression of memory and synaptic plasticity. Brain 131:651–664. https://doi.org/10.1093/brain/awn008

Kudo T, Takuwa H, Takahashi M et al (2023) Selective dysfunction of fast-spiking inhibitory interneurons and disruption of perineuronal nets in a tauopathy mouse model. iScience 26:106342. https://doi.org/10.1016/j.isci.2023.106342

Kvartsberg H, Hanes J, Benedet AL et al (2020) Quantification of tau phosphorylated at threonine 217 using a novel ultrasensitive immunoassay distinguishes Alzheimer's disease from healthy controls. Alzheimers Dement 16. https://doi.org/10.1002/alz.043467

Lam AD, Sarkis RA, Pellerin KR et al (2020) Association of epileptiform abnormalities and seizures in Alzheimer disease. Neurology 95:e2259–e2270. https://doi.org/10.1212/wnl.0000000000010612

Langballe EM, Ask H, Holmen J et al (2015) Alcohol consumption and risk of dementia up to 27 years later in a large, population-based sample: the HUNT study, Norway. Eur J Epidemiol 30:1049–1056. https://doi.org/10.1007/s10654-015-0029-2

Lanoiselée HM, Nicolas G, Wallon D et al (2017) APP, PSEN1, and PSEN2 mutations in early-onset Alzheimer disease: A genetic screening study of familial and sporadic cases. PLoS Med 14:e1002270. https://doi.org/10.1371/journal.pmed.1002270

Lauterborn JC, Scaduto P, Cox CD et al (2021) Increased excitatory to inhibitory synaptic ratio in parietal cortex samples from individuals with Alzheimer's disease. Nat Commun 12:2603. https://doi.org/10.1038/s41467-021-22742-8

Lee JH, Yang DS, Goulbourne CN et al (2022) Faulty autolysosome acidification in Alzheimer's disease mouse models induces autophagic build-up of Abeta in neurons, yielding senile plaques. Nat

Neurosci 25:688–701. https://doi.org/10.1038/s41593-022-01084-8

Lees B, Meredith LR, Kirkland AE et al (2020) Effect of alcohol use on the adolescent brain and behavior. Pharmacol Biochem Behav 192:172906. https://doi.org/10.1016/j.pbb.2020.172906

Leuzy A, Mattsson-Carlgren N, Palmqvist S et al (2022) Blood-based biomarkers for Alzheimer's disease. Embo Mol Med 14:e14408. https://doi.org/10.15252/emmm.202114408

Li D, Mielke MM (2019) An update on blood-based markers of Alzheimer's disease using the SiMoA platform. Neurology Ther 8:73–82. https://doi.org/10.1007/s40120-019-00164-5

Li X-L, Hu N, Tan M-S et al (2014) Behavioral and psychological symptoms in Alzheimer's disease. Biomed Res Int 2014:1–9. https://doi.org/10.1155/2014/927804

Li H, Guglielmetti C, Sei YJ et al (2023) Neurons require glucose uptake and glycolysis in vivo. Cell Rep 42. https://doi.org/10.1016/j.celrep.2023.112335

Lisdahl KM, Sher KJ, Conway KP et al (2018) Adolescent brain cognitive development (ABCD) study: Overview of substance use assessment methods. Dev Cogn Neurosci 32:80–96. https://doi.org/10.1016/j.dcn.2018.02.007

Lisgaras CP, Scharfman HE (2022) High frequency oscillations (250-500Hz) in animal models of Alzheimer's disease and two animal models of epilepsy. Epilepsia. https://doi.org/10.1111/epi.17462

Lisgaras CP, Scharfman HE (2023) Interictal spikes in Alzheimer's disease: Preclinical evidence for dominance of the dentate gyrus and cholinergic control by the medial septum. Neurobiol Dis 187:106294. https://doi.org/10.1016/j.nbd.2023.106294

Liu F, Gong C-X (2008) Tau exon 10 alternative splicing and tauopathies. Mol Neurodegener 3:8. https://doi.org/10.1186/1750-1326-3-8

Liu L, Drouet V, Wu JW et al (2012) Trans-synaptic spread of tau pathology in vivo. PLoS One 7:e31302. https://doi.org/10.1371/journal.pone.0031302

Liu C-C, Liu C-C, Kanekiyo T et al (2013) Apolipoprotein E and Alzheimer disease: risk, mechanisms and therapy. Nat Rev Neurol 9:106–118. https://doi.org/10.1038/nrneurol.2012.263

Liu W, Vetreno RP, Crews FT (2021) Hippocampal TNF-death receptors, caspase cell death cascades, and IL-8 in alcohol use disorder. Mol Psychiatry 26:2254–2262. https://doi.org/10.1038/s41380-020-0698-4

Livingston G, Huntley J, Sommerlad A et al (2020) Dementia prevention, intervention, and care: 2020 report of the Lancet Commission. Lancet 396:413–446. https://doi.org/10.1016/s0140-6736(20)30367-6

Long JM, Holtzman DM (2019) Alzheimer disease: an update on pathobiology and treatment strategies. Cell 179:312–339. https://doi.org/10.1016/j.cell.2019.09.001

Luciunaite A, McManus RM, Jankunec M et al (2020) Soluble Abeta oligomers and protofibrils induce NLRP3 inflammasome activation in microglia. J Neurochem 155:650–661. https://doi.org/10.1111/jnc.14945

Lue L-F, Walker DG, Brachova L et al (2001) Involvement of microglial receptor for advanced glycation end-products (RAGE) in Alzheimer's disease: identification of a cellular activation mechanism. Exp Neurol 171:29–45. https://doi.org/10.1006/exnr.2001.7732

Mace S, Cousin E, Ricard S et al (2005) ABCA2 is a strong genetic risk factor for early-onset Alzheimer's disease. Neurobiol Dis 18:119–125. https://doi.org/10.1016/j.nbd.2004.09.011

Maghsudi H, Schütze M, Maudsley AA et al (2020) Age-related brain metabolic changes up to seventh decade in healthy humans. Clin Neuroradiol 30:581–589. https://doi.org/10.1007/s00062-019-00814-z

Majerova P, Zilkova M, Kazmerova Z et al (2014) Microglia display modest phagocytic capacity for extracellular tau oligomers. J Neuroinflammation 11:161. https://doi.org/10.1186/s12974-014-0161-z

Malpetti M, Kievit RA, Passamonti L et al (2020) Microglial activation and tau burden predict cognitive decline in Alzheimer's disease. Brain 143:awaa088. https://doi.org/10.1093/brain/awaa088

Maphis NM, Morningstar MD, Linsenbardt DN (2024) Binge drinking in a mouse model of AD leads to unique cortical neurophysiological alterations. Poster presented at the Research Society on Alcohol conference, Minneapolis, MN, Tuesday, June 25, 2024. https://onlinelibrary.wiley.com/doi/epdf/10.1111/acer.15318

Marinkovic K, Myers ABA, Arienzo D et al (2022) Cortical GABA levels are reduced in young adult binge drinkers: Association with recent alcohol consumption and sex. NeuroImage: Clin 35:103091. https://doi.org/10.1016/j.nicl.2022.103091

Markram H, Toledo-Rodriguez M, Wang Y et al (2004) Interneurons of the neocortical inhibitory system. Nat Rev Neurosci 5:793–807. https://doi.org/10.1038/nrn1519

Marshall SA, McClain JA, Kelso ML et al (2013) Microglial activation is not equivalent to neuroinflammation in alcohol-induced neurodegeneration: The importance of microglia phenotype. Neurobiol Dis 54:239–251. https://doi.org/10.1016/j.nbd.2012.12.016

Marsland P, Vore AS, DaPrano E et al (2022) Sex-specific effects of ethanol consumption in older Fischer 344 rats on microglial dynamics and Aβ(1-42) accumulation. Alcohol. https://doi.org/10.1016/j.alcohol.2022.08.013

Martínez A, Carmona M, Portero-Otin M et al (2008) Type-dependent oxidative damage in frontotemporal lobar degeneration. J Neuropathol Exp Neurol 67:1122–1136. https://doi.org/10.1097/nen.0b013e31818e06f3

Mathys H, Adaikkan C, Gao F et al (2017) Temporal tracking of microglia activation in neurodegeneration at single-cell resolution. Cell Rep 21:366–380. https://doi.org/10.1016/j.celrep.2017.09.039

Matsushita S, Miyakawa T, Maesato H et al (2008) Elevated cerebrospinal fluid tau protein levels in Wernicke's encephalopathy. Alcohol Clin Exp Res 32:1091–1095. https://doi.org/10.1111/j.1530-0277.2008.00671.x

Mattson MP (2020) Involvement of GABAergic interneuron dysfunction and neuronal network hyperexcitability in Alzheimer's disease: Amelioration by metabolic switching. Int Rev Neurobiol 154:191–205. https://doi.org/10.1016/bs.irn.2020.01.006

McAlpine FE, Lee J-K, Harms AS et al (2009) Inhibition of soluble TNF signaling in a mouse model of Alzheimer's disease prevents pre-plaque amyloid-associated neuropathology. Neurobiol Dis 34:163–177. https://doi.org/10.1016/j.nbd.2009.01.006

McDonald TS, Lerskiatiphanich T, Woodruff TM et al (2023) Potential mechanisms to modify impaired glucose metabolism in neurodegenerative disorders. J Cereb Blood Flow Metab 43:26–43. https://doi.org/10.1177/0271678x221135061

McMillan P, Korvatska E, Poorkaj P et al (2008) Tau isoform regulation is region- and cell-specific in mouse brain. J Comp Neurol 511:788–803. https://doi.org/10.1002/cne.21867

Mergenthaler P, Lindauer U, Dienel GA, Meisel A (2013) Sugar for the brain: the role of glucose in physiological and pathological brain function. Trends Neurosci 36:587–597. https://doi.org/10.1016/j.tins.2013.07.001

Meyer P-F, Tremblay-Mercier J, Leoutsakos J et al (2019) INTREPAD: A randomized trial of naproxen to slow progress of presymptomatic Alzheimer disease. Neurology 92:e2070–e2080. https://doi.org/10.1212/wnl.0000000000007232

Meyer SD, Schaeverbeke J, Gille B et al (2020a) Comparison of two analytical platforms for blood-based surrogate biomarkers of amyloid pathology. Alzheimers Dement 16. https://doi.org/10.1002/alz.045110

Meyer SD, Schaeverbeke JM, Verberk IMW et al (2020b) Comparison of ELISA- and SIMOA-based quantification of plasma Aβ ratios for early detection of cerebral amyloidosis. Alzheimers Res Ther 12:162. https://doi.org/10.1186/s13195-020-00728-w

Miao J, Ma H, Yang Y et al (2023) Microglia in Alzheimer's disease: pathogenesis, mechanisms, and therapeutic potentials. Front Aging Neurosci 15:1201982. https://doi.org/10.3389/fnagi.2023.1201982

Mielke MM, Hagen CE, Xu J et al (2018) Plasma phospho-tau181 increases with Alzheimer's disease clinical severity and is associated with tau- and amyloid-positron emission tomography. Alzheimers Dement J Alzheimers Assoc 14:989–997. https://doi.org/10.1016/j.jalz.2018.02.013

Mihic SJ, Harris RA (1995) Alcohol actions at the GABAA receptor/chloride channels complex. In: Deitrich R, Erwin G (eds) Pharmacological effects of ethanol on the nervous system. CRC, Boca Raton, FL, pp 51–71

Mintz CM, Knox J, Hartz SM et al (2023) Demographic differences in the cascade of care for unhealthy alcohol use: A cross-sectional analysis of data from the 2015–2019 National Survey on Drug Use and Health. Alcohol Clin Exp Res 47:1890–1903. https://doi.org/10.1111/acer.15176

Moceri VM, Kukull WA, Emanuel I et al (2000) Early-life risk factors and the development of Alzheimer's disease. Neurology 54:415–415. https://doi.org/10.1212/wnl.54.2.415

Mohamed N, Herrou T, Plouffe V et al (2013) Spreading of tau pathology in Alzheimer's disease by cell-to-cell transmission. Eur J Neurosci 37:1939–1948. https://doi.org/10.1111/ejn.12229

Montesinos J, Alfonso-Loeches S, Guerri C (2016) Impact of the innate immune response in the actions of ethanol on the central nervous system. Alcohol Clin Exp Res 40:2260–2270. https://doi.org/10.1111/acer.13208

Montoliu C, Sancho-Tello M, Azorin I et al (1995) Ethanol increases cytochrome P4502E1 and induces oxidative stress in astrocytes. J Neurochem 65:2561–2570

Moriyama Y, Mimura M, Kato M, Kashima H (2006) Primary alcoholic dementia and alcohol-related dementia. Psychogeriatrics 6:114–118. https://doi.org/10.1111/j.1479-8301.2006.00168.x

Mosconi L, Tsui WH, Herholz K et al (2008) Multicenter standardized 18F-FDG PET diagnosis of mild cognitive impairment, Alzheimer's disease, and other dementias. J Nucl Med 49:390–398. https://doi.org/10.2967/jnumed.107.045385

Moscoso A, Grothe MJ, Ashton NJ et al (2021) Longitudinal associations of blood phosphorylated Tau181 and neurofilament light chain with neurodegeneration in Alzheimer disease. JAMA Neurol 78. https://doi.org/10.1001/jamaneurol.2020.4986

Möykkynen T, Korpi ER (2012) Acute effects of ethanol on glutamate receptors. Basic Clin Pharmacol Toxicol 111:4–13. https://doi.org/10.1111/j.1742-7843.2012.00879.x

Mufson EJ, Malek-Ahmadi M, Snyder N et al (2016) Braak stage and trajectory of cognitive decline innoncognitively impaired elders. Neurobiol Aging43:101–110. https://doi.org/10.1016/j.neurobiolaging.2016.03.003

Mukamal KJ, Kuller LH, Fitzpatrick AL et al (2003) Prospective study of alcohol consumption and risk of dementia in older adults. JAMA 289:1405–1413. https://doi.org/10.1001/jama.289.11.1405

Nakamura K, Iwahashi K, Furukawa A et al (2003) Acetaldehyde adducts in the brain of alcoholics. Arch Toxicol 77:591–593

National Institute on Alcohol Abuse and Alcoholism (NIAAA) (2024) Alcohol's effects on health. US Department of Health and Human Services. https://www.niaaa.nih.gov/alcohols-effects-health. Accessed 15 July 2024

National Survey on Drug Use and Health (NSDUH) (2021) Substance Abuse and Mental Health Services Administration, US Government. https://www.samhsa.gov/data/release/2021-national-survey-drug-use-and-healthnsduh-releases. Accessed 15 July 2024

National Survey on Drug Use and Health (NSDUH) (2022/2012) Substance Abuse and Mental Health

Services Administration, US Government. https://www.samhsa.gov/data/report/2022-nsduh-detailed-tables. Accessed 15 July 2024

National Survey on Drug Use and Health (NSDUH) (2023) Substance Abuse and Mental Health Services Administration, US Government. https://www.samhsa.gov/data/report/2023-nsduh-detailed-tables. Accessed 15 July 2024

Nilssen ES, Doan TP, Nigro MJ et al (2019) Neurons and networks in the entorhinal cortex: A reappraisal of the lateral and medial entorhinal subdivisions mediating parallel cortical pathways. Hippocampus 29:1238–1254. https://doi.org/10.1002/hipo.23145

Nixon RA (2013) The role of autophagy in neurodegenerative disease. Nat Med 19:983–997. https://doi.org/10.1038/nm.3232

Nixon K, Kim DH, Potts EN et al (2008) Distinct cell proliferation events during abstinence after alcohol dependence: Microglia proliferation precedes neurogenesis. Neurobiol Dis 31:218–229. https://doi.org/10.1016/j.nbd.2008.04.009

Noble W, Garwood C, Stephenson J et al (2009) Minocycline reduces the development of abnormal tau species in models of Alzheimer's disease. FASEB J 23:739–750. https://doi.org/10.1096/fj.08-113795

Noble W, Hanger DP, Miller CCJ, Lovestone S (2013) The importance of tau phosphorylation for neurodegenerative diseases. Front Neurol 4:83. https://doi.org/10.3389/fneur.2013.00083

Norden DM, Godbout JP (2013) Review: Microglia of the aged brain: primed to be activated and resistant to regulation. Neuropathol Appl Neurobiol 39:19–34. https://doi.org/10.1111/j.1365-2990.2012.01306.x

Novikova G, Kapoor M, Tcw J et al (2019) Integration of Alzheimer's disease genetics and myeloid cell genomics identifies novel causal variants, regulatory elements, genes and pathways. bioRxiv. https://doi.org/10.1101/694281

Ohara S, Rannap M, Tsutsui K-I et al (2023) Hippocampal-medial entorhinal circuit is differently organized along the dorsoventral axis in rodents. Cell Rep 42:112001. https://doi.org/10.1016/j.celrep.2023.112001

Olsen RW, Liang J (2017) Role of GABAA receptors in alcohol use disorders suggested by chronic intermittent ethanol (CIE) rodent model. Mol Brain 10:45. https://doi.org/10.1186/s13041-017-0325-8

Olsson B, Lautner R, Andreasson U et al (2016) CSF and blood biomarkers for the diagnosis of Alzheimer's disease: a systematic review and meta-analysis. Lancet Neurol 15:673–684. https://doi.org/10.1016/s1474-4422(16)00070-3

Orihuela R, McPherson CA, Harry GJ (2016) Microglial M1/M2 polarization and metabolic states. Br J Pharmacol 173:649–665. https://doi.org/10.1111/bph.13139

Ota M, Sato N, Nakata Y et al (2011) Relationship between apathy and diffusion tensor imaging metrics of the brain in Alzheimer's disease: Relationship between apathy and DTI in AD. Int J Geriatr Psychiatry 27:722–726. https://doi.org/10.1002/gps.2779

Oudman E, Oey MJ, Batjes D et al (2022) Wernicke-Korsakoff syndrome diagnostics and rehabilitation in the post-acute phase. Addict Neurosci 4:100043. https://doi.org/10.1016/j.addicn.2022.100043

Palop JJ, Chin J, Roberson ED et al (2007) Aberrant excitatory neuronal activity and compensatory remodeling of inhibitory hippocampal circuits in mouse models of Alzheimer's disease. Neuron 55:697–711. https://doi.org/10.1016/j.neuron.2007.07.025

Pasantes-Morales H, Tuz K (2006) Volume changes in neurons: hyperexcitability and neuronal death. Contrib Nephrol 152:221–240. https://doi.org/10.1159/000096326

Pascale A, Stephenson M, Barr P et al (2022) Exploring the relationships between adolescent alcohol misuse and later life health outcomes. Alcohol Clin Exp Res 46:1753–1765. https://doi.org/10.1111/acer.14917

Pascual M, Calvo-Rodriguez M, Núñez L et al (2021) Toll-like receptors in neuroinflammation, neurodegeneration, and alcohol-induced brain damage. IUBMB Life 73:900–915

Pati D, Marcinkiewcz CA, DiBerto JF et al (2020) Chronic intermittent ethanol exposure dysregulates a GABAergic microcircuit in the bed nucleus of the stria terminalis. Neuropharmacology 168:107759. https://doi.org/10.1016/j.neuropharm.2019.107759

Pati D, Downs AM, McElligott ZA, Kash TL (2022) Chronic ethanol exposure modulates periaqueductal gray to extended amygdala dopamine circuit. J Neurosci 43:709–721. https://doi.org/10.1523/jneurosci.1219-22.2022

Pedditzi E, Peters R, Beckett N (2016) The risk of overweight/obesity in mid-life and late life for the development of dementia: a systematic review and meta-analysis of longitudinal studies. Age Ageing 45:14–21. https://doi.org/10.1093/ageing/afv151

Pelicao R, Santos MC, Freitas-Lima LC et al (2016) URB597 inhibits oxidative stress induced by alcohol binging in the prefrontal cortex of adolescent rats. Neurosci Lett 624:17–22. https://doi.org/10.1016/j.neulet.2016.04.068

Peric A, Annaert W (2015) Early etiology of Alzheimer's disease: tipping the balance toward autophagy or endosomal dysfunction? Acta Neuropathol 129:363–381. https://doi.org/10.1007/s00401-014-1379-7

Peters R, Poulter R, Warner J et al (2008) Smoking, dementia and cognitive decline in the elderly, a systematic review. BMC Geriatr 8:36. https://doi.org/10.1186/1471-2318-8-36

Petrache AL, Rajulawalla A, Shi A et al (2019) Aberrant excitatory–inhibitory synaptic mechanisms in entorhinal cortex microcircuits during the pathogenesis of Alzheimer's disease. Cereb Cortex 29:bhz016. https://doi.org/10.1093/cercor/bhz016

Phillips RD, Bellis MDD, Brumback T et al (2021) Volumetric trajectories of hippocampal subfields and amygdala nuclei influenced by adolescent alcohol use and lifetime trauma. Transl Psychiatry 11:154. https://doi.org/10.1038/s41398-021-01275-0

Pinti M, Appay V, Campisi J et al (2016) Aging of the immune system: Focus on inflammation and vaccination. Eur J Immunol 46:2286–2301. https://doi.org/10.1002/eji.201546178

Polidori MC, Nelles G (2014) Antioxidant clinical trials in mild cognitive impairment and Alzheimer's disease–challenges and perspectives. Curr Pharm Des 20:3083–3092. https://doi.org/10.2174/13816128113196660706

Pooler AM, Phillips EC, Lau DH et al (2013) Physiological release of endogenous tau is stimulated by neuronal activity. EMBO Rep 14:389–394. https://doi.org/10.1038/embor.2013.15

Popova D, Sun J, Chow H, Hart RP (2024) A critical review of ethanol effects on neuronal firing: A metabolic perspective. Alcohol Clin Exp Res 48:450–458. https://doi.org/10.1111/acer.15266

Provenzano F, Pérez MJ, Deleidi M (2021) Redefining microglial identity in health and disease at single-cell resolution. Trends Mol Med 27:47–59. https://doi.org/10.1016/j.molmed.2020.09.001

Qin L, Crews FT (2012a) NADPH oxidase and reactive oxygen species contribute to alcohol-induced microglial activation and neurodegeneration. J Neuroinflammation 9:5. https://doi.org/10.1186/1742-2094-9-5

Qin L, Crews FT (2012b) Chronic ethanol increases systemic TLR3 agonist-induced neuroinflammation and neurodegeneration. J Neuroinflammation 9:130. https://doi.org/10.1186/1742-2094-9-130

Qin L, Zou J, Barnett A et al (2021) TRAIL mediates neuronal death in AUD: a link between neuroinflammation and neurodegeneration. Int J Mol Sci 22:2547. https://doi.org/10.3390/ijms22052547

Qin W, Li F, Jia L et al (2022) Phosphorylated tau 181 serum levels predict Alzheimer's disease in the preclinical stage. Front Aging Neurosci 14:900773. https://doi.org/10.3389/fnagi.2022.900773

Radulescu CI, Doostdar N, Zabouri N et al (2023) Age-related dysregulation of homeostatic control in neuronal microcircuits. Nat Neurosci 26:2158–2170. https://doi.org/10.1038/s41593-023-01451-z

Rajan KB, Weuve J, Barnes LL et al (2021) Population estimate of people with clinical Alzheimer's disease and mild cognitive impairment in the United States (2020–2060). Alzheimers Dement 17:1966–1975. https://doi.org/10.1002/alz.12362

Rapoport SI (2003) Coupled reductions in brain oxidative phosphorylation and synaptic function can be quantified and staged in the course of Alzheimer disease. Neurotox Res 5:385–397. https://doi.org/10.1007/bf03033167

Reddy SK, Husain K, Schlorff E et al (1999) Dose response of ethanol ingestion on antioxidant defense system in rat brain subcellular fractions. Neurotoxicology 20:977–988

Rehm J, Hasan OSM, Black SE et al (2019) Alcohol use and dementia: a systematic scoping review. Alzheimers Res Ther 11:1. https://doi.org/10.1186/s13195-018-0453-0

Rivers-Auty J, Mather AE, Peters R et al (2020) Anti-inflammatories in Alzheimer's disease—potential therapy or spurious correlate? Brain Commun 2:fcaa109. https://doi.org/10.1093/braincomms/fcaa109

Roberson ED, Halabisky B, Yoo JW et al (2011) Amyloid-β/Fyn-induced synaptic, network, and cognitive impairments depend on tau levels in multiple mouse models of Alzheimer's disease. J Neurosci 31:700–711. https://doi.org/10.1523/jneurosci.4152-10.2011

Robert A, Schöll M, Vogels T (2021) Tau seeding mouse models with patient brain-derived aggregates. Int J Mol Sci 22:6132. https://doi.org/10.3390/ijms22116132

Rorbach-Dolata A, Piwowar A (2019) Neurometabolic evidence supporting the hypothesis of increased incidence of type 3 diabetes mellitus in the 21st century. Biomed Res Int 2019:1435276. https://doi.org/10.1155/2019/1435276

Rossano SM, Johnson AS, Smith A et al (2024) Microglia measured by TSPO PET are associated with Alzheimer's disease pathology and mediate key steps in a disease progression model. Alzheimers Dement 20:2397–2407. https://doi.org/10.1002/alz.13699

Ruan Z, Pathak D, Kalavai SV et al (2020) Alzheimer's disease brain-derived extracellular vesicles spread tau pathology in interneurons. Brain 144:awaa376. https://doi.org/10.1093/brain/awaa376

Sabia S, Fayosse A, Dumurgier J et al (2018) Alcohol consumption and risk of dementia: 23 year follow-up of Whitehall II cohort study. BMJ 362:k2927. https://doi.org/10.1136/bmj.k2927

Salling MC, Skelly MJ, Avegno E et al (2018) Alcohol consumption during adolescence in a mouse model of binge drinking alters the intrinsic excitability and function of the prefrontal cortex through a reduction in the hyperpolarization-activated cation current. J Neurosci 38:6207–6222. https://doi.org/10.1523/jneurosci.0550-18.2018

SAMHSA Center for Behavioral Health Statistics and Quality, National Survey on Drug Use and Health (2022) Table 5.9A—Alcohol use disorder in past year: among people aged 12 or older; by age group and demographic characteristics, numbers in thousands, 2021 and 2022

Sanna PP, Cabrelle C, Kawamura T et al (2023) A history of repeated alcohol intoxication promotes cognitive impairment and gene expression signatures of disease progression in the 3xTg mouse model of Alzheimer's disease. eneuro. https://doi.org/10.1523/eneuro.0456-22.2023

Sarnowski C, Ghanbari M, Bis JC et al (2022) Meta-analysis of genome-wide association studies identifies ancestry-specific associations underlying circulating total tau levels. Commun Biol 5:336. https://doi.org/10.1038/s42003-022-03287-y

Scaduto P, Lauterborn JC, Cox CD et al (2023) Functional excitatory to inhibitory synaptic imbalance and loss of cognitive performance in people with Alzheimer's disease neuropathologic change. Acta Neuropathol 145:303–324. https://doi.org/10.1007/s00401-022-02526-0

Scarmeas N, Honig LS, Choi H et al (2009) Seizures in Alzheimer disease: who, when, and how common? Arch Neurol (Chicago) 66:992–997. https://doi.org/10.1001/archneurol.2009.130

Scharfman HE (2012) Alzheimers disease and epilepsy: insight from animal models. Future Neurol 7:177–192. https://doi.org/10.2217/fnl.12.8

Scheltens P, Strooper BD, Kivipelto M et al (2021) Alzheimer's disease. Lancet 397:1577–1590. https://doi.org/10.1016/s0140-6736(20)32205-4

Schwarzinger M, Pollock BG, Hasan OSM et al (2018) Contribution of alcohol use disorders to the burden of dementia in France 2008-13: a nationwide retrospective cohort study. Lancet Public Health 3:e124–e132. https://doi.org/10.1016/s2468-2667(18)30022-7

Seabrook TJ, Jiang L, Maier M, Lemere CA (2006) Minocycline affects microglia activation, Aβ deposition, and behavior in APP-tg mice. Glia 53:776–782. https://doi.org/10.1002/glia.20338

Sears SM, Hewett SJ (2021) Influence of glutamate and GABA transport on brain excitatory/inhibitory balance. Exp Biol Med 246:1069–1083. https://doi.org/10.1177/1535370221989263

Shafighi K, Villeneuve S, Neto PR et al (2023) Social isolation is linked to classical risk factors of Alzheimer's disease-related dementias. PLoS One 18:e0280471. https://doi.org/10.1371/journal.pone.0280471

Shimojo M, Takuwa H, Takado Y et al (2020) Selective disruption of inhibitory synapses leading to neuronal hyperexcitability at an early stage of tau pathogenesis in a mouse model. J Neurosci 40:3491–3501. https://doi.org/10.1523/jneurosci.2880-19.2020

Sienski G et al (2021) APOE4 disrupts intracellular lipid homeostasis in human iPSC-derived glia. Sci Transl Med 13(583):eaaz4564. https://doi.org/10.1126/scitranslmed.aaz4564

Šimić G, Leko MB, Wray S et al (2016) Tau protein hyperphosphorylation and aggregation in Alzheimer's disease and other tauopathies, and possible neuroprotective strategies. Biomolecules 6:6. https://doi.org/10.3390/biom6010006

Smailagic N, Lafortune L, Kelly S et al (2018) 18F-FDG PET for prediction of conversion to Alzheimer's disease dementia in people with mild cognitive impairment: an updated systematic review of test accuracy. J Alzheimers Dis 64(4):1175–1194. https://doi.org/10.3233/jad-171125

Solito E, Sastre M (2012) Microglia function in Alzheimer's disease. Front Pharmacol 3:14. https://doi.org/10.3389/fphar.2012.00014

Song M, Jin J, Lim JE et al (2011) TLR4 mutation reduces microglial activation, increases Abeta deposits and exacerbates cognitive deficits in a mouse model of Alzheimer's disease. J Neuroinflammation 8:92. https://doi.org/10.1186/1742-2094-8-92

Soula M, Maslarova A, Harvey RE et al (2023) Interictal epileptiform discharges affect memory in an Alzheimer's disease mouse model. bioRxiv. https://doi.org/10.1101/2023.02.15.528683

Spear LP (2000) The adolescent brain and age-related behavioral manifestations. Neurosci Biobehav Rev 24:417–463. https://doi.org/10.1016/s0149-7634(00)00014-2

Spear L (2013) The teenage brain. Curr Dir Psychol Sci 22:152–157. https://doi.org/10.1177/0963721412472192

Spear LP (2018) Effects of adolescent alcohol consumption on the brain and behaviour. Nat Rev Neurosci 19:197–214. https://doi.org/10.1038/nrn.2018.10

Stamer K, Vogel R, Thies E et al (2002) Tau blocks traffic of organelles, neurofilaments, and APP vesicles in neurons and enhances oxidative stress. J Cell Biol 156:1051–1063. https://doi.org/10.1083/jcb.200108057

Stirling DP, Koochesfahani KM, Steeves JD, Tetzlaff W (2005) Minocycline as a neuroprotective agent. Neuroscientist 11:308–322. https://doi.org/10.1177/1073858405275175

Strang KH, Golde TE, Giasson BI (2019) MAPT mutations, tauopathy, and mechanisms of neurodegeneration. Lab Investig 99:912–928. https://doi.org/10.1038/s41374-019-0197-x

Suh SW, Aoyama K, Chen Y et al (2003) Hypoglycemic neuronal death and cognitive impairment are prevented by poly(ADP-ribose) polymerase inhibitors administered after hypoglycemia. J Neurosci 23:10681–10690. https://doi.org/10.1523/jneurosci.23-33-10681.2003

Szekely CA, Green RC, Breitner JCS et al (2008) No advantage of Ab42-lowering NSAIDs for prevention of Alzheimer dementia in six pooled cohort studies. Neurology 70:2291–2298. https://doi.org/10.1212/01.wnl.0000313933.17796.f6

Tabakoff B, Hoffman PL (1996) Alcohol addiction: an enigma among us. Neuron 16:909–912. https://doi.org/10.1016/s0896-6273(00)80113-0

Tábuas-Pereira M, Durães J, Lopes J et al (2019) Increased CSF tau is associated with a higher risk of seizures in patients with Alzheimer's disease. Epilepsy Behav 98:207–209. https://doi.org/10.1016/j.yebeh.2019.06.033

Tai LM, Weng JM, LaDu MJ, Brady ST (2020) Relevance of transgenic mouse models for Alzheimer's disease. Prog Mol Biol Transl Sci 177:1–48. https://doi.org/10.1016/bs.pmbts.2020.07.007

Tamagno E, Guglielmotto M, Vasciaveo V, Tabaton M (2021) Oxidative stress and beta amyloid in Alzheimer's disease. Which comes first: the chicken or the egg? Antioxidants 10:1479. https://doi.org/10.3390/antiox10091479

Tang C, Ma Y, Lei X et al (2023) Hypertension linked to Alzheimer's disease via stroke: Mendelian randomization. Sci Rep 13:21606. https://doi.org/10.1038/s41598-023-49087-0

Tapia-Rojas C, Carvajal FJ, Mira RG et al (2018) Adolescent binge alcohol exposure affects the brain function through mitochondrial impairment. Mol Neurobiol 55:4473–4491. https://doi.org/10.1007/s12035-017-0613-4

Teunissen CE, Verberk IMW, Thijssen EH et al (2021) Blood-based biomarkers for Alzheimer's disease: towards clinical implementation. Lancet Neurol 21:66–77. https://doi.org/10.1016/s1474-4422(21)00361-6

Thanos PK, Wang G-J, Volkow ND (2008) Positron emission tomography as a tool for studying alcohol abuse. Alcohol Res Health 31:233–237

The Alzheimer's Association (2024) 2024 Alzheimer's disease facts and figures. Alzheimers Dement 20(5):3708–3821. https://doi.org/10.1002/alz.13809

Thijssen EH, Verberk IMW, Stoops E et al (2020) Amyloid, pTau, NfL, and GFAP as biomarkers for Alzheimer's disease. Alzheimers Dement 16. https://doi.org/10.1002/alz.038179

Thijssen EH, Joie RL, Strom A et al (2021) Plasma phosphorylated tau 217 and phosphorylated tau 181 as biomarkers in Alzheimer's disease and frontotemporal lobar degeneration: a retrospective diagnostic performance study. Lancet Neurol 20:739–752. https://doi.org/10.1016/s1474-4422(21)00214-3

Thomas AJ, Hamilton CA, Heslegrave A et al (2022) A longitudinal study of plasma pTau181 in mild cognitive impairment with Lewy bodies and Alzheimer's disease. Mov Disord 37:1495–1504. https://doi.org/10.1002/mds.28994

Thorlakur J, Hreinn S, Stacy S et al (2013) Variant of TREM2 associated with the risk of Alzheimer's disease. N Engl J Med 368:107–116. https://doi.org/10.1056/nejmoa1211103

Tomasi DG, Wiers CE, Shokri-Kojori E et al (2019) Association between reduced brain glucose metabolism and cortical thickness in alcoholics: evidence of neurotoxicity. Int J Neuropsychopharmacol 22:548–559. https://doi.org/10.1093/ijnp/pyz036

Topiwala A, Ebmeier KP, Maullin-Sapey T, Nichols TE (2022) Alcohol consumption and MRI markers of brain structure and function: Cohort study of 25,378 UK Biobank participants. NeuroImage: Clin 35: 103066. https://doi.org/10.1016/j.nicl.2022.103066

Tucker AE, Pauneto CDMA, Barnett AM, Coleman LG (2022) Chronic ethanol causes persistent increases in Alzheimer's tau pathology in female 3xTg-AD mice: a potential role for lysosomal impairment. Front Behav Neurosci 16:886634. https://doi.org/10.3389/fnbeh.2022.886634

Tyas SL (2001) Alcohol use and the risk of developing Alzheimer's disease. Alcohol Res Health 25:299

Tzartos JS, Boufidou F, Stergiou C et al (2022) Plasma P-Tau181 for the discrimination of Alzheimer's disease from other primary dementing and/or movement disorders. Biomolecules 12:1099. https://doi.org/10.3390/biom12081099

Valenzuela CF (1997) Alcohol and neurotransmitter interactions. Alcohol Health Res World 21:144–148

Valenzuela CF, Harris RA (1997) Alcohol: neurobiology. In: Lowinson JH, Ruiz P, Millman RB, Langrod JG (eds) Substance abuse: A comprehensive textbook. Williams & Wilkins, Baltimore, pp 119–142

Valenzuela CF, Jotty K (2015) Mini-review: effects of ethanol on gabaa receptor-mediated neurotransmission in the cerebellar cortex—recent advances. Cerebellum 14:438–446. https://doi.org/10.1007/s12311-014-0639-3

van der Flier WM, Scheltens P (2022) The ATN framework—moving preclinical Alzheimer disease to clinical relevance. JAMA Neurol 79:968–970. https://doi.org/10.1001/jamaneurol.2022.2967

van Groen T, Miettinen P, Kadish I (2003) The entorhinal cortex of the mouse: Organization of the projection to the hippocampal formation. Hippocampus 13:133–149. https://doi.org/10.1002/hipo.10037

Vermunt L, Sikkes SAM, van den Hout A et al (2019) Duration of preclinical, prodromal, and dementia stages of Alzheimer's disease in relation to age, sex, and APOE genotype. Alzheimers Dement 15:888–898. https://doi.org/10.1016/j.jalz.2019.04.001

Verret L, Mann EO, Hang GB et al (2012) Inhibitory interneuron deficit links altered network activity and cognitive dysfunction in Alzheimer model. Cell 149:708–721. https://doi.org/10.1016/j.cell.2012.02.046

Vetreno RP, Crews FT (2012) Adolescent binge drinking increases expression of the danger signal receptor agonist HMGB1 and Toll-like receptors in the adult prefrontal cortex. Neuroscience 226:475–488. https://doi.org/10.1016/j.neuroscience.2012.08.046

Vetreno RP, Broadwater M, Liu W et al (2014) Adolescent, but not adult, binge ethanol exposure leads to persistent global reductions of choline acetyltransferase expressing neurons in brain. PLoS One 9:e113421. https://doi.org/10.1371/journal.pone.0113421

Vlad SC, Miller DR, Kowall NW, Felson DT (2008) Protective effects of NSAIDs on the development of Alzheimer disease. Neurology 70:1672–1677. https://doi.org/10.1212/01.wnl.0000311269.57716.63

Vogels T, Leuzy A, Cicognola C et al (2019) Propagation of tau pathology: integrating insights from post mortem and in vivo studies. Biol Psychiatry. https://doi.org/10.1016/j.biopsych.2019.09.019

Vogels T, Vargová G, Brezováková V et al (2020) Viral delivery of non-mutated human truncated tau to neurons recapitulates key features of human tauopathy in wild-type mice. J Alzheimers Dis 77:551–568. https://doi.org/10.3233/jad-200047

Volkow ND, Wang G-J, Kojori ES et al (2015) Alcohol decreases baseline brain glucose metabolism more in heavy drinkers than controls but has no effect on stimulation-induced metabolic increases. J Neurosci 35:3248–3255. https://doi.org/10.1523/jneurosci.4877-14.2015

Vossel KA, Beagle AJ, Rabinovici GD et al (2013) Seizures and epileptiform activity in the early stages of Alzheimer disease. JAMA Neurol 70:1158–1166. https://doi.org/10.1001/jamaneurol.2013.136

Vossel KA, Ranasinghe KG, Beagle AJ et al (2016) Incidence and impact of subclinical epileptiform activity in Alzheimer's disease. Ann Neurol 80:858–870. https://doi.org/10.1002/ana.24794

Vossel KA, Tartaglia MC, Nygaard HB et al (2017) Epileptic activity in Alzheimer's disease: causes and clinical relevance. Lancet Neurol 16:311–322. https://doi.org/10.1016/s1474-4422(17)30044-3

Walsh DM, Selkoe DJ (2016) A critical appraisal of the pathogenic protein spread hypothesis of neurodegeneration. Nat Rev Neurosci 17:251–260. https://doi.org/10.1038/nrn.2016.13

Walter S, Letiembre M, Liu Y et al (2007) Role of the toll-like receptor 4 in neuroinflammation in Alzheimer's disease. Cell Physiol Biochem 20:947–956. https://doi.org/10.1159/000110455

Walter TJ, Vetreno RP, Crews FT (2017) Alcohol and stress activation of microglia and neurons: brain regional effects. Alcohol Clin Exp Res 41:2066–2081. https://doi.org/10.1111/acer.13511

Walter KR, Ricketts DK, Presswood BH et al (2022) Prenatal alcohol exposure causes persistent microglial activation and age- and sex-specific effects on cognition and metabolic outcomes in an Alzheimer's disease mouse model. Am J Drug Alcohol Abus 1–19. https://doi.org/10.1080/00952990.2022.2119571

Wang Y, Mandelkow E (2016) Tau in physiology and pathology. Nat Rev Neurosci 17:22–35. https://doi.org/10.1038/nrn.2015.1

Wang J-Z, Xia Y-Y, Grundke-Iqbal I, Iqbal K (2013) Abnormal hyperphosphorylation of tau: sites, regulation, and molecular mechanism of neurofibrillary degeneration. J Alzheimers Dis 33:S123–S139. https://doi.org/10.3233/jad-2012-129031

Wang Y, Ulland TK, Ulrich JD et al (2016) TREM2-mediated early microglial response limits diffusion and toxicity of amyloid plaques. J Exp Med 213:667–675. https://doi.org/10.1084/jem.20151948

Wang Z-T, Li K-Y, Tan C-C et al (2021) Associations of alcohol consumption with cerebrospinal fluid biomarkers of Alzheimer's disease pathology in cognitively intact older adults: the CABLE Study. J Alzheimers Dis 82:1045–1054. https://doi.org/10.3233/jad-210140

Wegmann S, Bennett RE, Delorme L et al (2019) Experimental evidence for the age dependence of tau protein spread in the brain. Sci Adv 5:eaaw6404. https://doi.org/10.1126/sciadv.aaw6404

Wesseling H, Mair W, Kumar M et al (2020) Tau PTM profiles identify patient heterogeneity and stages of Alzheimer's disease. Cell. https://doi.org/10.1016/j.cell.2020.10.029

Weston LL, Jiang S, Chisholm D et al (2021) Interleukin-10 deficiency exacerbates inflammation-induced tau pathology. J Neuroinflammation 18:161. https://doi.org/10.1186/s12974-021-02211-1

Wieckowska-Gacek A, Mietelska-Porowska A, Wydrych M, Wojda U (2021) Western diet as a trigger of Alzheimer's disease: From metabolic syndrome and systemic inflammation to neuroinflammation and neurodegeneration. Ageing Res Rev 70:101397. https://doi.org/10.1016/j.arr.2021.101397

Wilcock DM, Lamb BT (2024) The importance of continuing development of novel animal models of Alzheimer's disease and Alzheimer's disease and related dementias. Alzheimers Dement 20(7):5078–5079. PMID: 38923143. PMCID: PMC11247659. https://doi.org/10.1002/alz.14105. Epub 2024 Jun 26

Wilson DF, Matschinsky FM (2020) Ethanol metabolism: The good, the bad, and the ugly. Méd Hypotheses 140:109638. https://doi.org/10.1016/j.mehy.2020.109638

Wisch JK, Gordon BA, Barthélemy NR et al (2024) Predicting continuous amyloid PET values with CSF tau phosphorylation occupancies. Alzheimers Dement. https://doi.org/10.1002/alz.14132

Wolf SA, Boddeke HWGM, Kettenmann H (2016) Microglia in physiology and disease. Annu Rev Physiol 79:619–643. https://doi.org/10.1146/annurev-physiol-022516-034406

Wolfe DM, Lee JH, Kumar A et al (2013) Autophagy failure in Alzheimer's disease and the role of defective lysosomal acidification. Eur J Neurosci 37:1949–1961. https://doi.org/10.1111/ejn.12169

Wu JW, Hussaini SA, Bastille IM et al (2016) Neuronal activity enhances tau propagation and tau pathology in vivo. Nat Neurosci. https://doi.org/10.1038/nn.4328

Xiang X, Wind K, Wiedemann T et al (2021) Microglial activation states drive glucose uptake and FDG-PET alterations in neurodegenerative diseases. Sci Transl Med 13:eabe5640. https://doi.org/10.1126/scitranslmed.abe5640

Xu Y, Zhao M, Han Y, Zhang H (2020) GABAergic inhibitory interneuron deficits in Alzheimer's disease: implications for treatment. Front Neurosci 14:660. https://doi.org/10.3389/fnins.2020.00660

Yamada K, Holth JK, Liao F et al (2014) Neuronal activity regulates extracellular tau in vivo. J Exp Med 211:387–393. https://doi.org/10.1084/jem.20131685

Yan R, Wang W, Yang W et al (2024) Mitochondria-related candidate genes and diagnostic model to predict late-onset Alzheimer's disease and mild cognitive impairment. J Alzheimers Dis 99:S299–S315. https://doi.org/10.3233/jad-230314

Yang J-Y, Xue X, Tian H et al (2014) Role of microglia in ethanol-induced neurodegenerative disease: Pathological and behavioral dysfunction at different developmental stages. Pharmacol Ther 144:321–337. https://doi.org/10.1016/j.mcn.2019.103409

Yang T, Li S, Xu H et al (2017) Large soluble oligomers of amyloid beta-protein from Alzheimer brain are far less neuroactive than the smaller oligomers to which they dissociate. J Neurosci 37:152–163. https://doi.org/10.1523/jneurosci.1698-16.2016

Yeh W-C, Hsu C-Y, Li K-Y et al (2022) Association between subclinical epileptiform discharge and the severity of cognitive decline in Alzheimer's disease: a longitudinal cohort study. J Alzheimers Dis 90:305–312. https://doi.org/10.3233/jad-220567

Young ML, Franklin JL (2019) The mitochondria-targeted antioxidant MitoQ inhibits memory loss, neuropathol-

ogy, and extends lifespan in aged 3xTg-AD mice. Mol Cell Neurosci 101:103409. https://doi.org/10.1016/j.pharmthera.2014.07.002
Yu Y, Run X, Liang Z et al (2009) Developmental regulation of tau phosphorylation, tau kinases, and tau phosphatases. J Neurochem 108:1480–1494. https://doi.org/10.1111/j.1471-4159.2009.05882.x
Yu JT, Xu W, Tan CC et al (2020) Evidence-based prevention of Alzheimer's disease: systematic review and meta-analysis of 243 observational prospective studies and 153 randomised controlled trials. J Neurol Neurosurg Psychiatry 91:1201–1209. https://doi.org/10.1136/jnnp-2019-321913
Zandi PP, Anthony JC, Hayden KM et al (2002) Reduced incidence of AD with NSAID but not H2 receptor antagonists: the Cache County Study. Neurology 59:880–886. https://doi.org/10.1212/wnl.59.6.880
Zhang L, Gai Y, Liu Y et al (2024) Tau induces inflammasome activation and microgliosis through acetylating NLRP3. Clin Transl Med 14:e1623. https://doi.org/10.1002/ctm2.1623
Zhao S, Widman L, Hagström H, Shang Y (2024) Disentangling the contributions of alcohol use disorder and alcohol-related liver disease towards dementia: A population-based cohort study. Addiction 119:706–716. https://doi.org/10.1111/add.16395
Zheng L, Roberg K, Jerhammar F et al (2006) Oxidative stress induces intralysosomal accumulation of alzheimer amyloid β-protein in cultured neuroblastoma cells. Ann N Y Acad Sci 1067:248–251. https://doi.org/10.1196/annals.1354.032
Zimatkin SM, Deitrich RA (1997) Ethanol metabolism in the brain. Addict Biol 2:387–400. https://doi.org/10.1080/13556219772444

11 Recent Investigations Designed to Unravel the Interaction of Age and Alcohol on Behavior and Cognition: Potential Neurobiological Mechanisms

Douglas B. Matthews, Emily Kerr, Aeda Zank, Jadyn Hartwig, Averi Garscia, Samuel Stumo, and Pravesh Sharma

Abstract

Understanding factors that alter the effect of alcohol in biological systems has been an area of active investigation for several decades. Recently, it has become clear that age is one of the most salient factors influencing how both acute and chronic alcohol exposure alters behavioral function. The following book chapter discusses how alcohol produces differential effects in adolescent animals in comparison to adult and aged (i.e., older) animals. Furthermore, where possible, relevant research identifying possible brain mechanisms mediating the differential effects of alcohol will be discussed. Finally, we highlight a small number of studies where sex and age of the subject interact to modify cognitive impairments produced by alcohol. We conclude that much work still needs to be done to fully understand how age, sex, and alcohol interact to produce the wide range of effects caused by consumption of the drug.

D. B. Matthews (✉) · E. Kerr · A. Zank · J. Hartwig
A. Garscia · S. Stumo
Department of Psychology, University of Wisconsin – Eau Claire, Eau Claire, WI, USA
e-mail: matthedb@uwec.edu

P. Sharma
Mayo Clinic Health System, Eau Claire, WI, USA

Keywords

Cognition · Alzheimer's disease · Ataxia · Alcohol · Aging · Adolescent

11.1 Introduction

Understanding the interaction of factors that led to excessive alcohol use and misuse is a critical public health concern due to the dramatic and often negative effects that alcohol has on both people and society. For example, in the United States, over 174 million individuals used alcohol in the previous year, and 61 million people report binge drinking (5+ drinks in males, 4+ drinks in females over 2 h that produce a blood alcohol level greater than 0.08 mg/dl) in the previous month while greater than 16 million people report heavy alcohol drinking in the past month (NSDUH survey 2022). In addition, the onset and continued impact of the COVID-19 pandemic has only exacerbated the increase in alcohol consumption (Castaldelli-Maia et al. 2021) magnifying unhealthy life states including increases in death rates due to alcohol misuse (White et al. 2022). The impact of high levels of alcohol use and misuse are staggering in that alcohol misuse results in 178,000 deaths each year reducing the average lifespan of those who die by greater than

C. F. Valenzuela et al. (eds.), *Effects of Alcohol on the Brain across the Lifespan*, Advances in Experimental Medicine and Biology 1473, https://doi.org/10.1007/978-3-031-81908-7_11

24 years in the United States (CDC fact sheet 2021). Unfortunately, the problems produced by alcohol are not limited only to the United States as over 2.6 million deaths in 2019 occurred worldwide due to alcohol misuse. Finally, alcohol misuse results in a loss of over $249 billion in the United States (Sacks et al. 2015).

Most people initiate alcohol use and misuse during adolescence, a life stage characterized by exploration and risk-taking behavior (Witt 2010; Patrick and Terry-McElrath 2019; Patton et al. 2018; US Department of Health and Human Services 2019). A large body of research has investigated factors that increase the likelihood of alcohol use and misuse in younger, adolescent populations including alcohol use and misuse in different racial and ethnic groups (Tam et al. 2024; Chung et al. 2018; Tapert and Eberson-Shumate 2022). Recently, it has been reported that overall alcohol use rates in adolescents are declining, (NSDUH data) even though use rates in the female population are increasing (White 2020). Furthermore, the overall decrease in alcohol consumption in adolescence has not been universally reported in all age demographics. Somewhat surprisingly, a consistent body of literature has revealed that in the older, aged demographic, alcohol use and misuse is increasing (Breslow et al. 2017; Calvo et al. 2020; Han et al. 2017, 2019; Keyes 2023; White et al. 2023). Furthermore, the increase in alcohol consumption in the older population is sex-dependent with older females without children identified as a possible at-risk group of alcohol users (McKetta and Keyes 2019, for review see White 2020).

The increased alcohol use and misuse in the older population is particularly concerning. Due to various medical, health, and nutritional factors, the average age of the population in most countries is increasing. It is predicted that by 2050 the number of people over the age of 60 will increase from 900 million to 2.1 billion people (United Nations 2022). In addition, the percentage of the global population older than 65 years will likely increase from 10% in 2022 to 16% by 2050 (United Nations 2022). Furthermore, older adults use healthcare systems significantly more than younger adults as represented by dollars spent on healthcare (Lassman et al. 2014). This increase in the percentage of older adults threatens to produce a "silver tsunami" that may overwhelm healthcare systems (Tampi et al. 2015). In addition, alcohol use and misuse has been linked to a variety of disease states that further necessitates investigating how the drug impacts the older population. Given these factors, it is critical to further explore how alcohol affects the aged population (Matthews and Koob 2023).

The focus of the current review is to detail the preclinical animal research that highlights how animals of different ages are uniquely impacted by alcohol exposure. Our focus will be on both alcohol exposure that occurs at different ages (e.g., aged vs. adult vs. adolescent exposure) as well as alcohol exposure that occurs early in life (during adolescence) and the longitudinal effects of this exposure that manifest across the lifespan. In doing so, we hope to highlight neurobiological changes that occur due to aging, particularly late-in-life changes, as a factor that is critical to understanding the effect of alcohol on behavior.

11.2 Effect of Acute Alcohol Exposure

Acute alcohol exposure produces a variety of behavioral effects that are age-dependent (see Table 11.1). For example, administration of a high dose of alcohol produces a loss of consciousness which is termed "sleep time" or loss of the righting reflex (Majchrowicz 1975; Matthews et al. 2008). Lower doses of alcohol can induce hypothermia (Abel and York 1979; Freund 1973; York 1982), produce ataxia (Bogo et al. 1981; Frye et al. 1981; Rustay et al. 2003) and an anxiolytic effect that can be modeled in rodents using a variety of techniques (Bertoglio and Carobrez 2002; Criswell et al. 1994). Finally, acute alcohol exposure can selectively impair hippocampal-dependent learning and memory while leaving hippocampal independent cognition relatively unaffected (Matthews et al. 1995; Hoffmann and Matthews 2001; for review see Van Skike et al. 2019). As expected, chronic alco-

Table 11.1 Summary of differential effects of acute alcohol on a variety of behaviors as a function of the subject's age. The down arrows indicate the magnitude of the effect while a flat arrow indicates no effect

Behavioral effect	Adolescent	Adult	Aged	Reference
Loss of righting reflex	↓	↓↓	↓↓↓	Novier et al. (2015); Gano et al. (2017); Perkins et al. (2018)
Hypothermia	↓	↓	↓↓	Watson et al. (2020)
Ataxia	↓	↓↓	↓↓↓	Van Skike et al. (2010); Matthews and Mittleman (2017)
Hippocampal-dependent memory	↓↓	↓↓	↓↓	Chin et al. (2011); Matthews et al. (2020)
Non-hippocampal-dependent memory	→	→	↓	Chin et al. (2011); Matthews et al. (2020)
Anxiolytic				
Female	→	↑	↑↑↑	Matthews et al. (2024)
Male	→	↑	→	
Motor simulating				
Female	↑↑	→	↓↓	Matthews et al. (2024)
Male	→	→	↑↑	

hol exposure impacts each of these behavioral endpoints, often by producing a tolerance state to an acute challenge as seen in the loss of righting reflex task (Swartzwelder et al. 1998; Matthews et al. 2008), hypothermia (Swartzwelder et al. 1998), ataxia (White et al. 2002), and hippocampal-dependent cognition (Silvers et al. 2006). Furthermore, termination of chronic alcohol exposure can produce a withdrawal state that causes an anxiogenic-like effect in rodents (Van Skike et al. 2015). Interestingly, age of the subject is a critical factor to understanding the magnitude of behavioral alterations produced in each of these endpoints, and, with many of the endpoints, central nervous system mechanisms have been proposed that mediate these alterations.

Research has focused on investigating the differential effects of acute alcohol exposure between adolescent and adult rodents, mostly rats, and identified biological and brain factors that may produce differential effects between the two ages (see Chin et al. 2010; Spear 2018). As previously stated, most individuals first begin to experiment with alcohol consumption during adolescence, a developmental time point that can lead to severe consequences for drug use (National Survey on Drug Use and Health, SAMHSA 2022; Green et al. 2024), and as such, it is important to understand the effects of alcohol in this age group (for review of this literature see Novier et al. 2015). After many years of investigation, we concluded that adolescent animals are *less* sensitive to the effects of acute alcohol compared to adult animals. For example, adolescent rats have significantly shorter loss of righting reflex due to a high-dose alcohol challenge compared to adult animals (Little et al. 1996; Matthews et al. 2008; Silveri and Spear 1998; Vetreno et al. 2023) and adolescent animals regain the righting reflex at significantly higher blood alcohol concentrations compared to adult animals (Silveri and Spear 1998) demonstrating the lack of sensitivity in adolescent rats to alcohol is likely due to central nervous system mechanism(s). However, adolescent rats can be significantly more sensitive to the hypothermic effects of alcohol compared to adults (Ristuccia et al. 2007; Watson et al. 2020) demonstrating the effect of alcohol in adolescence is task and/or behavior dependent.

Ataxia due to acute alcohol administration has been more extensively studied, and, once again results consistently demonstrate that adolescent rats are significantly less sensitive to the motor impairing effects of alcohol compared to adult animals. For example, adolescent rats are less impaired by acute alcohol administration on the tilting plane task (White et al. 2002; Vetreno et al. 2023) and accelerating rotarod compared to adult animals (Ornelas et al. 2015). In our laboratory we have frequently used the aerial righting reflex to investigate the ataxic effects produced by acute alcohol administration (e.g., see Van Skike et al. 2010). The aerial righting reflex is a valid and reliable behavioral task to investigate the effect of acute alcohol on general motor ability. In the task, an initial baseline score is determined by holding the subject in a supine position 5″ over a soft foam pad and released. If the subject successfully rights itself two out of three times the subject is given a score of 5″; however, if the subject does not successfully right itself, the height the animal is held over the foam pad is increased by 5″ and the animal is retested. Animals are never tested at a height greater than 20 in. Following determination of the baseline score, animals are injected with either saline or an alcohol dose and retested several times over a 60–90-min post-injection window to determine the effect of alcohol on ataxia. While quite simplistic, the aerial righting reflex task has several advantages. First, as a reflex it takes no training and therefore data can be quickly collected in a single day; second, age and sex do not produce baseline differences and hence any resultant differential effect in subjects' performance is most likely due to the alcohol challenge; and third, minimal experimental training of researchers is necessary for data collection.

Employing the aerial righting reflex, we have consistently found that adolescent animals are less sensitive to the motor impairing effects of acute alcohol administration (Van Skike et al. 2010; Ornelas et al. 2015; Matthews and Mittleman 2017). Initially we sought to determine if the effect was primarily driven by differential blood alcohol levels which would suggest the behavioral difference seen between adolescent and adult animals was not due to a central nervous system mechanism. Interestingly, we found that differential blood alcohol levels between adolescent and adult animals could not explain the differential aerial righting reflex data (Van Skike et al. 2010). We then sought to determine if we could identify a neuronal effect that mirrors the behavioral impairment produced by acute alcohol exposure. Given we were measuring a behavior that was dependent on the subjects' motor function and balance, we investigated the effect of an acute alcohol challenge on the spontaneous activity of individual Purkinje neurons from adolescent and adult rats while under urethane anesthetic. Interestingly, we found that acute alcohol, at the doses used in the aerial righting reflex task, inhibited the spontaneous activity of cerebellar Purkinje neurons significantly less in adolescent rats than it did in adult rats thereby mirroring the reduced behavioral effect found in adolescent rats compared to adult rats (Van Skike et al. 2010). The complementary behavioral and single unit, in vivo electrophysiological data supported further investigation into the cerebellum as the site of action impacting age-dependent effects of acute alcohol on ataxia. Consequently, we next investigated if differential expression of protein kinase C gamma (PKCγ) due to age existed in the cerebellum. We selected PKCγ to investigate due to data from targeted genetic research demonstrating that PKCγ knockout mice are significantly less sensitive to the ataxic effects of acute alcohol (Harris et al. 1995). We therefore predicted differential PKCγ expression in cerebellum due to biological aging where adolescent rats would have less PKCγ compared to adult rats. This is what was found via Western blot analysis. Specifically, adolescent rats were found to have modest, but significantly less, PKCγ in their cerebellum compared to adult animals (Van Skike et al. 2010). As such we proposed that the normal aging process produces changes in protein expression in the cerebellum that alters the electrophysiological effect of acute alcohol in the same brain region. The alteration in the electrophysiological effect of alcohol consequently altered the behavioral effect of the drug.

Acute alcohol administration also selectively impairs hippocampal-dependent memory as exhaustively reviewed by Van Skike et al. (2019). Briefly, when tested approximately 30 minutes following an acute alcohol challenge, adult rats have impaired spatial memory (Matthews et al. 1995, 1999; Hoffmann and Matthews 2001; García-Moreno and Cimadevilla 2012), contextual memory (Hefner and Holmes 2007; Melia et al. 1996; Weitemier and Ryabinin 2003) which may be sex- and age-dependent (Sircar 2019), trace conditioned memory in rats (Weitemier and Ryabinin 2003; Hunt et al. 2009) and in mice (Tipps et al. 2015), and novel object recognition in both rats and mice (Ryabinin et al. 2002; Swartzwelder et al. 2012). To identify potential electrophysiological correlates of the cognitive impairment, we investigated if similar doses of alcohol (1.5 g/kg alcohol) degrade the spatial specificity of hippocampal place cells. Place cells are individual hippocampal CA1 or CA3 cells that display higher firing rates when an animal is in a specific location as opposed to the activity the animal is engaging in the location (Best et al. 2001). We found that similar doses of alcohol that impair hippocampal-dependent cognitive performance also degrade the spatial specificity of hippocampal place cells recorded in awake freely behaving animals (Matthews et al. 1996) primarily by reducing the spontaneous activity of the neurons as the animal navigates an environment (White and Best 2000). Furthermore, recent investigations using tetrode arrays confirmed these early findings by demonstrating that the spatial representation of place cells is reduced by two components, a significant reduction in "in-field" place cell firing and a significant increase in place fields that disappear (i.e., a dramatic reduction in the location-specific firing of a particular place cell) while under the effects of alcohol (Miyake et al. 2020). This and related work allow for the leveraging of behavioral and electrophysiological studies to begin providing insights into how alcohol alters cognitive processing (see Lapish 2024).

Given the reduced sensitivity to the effects of alcohol in adolescents, like what we reviewed for ataxia, one might predict that hippocampal-dependent memory would be less degraded in adolescent animals compared to adult animals following an acute alcohol challenge. In fact, it was initially reported using the Morris water maze that adolescent rats were *more* sensitive to the cognitive impairing effects of acute alcohol administration compared to adult animals (Markwiese et al. 1998). However, follow-up research has shown that adolescent and adult rats have similar impaired spatial memory by acute alcohol (Chin et al. 2011) and additional research from several other laboratories suggests that the memory impairing effects of acute alcohol are not age-dependent but instead dependent on several experimental factors (see Novier et al. 2015 for an extensive discussion). While hippocampal place cell analysis has been applied to prenatal alcohol-exposed animals (Harvey et al. 2020), unfortunately, the effect of acute alcohol administration on hippocampal place cell spatial specificity in adolescent subjects has yet to be conducted. Consequently, it is still an open question as to whether hippocampal function is differentially affected by the age of subjects following an alcohol challenge.

Recently, our lab and others have been investigating the effect of acute alcohol exposure in aged animals and comparing the results to younger age groups to determine if changes during normal aging continue to alter the effect of the drug across the lifespan. As previously mentioned, the older demographic is continuing to consume alcohol, often in dangerous binge-like patterns (Blazer and Wu 2009; Breslow et al. 2017; Calvo et al. 2020; Han et al. 2019; Laberge et al. 2020; Keyes 2023) and this demographic uses healthcare significantly more than younger demographics (Tampi et al. 2015). Therefore, understanding behavioral health risks that impact the older demographic is important. However, little is known if the age-dependent effects of acute alcohol that have been reported between adolescent animals and adult animals extends to aged animals or if older animals respond similarly to alcohol as younger animals.

Aged rats are significantly more sensitive to the effects of acute alcohol administration compared to both adolescent and adult rats. For example, it has been demonstrated that aged rats, both male and female subjects, are significantly more sensitive to the hypnotic effect (i.e., the loss of righting reflex) produced by acute alcohol administration compared to adult rats (Ornelas et al. 2015; Gano et al. 2017; Perkins et al. 2018). In addition, aged rats regain the loss of righting reflex at significantly lower blood alcohol concentrations than do younger animals further supporting the conclusion of increased sensitivity (Perkins et al. 2018). In agreement with the loss of righting reflex studies are additional experiments that demonstrate that aged male rats are significantly more sensitive to the hypothermic effects of an acute alcohol challenge compared to younger animals (Watson et al. 2020). In addition, acute alcohol administration produces significantly more motor impairment in aged male rats compared to younger male rats as measured via the aerial righting reflex (Van Skike et al. 2010; Ornelas et al. 2015; Matthews and Mittleman 2017). Our laboratory is currently investigating potential molecular mechanisms that may mediate the increased ataxic effect of acute alcohol in aged subjects, once again focusing on differential expression of various PKC isoforms in the cerebellum (gamma, delta, and epsilon) (Matthews et al. 2023b).

Acute alcohol administration produces an anxiolytic-like response in most studies, a response that is often investigated in rats and mice using the elevated plus maze, where subjects administered low to moderate doses of alcohol enter the open arms of the maze significantly more than control animals (Criswell et al. 1994; LaBuda and Hale 2000; LaBuda and Fuchs 2001; Wilson et al. 2004). In addition, administration of low-dose alcohol can also facilitate stimulatory movement in rodent models (Rodd et al. 2004; Da Silva et al. 2005; Karlsson and Roman 2016). A recent study investigated the impact of subjects' age (adolescent, adult, or aged) and sex on the anxiolytic and motor stimulator effect of 1.0 g/kg alcohol in rats and found the effect of acute alcohol was both age- and sex-dependent. Specifically, aged female rats showed a significantly greater anxiolytic-like response to acute alcohol compared to aged male rats or younger animals while aged males had a significantly greater motor stimulating effect compared to aged females or younger animals (Matthews et al. 2024).

As previously discussed, acute alcohol selectively impairs hippocampal-dependent memory but does not impair hippocampal-independent memory such as procedural or cue-response memory (reviewed in Van Skike et al. 2019). We recently investigated the effect of an acute alcohol challenge on spatial and nonspatial memory via the Morris water maze in "cognitively spared" aged animals compared to younger animals. In this experiment, we defined cognitively spared aged subjects as subjects that could learn the nonspatial or spatial task to a level similar to young adult animals. Interestingly, we found that while acute alcohol administration did not produce a significant impairment in nonspatial cognitive performance, the alcohol challenge also did not interact with age to increase the spatial memory impairment in 18–24-month-old male rats. In addition, when we tested 29–33-month-old male rats, an acute ethanol challenge produced significantly greater nonspatial and spatial memory impairments compared to younger animals (Matthews et al. 2020).

The increased sensitivity by aged, older animals to the effects of acute alcohol administration can be due to a variety of factors including central nervous system mechanisms. However, it is also important to determine if the increased sensitivity can be due to altered (typically slowed) metabolism of alcohol in aged animals thereby resulting in higher blood alcohol levels compared to younger animals. If such was the case, then aged animals are not more sensitive to alcohol but instead simply break down the drug slower resulting in elevated blood alcohol levels and increased behavioral effects. However, it has been shown that aged rats have similar blood alcohol concentrations following intraperitoneal injections if the amount of alcohol is less than ~2.5 g/kg and if blood alcohol levels are taken within 60 min of the injection. Given all the previously reported behavioral data occurred within this period, it is likely that the central nervous system is altered due to biological aging such that older animals are significantly more sensitive to alcohol compared to younger animals.

These data demonstrate several important facts related to the effect of acute alcohol in aged animals. First, aged animals are clearly more sensitive to the behavioral effects of alcohol compared to younger animals. Second, given that adolescent animals are typically less sensitive to the effects of alcohol than adults, but aged animals are more sensitive to the effects of alcohol than adults it appears that the natural aging process produces changes that increase the sensitivity to alcohol. Third, the magnitude of the increased effect in aged animals to alcohol varies based on behavior with ataxic effects being sensitive while cognitive effects are less sensitive. Fourth, sex and different lifespan ages interact to impact the effect of alcohol on behavior. Fifth, it is important to remember that subjects (rats or mice) may respond differently to the stress of injection or intoxication based on age and sex and confounds such as these may impact findings and conclusions. Finally, it is important to determine when conducting cognitive studies if the subjects are, or are not, cognitively impaired prior to an alcohol challenge. Given cognition can be compromised in aging and the continued increase in cases of mild cognitive impairment and Alzheimer's disease and related dementia (ADRD), this is likely to be an important field of study to fully investigate.

11.3 Effect of Chronic Intermittent Alcohol Exposure During Adolescence on Behavior Across the Lifespan

Most individuals begin drinking alcohol during adolescence (National Institute on Alcohol Abuse and Alcoholism 2025) and age of drinking onset is a strong predictor of alcohol-related problems later in life (DeWit et al. 2000). As reviewed in the current paper and in other chapters in the current issue (Match et al in this volume; Sicher and Crowley, in this volume), much is known about the behavioral effects and underlying brain changes due to adolescent alcohol use. However, little has been investigated concerning the impact of alcohol administration during adolescence on behavioral changes over a majority of the lifespan. To address this, research has begun investigating the behavioral and central nervous system alterations produced by adolescent alcohol exposure over the lifespan. Specifically, in our first study on this topic (Matthews et al. 2017), we exposed adolescent male rats to either saline or 4.0 g/kg alcohol every other day during adolescence (postnatal day 30–48) for a total of ten alcohol exposures and ten alcohol withdrawals. After this exposure period, animals were administered alcohol or saline prior to a series of behavioral tests at specific time points over the next ~500 days. A final experimental group was exposed to saline during adolescence and before each test session except the final test session at postnatal days 530–532 where they received alcohol to test the effect of the drug late in life. This project demonstrated that adolescent alcohol exposure had long-lasting effects on the subjects' later responses to an alcohol challenge. Specifically, we reported long-lasting tolerance to a high-dose alcohol challenge as measured by the loss of righting reflex, and potentiated spatial memory impairments to a moderate alcohol challenge. In fact, these effects were found approximately 500 days following the chronic intermittent alcohol-exposure period during adolescence (Matthews et al. 2017). These data demonstrate that life-long changes in biological function occur following adolescent alcohol exposure and highlight the limbic system, particularly the hippocampus and related structures, as brain areas that may be compromised by chronic intermittent alcohol exposure during adolescence.

In our follow-up longitudinal studies, we sought to further explore cognitive deficits that may exist across the lifespan in animals exposed to alcohol during adolescence. In addition, we focused on directly investigating sex-dependent differences that may exist by treating and testing both female and male rats. To this end, we treated male and female rats on postnatal day 30–48 with either water, 3.0 g/kg, or 5.0 g/kg alcohol gavaged every other day for 20 days resulting in 10 intoxication and 10 withdrawal periods. The day following the completion of the alcohol-exposure period, animals' nonspatial learning, spatial learning, and behavioral flexibility were tested in the Morris water maze. We selected each of these tests for specific reasons: nonspatial learning tested for general motivation to escape the water maze and later tests to confirm visual ability in aged animals as recommended by Foster (2023); spatial learning tested the function of the hippocampal system; behavioral flexibility was accomplished by a 2-day reversal task in the Morris water maze which tested orbital frontal/prefrontal cortex functioning (Stalnaker et al. 2015; Rudebeck and Murray 2014). The chronic intermittent alcohol-exposure paradigm during adolescence did not impair nonspatial learning or spatial learning in the Morris water maze although aged animals (~postnatal day 630), particularly males, did show general impairments in spatial learning compared to their performance earlier in life. However, a significant impairment was found in behavioral flexibility performance due to alcohol exposure during adolescence and the impairment in behavioral flexibility was sex-dependent. Specifically, for male subjects the impairment in behavioral flexibility as a function of alcohol exposure was found primarily on the last test almost 600 days following the adolescent alcohol exposure. For female subjects the impairment in behavioral flexibility was observed at approximately 300 days post alcohol exposure (~10 months of age) and was present throughout the study until normal biological aging in the control animals began to impair performance (Matthews et al. 2022).

These data are important for a variety of reasons. First, they further extend our earlier results (Matthews et al. 2017) demonstrating that adolescent alcohol exposure can impair cognition in aged animals. In addition, the research supports the conclusion that sex is an important factor to investigate as it relates to the interaction of age and alcohol exposure. Finally, the impairment in behavioral flexibility is important. Significantly, recent research studies have begun to reveal a potential relationship between an alcohol use diagnosis and the development of Alzheimer's disease and related dementia (ADRD) (Schwarzinger et al. 2018; Rehm et al. 2019; Zhang et al. 2022; Anton et al, in this volume). Similarly, in clinical studies, research using tasks such as the Wisconsin Card Sorting Task often finds deficits in behavioral flexibility in patients diagnosed with Alzheimer's disease and related dementia (see Guarino et al. 2019 for a recent review of this literature).

Given the consistent cognitive deficits in aged animals following adolescent alcohol exposure and the finding that such cognitive deficits vary as a function of sex, we further investigated how adolescent alcohol exposure impacts cognition late in life and queried if chronic intermittent alcohol exposure during adolescence alters an anxiety-like state over the course of the lifespan. Once again, we exposed female and male rats to either water, 3.0 g/kg, or 5.0 g/kg alcohol during adolescence in a chronic intermittent fashion from postnatal day 30–48 via gavage. This resulted in ten alcohol intoxication episodes and ten alcohol withdrawal episodes. The day following the adolescent alcohol exposure period and two other times over the next ~16 months, animals' anxiety-like behavior was assessed in the elevated plus maze and the open field task. Finally, on postnatal day 582, animals were trained on the standard spatial version of the Morris water maze followed by a 2-day reversal task to measure behavioral flexibility as previously done. Chronic intermittent alcohol exposure during adolescence initially produced an anxiogenic effect in male rats that resolved over the course of the lifespan. In addition, and consistent with our previous results (Matthews et al. 2022), male rats exposed to alcohol during adolescence had significantly impaired spatial learning and significantly impaired behavioral flexibility (Matthews et al. 2023a). Furthermore, within 2 days of the last water maze test, brains were removed and [^{3}H]PK11195 binding was used to measure microglial reactivity in selected brain regions. We found that [^{3}H]PK11195 binding was not altered by previous alcohol exposure during adolescence but did display differential sex- and brain-region dependent expression

(Matthews et al. 2023a). These studies further solidify the conclusion that early life binge-like alcohol exposure produces cognitive deficits late in life that are (a) sex-dependent and (b) mirror cognitive deficits found in individuals diagnosed with Alzheimer's disease or related dementia.

11.4 Does Chronic Alcohol in Aged Animals Produce Dementia-Like Effects?

As previously stated, research is beginning to develop a possible relationship between chronic alcohol exposure and the development of Alzheimer's disease or related dementia (Schwarzinger et al. 2018; Rehm et al. 2019; Zhang et al. 2022). In fact, it has been proposed that being diagnosed with an alcohol use disorder is a behavioral predictor of developing early onset ADRD (Schwarzinger et al. 2018). In addition, the possible connection between alcohol exposure and ADRD has been supported by preclinical animal models (Downs et al. 2023; Tousley et al. 2023; Marsland et al. 2023; however see Kang et al. 2023). Based on these overlapping lines of evidence, we sought to further investigate in aged animals the impact of chronic intermittent alcohol exposure on behavioral flexibility in the Morris water maze. Specifically, we sought to test the hypothesis that chronic intermittent alcohol exposure would produce differential cognitive impairments in aged male rats compared to adult male rats and then, we wanted to investigate protein-specific pathways in the hippocampus via proteomics and phosphoproteomics that may be differentially altered by the interaction of the alcohol exposure and subjects' age (Ho et al. 2022).

We trained male aged (postnatal 19 months) and young male adult (postnatal 3 months) rats the standard spatial version of the Morris water maze task to a criterion of about 10 s average performance per day for 2–3 days to ensure we were testing cognitive spared rats as previously done (Matthews et al. 2020). Animals were then administered via gavage either 5.0 g/kg alcohol or water every other day for 20 days resulting in 10 intoxication and 10 withdrawal periods before a reversal task was administered to determine the effect of chronic intermittent alcohol exposure on behavioral flexibility. We found that aged male rats, but not adult male rats, had impaired behavioral flexibility following the alcohol exposure. In addition, proteomic analysis revealed altered relative expression of multiple proteins and protein pathways in the hippocampus that are compromised by the alcohol exposure paradigm including relative PKC delta expression and CamK2a expression, proteins that also have been linked to ADRD. In addition, iron transporter proteins were altered suggesting increased iron concentration in the hippocampus of aged subjects. Finally, KEGG and Reactome pathways for proteomic and phosphoproteomics were also altered in a manner like that found in ADRD patients (Ho et al. 2022). These data further support the hypothesis that chronic alcohol exposure can produce cognitive deficits and underlying changes in brain protein pathways that can mirror those that are found in Alzheimer's disease and related dementia. However, significant work is needed to determine if there is a causal link between chronic alcohol exposure and Alzheimer's disease or if chronic alcohol exposure produces a dementia state that is unique to alcohol due to accelerated brain changes (Zhao et al. 2020; Sullivan and Pfefferbaum 2023; Zahr 2024; Zillich et al. 2024)

11.5 Conclusion

The age of a subject is a primary factor impacting the effect of alcohol on behavior and brain systems. The current work reviews our research focused on understanding how alcohol produces differential effects in adolescent animals compared to adult animals and our recent work highlighting the differential effects of alcohol in aged animals compared to adult animals. In addition, we have reviewed recent neurological changes due to the interaction of age and alcohol exposure that may mediate the differential behavioral effects. The general conclusion of this work is that as an individual ages, their sensitivity to alcohol increases in most, if not all, behavioral and neurobiological effects. Stated more directly, the older population is more sensitive to alcohol compared to the adult population which is more sensitive to alcohol compared to the adolescent population. In addition, it does not appear as if this effect is driven by alterations in liver function and alcohol clearance if behavioral effects are measured within a short (<90 min) time periods following alcohol administration. As such, we have investigated, and are continuing to expand our investigations, into potential central nervous system mechanism(s) that may underlie the effect of alcohol in different age groups.

In addition to the normal aging process resulting in greater sensitivity to alcohol, we have also investigated the interactive effect of adolescent alcohol exposure and sex on later cognitive function. Specifically, adolescent alcohol exposure in both female and male rats results in impairments in behavioral flexibility across the lifespan but the impairments manifest themselves significantly faster in female subjects compared to male subjects. This differential effect on cognitive function by sex requires further investigation to understand biological factors that may cause the disparate deficits in cognition.

Finally, we and others have shown that chronic alcohol exposure in aged animals can alter relative protein expression and change protein pathways in a manner like what is found in patients with Alzheimer's disease or related dementia. Given there is some evidence that an alcohol use disorder may impact the development of ADRD, it is imperative to further investigate the causal link between chronic alcohol use and degenerative neurological diseases.

Alcohol has long been and still is the most used and misused drug in the world (Ferreira and Willoughby 2008; Sudhinaraset et al. 2016) and is consumed by almost every age demographic from adolescents to adults to the older population. As individuals age, their neurobiological and physical bodies change resulting in an increase in the effect of alcohol. Understanding the impact of the subjects' age and the interaction of sex and aging will allow researchers to better understand how alcohol can negatively impact people and reduce the harmful outcomes of alcohol use and misuse.

References

Abel EL, York JL (1979) Age-related differences in response to ethanol in the rat. Physiol Psychol 7:391–395

Bertoglio LJ, Carobrez AP (2002) Anxiolytic effects of ethanol and phenobarbital are abolished in test-experienced rats submitted to the elevated plus maze. Pharmacol Biochem Behav 73:963–969. https://doi.org/10.1016/s0091-3057(02)00958-9

Best PJ, White AM, Minai A (2001) Spatial processing in the brain: the activity of hippocampal place cells. Annu Rev Neurosci 24:459–486

Blazer DG, Wu LT (2009) The epidemiology of at-risk and binge drinking among middle-aged and elderly community adults: National Survey on Drug Use and Health. Am J Psychiatry 166:1162–1169

Bogo V, Hill TA, Young RW (1981) Comparison of accelerod and rotarod sensitivity in detecting ethanol-induced and acrylamide-induced performance decrement in rats—review of experimental considerations of rotating rod systems. Neurotoxicology 2:765–787

Breslow RA, Castle IP, Chen CM et al (2017) Trends in alcohol consumption among older Americans: National Health Interview Surveys, 1997 to 2014. Alcohol Clin Exp Res 41:976–986. https://doi.org/10.1111/acer.13365

Calvo E, Allel K, Staudinger UM et al (2020) Cross-country differences in age trends in alcohol consumption among older adults: a cross-sectional study of individuals aged 50 years and older in 22 countries. Addiction 1399–1412. https://doi.org/10.1111/add.15292

Castaldelli-Maia JM, Segura LE, Martins SS (2021) The concerning increasing trend of alcohol beverage sales in the U.S. during the COVID-19 pandemic. Alcohol 96:37–42

Centers for Disease Control and Prevention (CDC) (2021) https://www.cdc.gov/alcohol/fact-sheets/states/excessive-alcohol-use-united-states.html

Chin VS, Van Skike CE, Matthews DB (2010) Effects of ethanol on hippocampal function during adolescence: a look at the past and thoughts on the future. Alcohol 44:3–14

Chin VS, Van Skike CE, Berry RB et al (2011) Effect of acute ethanol and acute allopregnanolone on spatial memory in adolescent and adult rats. Alcohol 45:473–483. https://doi.org/10.1016/j.alcohol.2011.03.001. Epub 2011 May 19

Chung T, Creswell KG, Bachrach R, Clark DB, Martin CS (2018) Adolescent binge drinking. Alcohol Res 39(1):5–15. PMID: 30557142; PMCID: PMC6104966

Criswell HE, Knapp DJ, Overstreet DH et al (1994) Effects of ethanol, chlordiazepoxide, and MK-801 on performance in the elevated-plus maze and on locomotor activity. Alcohol Clin Exp Res 18:596–601. https://doi.org/10.1111/j.1530-0277.1994.tb00916.x

Da Silva GE, Vendruscolo LF, Takahashi RN (2005) Effects of ethanol on locomotor and anxiety-like behaviors and the acquisition of ethanol intake in Lewis and spontaneously hypertensive rats. Life Sci 77:693–706

DeWit DJ, Adlaf EM, Offord DR et al (2000) Age at first alcohol use: a risk factor for the development of alcohol disorders. Am J Psychiatry 157:745–750. https://doi.org/10.1176/appi.ajp.157.5.745. PMID: 10784467

Downs AM, Catavero CM, Kasten MR et al (2023) Tauopathy and alcohol consumption interact to alter locus coeruleus excitatory transmission and excitability in male and female mice. Alcohol 107:97–107

Ferreira MP, Willoughby D (2008) Alcohol consumption: the good, the bad, and the indifferent. Appl Physiol Nutr Metab 33:12–20. https://doi.org/10.1139/H07-175

Foster TC (2023) Animal models for studies of alcohol effects on the trajectory of age-related cognitive decline. Alcohol 107:4–11

Freund G (1973) Hypothermia after acute ethanol and benzyl alcohol administration. Life Sci 13:345–349. https://www.ncbi.nlm.nih.gov/pubmed/4752488

Frye GD, Chapin RE, Vogel RA et al (1981) Effects of acute and chronic 1,3-butanediol treatment on central nervous-system function—a comparison with ethanol. J Pharm Exper Ther 216:306–314

Gano A, Doremus-Fitzwater TL, Deak T (2017) A cross-sectional comparison of ethanol-related cytokine expression in the hippocampus of young and aged Fischer 344 rats. Neurobiol Aging 54:40–53. https://doi.org/10.1016/j.neurobiolaging.2017.01.025. Epub 2017 Mar 17. PMID: 28319836; PMCID: PMC5401774

García-Moreno LM, Cimadevilla JM (2012) Acute and chronic ethanol intake: effects on spatial and non-spatial memory in rats. Alcohol 46(8):757–762. https://doi.org/10.1016/j.alcohol.2012.08.001. Epub 2012 Sept 1. PMID: 22944615

Green R, Wolf BJ, Chen A, Kirkland AE, Ferguson PL, Browning BD, Bryant BE, Tomko RL, Gray KM, Mewton L, Squeglia LM (2024) Predictors of substance use initiation by early adolescence. Am J Psychiatry 181(5):423–433. https://doi.org/10.1176/appi.ajp.20230882. PMID: 38706327

Guarino A, Favieri F, Boncompagni I et al (2019) Executive functions in Alzheimer disease: A systematic review. Front Aging Neurosci 10:Article 437

Han BH, Moore AA, Sherman S et al (2017) Demographic trends of binge alcohol use and alcohol use disorders among older adults in the United States, 2005-2014. Drug Alcohol Depend 170:198–207. https://doi.org/10.1016/j.drugalcdep.2016.11.003

Han BH, Moore AA, Ferris R et al (2019) Binge drinking among older adults in the United States, 2015 to 2017. J Am Geriatr Soc 67:2139–2144. https://doi.org/10.1111/jgs.16071

Harris RA, McQuilkin SJ, Paylor R (1995) Mutant mice lacking the gamma isoform of protein kinase C show decreased behavioral actions of ethanol and altered function of gamma-aminobutyrate type A receptors. Proc Natl Acad Sci USA 92:3658–3662. https://doi.org/10.1073/pnas.92.9.3658. PMID: 7731960; PMCID: PMC42020

Harvey RE, Berkowitz LE, Savage DD et al (2020) Altered hippocampal place cell representation and theta rhythmicity following moderate prenatal alcohol exposure. Curr Biol 30:3556–3569.e5

Hefner K, Holmes A (2007) An investigation of the behavioral actions of ethanol across adolescence in mice. Psychopharmacology 191:311–322

Ho AMC, Peyton MP, Scaletty S et al (2022) Chronic intermittent ethanol exposure alters behavioral flexibility in aged rats compared to adult rats and modifies protein and protein pathways related to Alzheimer's disease. ACS Omega 7:46260–44627

Hoffmann SE, Matthews DB (2001) Ethanol-induced impairments in spatial working memory are not due to deficits in learning. Alcohol Clin Exp Res 25:856–861

Hunt PS, Levillain ME, Spector BM et al (2009) Post-training ethanol disrupts trace conditioned fear in rats: Effects of timing of ethanol, dose and trace interval duration. Neurobiol Learn Mem 91:73–80

Kang S, Lee J, Choi S, Nesbitt J, Min PH, Trushina E, Choi DS (2023) Moderate ethanol exposure reduces astrocyte-induced neuroinflammatory signaling and cognitive decline in presymptomatic APP/PS1 mice. Res Sq [Preprint]. https://doi.org/10.21203/rs.3.rs-3627637/v1. PMID: 38077051; PMCID: PMC10705690

Karlsson O, Roman E (2016) Dose-dependent effects of alcohol administration on behavioral profiles in the MCSF test. Alcohol 50:51–56. https://doi.org/10.1016/j.alcohol.2015.10.003

Keyes KM (2023) Alcohol use in the older adult US population: Trends, causes, and consequences. Alcohol 107:28–31

Laberge S, Bigelow P, Lagarde E et al (2020) Examining the association between alcohol consumption and health conditions in community dwelling older adults. J Community Health. https://doi.org/10.1007/s10900-020-00842-8

LaBuda CJ, Fuchs PN (2001) The anxiolytic effect of acute ethanol or diazepam exposure is unaltered in mu-opioid receptor knockout mice. Brain Res Bull 55(6):755–760. https://doi.org/10.1016/s0361-9230(01)00569-x. PMID: 11595359

LaBuda CJ, Hale RL (2000) Anxiety in mice following acute aspartame and ethanol exposure. Alcohol 20(1):69–74. https://doi.org/10.1016/s0741-8329(99)00060-9. PMID: 10680719

Lapish CC (2024) Understanding how acute alcohol impacts neural encoding in the rodent brain. Curr Top Behav Neurosci. https://doi.org/10.1007/7854_2024_479. Epub ahead of print. PMID: 38858298

Lassman D, Hartman M, Washington B et al (2014) US health spending trends by age and gender: selected years 2002-10. Health Aff 33:815–822. https://doi.org/10.1377/hlthaff.2013.1224
Little PJ, Kuhn CM, Wilson WA et al (1996) Differential effects of ethanol in adolescent and adult rats. Alcohol Clin Exp Res 20:1346–1351
Majchrowicz E (1975) Induction of physical dependence upon ethanol and the associated behavioral changes in rats. Psychopharmacologia 43:245–254. https://www.ncbi.nlm.nih.gov/pubmed/1237914
Markwiese BJ, Acheson SK, Levin ED et al (1998) Differential effects of ethanol on memory in adolescent and adult rats. Alcohol Clin Exp Res 22:416–421
Marsland P, Vore AS, Daprano EM et al (2023) Sex-specific effects of ethanol consumption in older Fischer 344 rats on microglial dynamics and AB(1-42) accumulation. Alcohol 107:108–118
Matthews DB, Koob GF (2023) On the critical need to investigate the effect of alcohol in the older population. Alcohol 107:2–3
Matthews DB, Mittleman G (2017) Age-dependent effects of chronic intermittent ethanol treatment: Gross motor behavior and body weight in aged, adult and adolescent rats. Neurosci Lett 657:146–150. https://doi.org/10.1016/j.neulet.2017.08.012. Epub 2017 Aug 5. PMID: 28789984
Matthews DB, Simson PE, Best PJ (1995) Acute ethanol impair spatial memory but not stimulus/response memory in the rat. Alcohol Clin Exp Res 19:902–909
Matthews DB, Simson PE, Best PJ (1996) Ethanol alters spatial processing of hippocampal place cells: A mechanism for impaired navigation when intoxicated. Alcohol Clin Exp Res 20:404–407
Matthews DB, Ilgen M, White AM et al (1999) Acute ethanol administration impairs spatial performance while facilitating nonspatial performance in rats. Neurobiol Learn Mem 72:169–179. https://doi.org/10.1006/nlme.1998.3900
Matthews DB, Tinsley KL, Diaz-Granados JL et al (2008) Chronic intermittent exposure to ethanol during adolescence produces tolerance to the hypnotic effects of ethanol in male rats: a dose-dependent analysis. Alcohol 42:617–621. https://doi.org/10.1016/j.alcohol.2008.09.001. PMID: 19038695
Matthews DB, Novier A, Diaz-Granados JL et al (2017) Impact of adolescent alcohol use across the lifespan: Long-lasting tolerance to high-dose alcohol coupled with potentiated spatial memory impairments to moderate-dose alcohol. Alcohol 61:33–42. https://doi.org/10.1016/j.alcohol.2017.01.012
Matthews DB, Scaletty S, Schreiber A et al (2020) Acute ethanol administration produces larger spatial and nonspatial memory impairments in 29-33 month old rats compared to adult and 18-24 month old rats. Pharmacol Biochem Behav 199:173074. https://doi.org/10.1016/j.pbb.2020.173074
Matthews DB, Scaletty S, Trapp S et al (2022) Chronic intermittent ethanol administration during adolescence produces sex dependent impairments in behavioral flexibility and survivability. Brain Sci 12:606. https://doi.org/10.3390/brainsci12050606
Matthews DB, Scaletty S, Trapp S et al (2023a) Chronic intermittent ethanol exposure during adolescence produces sex- and age-dependent changes in anxiety and cognition without changes in microglia reactivity late in life. Front Behav Neurosci 17. https://doi.org/10.3389/fn-beh.2023.1223883
Matthews DB, Wieden T, Turunen A et al (2023b) Effect of chronic intermittent ethanol exposure and age on PKC-isoform expression in the cerebellum. Alcohol Clin Exp Res 47(S1):182
Matthews DB, Rossmann G, Matthews SJ, Zank A, Shult C, Turunen A, Sharma P (2024) Increased anxiolytic effect in aged female rats and increased motoric behavior in aged male rats to acute alcohol administration: Comparison to younger animals. Pharmacol Biochem Behav 239:173770. https://doi.org/10.1016/j.pbb.2024.173770. Epub 2024 Apr 16. PMID: 38636813
McKetta S, Keyes KM (2019) Heavy and binge alcohol drinking and parenting status in the United States from 2006 to 2018: An analysis of nationally representative cross-sectional surveys. PLoS Med 16:e1002954. https://doi.org/10.1371/journal.pmed.1002954
Melia KR, Ryabinin AE, Corodimas KP et al (1996) Hippocampal-dependent learning and experience-dependent activation of the hippocampus are preferentially disrupted by ethanol. Neuroscience 74:313–322
Miyake K, Yagi S, Aoki Y, Shikano Y, Ikegaya Y, Sasaki T (2020) Acute effects of ethanol on hippocampal spatial representation and offline reactivation. Front Cell Neurosci 14:571175. https://doi.org/10.3389/fncel.2020.571175. PMID: 33250711; PMCID: PMC7674284
National Institute on Alcohol Abuse and Alcoholism (2025) Get the facts about underage drinking. https://www.niaaa.nih.gov/publications/brochures-and-fact-sheets/underage-drinking
National Survey on Drug Use and Health, SAMHSA (2022) https://www.samhsa.gov/data/release/2022-national-survey-drug-use-and-health-nsduh-releases
Novier A, Diaz-Granados JL, Matthews DB (2015) Alcohol use across the lifespan: An analysis of adolescent and aged rodents and humans. Pharmacol Biochem Behav 133:65–82. https://doi.org/10.1016/j.pbb.2015.03.015
Ornelas LC, Novier A, Van Skike CE et al (2015) The effects of acute alcohol on motor impairments in adolescent, adult, and aged rats. Alcohol 49:121–126. https://doi.org/10.1016/j.alcohol.2014.12.002
Patrick ME, Terry-McElrath YM (2019) Prevalence of high-intensity drinking from adolescence through young adulthood: National data from 2016-2017. Sub Abuse Res Treat 13:1178221818822976. https://doi.org/10.1177/1178221818822976
Patton G, Olsson C, Skirbekk V et al (2018) Adolescence and the next generation. Nature 554:458–466. https://doi.org/10.1038/nature25759

Perkins AE, Vore AS, Lovelock D, Varlinskaya E, Deak T (2018) Late aging alters behavioral sensitivity to ethanol in a sex-specific manner in Fischer 344 rats. Pharmacol Biochem Behav 175:1–9. https://doi.org/10.1016/j.pbb.2018.07.012. Epub 2018 Aug 29. PMID: 30171932; PMCID: PMC6240477

Rehm J, Hasan OSM, Black SE et al (2019) Alcohol use and dementia: a systematic scoping review. Alzheimers Res Ther 11:1. https://doi.org/10.1186/s13195-018-0453-0. PMID: 30611304; PMCID: PMC6320619

Ristuccia RC, Hernandez M, Wilmouth CE et al (2007) Differential expression of ethanol-induced hypothermia in adolescent and adult rats induced by pretest familiarization to the handling/injection procedure. Alcohol Clin Exp Res 31:575–581. https://doi.org/10.1111/j.1530-0277.2007.00341.x

Rodd ZA, Bell RL, McKinzie DL et al (2004) Low-dose stimulatory effects of ethanol during adolescence in rat lines selectively bred for high alcohol intake. Alcohol Clin Exp Res 28(4):535–543. https://doi.org/10.1097/01.alc.0000122107.08417.d0

Rudebeck PH, Murray EA (2014) The orbitofrontal oracle: cortical mechanisms for the prediction and evaluation of specific behavioral outcomes. Neuron 84:1143–1156. https://doi.org/10.1016/j.neuron.2014.10.049. PMID: 25521376; PMCID: PMC4271193

Rustay NR, Wahlsten D, Crabbe JC (2003) Assessment of genetic susceptibility to ethanol intoxication in mice. Proc Natl Acad Sci USA 100:2917–2922. https://doi.org/10.1073/pnas.0437273100

Ryabinin AE, Miller MN, Durrant S (2002) Effects of acute alcohol administration on object recognition learning in C57BL/6J mice. Pharmacol Biochem Behav 71:307–331

Sacks JJ, Gonzales KR, Bouchery EE et al (2015) 2010 national and state costs of excessive alcohol consumption. Am J Prev Med 49:e73–e79

Schwarzinger M, Pollock BG, Hasan OSM et al (2018) Contribution of alcohol use disorders to the burden of dementia in France 2008-13: a nationwide retrospective cohort study. Lancet Public Health 3:e124–e132. https://doi.org/10.1016/S2468-2667(18)30022-7. Epub 2018 Feb 21. PMID: 29475810

Silveri MM, Spear LP (1998) Decreased sensitivity to the hypnotic effects of ethanol early in ontogeny. Alcohol Clin Exp Res 22:670–676

Silvers JM, Tokunaga S, Mittleman G et al (2006) Chronic intermittent ethanol exposure during adolescence reduces the effect of ethanol challenge on hippocampal allopregnanolone levels and Morris water maze task performance. Alcohol 39:151–158. https://doi.org/10.1016/j.alcohol.2006.09.001. PMID: 17127134

Sircar R (2019) Estrogen modulates ethanol-induced memory deficit in postpubertal adolescent rats. Alcohol Clin Exp Res 43(1):61–68. https://doi.org/10.1111/acer.13921. Epub 2018 Dec 9. PMID: 30408223

Spear LP (2018) Effects of adolescent alcohol consumption on the brain and behaviour. Nat Rev Neurosci 19:197–214. https://doi.org/10.1038/nrn.2018.10

Stalnaker TA, Cooch NK, Schoenbaum G (2015) What the orbitofrontal cortex does not do. Nat Neurosci 18:620–627. https://doi.org/10.1038/nn.3982. PMID: 25919962; PMCID: PMC5541252

Sudhinaraset M, Wigglesworth C, Takeuchi DT (2016) Social and cultural contexts of alcohol use: influences in a social-ecological framework. Alcohol Res 38(1):35–45. PMID: 27159810; PMCID: PMC4872611

Sullivan EV, Pfefferbaum A (2023) Alcohol use disorder: Neuroimaging evidence for accelerated aging of brain morphology and hypothesized contribution to age-related dementia. Alcohol 107:44–55. https://doi.org/10.1016/j.alcohol.2022.06.002. Epub 2022 July 1. PMID: 35781021

Swartzwelder HS, Richardson RC, Markwiese-Foerch B et al (1998) Developmental differences in the acquisition of tolerance to ethanol. Alcohol 15:311–314. https://doi.org/10.1016/s0741-8329(97)00135-3. PMID: 9590516

Swartzwelder NA, Risher ML, Abdelwahab SH et al (2012) Effects of ethanol, Delta-Tetrahydrocannabinol, or their combination on object recognition memory and object preference in adolescent and adult male rats. Neurosci Lett 527:11–15

Tam CC, Li L, Lui CK, Cook WK (2024) Drinking patterns among US men and women: Racial and ethnic differences from adolescence to early midlife. Alcohol Clin Exp Res (Hoboken) 48(6):1076–1087. https://doi.org/10.1111/acer.15308. Epub 2024 June 3. PMID: 38829485; PMCID: PMC11178455

Tampi RR, Tampi DJ, Durning M (2015) Substance use disorders in late life: A review of current evidence. Heal Aging Res 4

Tapert SF, Eberson-Shumate S (2022) Alcohol and the adolescent brain: what we've learned and where the data are taking us. Alcohol Res 42(1):07. https://doi.org/10.35946/arcr.v42.1.07. PMID: 35465194; PMCID: PMC8999519

Tipps ME, Raybuck JD, Buck KJ, Lattal KM (2015) Acute ethanol withdrawal impairs contextual learning and enhances cued learning. Alcohol Clin Exp Res 39(2):282–290. https://doi.org/10.1111/acer.12614. PMID: 25684050; PMCID: PMC4331355

Tousley AR, Yeh PWL, Yeh HH (2023) Precocious emergence of cognitive and synaptic dysfunction in 3xTg-AD mice exposed prenatally to ethanol. Alcohol 107:56–72

United Nations, Department of Economic and Social Affairs, World population prospects, 2022. https://www.un.org/development/desa/pd/sites/www.un.org.development.desa.pd/files/wpp2022_summary_of_results.pdf

US Department of Health and Human Services, 2019 fact sheet. https://www.niaaa.nih.gov/alcohols-effects-health/alcohol-topics/alcohol-facts-and-statistics

Van Skike CE, Botta P, Chin VS et al (2010) Behavioral effects of ethanol in cerebellum are age dependent: potential system and molecular mechanisms. Alcohol Clin Exp Res 34:2070–2080. https://doi.org/10.1111/j.1530-0277.2010.01303.x

Van Skike CE, Diaz-Granados JL, Matthews DB (2015) Chronic intermittent ethanol exposure produces persistent anxiety in adolescent and adult rats. Alcohol Clin Exp Res 39:262–271. https://doi.org/10.1111/acer.12617

Van Skike CE, Goodlett C, Matthews DB (2019) Acute alcohol and cognition: Remembering what it causes us to forget. Alcohol 79:105–125

Vetreno RP, Campbell J, Crews FT (2023) A multicomponent ethanol response battery across a cumulative dose ethanol challenge reveals diminished adolescent rat ethanol responsivity relative to adults. Adv Drug Alcohol Res 3:11888. https://doi.org/10.3389/adar.2023.11888. PMID: 38389807; PMCID: PMC10880770

Watson MR, James K, Mittleman G et al (2020) Impact of acute ethanol exposure on body temperatures in aged, adult and adolescent male rats. Alcohol 82:81–89. https://doi.org/10.1016/j.alcohol.2019.08.001

Weitemier AZ, Ryabinin AE (2003) Alcohol-induced memory impairments in trace fear conditioning: A Hippocampus-specific effect. Hippocampus 13:305–315

White AM (2020) Gender differences in the epidemiology of alcohol use and related harms in the United States. Alcohol Res 40(2):01. https://doi.org/10.35946/arcr.v40.2.01. PMID: 33133878; PMCID: PMC7590834

White AM, Best PJ (2000) Effects of ethanol on hippocampal place-cell and interneuron activity. Brain Res 876:154–165

White AM, Bae JG, Truesdale MC et al (2002) Chronic-intermittent ethanol exposure during adolescence prevents normal developmental changes in sensitivity to ethanol-induced motor impairments. Alcohol Clin Exp Res 26:960–968

White AM, Castle IJP, Powell PA et al (2022) Alcohol-related deaths during the COVID-19 pandemic. JAMA 321:1704–1706

White AM, Orosz A, Powell PA et al (2023) Alcohol and aging—an area of increasing concern. Alcohol 107:19–27

Wilson MA, Burghardt PR, Ford KA, Wilkinson MB, Primeaux SD (2004) Anxiolytic effects of diazepam and ethanol in two behavioral models: comparison of males and females. Pharmacol Biochem Behav 78(3):445–458. https://doi.org/10.1016/j.pbb.2004.04.017

Witt ED (2010) Research on alcohol and adolescent brain development: opportunities and future directions. Alcohol 44:119–124

York JL (1982) Body water content, ethanol pharmacokinetics, and the responsiveness to ethanol in young and old rats. Dev Pharmacol Ther 4:106–116. https://www.ncbi.nlm.nih.gov/pubmed/7117085

Zahr NM (2024) Alcohol use disorder and dementia: a review. Alcohol Res 44(1):03. https://doi.org/10.35946/arcr.v44.1.03. PMID: 38812709; PMCID: PMC11135165

Zhang P, Edenberg HJ, Nurnberger J et al (2022) Alcohol use disorder is associated with higher risks of Alzheimer's and Parkinson's diseases: A study of US insurance claims data. Alzheimers Dement (Amst) 14:e12370. https://doi.org/10.1002/dad2.12370. PMID: 36419637; PMCID: PMC9677510

Zhao Q, Pfefferbaum A, Podhajsky S, Pohl KM, Sullivan EV (2020) Accelerated aging and motor control deficits are related to regional deformation of central cerebellar white matter in alcohol use disorder. Addict Biol 25(3):e12746. https://doi.org/10.1111/adb.12746. Epub 2019 Apr 1. PMID: 30932270; PMCID: PMC6773538

Zillich L, Cetin M, Hummel EM, Poisel E, Fries GR, Frank J, Streit F, Foo JC, Sirignano L, Friske MM, Lenz B, Hoffmann S, Adorjan K, Kiefer F, Bakalkin G, Hansson AC, Lohoff FW, Kärkkäinen O, Kok E, Karhunen PJ, Sutherland GT, Walss-Bass C, Spanagel R, Rietschel M, Moser DA, Witt SH (2024) Biological aging markers in blood and brain tissue indicate age acceleration in alcohol use disorder. Alcohol Clin Exp Res (Hoboken) 48(2):250–259. https://doi.org/10.1111/acer.15241. Epub 2024 Jan 26. PMID: 38276909; PMCID: PMC10922212

12 Adolescent Alcohol and the Spectrum of Cognitive Dysfunction in Aging

Terrence Deak, Hannah E. Burzynski, Polliana T. Nunes, Stephen M. Day, and Lisa M. Savage

Abstract

Among the many changes associated with aging, inflammation in the central nervous system (CNS) and throughout the body likely contributes to the constellation of health-related maladies associated with aging. Genetics, lifestyle factors, and environmental experiences shape the trajectory of aging-associated inflammation, including the developmental timing, frequency, and intensity of alcohol consumption. This chapter posits that neuroinflammatory processes form a critical link between alcohol exposure and the trajectory of healthy aging, at least in part through direct or indirect interactions with cholinergic circuits that are crucial to cognitive integrity. In this chapter, we begin with a discussion of how inflammation changes from early development through late aging; discuss the role of inflammation and alcohol in the emergence of mild cognitive impairment (MCI); elaborate on critical findings on the contribution of alcohol-related thiamine deficiency to the loss of cholinergic function and subsequent development of Wernicke–Korsakoff syndrome (WKS); and present emerging findings at the intersection of alcohol and Alzheimer's disease and related dementias (ADRD). In doing so, our analysis points toward inflammation-mediated compromise of basal forebrain cholinergic function as a key culprit in cognitive dysfunction associated with chronic alcohol exposure, effects that may be rescuable through either pharmacological or behavioral approaches. Furthermore, our chapter reveals an interesting dichotomy in the effects of alcohol on neuropathological markers of ADRD that depend upon both biological sex and genetic vulnerability.

Keywords

Ethanol · Aging · Cognition · Adolescence · Cholinergic · Dementia · Neuroinflammation

12.1 Introduction

Excessive alcohol use is a major risk factor for early cognitive decline and dementia (Rehm et al. 2019), the emergence of neurodegenerative states (Crews and Nixon 2009), and all-cause mortality (World Health Organization 2019). In particular, recent studies have shown that early-onset dementia was especially prevalent among individuals diagnosed with alcohol use disorder (AUD) (Schwarzinger et al. 2018). In contrast, a

T. Deak (✉) · H. E. Burzynski · P. T. Nunes
S. M. Day · L. M. Savage
Developmental Exposure Alcohol Research Center (DEARC), Behavioral Neuroscience Program, Department of Psychology, Binghamton University-State University of New York, Binghamton, NY, USA
e-mail: tdeak@binghamton.edu

C. F. Valenzuela et al. (eds.), *Effects of Alcohol on the Brain across the Lifespan*, Advances in Experimental Medicine and Biology 1473, https://doi.org/10.1007/978-3-031-81908-7_12

systematic review of studies on light-to-moderate drinking reported a slightly reduced risk of dementia relative to individuals who largely abstained from alcohol consumption (Ilomaki et al. 2015). Despite these long-standing associations, alcohol misuse was historically an exclusion criterion for evaluating dementia prevalence worldwide but is now listed as one of the top three modifiable risk factors for dementia (Livingston et al. 2020). These findings have motivated extensive research into the mechanisms contributing to alcohol-mediated disruption in cognitive dysfunction across the lifespan, which can manifest in a wide spectrum of neuro-compromised states.

The immune system transforms across early development and later aging, showing progressive deterioration throughout the lifespan (Lynch et al. 2010). Even in healthy individuals, aging produces sustained inflammation due to exacerbated inflammatory responses of immune cells and their eventual senescence (Rozovsky et al. 1998), which often takes the form of delayed or impaired "shutoff" by late-acting, anti-inflammatory mechanisms (Norden and Godbout 2013; Norden et al. 2015). Though this transformation occurs throughout the body, it is especially harmful in the central nervous system (CNS) as neuroinflammation is a key driver of age-related cognitive decline. While natural aging is inevitable, certain lifestyle choices, such as alcohol use, can exacerbate these sensitized immune responses (Carlson et al. 2023). Indeed, a substantial challenge for the field is to identify how the developmental timing, frequency, and intensity of alcohol consumption might influence the trajectory of healthy brain aging and cognitive integrity (Deak et al. 2022; Deak and Savage 2019; Nunes et al. 2019). The overarching goal of this chapter is to integrate what is known about the relationship between binge-like alcohol consumption (or exposure models) and the spectrum of cognitive dysfunction that is commonly associated with prolonged alcohol misuse (see Fig. 12.1). To do this, we will first discuss consequences of natural aging, specifically age-related impairments in neuroimmune function, and then highlight how alcohol can aggravate and advance such dysfunction. Subsequent sections will describe cholinergic mechanisms by which alcohol can accelerate normal, age-related cognitive decline to progress into mild cognitive impairment (MCI) and more severe pathological states such as Wernicke–Korsakoff syndrome (WKS) and Alzheimer's disease and related dementias (ADRD).

12.2 Alcohol Interactions with Natural Aging

12.2.1 The Neuroimmunology of Aging

Natural aging is accompanied by a decline in the efficiency and accuracy of the neuroimmune system that first emerges in middle age and amplifies in later ages (Moca et al. 2022). Such impairments have been observed in a variety of immune-derived cells within the CNS, and extensive work has described age-related changes in microglia and astrocytes (Lynch et al. 2010; Fonken et al. 2016; Frank et al. 2016; Clarke et al. 2018). For example, multiple studies have suggested that aged microglia adopt a sensitized or "primed" state that is characterized by excessive and prolonged inflammatory responses (Fonken et al. 2016; Norden and Godbout 2013). Under normal conditions, microglia are extremely dynamic and rapidly transition from their homeostatic resting state to a reactive, pro-inflammatory state when stimulated by an injury or pathogen (Yirmiya et al. 2015). Reactive microglia secrete a host of inflammatory cytokines, chemokines, and reactive oxygen species (ROS) in order to remove the insult and limit damage to surrounding tissue (Safaiyan et al. 2016; Eggen et al. 2013). This process differs based on the type of insult, and activated microglia respond differently to an antigen compared to cerebral infarct or injury. When a pathogen-associated antigen is detected, activated microglia initiate a signaling cascade of inflammatory mediators that may recruit peripheral leukocytes to the CNS, which then remove the pathogen through phagocytosis (Rock et al.

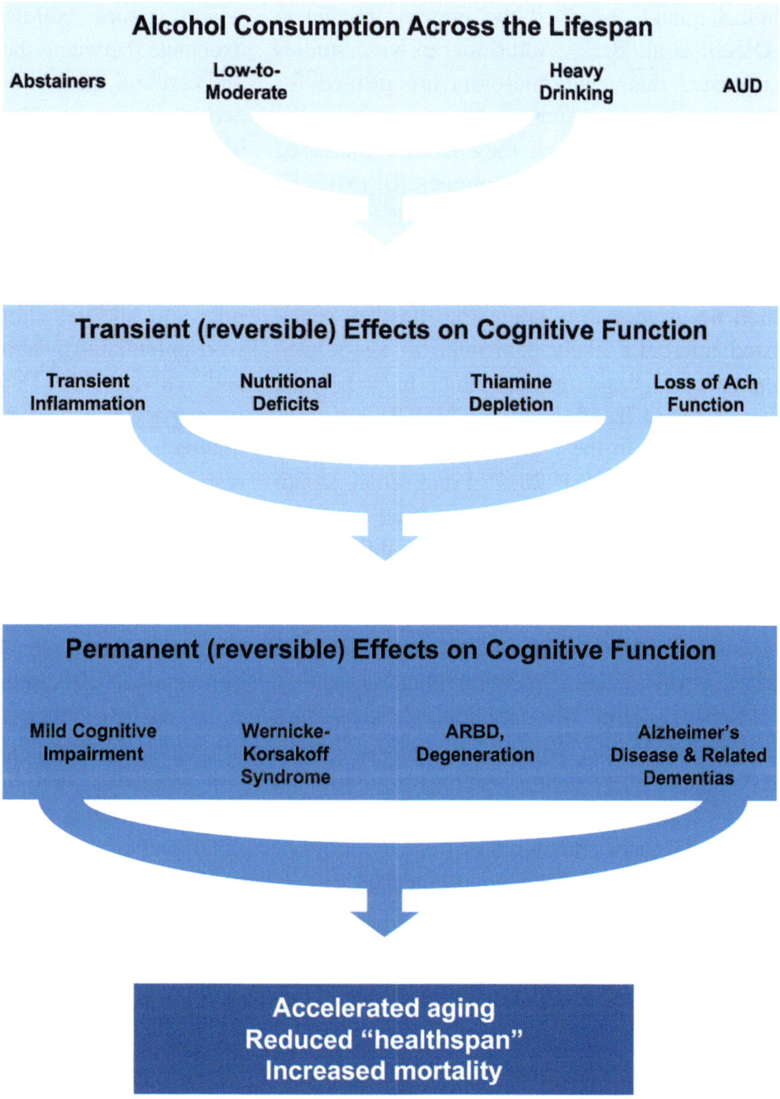

Fig. 12.1 Cascading impact of chronic alcohol use across the lifespan

2004). Microglia themselves are phagocytic, meaning they are also able to clear cellular waste and debris, including dead and dying cells (Sierra et al. 2010; Neumann et al. 2009). This is especially important after CNS injuries and infarcts as reactive microglia can remove damaged cells, including neurons that are secreting toxic amounts of excitatory signals, and limit damage to surrounding cells (Loane and Byrnes 2010). Importantly, healthy microglia can quickly transition back to their resting state when the threat is resolved and the anti-inflammatory response is initiated (Eggen et al. 2013).

Unlike activated microglia found in healthy adults, microglia in aged brains are much less efficient at transitioning back to a resting state, resulting in their sustained activation and release of pro-inflammatory mediators (Norden and Godbout 2013). Preclinical studies reported significant elevations in the pro-inflammatory cytokines interleukin (IL)-1β and IL-6 in the brains of aged rodents relative to adults after they were challenged with the endotoxin lipopolysaccharide (LPS) (Godbout et al. 2005; Henry et al. 2009). Furthermore, this heightened inflammation persisted in the aged brains while adult brains quickly resolved the immune challenge (O'Neil et al. 2022). Additional ex vivo studies suggested that aged microglia are primed for these heightened responses because even in the absence of a stimulus, they exhibit increased major histocompatibility complex II (MHCII) expression, which is indicative of a reactive, pro-inflammatory phenotype (Henry et al. 2009; Frank et al. 2010). As immune threats are a common occurrence throughout the lifespan, sensitized microglia likely contribute to age-related increases in basal inflammation. Indeed, basal expression of IL-1β, IL-6, and MHCII is significantly higher in the brains of aged rodents and humans (Gano et al. 2017; Frank et al. 2006; Streit et al. 2004). Similar age-related increases have been observed in the pro-inflammatory cytokine, high mobility group box 1 (HMGB1), and toll-like receptors (TLRs), which are also considered a result of primed microglia (Fonken et al. 2016). Along with aberrant inflammatory responses, aging also influences microglia's phagocytic activity as aged microglia express less phagocytic receptors and exhibit reduced debris clearance (Thomas et al. 2022). It has been proposed that such dysfunction is due to increased lipid burden within aged microglia that inhibits phagocytic activity (Marschallinger et al. 2020). While the mechanisms driving this impairment have not been fully elucidated, the accumulation of waste within the CNS is known to produce cytotoxic amounts of inflammatory mediators and cause widespread cell death (Neumann et al. 2009). Thus, age-related increases in neuroinflammation are not solely due to microglial priming but are a product of multifaceted alterations in the functional state of the neuroimmune defense network.

While dysfunctional microglia greatly contribute to the sustained neuroinflammation seen in aging, astrocytes undergo a similar deterioration that further drives this inflammatory state (Jyothi et al. 2015). Astrocytes regulate a number of processes within the CNS including neurotransmitter release and blood flow, as well as synapse formation and elimination (Sofroniew and Vinters 2010). Additionally, astrocytes play an important role in the glymphatic system, which clears cellular waste through fluid exchange between the brain's interstitial fluid and cerebral spinal fluid (Iliff et al. 2012). Such actions are impaired in aging brains as there is an age-associated reduction in the expression of aquaporin 4 (AQP4), the astrocytic water channel responsible for this exchange (Kress et al. 2014). As a result, the accumulation of cellular debris and neuroinflammation produced by aged microglia and their impaired phagocytic activity were potentiated by aging astrocytes (Verkhratsky and Semyanov 2023). Lastly, much like microglia, astrocytes can adopt a pro-inflammatory phenotype to respond to an immune threat, and such responses can become dysregulated in aging (Clarke et al. 2018). Reactive astrocytes are often characterized by an upregulation in glial fibrillary acidic protein (GFAP), and numerous preclinical and clinical studies report elevated GFAP expression in aging brains (Verkhratsky and Semyanov 2023; Nichols et al. 1993; Robillard et al. 2016). Other studies suggest that like microglia, aging astrocytes adopt a primed state that sensitizes their inflammatory responses to immune threats and increases basal inflammation (Clarke et al. 2018).

Another shared consequence of aging that affects both microglia and astrocytes is that these cells become resistant to anti-inflammatory mechanisms (O'Neil et al. 2022). Normally, once an immune threat is contained, anti-inflammatory mediators regulate the transition of reactive microglia to their homeostatic resting states or an anti-inflammatory, reparative state (Eggen et al. 2013). However, activated microglia from aged brains show reduced sensitivity to these markers, especially the prominent anti-inflammatory mediators IL-4 and transforming growth factor beta (TGFβ) (Fenn et al. 2012, 2014; Tichauer et al. 2014; Rozovsky et al. 1998). Reactive microglia are also regulated by astrocytes through their release of the anti-inflammatory cytokine IL-10 which stimulates the production of TGFβ and inhibits microglial activation (Norden et al. 2014). Unfortunately, aged astrocytes exhibit a similar decline in anti-inflammatory actions as reduced IL-10 expression was observed in astrocytes from aged rats both 4 and 24 h after LPS challenge and they were unable to prevent microglial activation (O'Neil et al. 2022; Norden et al. 2016). These age-related resistant and erroneous immune responses are referred to as immunosenescence, which is thought to propagate neuroinflammation as it is no longer contained by critical regulatory feedback systems (O'Neil et al. 2022).

Interestingly, the priming and eventual immunosenescence of aging astrocytes and microglia appear to be region specific, with some areas showing extreme sensitivity to these age-related effects. One such region is the hippocampus, which plays an integral role in cognitive function, particularly learning and memory (Golomb et al. 1993). Aging studies consistently report exaggerated levels of pro-inflammatory mediators, including IL-1β and IL-6, in the hippocampus of aged rodents relative to adults (Gano et al. 2017; Sierra et al. 2007). There is extensive evidence that these elevations are due to increased microglial priming in this brain region as an LPS challenge elicited much larger inflammatory responses in the hippocampi of aged rats (Barrientos et al. 2006; Frank et al. 2010). Similarly, there is evidence from both rodent and human studies that hippocampal astrocytes are also more susceptible to age-related dysfunction as they show the largest upregulation of GFAP and undergo the most transcriptomic changes that promote a primed, pro-inflammatory state (Nichols et al. 1993; Soreq et al. 2017; Clarke et al. 2018; David et al. 1997). As the hippocampus is a critical integration center for learning and memory, it is unsurprising that sustained neuroinflammation in this region has been identified as a key contributor to age-related cognitive decline. Beyond the hippocampus, neuroinflammation in other regions such as the cortex, thalamus, and cerebellum are thought to contribute to age-associated impairments in executive function, sensory processing, speech, and coordination (Carlson et al. 2023; Ownby 2010). Given the growing number of aged individuals in the global population, it is important to identify which lifestyle choices may accelerate or prevent such detrimental changes in neuroimmune function.

12.2.2 Interactive Effects of Alcohol and Aging

One lifestyle/experiential variable that can greatly accelerate age-related inflammation, especially within the CNS, is alcohol exposure across the lifespan (Carlson et al. 2023; Deak et al. 2022). A major challenge facing the alcohol field is to better understand the relationship between alcohol consumption patterns and its influence on overall health outcomes, including and especially for CNS function. Not only is it difficult for individuals to accurately reconstruct their history of alcohol consumption across prolonged periods of time, but self-report of alcohol intake is also influenced by a variety of factors such as individual expectancies, positive self-presentation, and memory impairments associated with drinking (Maisto et al. 1995; Tevik et al. 2021). Although objective biomarkers of alcohol exposure such as phosphatidylethanol (PETH) concentrations offer some promise (Harris et al. 2021), PETH levels persist in blood and organs on the timescale of days to weeks, which is probably not adequate for realistic assessments of lifetime alcohol exposure. PETH is also not an effective measure in preclinical (rodent) models, further limiting its use in research and development (Aradóttir et al. 2004). For these reasons, many investigators have now changed tactics to incorporate objective measures of senescence markers as a means to determine the influence of alcohol exposure and/or consumption on the trajectory of aging, further cementing a transition in focus from chronological aging to biological aging (e.g., Zillich et al. 2024).

A second significant challenge in evaluating the influence of alcohol consumption patterns across the lifespan on aging-related CNS function is the developmental period in which alcohol exposure was incurred. As discussed in Chap. 11 by Matthews and colleagues, developmental stage is a critical issue because alcohol exposure is not uniform across the lifespan, nor are the biological substrates on which alcohol acts. For instance, it is estimated that between 1 and 9% of individuals worldwide have been exposed to alcohol in utero, making fetal alcohol spectrum disorders (FASD) the leading cause of preventable intellectual disabilities (May et al. 2018). Prenatal development is undoubtedly a period during which the organism may be especially vulnerable to accelerated aging outcomes (e.g., Smith et al. 2022; Church et al. 1996) as well as neurodegenerative diseases (Araujo et al. 2021), with many of these effects involving long-lasting changes in microglial activity and other features of neuroimmune signaling later in life (Walter et al. 2023; Bake et al. 2023; Reid et al. 2015; Boots et al. 2023; Pinson et al. 2021; Gano et al. 2020). These conclusions are built upon a long history and strong foundation of preclinical and clinical studies elaborating hippocampal and cortical deficits associated with prenatal alcohol exposure (PAE) and form the basis for broader concerns about early alcohol effects on later cognitive function as a result.

Similarly, binge drinking peaks among adolescents and emerging adults, a developmental period marked by dynamic re-architecture of the CNS that involves substantial changes in glial cell activity and function (Doremus-Fitzwater and Deak 2022). For instance, synaptic pruning and paring mechanisms involve the active clearance of complement-tagged synapses through microglial phagocytic activity (Kopec et al. 2019; Germann et al. 2021), which thereby increases the strength of remaining synaptic contacts in an experience-dependent manner. Ongoing angiogenesis formalizes its delivery of blood perfusion to gross anatomical regions of the CNS (Rowan and Maxwell 1981; Ogunshola et al. 2000), while astrocytes envelop microcapillaries in a final sealing of the blood–brain barrier (BBB) (Molofsky and Deneen 2015). Similarly, progressive myelination of axons by oligodendrocytes during adolescence is critical to improvements in processing speed and connectivity across distal regions of the neuroaxis (Jessen et al. 2015; Rice and Barone 2000). The extent to which these processes are immediately responsive to acute alcohol consumption/exposure during adolescence is a somewhat open question for the field because most studies fail to include critical tests of alcohol action at multiple ages. However, the few studies comparing adolescent to adult neuroimmune sensitivity suggest that induction of neuroimmune genes is severely muted in adolescents (relative to adults), regardless of whether they have been challenged with alcohol, LPS, or an acute stress (Doremus-Fitzwater and Deak 2022; Doremus-Fitzwater et al. 2015; Marsland et al. 2022). Thus, it will be critical for the alcohol field to determine how and why the adolescent brain seems to show such categorically distinct sensitivity to alcohol relative to the mature adult brain.

Nevertheless, because binge drinking peaks during late adolescence and early adulthood, a considerable number of studies have examined long-lasting effects of adolescent alcohol exposure in preclinical models. By and large, studies suggest adolescent intermittent ethanol (AIE) substantially influences neuroimmune reactivity even after a protracted period of alcohol abstinence. For instance, a multi-day binge alcohol exposure in rodents sent microglia into a dystrophic state (Marshall et al. 2020; McClain et al. 2011; Marshall et al. 2016). Consistent with this, several studies have shown prolonged microglial activation after AIE, with at least some of these studies showing that the adverse effects of AIE were prevented by CSF1 inhibitors (Coleman et al. 2020), minocycline (Khan et al. 2023; Hu et al. 2020; Barnett et al. 2022), or other drugs with anti-inflammatory properties such as indomethacin (Pascual et al. 2007; Vetreno and Crews 2018; Vetreno et al. 2018; Macht et al. 2023; Monleón et al. 2020, 2022). Additionally, recent studies have shown sensitized neuroimmune gene induction in adult rats with a history of AIE (Vore et al. 2021) as well as a sensitized fever response to Poly I:C, a synthetic form of double-stranded RNA that mimics a viral infection (Gano et al. 2024). Additional findings on the neuroimmune consequences of adolescent alcohol are discussed in Chap. 9 by Macht and coworkers. These findings need to be kept in perspective, however, as other studies have shown reduced extracellular cytokine concentrations using large molecule microdialysis (Gano et al. 2019) and impaired fever responses provoked by LPS in AIE-exposed males (Telles et al. 2017; Cruz et al. 2020). Together, these studies paint a more complex portrait of neuroimmune reorganization after adolescent alcohol, with many of these effects persisting for 30 days or more after the final ethanol exposure.

Historically, studies examining patterns of alcohol intake across the lifespan have shown that for most individuals, alcohol intake (and binge drinking in particular) peaks in adolescents and emerging adults and then decreases gradually as individuals advance in age from adulthood into later life (Deak et al. 2022). However, some studies have found that rates of alcohol consumption among elderly individuals are on the rise and at a historic high (Grant et al. 2017; Blazer and Wu 2009; Breslow et al. 2017). Although the reasons for this shift toward greater alcohol consumption later in life remain unclear, several factors may contribute. For instance, social isolation tends to increase for many aging individuals as they lose friends/companions and become less mobile (Perkins et al. 2019; Perkins et al. 2016; Ravenel et al. 2024). Social isolation is especially challenging, and even rodents will increase alcohol intake during prolonged periods of social isolation (Karkhanis et al. 2016; McCool and Chappell 2009). As social occasions become less frequent, especially through the recent pandemic, older adults tend to consume more alcohol during such functions and adopt binge-like drinking behaviors (Kuerbis et al. 2014). Related to this, many elderly individuals experience painful bereavement and consume larger amounts of alcohol for its euphoric, relaxing, and anxiolytic effects (Kuerbis et al. 2014). Alternatively, it is possible that the rise in alcohol consumption among aged individuals reflects overall better health and longevity compared to previous decades. Indeed, both the overall lifespan and quality of life in later aging (termed the "healthspan") have increased substantially over the past five decades, perhaps setting the stage for alcohol consumption patterns to mirror what would previously have been observed at younger ages. Regardless, this troubling pattern of heightened alcohol intake among aging individuals requires additional studies to clarify the psychosocial factors that contribute.

As mentioned above, alcohol is a powerful toxicant, making increased alcohol consumption among older adults especially dangerous as alcohol produces many of the same neuroimmune deficits as aging (Carlson et al. 2023). In adult non-aged rodents, chronic ethanol has been shown to stimulate a pro-inflammatory state within the CNS by increasing levels of IL-1β, IL-6, and tumor necrosis factor alpha (TNF-α) (Qin et al. 2008; Lippai et al. 2013; Alfonso-Loeches et al. 2010). Similarly, elevated HMGB1 and TLR3 expression was observed in non-aged rodent brains following chronic ethanol exposure and in post-mortem brain tissue of individuals with alcohol use disorder (AUD) (Crews et al. 2013). Since these elevations in pro-inflammatory mediators in non-aged adults mirror the neuroimmune alterations seen in aging, it has been suggested that ethanol may produce a similar primed phenotype within microglia. For example, microglia isolated from non-aged rodents after a 4-day binge-like ethanol exposure displayed increased MHCII expression, a marker of activation, and similar findings have been reported in microglia isolated from aged rodents (Peng et al. 2017; Frank et al. 2006; Henry et al. 2009). Furthermore, mice exposed to chronic ethanol and then challenged with LPS showed exaggerated levels of brain IL-1β and TNF-α, which is reminiscent of the sensitized responses to LPS observed in the brains of aged rodents (Qin et al. 2008; Godbout et al. 2005; Henry et al. 2009). Beyond microglial priming, ethanol also mirrors age-related changes in the phagocytic activity of these cells as studies report that microglia isolated from mice exposed to chronic ethanol exhibit reduced expression of the phagocytic marker, CD68 (Lowe et al. 2020). Additionally, in vitro studies reported a decline in cultured microglia's ability to phagocytose amyloid beta after being exposed to ethanol for 24 h (Kalinin et al. 2018).

Though there is substantial evidence supporting the hypothesis that alcohol primes microglia and exaggerates immune responses, other studies have reported contrasting findings suggesting that alcohol may be better characterized as immunomodulatory, not purely neuroinflammatory. Specifically, studies have shown that relative to adult rats, adolescent rats display a blunted pro-inflammatory response within the CNS (IL-1β,

TNF-α) after an acute ethanol challenge (Doremus-Fitzwater et al. 2015). The same group exposed adult rats to multiple ethanol challenges and again observed a suppression of IL-1β and TNF-α in the CNS (Gano et al. 2016). Proteomic studies support this immunomodulatory role of alcohol as cultured microglia exposed to ethanol express more anti-inflammatory mediators compared to microglia exposed to LPS which primarily express pro-inflammatory signals (Guergues et al. 2020). These findings suggest that instead of priming microglia, alcohol may be causing these cells to become senescent as they are not producing the expected immune responses. For example, young rats given an acute ethanol challenge show increases in hippocampal IL-6 expression, whereas aged rats exhibited exaggerated basal levels of hippocampal IL-6 but no change after ethanol administration (Gano et al. 2017). A similar debate has occurred over the effects of alcohol on astrocytes as some studies report increased GFAP expression and astrocyte reactivity in the hippocampus and prefrontal cortex after chronic ethanol exposure (Evrard et al. 2006; Vongvatcharanon et al. 2010; Dalçik et al. 2009), while others report reduced GFAP expression in these regions which is more indicative of senescent cells (Franke 1995; Rintala et al. 2001). It is important to note that these contrasting studies used various alcohol exposure procedures, highlighting that consequences of alcohol on neuroimmune function differ greatly based on the age of consumption, route of administration, and the length of exposure (Deak and Savage 2019). Thus, more consistent studies are needed before the consequences of alcohol within the CNS and how they mirror or potentially accelerate age-related impairments can be fully understood.

Regardless of the mechanisms driving alcohol's deleterious actions in the CNS, aging can greatly increase an individual's alcohol sensitivity, making older adults more susceptible to its neurotoxic effects. This heightened sensitivity and lower tolerance to alcohol is likely due, at least in part, to age-related reductions in alcohol pharmacokinetics, including changes in body composition and ethanol metabolism. Aged individuals show reduced activity of acetaldehyde dehydrogenase and cytochrome P-4502E1, two major enzymes responsible for metabolizing alcohol (Meier and Seitz 2008). Additionally, aging is often associated with changes in body composition and lower water content, creating less volume for alcohol distribution (Vestal et al. 1977). As a result, the blood ethanol content (BEC) of an elderly individual will likely be larger than the BEC of a younger adult who consumed an equal amount of alcohol (Cederbaum 2012). This enhanced sensitivity increases the likelihood that elderly individuals will experience impaired motor coordination and greater confusion while intoxicated which can be extremely dangerous (Kalant 1998). Such impairments have been recapitulated in rodent studies that report greater ethanol-induced impairments in motor function and Morris water maze performance in aged rodents relative to adults (Ornelas et al. 2015; Perkins et al. 2018; Novier et al. 2013).

Interestingly, the cognitive impairments observed after ethanol exposure in rodents and humans closely resemble those seen in aging, which suggests that certain brain regions may be more sensitive to both. The hippocampus is one region that is vulnerable to age-related neurodegeneration as well as alcohol toxicity and both are associated with cognitive decline (Beresford et al. 2006; Golomb et al. 1993; Wilson et al. 2017; Jack et al. 2000). Though the hippocampus is the most widely studied, similar effects of aging and alcohol have been implicated in the frontal cortex, parietal cortex, temporal cortex, and the hypothalamus, which are also important mediators of cognitive function (Toledo Nunes et al. 2019; Harper and Kril 1989). These studies highlight an intersection of aging and alcohol, as any alcohol-induced impairments in these regions, whether a result of primed or immunosenescent cells, are likely amplified in the aged brain and vice versa. In sum, alcohol and aging have synergistic effects on brain health, especially neuroimmune function, which likely accelerates age-related cognitive decline. The following sections will discuss how this acceleration creates a sensitized environment that can rapidly progress into more severe neurodegenerative pathologies.

12.3 Developmental Alcohol Exposure and the Progression of Pathological Age-Related Cognitive Impairment

12.3.1 Alcohol Use Disorder (AUD) Is a Risk Factor for Pathological Aging

Excessive alcohol use over the lifespan can increase the risk for alcohol-related brain damage (ARBD) and cognitive decline. Over 70% of individuals with chronic AUD display some degree of brain pathology (Goldstein and Shelly 1980; Harper 1998), and many of the brain regions implicated in AUDs mirror those vulnerable to degeneration during advanced aging and Alzheimer's disease and related dementias (ADRD). This includes reductions in brain volume in several regions critical for cognition, including the frontal, temporal, parietal, cingulate, and insular cortices, cerebellum, thalamus, and hippocampus, and this loss can be more pronounced in adults with AUDs who are 65 years and older (Sullivan et al. 2018; Zahr et al. 2019).

Several cross-sectional human studies support the idea that chronic alcohol misuse alters normal aging as patients with AUD exhibit accelerated brain shrinkage by middle age, and other longitudinal studies revealed age–AUD neuropathological interactions in the frontal cortex and hippocampus (Zahr et al. 2019; Sullivan et al. 2018). Such studies have developed the hypothesis that heavy alcohol use (greater than 5 drinks per day) is associated with accelerated cognitive aging (Woods et al. 2016) and can also increase an individual's risk of developing Alzheimer's disease (AD) if they carry the ApoE e4 allele (Anttila et al. 2004; Kivipelto et al. 2008). Thus, AUD is an enduring, complex disease that continues to evolve during the aging process and likely exacerbates cognitive decline.

12.3.2 Mild Cognitive Impairment (MCI) as Pathological Aging

A diagnosis of mild cognitive impairment (MCI) is often made when cognitive difficulties are significantly greater than typical age-related cognitive decline, often over the age of 65. Both cross-sectional and quasi-longitudinal studies indicate modest declines in memory and reasoning abilities until about age 65, after which the decline accelerates (Salthouse 2019). However, there is individual variability in cognitive aging in humans and other species, and different domains are affected by aging. Specifically, aging has been associated with declines in processing speed, short-term memory, language, and visuospatial and executive functions (Harada et al. 2013).

The Mayo Clinic's Petersen/Winblad criteria for MCI include four aspects (Petersen et al. 2014; Winblad et al. 2004): (1) reported change in cognition; (2) impairment in one or more cognitive domains relative to a person's age and education; (3) spared independence in functional abilities; and (4) lack of dementia. In addition, MCI is classified into four subtypes: single-domain amnestic MCI, multi-domain amnestic MCI, single-domain non-amnestic MCI, and multi-domain non-amnestic MCI (Winblad et al. 2004). The amnestic subtypes are associated with higher risk of progression to AD, compared with non-amnestic subtypes.

Hippocampal volume has been identified as a significant predictor of cognitive decline in many cases of MCI (Mieling et al. 2023; Morrison et al. 2023; Richter et al. 2022). Recent research has extended this predictive capacity to basal forebrain volume, particularly in patients undergoing pharmacotherapy with cholinergic drugs to treat cognitive and memory dysfunction. Consistent with this, larger basal forebrain volume was associated with slower global cognitive decline, while greater left hippocampus volume was linked to slower memory decline.

In the transition from prodromal MCI to AD, the volume of the basal forebrain region has been found to predict cognitive decline (Richter et al. 2022; Tiernan et al. 2018). Basal forebrain structural changes seem to precede cortical atrophy in the progression of AD (Schmitz and Nathan Spreng 2016). Thus, alterations in basal forebrain structure serve as presymptomatic biomarkers for MCI and progression to AD (Richter et al. 2022; Tiernan et al. 2018; Nicolas et al. 2020). Dysregulation of basal forebrain circuitry in MCI and AD is not limited to cognitive decline;

it is also associated with dysregulation of the default mode network, which is critical for executive function and episodic memory (Nair et al. 2018). In addition, basal forebrain degeneration predicts AD-specific pathology. In vivo MRI studies have demonstrated a correlation between the degree of basal forebrain atrophy and amyloid-beta burden (Grothe et al. 2013; Kerbler et al. 2015). Notably, baseline volumes of the nucleus basalis of Meynert (NbM) predict progressive cortical degeneration and a trans-synaptic spread of amyloid-β that starts in the NbM (Kerbler et al. 2015; Schmitz and Nathan Spreng 2016). Furthermore, changes in cognition were associated with the development of pre-tangle markers in basal forebrain cholinergic neurons before frank tau deposition occurred throughout the brain (Vana et al. 2011). Such results suggest that cholinergic basal forebrain neurons are an early vulnerable target to AD-related pathology.

Lower cortical cholinergic innervation is also associated with cognitive decline and MCI (Xia et al. 2022). However, a higher educational level in MCI patients is linked to increased cholinergic activity, suggesting an initial compensatory effect, but it is unknown whether this prevails in later AD stages (Xia et al. 2022). Recent resting-state functional MRI studies have found reduced basal forebrain functional connectivity in several cholinergic projection sites (cortex, hippocampus, amygdala) during the early phases of AD (Li et al. 2017; Qi et al. 2021). NbM disconnectivity in early AD specifically targets cortical regions enriched with astrocytes/microglia cells and/or immune process-related genes (Ren et al. 2023).

In addition, there is a significant reduction in the number of adult-born neurons within the hippocampus in both MCI and AD patients, compared to normal age-matched controls (Salta et al. 2023). There is also a significant reduction in the number of neural progenitor cells in MCI and AD (Moreno-Jiménez et al. 2019; Tobin et al. 2019). Importantly, the number of new astrocytes increased in MCI and AD, compared to NCI, suggesting changes in the cell fate for newly born cells within the hippocampus during pathological aging (Ginsberg et al. 2019).

Thus, degeneration within the septohippocampal pathway appears critical to the transition from healthy aging toward MCI and ultimately in the development of AD. There are several factors such as genetics, sex, and the environment, including drug and alcohol exposure, which can delay or accelerate the progression of pathological aging (McQuail et al. 2020). Next, we will review the role of early developmental exposure to alcohol in the progression of pathological aging and the use of animal models to reveal critical behavioral changes and neurobiological mechanisms, with a focus on the septohippocampal circuit.

12.3.3 Aging and the Septohippocampal Circuit

Rodent models have shown that the septohippocampal circuit is vulnerable to both aging and chronic ethanol exposure, in particular AIE. It is important to note that age-related cognitive decline is not absolute; only a subset of aged rats display cognitive impairment as a consequence of natural aging (Fischer et al. 1989; Gage et al. 1989). In the rodent, cognitive deficits emerge in middle age (12–18 months), and impaired spatial memory begins to appear around 12 months of age, but cognitive performance can be highly variable (Guidi et al. 2014; Bizon et al. 2009). A longitudinal study also found impairments in episodic spatial memory using the water maze in middle-aged male rats (Febo et al. 2020).

Several studies have shown that hippocampal ACh efflux was reduced as a result of aging (Shao et al. 2019; Chang and Gold 2008; Stanley and Fadel 2012). Age-related declines in ACh efflux have been found as early as 10 months old (Chang and Gold 2008). Aged rats that are cognitively impaired also show a significant loss of cholinergic neurons in the MS/DB compared to young rats, but show no change in the pontine reticular cholinergic neurons, implicating the basal forebrain as a key structure in age-associated memory impairment (Baskerville et al. 2006). There is also evidence that the age-related loss of integrity of basal forebrain cholinergic neurons occurs to a greater extent in aged rats with impaired spatial learning (Sugaya et al. 1998). Studies have also reported age-related reductions in the density of cholinergic fibers within the dentate gyrus, which was more profound in aged male rats relative to aged female rats (Lukoyanov et al. 1999). In addition, cortical cholinergic innervation became less dense as rats aged, and this was further augmented when the expression of tropomyosin-related kinase A receptors was suppressed by viral vector-based RNA interference (Parikh et al. 2013). This suppression also decreased the ability of cholinergic neurons to release ACh, more so in aged-suppressed rats compared to aged or young controls (Parikh et al. 2013), suggesting that cholinergic transmission declines slowly over the lifespan.

Drugs that enhance ACh levels increase hippocampal adult neurogenesis (Kotani et al. 2006), and the survival of newly born neurons is regulated in part by nicotinic α7 nicotinic ACh receptors (AChR α7) (Kita et al. 2014). However, the neurogenic effect of AChRα7 may be limited to males (Otto and Yakel 2019). In a reciprocal fashion, neurogenesis appears to be critical for maintaining and stabilizing the cholinergic septohippocampal circuit, which is critical to successful spatial memory during normal aging (Kirshenbaum et al. 2023). Although hippocampal neurogenesis was decreased as rodents aged (as much as 80% by middle age) (Kuhn et al. 1996; Wu et al. 2023), there does not appear to be an association between decreased neurogenesis and degree of spatial memory impairment (Stepanichev et al. 2023). Although some studies reported that increased neurogenesis led to improvements in age-related cognitive decline (Marlatt et al. 2012), other studies reported that the decreased hippocampal neurogenesis that occurs with age did not predict the presence or severity of cognitive impairment (Bizon et al. 2004). Neurogenesis is not just involved in initial learning and pattern separation but is also a mechanism for ameliorating proactive memory interference (Akers et al. 2014; Scott et al. 2021), so this issue needs to be considered carefully in aging studies.

12.3.4 Adolescent Ethanol Exposure and the Septohippocampal Circuit

Heavy intermittent alcohol exposure during early adolescence is associated with persistent changes in brain structure and connectivity (Spear 2018). Two inter-related pathologies that have been consistently observed in rodent models of adolescent binge ethanol exposure (AIE) are a suppression of hippocampal neurogenesis (30–60%) (Macht et al. 2021; Vetreno and Crews 2015) and a decrease in the number of cholinergic basal forebrain neurons (25–30%) (Fernandez and Savage 2017; Vetreno et al. 2020). The loss of hippocampal neurogenesis and the reduction in cholinergic basal forebrain neurons seen following AIE were persistent into advanced adulthood (Reitz et al. 2021), but were not evident if the alcohol exposure occurred during adulthood (Broadwater et al. 2014; Vetreno et al. 2014). These findings suggest that the adolescent brain may be especially sensitive and/or vulnerable to long-lasting effects of AIE on cholinergic function. Despite the evidence surrounding the loss of forebrain cholinergic phenotype and the decrease in hippocampal neurogenesis following AIE, hippocampal-dependent spatial behavioral deficits are not consistently observed (Vetreno and Crews 2012; Swartzwelder et al. 2015; Macht et al. 2021). Indeed, there are certain cognitive tasks that are more sensitive to AIE than others (Crews et al. 2019).

12.3.5 Interactions Between Developmental Ethanol Exposure and Pathological Aging

Some studies have revealed that heavy but not light-to-moderate alcohol consumption increased the risk for dementia in humans after a diagnosis of MCI (Lao et al. 2021; Rehm et al. 2019; Xu et al. 2009). Thus, heavy alcohol use has emerged as a risk factor for progression from pathological aging to dementia. Recently, interest in the interactions between chronic alcohol exposure and pathological aging has intensified (White et al. 2023). This includes an interest in whether early developmental alcohol exposure alters the trajectory of healthy aging and animal models can be used to directly review the role of developmental alcohol exposure in the progression to pathological aging. An emerging hypothesis is that the septohippocampal pathway is a critical neural circuit that is especially sensitive to early developmental ethanol exposure and aging (Reitz et al. 2024).

An early study found that learning of place location was not impaired in young mice, but a deficit emerged when mice that underwent prenatal alcohol exposure (PAE) reached middle age (White et al. 2023). Another PAE study in male rats found spatial navigation deficits in the Morris water maze in both young and middle-aged animals (Gabriel et al. 2002). However, a progressive change in age-related cognitive decline (spontaneous alternation, novel object recognition, and fear conditioning) were not found in C57BL/6J mice exposed to PAE (Smith et al. 2022), and no aging effects were observed on these behaviors at 17 months of age. These data suggest that PAE may accelerate pathological aging, dependent on the PAE model and the behavioral paradigms used to assess cognitive function.

A few studies have examined the effects of adolescent ethanol exposure on successful cognitive aging. Intermittent alcohol exposure during adolescence (from PD 30 to PD 48) facilitated spatial memory impairments to acute ethanol challenges at 18 months of age (Matthews et al. 2017). Another study (Matthews et al. 2023) found that when rats exposed to AIE were tested for spatial memory at 19.5 months, male but not female rats exposed to AIE displayed heightened anxiety and significantly longer swim path lengths on several training days. In addition, changing the platform location to test behavioral flexibility following spatial learning revealed that AIE impaired performance in a sex-dependent manner (males not females) over age-matched controls. However, in a longitudinal design assessing cognition multiple times across 20 months following AIE, spatial memory in the water maze was not affected by alcohol, whereas reversal learning was impaired by AIE—but without alcohol–age-dependent interactions (Matthews et al. 2022). One concern with longitudinal studies is that repeated cognitive testing can serve to "boost" later cognitive performance as either a "learning to learn" or "practice effect" or a form of enrichment (Cnops et al. 2022; Zheng et al. 2022). This needs to be balanced with the benefits of longitudinal aging studies, which can better identify the onset of cognitive impairment and control for cohort effects (McQuail et al. 2020).

In a cross-sectional study, male and female rats underwent AIE and were tested as either young adults (4 months old) or at middle age (14 months old) on two hippocampal behavioral assays. These assays were concurrent with the assessment of behaviorally evoked Ach efflux and followed by histological assessment of cholinergic neurons in the medial septum-diagonal band and hippocampal neurogenesis (Reitz et al. 2021). We found that AIE-induced impairments on an object location task resembled age-related cognitive decline. Aging, regardless of AIE status, led to suppression of hippocampal cholinergic tone during the object location task. In contrast, hippocampal neurogenesis was sensitive to the synergistic interaction between AIE and aging, as young adult AIE rats had a higher number of doublecortin staining cells than middle-aged AIE rats. On a different hippocampal-dependent task, spontaneous alternation, we found a sex-dependent effect: Middle-aged males were impaired relative to young males, but middle-aged females performed similar to young females. As in previous studies (Kipp et al. 2021a; Fernandez and Savage 2017), AIE had no effect on spontaneous alternation. However, regardless of age, male AIE rats had lower behaviorally evoked ACh during spontaneous alternation, compared to male control rats: an effect not seen in female rats and not observed previously when only young males were assessed (Kipp et al. 2021a; Fernandez and Savage 2017). Middle age did not lead to the suppression of the cholinergic phenotype, which other studies have not observed until advanced age (20–24 months; Martínez-Serrano and Björklund 1998; Pitkin and Savage 2004). Thus, our findings support the hypothesis that developmental ethanol exposure may lead to accelerated age-related cognitive decline in spatial location memory and that this may be associated with reduced hippocampal neurogenesis and Ach deficits.

12.3.6 Conclusions: Developmental Alcohol Is a Factor That Drives Pathological Aging

Emerging evidence suggests that binge-like alcohol exposure during adolescence contributes to accelerated pathological aging—particularly in male rodents starting at middle age and progressing into advanced age. A dysfunctional septohippocampal system may make an organism more vulnerable to normal and pathological aging processes, and whether adaptive aging responses are able to compensate under such disease states is unknown (Gray and Barnes 2015). In contrast, in rodents with AD transgenes, the progression of cognitive decline and AD-associated pathology occurred more in females than males, raising important questions about mechanisms contributing to sex-specific vulnerabilities to alcohol, aging, and the progression of ADRD (more on this below).

In humans, there is evidence that basal forebrain pathology is critical for the transition from healthy aging to pathological aging, including the development of MCI and ultimately AD. Degeneration of cholinergic neurons in the basal forebrain is a hallmark of pathological aging, contributing to the reduction of cortical innervation and ACh activity—thereby reducing activity of cortical regions involved in memory and cognition. Basal forebrain volume predicts longitudinal limbic cortical degeneration and the spread of amyloid biomarkers in MCI and AD (Schmitz and Nathan Spreng 2016). Cholinergic neurons in the basal forebrain accumulate both intraneuronal tau and Aβ, and this becomes more profound with the transition to MCI (Schmitz et al. 2018).

A key pathology that arises from adolescent alcohol exposure is loss of the basal forebrain cholinergic phenotype, suppression of behaviorally activated ACh activity in the hippocampus and frontal cortex, and reduced hippocampal neurogenesis (Kipp et al. 2021b; Reitz et al. 2021). Given the overlap in septohippocampal pathology between pathological aging and AUD, it is critical to understand how these neurobiological changes associated with developmental exposure to alcohol drive the brain into pathological aging.

12.4 Thiamine Deficiency as a Driver of Adult Alcohol-Related Brain Damage and Memory Impairment

12.4.1 Thiamine Deficiency Related and Unrelated to Alcohol Use Disorders in Humans

Prolonged and excessive alcohol use has been related to different forms of dementia as well as structural and functional changes in the brain. One of the critical drivers of alcohol-related brain damage (ARBD) is thiamine deficiency (TD). Thiamine (vitamin B1) is an important nutrient required by all tissue with an essential role in the development and maintenance of brain function. During TD, the metabolism of lipids, glucose, and amino acids and syntheses of neurotransmitters—aminobutyric acid (GABA), ACh, and glutamate—are affected in both neurons and glial cells (Abdou and Hazell 2015; Gibson et al. 2016). These effects in the brain (human and experimental models) have been associated with development of ARBD (Butterworth 2003; Nardone et al. 2013). It is estimated that TD is prevalent in 15–80% of patients with AUD (Li et al. 2008; Morgan 1982). Alcohol-related dementia (ARD), a chronic AUD in the absence of TD, or other complicating factors, has a less distinct pathophysiological profile than TD (Ridley et al. 2013). Furthermore, it has been questioned whether ARD is a distinct neurocognitive disorder, or has a multifactorial etiopathology that includes TD, head injury, or liver disease (Palm et al. 2022). Given this, we will focus on AUD with TD. Among individuals with AUD, TD occurs due to a combination of poor dietary intake and a decrease in gastrointestinal absorption of thiamine. The high calories and low nutrients in alcoholic beverages require more thiamine to maintain metabolic function, affecting glucose metabolism by inhibiting phosphorylation of thiamine and its incorporation into enzymes leading to TD (Butterworth et al. 1993; Chandrakumar et al. 2018; Donnino et al. 2007; Harper 2006). TD is common in developing countries due to the prevalence of acute malnutrition (Attias et al. 2012); however, TD is also reported in developed countries and occurs as a result of many health conditions, including diseases associated with aging (e.g., Alzheimer's disease, Parkinson's disease), bariatric surgery, eating disorders, HIV, diabetes, and hyperemesis gravidarum (for review: Dhir et al. 2019).

If left untreated, TD can manifest into Wernicke's encephalopathy (WE) and if thiamine is not restored the chronic neurological sequela, Wernicke–Korsakoff syndrome (WKS), may develop (Nardone et al. 2013). The classic triad of signs and symptoms associated with WE are abnormal eye movements, gait ataxia, and cognitive impairment (Harper et al. 1986). Studies have shown that if WE patients are treated with thiamine before the development of significant brain damage, the associated cognitive dysfunctions may be reversible. However, irreversible lesions can develop and persist if TD continues (Isenberg-Grzeda et al. 2012; Ogershok et al. 2002), and the disorder can progress to a long-lasting amnestic state, Korsakoff's amnesic syndrome, or WKS. This progression happens in approximately 56–84% of individuals with AUD, with less frequency in cases unrelated to alcohol misuse (Arts et al. 2017; Harper 2006; Kopelman et al. 2009; Victor et al. 1971).

The primary cognitive signs of WKS include psychosis, confabulation, memory loss, and learning deficits (Marrs and Lonsdale 2021). Individuals with WKS have difficulty in forming new memories, as well as retaining or recalling new information, due to the profound anterograde amnestic state. Although the anterograde memory is usually more affected than retrograde memory in WKS patients, retrograde amnesia is observed with deficits in temporal gradient, where the memory impairments could extend retrospectively for up to 30 years (Arts et al. 2017; Kopelman et al. 1999, 2009; Victor et al. 1971).

The neural underpinnings of the profound amnesia in WKS have been studied in both post-mortem studies and imaging studies with damage to the mammillary bodies and the thalamic nuclei as key regions of interest (Harding et al. 2000; Mayes et al. 1988; Kopelman 1995; Pitel et al. 2012; Sullivan and Pfefferbaum 2009). Although shrinkage of frontal cortical structures (gray and white matter), hippocampus, and cerebellar cell loss is seen in WKS, it occurs to a lesser degree and with less frequency than diencephalic damage (Nunes et al. 2019; Sambon et al. 2021; Sullivan and Pfefferbaum 2009). Importantly, pathology of the thalamus has been considered one of the critical predictors of memory dysfunction in alcohol-related WKS (Harding et al. 2000; Pitel et al. 2012, 2015; for review: Savage et al. 2021; Sullivan and Pfefferbaum 2009). The thalamus plays an important role as a node of two networks that are implicated in AUD patients with WKS: The Papez circuit that subserves episodic memory and the frontocerebellar circuit involved in working memory and executive functions (Fadda and Rossetti 1998; Nunes et al. 2019; Pitel et al. 2011; Pitel et al. 2015). Segobin et al. (2019) showed that atrophy of different thalamic nuclei is dissociated in these two brain circuits where atrophy of the anterior thalamus was found uniquely impacted in WKS patients. Decreased connectivity between anterior thalamic nuclei and hippocampus was also observed in WKS (Segobin et al. 2019), likely a consequence of disruption of the fornix. This finding supports the hypothesis that amnesia in WKS is associated with a disrupted neural circuit involving medial temporal lobe and diencephalic regions (Nahum et al. 2015).

Investigating cognitive and brain changes in patients with WKS over months and up to 10 years after the diagnosis, Maillard and colleagues (Maillard et al. 2021) showed that even after years of alcohol abstinence, only a mild recovery of the volume in three brain regions of frontocerebellar circuit (cerebellum, pontine crossing tract, and middle cerebellar peduncle) was observed, and there was no significant improvement in cognitive performance (especially episodic memory). In addition, structural and metabolic alterations of the Papez circuit in WKS persisted: the thalamus, hypothalamus, and fornix were severely atrophied at all times (early, 1, and 10 years after diagnosis; Maillard et al. 2021). In contrast, studies have shown that in individuals with AUD, without TD, prolonged abstinence led to the recovery of memory and restoration of brain volume (Segobin et al. 2014; Pfefferbaum et al. 1995). These results emphasize that irreversible anterograde amnesia is observed in WKS from TD and not the alcohol damage per se (Arts et al. 2017; Maillard et al. 2021).

Considering that WE is underdiagnosed and the existence of high mortality rates of untreated patients (20%; (Harper et al. 1986)), thiamine supplementation has been considered as a treatment for suspected WE. It should also be considered as a treatment for AUD patients with or without diagnosis of WE, to avoid the progression to WE (Pruckner et al. 2019). Praharaj et al. (2021) also suggested that other neuropsychiatric syndromes associated with TD—in the context of AUD—such as alcohol cerebellar syndrome, Marchiafava–Bignami syndrome, alcoholic polyneuropathy, and delirium tremens should also receive thiamine replacement therapy (Praharaj et al. 2021).

Oral thiamine supplementation has also been suggested to prevent the development of WE in AUD as well as to improve alcohol-induced cognitive impairment (Dhir et al. 2019; Lagercrantz et al. 1986; Smith et al. 2021). Examining alcohol-dependent persons during routine inpatient detoxification with oral thiamine supplementation (oral dose of 200 mg thiamine/day, scheduled for 14 days), a recent study showed that thiamine replacement together with alcohol abstinence improved thiamine blood levels and cognitive function (Bonnet et al. 2023). Listabarth et al. (2023) also revealed an association between improvements in memory and thiamine replacement, showing that both oral and intravenous thiamine administration had equal efficiency in increasing the blood thiamine pyrophosphate levels (Listabarth et al. 2023). These findings suggest that thiamine supplementation should be implemented as a clinical practice to prevent cognitive impairment in AUD. Aside from the intravenous thiamine replacement therapy and oral supplements, there is no optimal treatment for AUD, WE, or WKS since the treatment varies depending on symptoms and severity (Alosco and Stern 2019). Furthermore, the thiamine dose, duration of treatment, and type or frequency of administration are still unclear (Isenberg-Grzeda et al. 2012; Latt and Dore 2014; Pruckner et al. 2019).

12.4.2 Developmental Thiamine Is Rare and Understudied

As described earlier, TD is frequently associated with alcohol consumption; however, TD can also occur during pregnancy and in the months following parturition (Fattal et al. 2011; Guerrini et al. 2007). Pregnant women are at risk of TD development when there is severe malnutrition (especially reported in low- and middle-income countries), excessive alcohol use, or hyperemesis gravidarum. In a recent review, Kareem et al. (2023) presented a broad range of clinical manifestations of TD during pregnancy and their consequences to the mother and fetus, e.g.: subclinical TD, wet beriberi, dry beriberi (WE and WKS), and infantile beriberi (Kareem et al. 2023).

Scattered clinical cases have reported WE's symptoms and the risk of the WKS development in pregnant and lactating women due to hyperemesis gravidarum, characterized by persistent severe nausea and vomiting during pregnancy that causes a rapid loss of thiamine (Hillbom et al. 1999; Ismail and Kenny 2007; Oudman et al. 2019; Rane et al. 2022). Fetal and maternal mortality was reported in half of the patients diagnosed with WE from hyperemesis gravidarum; however, there was not a relation between fetus survival and thiamine dose used to treat WE symptoms (Oudman et al. 2019). Breastfed infants of TD mothers are at highest risk for developing TD, resulting in infantile beriberi. If left untreated, infantile beriberi can result in death or the emergence of neurodevelopmental problems later in life (Allen 2012; Mimouni-Bloch et al. 2014). For instance, in 2003, several infants were hospitalized in Israel with severe neurological symptoms, prolonged vomiting, nystagmus, seizures, and coma. After investigation, they were diagnosed with infantile TD, which was caused by a non-dairy, soy-based infant formula that was not supplemented with thiamine. These TD infants were responsive to thiamine treatment; however, years later, some children who had infantile TD had persistent cognitive and language impairments (Fattal-Valevski et al. 2005, 2009; Fattal et al. 2011).

In 1990, a clinical study showed that mothers with fetuses with severe intrauterine growth retardation exhibited lower blood cell thiamine content compared to women with normal pregnancies, suggesting that thiamine supplementation should be included during pregnancy (Heinze and Weber 1990). The thiamine-depleting effects of alcohol are expected to result in adverse health outcomes since thiamine levels are crucial for neurodevelopment and function, and alcohol is a potent teratogen in humans (Bâ 2017). Indeed, the high incidence of intrauterine growth retardation and other abnormalities observed in children born to alcohol-consuming mothers in the 1960s was later characterized as "fetal alcohol spectrum—FAS" (Butterworth 1993). Although both TD and early developmental alcohol exposure separately cause FAS characteristics such as intrauterine growth retardation, microcephaly, and language impairments, the lack of developmental data on TD-alcohol synergism during pregnancy and lactation in humans and how it can interfere in fetal brain development in human population is still unresolved (Bâ 2011, 2017; for review see: Kloss et al. 2018).

Experiments with TD in female rats during pregnancy support the idea that adequate intake of thiamine during pregnancy is essential for successful development of offspring. Maternal thiamine deficiency induces delayed development of the fetus, decreased brain weight, myelination, and neurochemical changes in the CNS (Fournier and Butterworth 1990; Roecklein et al. 1985; Trostler et al. 1977). Additionally, de Freitas-Silva et al. (2010) have shown that maternal thiamine restriction in the perinatal period induced spatial learning deficits that were observed in peri-adolescent but not adult rats (de Freitas-Silva et al. 2010). Extensive work with different patterns of maternal TD, with or without chronic ethanol intake, supports cellular differentiation as the critical period for alcohol–thiamine synergism (Bâ 2009, 2011). This synergism also provokes extensive cellular death and tissue necrosis related to FAS. Thus, understanding how apoptotic mechanisms compromise the trajectory of neurodevelopment and brain aging in cases of ethanol exposure and TD will be crucial for developing therapeutic strategies for FAS treatment (Bâ 2017).

12.4.3 Animal Models for Thiamine Deficiency and Pathological Aging

Rodent models of thiamine deficiency have been used to reproduce neurological, neuropathological, and neurochemical changes described in WE and WKS patients. TD is induced by feeding adult rats/mice with TD diet in combination with *i.p.* injections of pyrithiamine, a thiamine pyrophosphokinase inhibitor that accelerates thiamine depletion (review: Savage et al. 2012; Vetreno et al. 2012). As such, pyrithiamine-induced thiamine depletion (PTD) provides a useful and highly tractable model for examining the neurological consequences of prolonged thiamine depletion in preclinical (rodent) models. A few studies have also used a prolonged ethanol consumption model with or without PTD to understand better how the treatments interact and the independent effects of each treatment (review: Nunes et al. 2019; Vetreno et al. 2011).

Over the years, studies have shown that TD in rodents using the PTD model can affect the expression of genes/proteins related to energy metabolism, neuroinflammation, neurotransmitter synthesis (cholinergic, GABAergic, glutamatergic systems), and myelin production in the brain. Additionally, PTD increased oxidative stress and mitochondrial dysfunction, contributing to neurological symptoms like impaired cognitive function and motor incoordination (Liu et al. 2017; Nardone et al. 2013; Nunes et al. 2018, 2019; Vetreno et al. 2012). Given that molecular and behavioral changes related to TD have been observed in many aging-related neurodegenerative diseases, such as Alzheimer's disease and Parkinson's disease, animal models of TD are also being used in research of degenerative processes associated with aging (Ke and Gibson 2004; Nunes et al. 2019).

Significant changes in the thalamus observed in WKS patients have been confirmed by histological and immunohistochemical assessment in brains of PTD rats, with neuronal loss (44–83%) in many thalamic nuclei and thalamic mass loss (20–30%) (Anzalone et al. 2010; Hall and Savage 2016; Kipp et al. 2021b; Savage et al. 2021; Vemuganti et al. 2006). Although most studies using the PTD model have been conducted in young male rats (3 months), studies have shown that aging can potentiate the neuropathology associated with TD—where thalamic lesions were more prevalent in middle-aged (10 months) and aged (22–23 months) rats exposed to PTD treatment compared to young rats of 2–3 months (Pitkin and Savage 2001, 2004). In addition, studies have shown that thalamic shrinkage/lesions were prominent in middle-aged rats (8–10 months) after months of recovery from PTD treatment (Kipp et al. 2021b; Vedder et al. 2015). Furthermore, middle-aged rats (9–10 months) treated with PTD in combination with prolonged chronic ethanol consumption also displayed thalamic shrinkage (Kipp et al. 2021b). However, chronic ethanol alone did not result in thalamic lesions (Kipp et al. 2021b; Vedder et al. 2015). The induction of thalamic lesions and the severity of the lesions may be explained by the possible contribution of rapid neuroimmune changes observed in response to TD. Our previous study showed that PTD treatment led to a profound increase of neuroinflammation markers within the thalamus of middle-aged rats (9 months) while chronic ethanol exposure showed small fluctuations in inflammatory genes regardless of brain region examined. In addition, fluctuations in neuroimmune genes varied as a function of vulnerability of brain regions by TD where the thalamus showed distinct and rapid neuroinflammation profile during PTD treatment compared with other two regions of interest in ARBD (hippocampus and frontal cortex). The neuroimmune gene induction also varied significantly as a function of stage of TD and time of recovery (Nunes et al. 2019). These results suggested that the severity of TD, the order of the TD that co-occurs with ethanol exposure, the age of the animals, and the mechanisms that contribute to ARBD, such as neuroinflammation, are important determinants for the extent of thalamic pathology associated with AUD (Nunes et al. 2019; Savage et al. 2021).

There is a close relationship between thalamic pathology, learning and memory impairments, and loss of cholinergic neurons in the medial septum/diagonal band (MS/DB) in the PTD rodent model (Hall and Savage 2016; Langlais and Savage 1995; Nardone et al. 2013). However, aging appears to cause greater dysfunction in the MS/DB cholinergic system following PTD treatment; this neuropathology was intensified by the duration of PTD treatment where aged PTD rats exhibited the greatest decline in cholinergic cell numbers with increasing duration of PTD treatment (Pitkin and Savage 2001, 2004). In addition, thalamic pathology found in rats co-exposed to PTD during chronic ethanol (over 9–10 months) correlated with suppression of ACh levels in the frontal cortex measured through in vivo microdialysis during spontaneous alternation but did not correlate with ACh levels in the hippocampus (Kipp et al. 2021b). These findings suggest that aging with TD and chronic ethanol disrupted the thalamocortical circuits to the greatest degree. In order to explore conditions that contribute to the development WKS and the mechanisms underlying the pathological changes, recent studies have explored the effects of TD and chronic ethanol consumption alone or combined on behavior tasks dependent on the frontal cortex, cerebellum, and hippocampus. In a study by Moya and colleagues (Moya et al. 2022), behavioral tests were performed during TD and at the end of EtOH exposure, without a withdrawal period. Hyperactivity and deficits in recognition memory were observed in rats treated with EtOH independent of TD; however, spatial memory was not affected by any treatments. The combination of chronic ethanol and PTD resulted in disinhibited-like behavior evidenced by more time spent in the inner zone of an open-field arena and in the open arm of the elevated plus maze, suggestive of heightened anxiety. The measures used on the disinhibition tests were correlated with changes in markers of lipid peroxidation (4-hydroxynonenal), apoptosis (caspase 9), and cell damage (HSP70 and HMGB1) in the frontal cortex, as well as a marker of nitrosative stress (nitrite) in the plasma. These changes suggest a synergism between chronic ethanol and PTD aggravates the molecular and behavior changes dependent on the frontal cortex in middle-aged rats (Moya et al. 2022).

A recent study (Kipp et al. 2021b), which included both sexes, also found spatial working memory impairments associated with chronic ethanol with or without PTD in male and female rats as they approached middle age (10 months of age). The memory deficit was correlated with reductions in acetylcholine efflux in the prefrontal cortex and hippocampus. The combination of TD and ethanol appeared to affect frontal cortical activity, assessed by acetylcholine efflux, and decision times during attention set shifting. Although there was no sex difference on behavioral and neurochemical disruption, female rats displayed similar impairments even with lower BECs than males, suggesting that female rats may become more sensitive to ethanol toxicity as they advance into middle age.

In summary, data across several studies confirm that TD causes severe neurobehavioral impairments, including deficits in spatial and working memory, and behavioral inflexibility when compared with ethanol exposure alone. Consistent with this, chronic alcohol appears to exacerbate the effects of TD across the lifespan on behavioral deficits and suppression of frontal cortical activity. Overall, chronic ethanol consumption, TD, or both conditions together damage critical neural circuits, but to different degrees, and these differences can be modulated by the age of the animals, severity of treatment, ethanol concentration, timing, recovery, and sex.

12.4.4 Conclusions: Similarity Between WKS and Dementia

AUDs are associated with an elevated risk of all types of dementia (Rehm et al. 2019). Some studies have used the term alcohol-related brain damage to include neurocognitive disorders related to chronic alcohol use, including WKS and ARD. Similar to WKS, ARD seems to be a direct result of TD and possibly ethanol neurotoxicity, suggesting that ARD has a multifactorial etiopathology (Arts et al. 2017; Ridley et al. 2013). Although evidence suggesting that the two disorders have overlapping clinical symptoms, such as peripheral neuropathology and ataxia, the cognitive profile of ARD entails impairments in visuospatial function, memory, and executive tasks, whereas WKS patients show deficits on executive tasks in conjunction with memory deficits (Ridley et al. 2013; Smith and Atkinson 1995). A recent study showed that both disorders are more prevalent in men (Palm et al. 2022). In addition, WKS most commonly is first diagnosed in people aged 50–59 years, in contrast to ARD that occurs commonly in people aged 70–79 years. Both diseases tend to be diagnosed at a younger age (middle age to middle old) relative to other progressive neurocognitive disorders (Palm et al. 2022).

In summary, WKS and AD are two major neurocognitive disorders that lead to memory impairment. Diencephalic amnesia is most described in WKS patients (Segobin and Pitel 2021). In contrast, AD shows a medial temporal lobe amnesia, with hippocampal atrophy at the early stage and progressively extending to neocortical areas (Braak and Braak 1991). However, both disorders damaged to the same extent the anterior thalamic nuclei, cingulate cortex, and hippocampus (only in moderate AD), with the mediodorsal thalamic nuclei and mammillary bodies more severely damaged in WKS than AD (Segobin et al. 2023). These findings reinforce the importance of examining brain networks involved in memory function and how they are disrupted in these neurocognitive disorders.

According to Gibson et al. (2022), WKS and AD have similarities in clinical manifestations and molecular mechanisms, and both diseases have TD as a common factor. Neuronal loss, increased neuroinflammation, cholinergic dysfunction, alteration of neurofilaments, and exacerbation of amyloid plaques are common features of TD in rodents (review: Gibson et al. 2022; Hazell et al. 2001; Ke and Gibson 2004; Kopelman 1991) with aggravation of these effects in transgenic models of AD (Calingasan et al. 1996; Karuppagounder et al. 2009). In conclusion, thiamine deficiency, chronic ethanol consumption, and aging are key factors that increase susceptibility to ARBD and other dementias. Alone or combined, thiamine deficiency and ethanol become increasingly relevant as individuals age, causing serious health issues, including cognitive impairments and dementia, which appear to be more pronounced with aging. However, future research is needed to better understand the genetic, molecular, and neurobehavioral changes that emerge across the lifespan due to natural aging, TD, and chronic ethanol consumption.

12.5 Alcohol Use Disorder as a Risk for AD and Related Dementias

AD is the most common form of dementia, affecting ~30 million people worldwide (Holtzman et al. 2011). It is characterized by the aggregation of extracellular amyloid plaques, the accumulation of intracellular neurofibrillary tau tangles, and neurodegeneration, which begins to accumulate 15–20 years before the onset of clinical symptoms (Jack et al. 2010). Thus, it is important to identify factors that can reduce or prolong this presymptomatic period. While a few studies suggest that low-to-moderate ethanol consumption may reduce the risk of AD (Koch et al. 2019; Luchsinger et al. 2004), several epidemiological studies have identified alcohol use disorder (AUD) as a risk factor for AD and AD-related pathology (Harwood et al. 2010; Rehm et al. 2019; Schwarzinger et al. 2018). Preclinical studies provide additional evidence that chronic ethanol exposure drives AD-related pathology; however, the biological mechanisms linking the two conditions are poorly understood. In the following sections, we will review the current literature describing the relationship between AUD and AD. We will begin by describing the pathological biomarkers associated with AD, namely, amyloid-β (Aβ) and tau. We will then review clinical and preclinical studies characterizing how alcohol impacts AD pathology in humans and transgenic animal models. Lastly, we will review potential mechanisms by which chronic alcohol misuse promotes AD pathology.

12.5.1 AD-Related Pathological Markers

Amyloid plaques are primarily comprised of Aβ, a post-translational cleavage product of amyloid precursor protein (APP). APP is a transmembrane cell surface protein that is broken down through competing pathways that produce pathologically inert substrates (non-amyloidogenic) or Aβ (amyloidogenic) (Haass et al. 2012). In the non-amyloidogenic pathway, APP is cleaved midway through the Aβ domain by an α-secretase enzyme, which produces a truncated APP C-terminal fragment-α (CTF-α). CTF-α is subsequently broken down by γ-secretase (Haass et al. 2012). In the amyloidogenic pathway, APP is cleaved by a β-secretase enzyme at the N-terminus end of the Aβ domain, producing a CTF-β peptide. CTF-β is then processed by γ-secretase to produce Aβ peptide. Aβ40 and Aβ42 are the most abundant Aβ products, with this pathway favoring Aβ40 production over Aβ42 (Citron et al. 1992; Haass et al. 2012). Of the two, Aβ42 is more aggregate-prone than Aβ40 and CSF Aβ42 levels decrease with AD progression, indicating increased deposition in the brain; CSF Aβ40 levels are unchanged with AD progression (Blennow et al. 2015). In fact, a reduced CSF Aβ42/40 ratio is a biomarker of AD progression and is a predictor of elevated phosphorylated tau levels (Wiltfang et al. 2007; Hansson et al. 2007). In neurons, Aβ is produced in endosomes and released at the synapses in an activity-dependent manner (Bero et al. 2011; Cirrito et al. 2008, 2005; Hettinger et al. 2018; Verges et al. 2011). Once in the extracellular space Aβ forms oligomers and aggregates into extracellular plaques in a concentration-dependent manner (Yan et al. 2009). While the severity of Aβ pathology does not correspond to cognitive decline, plaque pathology precedes tau pathology and neurodegeneration by several years (Jack et al. 2010; Musiek and Holtzman 2015). Tau protein is primarily expressed in neurons, where it binds to tubulin and functions to stabilize microtubules and helps regulate axonal transport (Wang and Mandelkow 2016). Tau contains 85 phosphorylation sites and is phosphorylated by several kinases (e.g., GSK3, Cdk5, MAPK, PKA, CaMKII) (Guo et al. 2017). Normal phosphorylation of tau regulates its distribution and function along the axon. However, tau hyperphosphorylation decreases its binding affinity to microtubules, subsequently promoting aggregation into paired helical filaments and neurofibrillary tangles (Castellani and Perry 2019). Tau pathology is a better predictor of cognitive impairment than Aβ and correlates with cognitive decline and neurodegeneration (Giannakopoulos et al. 2003; Nelson et al. 2012).

In AD, tau pathology begins to accumulate in the entorhinal cortex, locus coeruleus, and medial temporal lobes (Beardmore et al. 2021; Crary et al. 2014). As AD progresses, pathological tau propagates trans-synaptically in a prion-like manner eventually spreading throughout the neocortex (Guo and Lee 2011). Recent studies have provided evidence that basal forebrain degradation precedes neurodegeneration in the entorhinal cortex. Longitudinal neuroimaging data showed that degradation in the nucleus basalis precedes neurodegeneration in the entorhinal cortex, and this degradation is exacerbated by the presence of Aβ and tau (Fernández-Cabello et al. 2020; Schmitz and Nathan Spreng 2016). According to the A/T/N (amyloid/tau/neurodegeneration) framework model of AD, Aβ, tau, and neurodegeneration are all required for a diagnosis of AD. Therefore, we will discuss how alcohol drives these pathological biomarkers in humans and rodent models of pathology in the following sections.

12.5.2 Human Studies: The Role of AUD as a Risk Factor for AD

While epidemiological studies have begun to identify AUD as a risk factor for AD, there are conflicting studies on the degree to which alcohol use and misuse impact dementia, and the mechanisms by which it promotes cognitive decline. A 2004 study conducted on the Washington Heights-Inwood Columbia Aging Project cohort, which included Medicare beneficiaries aged 65 and over, reported that light and moderate alcohol intake (1 serving/month to 3 servings/day) was associated with a lower risk of dementia in individuals without the ApoE4 allele (Luchsinger et al. 2004). Similarly, a meta-analysis evaluating alcohol intake and dementia risk from 15 epidemiological studies conducted across multiple countries found that moderate drinkers (<40 g/day) had a lower dementia risk than individuals who abstained from alcohol (Mewton et al. 2023). One study using Pittsburgh Compound-B ([^{11}C]PiB) PET imaging compared amyloid deposition and cortical thickness in healthy middle-aged adults (40–65 years old, $n = 20$), and those with a history of AUD (40–65 years old, $n = 19$). The researchers hypothesized that individuals with a history of AUD would show greater Aβ deposition; however, Aβ was not detected in either group (Flanigan et al. 2021). Despite this, the study reported that individuals with AUD showed reduced cortical thickness in the inferior temporal gyrus, middle temporal gyrus, and fusiform gyrus, as well as reduced gray matter volumes in the hippocampus (Flanigan et al. 2021). An MRI study conducted in 2021 reported that individuals with AUD showed smaller gray matter volume in the sensorimotor complex and hippocampal formation (Zhornitsky et al. 2021). While reduced cortical thickness and gray matter in these areas is associated with AD progression, it is typically preceded by the presence of amyloid and tau pathology (Dickerson et al. 2009; Mattsson et al. 2014). Thus, alcohol misuse may promote AD-related neurodegeneration before the onset of amyloid and tau pathology.

Conversely, several epidemiological studies have reported a stronger relationship between heavy alcohol misuse and dementia or AD. A nationwide cohort study conducted in France from 2008 to 2013 reported men and women with a history of AUD had a threefold increased risk of dementia, while 16.5% of men and 4% of women with dementia had a history of AUD (Schwarzinger et al. 2018). Another nationwide cohort study from Finland identified alcohol misuse as the strongest modifiable risk factor for dementia and found that it was associated with a 2.2-fold increased risk for vascular dementia and a 1.6-fold increased risk for AD (Kauko et al. 2024). In the DELIVER cohort (Decoding the Epidemiology of LIVER disease in Sweden; 1987–2020), individuals with AUD had a 4.6-fold increased risk of dementia (Zhao et al. 2023). Finally, in the Whitehall II study, 35–55-year-old participants were recruited in London between 1985 and 1988. Alcohol consumption was frequently assessed (8 times between 1985 and 2016) and dementia outcomes were reported through national databases until 2017. This study reported an interesting relationship between dementia outcomes and alcohol abstinence, moderate alcohol use (1–14 units/week), and heavy alcohol use (>14 units/week). The association between alcohol use and dementia appeared to follow a U-shaped curve, with moderate alcohol use conferring a lower risk for dementia than abstinence or heavy alcohol use (Sabia et al. 2018). This study links conflicting studies discussed above, demonstrating that moderate alcohol consumption may be protective against AD and dementia, whereas heavy alcohol misuse may exacerbate dementia. Collectively, these epidemiological, population cohort, and clinical studies have identified alcohol misuse as a novel risk factor for AD and AD-related dementias and have laid the foundation for a new field of research.

It is important to note, however, that these approaches have a few important limitations. First, many of these studies rely on self-reporting for alcohol intake assessment, which could be unreliable in some cases. Furthermore, individual differences in biological variables that impact ethanol metabolism (e.g., sex/gender, height, age, body composition, genetics) could influence AD risk and progression. Next, many of these studies did not report the diagnostic criteria with which an AD diagnosis was made. Thus, it is unknown whether individuals in these studies identified as AD patients truly met the appropriate diagnostic standard (i.e., A/T/N framework). Lastly, studies using human subjects are limited in their ability to manipulate environmental or genetic factors and their ability to characterize clinical and pathological outcomes in individuals. Thus, preclinical studies using animal models can provide further details on the genetic and environmental factors that drive AD-related pathology under the influence of alcohol. In turn, these studies can identify treatments or interventions that will limit the risk of AD in AUD patients.

12.5.3 Preclinical Evidence for AUD as a Driver of AD-Related Pathology

As discussed in Chap. 10 by Anton and colleagues, preclinical studies have begun to characterize and explore the mechanistic links between alcohol exposure and AD-related pathology. In one study, adult 3-month-old 3xTg-AD mice (Table 12.1) were exposed to voluntary ethanol intake for 4 months and then euthanized 1 month after ethanol cessation. Ethanol exposure was associated with increased Aβ42 levels in the lateral entorhinal cortex and prefrontal cortex (Hoffman et al. 2019). Tau levels were also increased in the lateral entorhinal cortex, medial prefrontal cortex, and amygdala, and there was increased p-tau in the CA1 region of the hippocampus (Hoffman et al. 2019). In another study, 10 weeks of ethanol exposure via a moderate two-bottle choice drinking paradigm, 8-month-old APP/PS1 mice (Table 12.1) showed an increased number of plaques in the hippocampus. Interestingly, there were a greater number of smaller plaques in the hippocampus and cortex in ethanol-treated mice, suggesting that ethanol may be either limiting plaque growth or promoting greater plaque proliferation. Despite this increased Aβ deposition and proliferation, ethanol exposure did not increase APP, CTF-β, or BACE-1 protein levels (Day et al. 2023). Ethanol-treated APP/PS1 mice also had reduced brain mass compared to APP/PS1 controls, indicating that long-term ethanol exposure may promote neurodegeneration in the presence of Aβ overexpression (Day et al. 2023). In another study, 5-month-old 3xTg-AD mice were treated with 5 g/kg of ethanol 5 days/week for 3 months and then aged to 14 months without any additional ethanol. Western blot experiments showed that hippocampal Aβ42 levels were increased in ethanol-exposed female mice, but not males. Ethanol exposure also increased cortical and hippocampal t-tau as well as cortical p-tau levels, but only in female mice (Tucker et al. 2022). Collectively, these studies demonstrate that ethanol exposure during adulthood can exacerbate amyloid burden, tau pathology, and neurodegeneration.

Adolescence is an especially vulnerable period of neurodevelopment and binge-like ethanol exposure during this period can have long-lasting neurobehavioral consequences (Crews et al. 2019; Spear 2018), which may extend to an increased risk for AD and AD-related pathology. Six-week-old APP23/PS45 mice (Table 12.1) were exposed to ethanol via a drinking-in-the dark paradigm for 4 weeks (4 h of ethanol exposure during their dark cycle). Mice were euthanized after two days of withdrawal and brains were homogenized for Western blot experiments. In these mice, ethanol exposure increased APP, BACE-1, and Aβ40 and Aβ42 protein levels, which translated to an increased number of amyloid plaques in the hippocampus (Huang et al. 2018). In another study, adolescent 3xTg-AD mice were exposed to an adolescent intermittent ethanol (AIE) exposure paradigm for 30 days (5 g/kg; 2 days on, 2 days off; P25–P55), and pathology was assessed when mice were 6 months old. AIE-exposed mice had greater intracellular Aβ42 staining in the subiculum, entorhinal cortex, and amygdala (Barnett et al. 2022). Western blot experiments also showed that AIE-treated mice had increased p-tau-181 in the hippocampus, but not in the cortex (Barnett et al. 2022). Lastly, adolescent ethanol exposure also led to increased hippocampal Aβ42 protein levels at 6 and 12 months of age (Ledesma et al. 2021).

There is also evidence that ethanol exposure during the adolescent period accelerated the appearance of cognitive decline in AD models. In female 3xTg-AD mice, AIE did not induce impairments in novel object memory, but did reduce recall of spatial memory in the Morris water maze and decreased exploration (Barnett et al. 2022). In the APP/PSEN model, AIE dramatically increased time to reach the goal box and errors made in the Hebb maze selectively at middle age (Ledesma et al. 2021). Prenatal alcohol exposure (PAE) may also accelerate AD-related cognitive decline. 3xTg-AD mice exposed to a PAE paradigm showed spatial memory impairments and behavioral inflexibility at 4 months of age, compared to unexposed 3xTg-AD mice. Interestingly, PAE and unexposed 3xTg-AD mice showed similar behavioral deficits by 6 months of age (Tousley et al. 2023). These cognitive deficits also corresponded with reduced GABAergic interneuron function. PAE-treated 3xTg-AD mice showed deceased spontaneous inhibitory postsynaptic current (sIPSC) frequency which corre-

Table 12.1 Selective list of transgenic rodent models of AD-related pathology used in alcohol studies described in this chapter. Included are the model names, strain background, species, and types of AD-related pathology expressed

Model	Strain, species	Amyloid pathology	Tau pathology	References
3xTg-AD	B6;129, mouse	Yes	Yes	Oddo et al. (2003)
APP23/PS45	C57BL/6, mouse	Yes	No	Huang et al. (2018)
APP/PS1	C57BL/6J, mouse B6C3, mouse	Yes	No	Jankowsky et al. (2004)
Tg2576	B6SJL, mouse	Yes	No	Hsiao et al. (1996)
P301S	C57BL/6J, mouse	No	Yes	Yoshiyama et al. (2007)
Tg-F344-AD	Fischer 344, rat	Yes	Yes	Cohen et al. (2013)

sponded with reduced PV+ GABAergic interneurons in the mPFC at 4 months of age, indicating a potential hyperexcitability phenotype (Tousley et al. 2023). In the same study, 4-month-old PAE 3xTg-AD mice showed elevated Aβ immunofluorescence in the medial prefrontal cortex (mPFC); however, these particular results should be viewed skeptically given a small *n* ($n = 3$ per group), as the findings are likely underpowered (Tousley et al. 2023). Future studies examining how PAE drives AD-related pathology should seek to expand these findings. Collectively, these emerging studies demonstrate that ethanol has a clear impact on Aβ deposition, and identify alcohol misuse as an early, modifiable risk factor for AD. Importantly, these studies also provide evidence that chronic ethanol exposure during multiple phases of life (i.e., prenatal, adolescence, early adulthood, middle adulthood) leads to increased AD-related pathology during late adulthood. Despite this growing body of evidence clearly showing a relationship between chronic ethanol exposure and AD-related pathology, the mechanisms linking the two remain unknown. The data suggest that alcohol exposure significantly influences the trajectory of AD-like brain pathology, effects that appear to compound with age. Therefore, any potential mechanism must be one that is both persistent after the cessation of ethanol and one that drives AD-related pathology. Thus, the following section will discuss two potential mechanisms that preclinical studies are currently focused on neuroinflammation and neuronal hyperexcitability.

12.5.4 Potential Mechanisms Driving AD-Related Pathology

12.5.4.1 Neuroinflammation

A few studies have begun to investigate how ethanol exposure impacts microglia and neuroinflammation in the context of AD-related pathology. In one study, Aβ42 phagocytosis was reduced in rat microglia cultures after ethanol exposure in vitro (Kalinin et al. 2018). Conversely, another study showed that ethanol exposure may promote Aβ phagocytosis when ethanol consumption occurred between the ages of 10–14 months in F344 rats. Specifically, immunofluorescent labeling showed that iba1+ cells colocalized with Aβ42 in female, but not male, F344 rats who consumed 10% ethanol following a 2-day on/2 day off, single-bottle procedure for 6 months (Marsland et al. 2023). These conflicting results may be due to differences in models (cell culture vs animal model) or ethanol exposure length (acute vs chronic) but are certainly in accord with human studies showing that patterns of drinking (moderate vs heavy consumption) may produce opposite outcomes (Livingston et al., 2020). Thus, further studies are needed to better understand how ethanol influences Aβ phagocytosis. Despite these differences there appears to be consensus on the long-term impact of chronic ethanol exposure on neuroinflammation. In one study, 3xTg-AD mice were exposed to a binge-like ethanol exposure paradigm (5 g/kg; 2 days on/2 days off) for 30 days during adolescence (P25–P55) and then aged to 6.5 months without any additional ethanol. Ethanol-exposed mice showed increased levels of the pro-inflammatory cytokines INFα, IL1β, MCP1, TNF-α, and TLR4. Interestingly, cytokine levels positively correlated with Aβ42 and p-tau-181 protein levels (Barnett et al. 2022). In another study, female 3xTg-AD mice were exposed to a binge-like ethanol exposure (5.0 g/kg; 5 days/2 days off) for 3 months (at age 5–8 months) and then aged to 14 months without any additional ethanol; t-tau and p-tau-214 were increased in the cortex while Aβ42 was increased in the hippocampus. Importantly, tau and Aβ levels were positively correlated with neuroimmune gene expression (Tucker et al. 2022).

Our laboratories have recently used the TgF344-AD model to examine whether adolescent ethanol, both chronic drinking and AIE, alters the progression of AD-related cognitive decline and neuropathological markers. In the chronic drinking models, with moderate BECs relatively minimal effects of lifelong alcohol consumption on cognitive function were found, despite signs of increased anxiety in multiple neurobehavioral tasks. When plaque pathology was examined, we observed sex-specific effects in which females showed increased colocalization of Aβ42 in iba1+ cells (Marsland et al., in prep). In contrast, heavy binge-type ethanol exposure using the AIE model found sex-specific effects of AD transgenes, AIE, and their interaction manifested in distinct behavioral and neurobiological outcomes. In male rats carrying AD transgenes, spatial navigation deficits were evident by 3 months of age that were unaffected by adolescent ethanol exposure. Furthermore, AD transgenes combined with AIE led to pathological changes characterized by increased p-Tau levels and a shift in the balance of proNGF/NGF receptors favoring cell death mechanisms. Conversely, in female TgF344-AD rats, AIE accelerated AD-induced cognitive decline. This was accompanied by a reduction in tropomyosin receptor kinase A (TrkA) receptors due to AIE and a decrease in vesicular acetylcholine transporter (VAChT) resulting from AD transgenes. The observed dysregulation in the cholinergic system suggests its potential contribution to the exacerbated behavioral impairments in female rats carrying AD transgenes exposed to adolescent ethanol.

Together, these studies provide evidence that chronic ethanol exposure has lasting effects on AD biomarkers and a pro-inflammatory phenotype (see also Sect. 12.2.2 in this chapter), especially in models of heavy ethanol exposure. However, another central problem facing the field is the extensive reliance on transgenic models of AD in determining the relationship between alcohol exposure and AD risk, which may overstate this association due to the relatively low incidence of familial AD. Therefore, future studies should investigate whether ethanol drives neuroinflammation through increased pathology or drives pathology through increased neuroinflammation in both genetically vulnerable (familial AD) and genetically typical (sporadic AD) rodent models.

12.5.4.2 Neuronal Excitability

AUD is characterized by neuronal hyperactivity during periods of withdrawal (Ariwodola and Weiner 2004; Cheaha et al. 2014; Slawecki et al. 2006; Wang and Mandelkow 2016). In one study, chronic intermittent ethanol (CIE) exposure increased the frequency of spontaneous interictal spikes and lowered seizure thresholds throughout the brain (Alberto et al. 2023). Ethanol exposure bidirectionally altered N-methyl-D-aspartate receptor (NMDAR)-mediated synaptic activity in multiple regions of the brain during periods of intoxication and withdrawal. In the agranular insular cortex, acute ethanol exposure inhibited NMDAR currents and disrupted long-term potentiation (LTP) (Shillinglaw et al. 2018). In CIE-treated mice, acute ethanol exposure reduced NMDAR-mediated excitatory synaptic transmission in the central amygdala (Roberto et al. 2004). During withdrawal from a 10-day CIE treatment, NMDAR function was increased in the basolateral amygdala during withdrawal (McGinnis et al. 2020). Withdrawal from CIE treatment also enhanced Glun2B-dependent LTP in the bed nucleus of the stria terminalis (Wills et al. 2012). These findings are relevant in the present context because neuronal activity also regulates Aβ and tau pathology in rodent models of Aβ and tau overexpression. In vivo microdialysis experiments demonstrated that pharmacological stimulation of neuronal activity increased tau levels in vivo, whereas pharmacological inhibition had no effect on tau (Yamada et al. 2014). Electrical, pharmacological, and optogenetic stimulation of neuronal activity also increased Aβ levels in vivo, whereas pharmacological inhibition decreased Aβ levels in Tg2576 mice (Bero et al. 2011; Cirrito et al. 2005, 2008; Yamamoto et al. 2015). Consequently, targeting neuronal activity may reduce Aβ deposition over time. Consistent with this, treatment with the NMDAR antagonist memantine decreased Aβ levels, amyloid plaque formation, and behavioral deficits in mouse models of Aβ overexpression (Dong et al. 2008; Stazi and Wirths 2021). In 7-month-old APP/PS1 mice, 28 days of unilateral vibrissal deprivation reduced amyloid plaque formation and growth in the barrel cortex of the corresponding hemisphere (Bero et al. 2011). Given the innervation between the vibrissae and the barrel cortex, this decrease was likely due to reduced neuronal activity as there were no differences between microglial and astrocytic activation (Bero et al. 2011). Thus, ethanol-induced disruptions in the brain's excitatory/inhibitory balance may ultimately exacerbate the activity-dependent production and propagation of Aβ and tau.

A few studies have begun to investigate how chronic ethanol alters brain excitability in rodent models of AD-like pathology. In one study, 5.5-month-old APP/PS1 mice were exposed to a moderate two-bottle choice drinking paradigm (20% ethanol, 12 h/day, 4 consecutive days/week) for 10 weeks. Ethanol-treated APP/PS1 mice had increased cortical GluN2B mRNA levels, compared to ethanol-treated wildtype mice. In the same study $GABA_AR$, α5 subunit mRNA levels were elevated in water-treated APP/PS1 mice, which were decreased in ethanol-exposed APP/PS1 mice (Day et al. 2023). This data suggests that even moderate levels of ethanol exposure may disrupt the brain's excitatory/inhibitor balance in APP/PS1 mice. In another study, 3-month-old P301S mice were exposed to an intermittent two-bottle choice drinking paradigm (20% ethanol, 24 h/day, Monday/Wednesday/Friday) for 16 weeks. In locus coeruleus (LC) neurons, the action potential threshold was decreased in ethanol-exposed males and females, in both wildtype and P301S mice. This translated to increased neuronal excitability in the LC of ethanol-exposed male and female, wildtype and P301S mice; ethanol-exposed P301S female mice showed the greatest increase in excitability (Downs et al. 2023). Thus, chronic ethanol exposure throughout early and middle adulthood rendered LC neurons to a hyperexcitable state, potentially driving AD-related pathology. Collectively, these studies demonstrate that chronic ethanol exposure during early and middle adulthood exacerbated brain excitability in the context of AD-related pathology. In these studies, measures of excitability were taken during an early withdrawal period (~72 h post-ethanol). Furthermore, these studies were also conducted in animals at an age when amyloid or tau pathology first begins to appear. It is unclear how ethanol-induced brain excitability changes over time and with age. Thus, future studies should consider the long-term consequences of ethanol exposure during different phases of development on brain excitability and AD-related pathology. Furthermore, as discussed with the ethanol-associated neuroinflammatory phenotype, it is unclear whether ethanol-induced hyperexcitability is the cause or consequence of increased pathology. Thus, future studies investigating this mechanism should seek to characterize the directionality of this relationship.

12.5.5 Concluding Remarks

Ongoing research continues to provide evidence that alcohol misuse or chronic ethanol exposure increases the risk of AD and drives AD-related pathology; however, there are a few key issues that need to be addressed in this growing field. While epidemiological and clinical studies provide evidence that heavy alcohol misuse increases the risk for dementia and AD, mild-to-moderate alcohol usage may offer a protective effect relative to complete abstinence. Here, we identified neuroinflammation and neuronal hyperexcitability as potential mechanisms by which alcohol misuse exacerbates AD-related pathology. It is important to note, however, that alcohol has widespread effects on the nervous system and peripheral organs and that AD is a multifaceted disease with many genetic and environmental risk factors. Thus, the relationship between AUD and AD may not be limited to a single mechanism. Moreover, it is also important to account for bidirectionality between the systems disrupted by alcohol misuse and those exacerbating AD-related pathology. Thus, future research in this field should be done from a multidisciplinary approach.

12.6 General Conclusions

Natural aging is associated with a wide range of cellular, structural, and circuit-level disruptions that collectively contribute to aging-related cognitive decline, mild cognitive impairments, and for some individuals, ADRD. The contribution of lifestyle factors such as patterns, frequency, and intensity of alcohol consumption appears to modify the trajectory of healthy brain aging, and the mechanisms contributing to these changes are just now beginning to emerge. The aged brain is associated with both heightened basal inflammation and delayed recovery of inflammatory processes, at least in part due to deficits in anti-inflammatory shutoff mechanisms that govern inflammation in the young brain. Emerging evidence suggests that natural aging-related changes in inflammation may contribute to earlier emergence of dementia, development of alcohol-related brain damage due to excessive inflammation (directly), or through thiamine deficiency (indirectly). Binge-like alcohol exposure during adolescence, a developmental period during which ethanol consumption typically peaks for most individuals, is also associated with reduced cholinergic output from basal forebrain cholinergic neurons. In this sense, it is perhaps noteworthy to mention that cholinergic signaling has been shown to have moderate anti-inflammatory effects itself. If caught early, loss of the cholinergic phenotype after chronic ethanol (with or without thiamine depletion) may be rescued by thiamine replacement, regular exercise, and other corrective measures that provoke a pro-neurotrophic response. However, once diencephalic damage associated with severe thiamine depletion is instantiated, cognitive deficits become irreversible. The same also appears to be true in the case of ADRD; as alcohol consumption and aging progress, the accumulation of cardinal neuropathological features of ADRD (amyloidopathy and tauopathy) appears to set the aging brain onto an irreversible course of cognitive dysfunction for which no effective treatments currently exist.

Some caveats to this framework should also be considered. For instance, substantial evidence from human epidemiological studies supports the notion that low to moderate levels of alcohol consumption may confer a slight protective benefit against dementia relative to non-drinkers. Furthermore, most preclinical studies examining alcohol-induced modulation of ADRD-like pathology have shown effects predominantly in transgenic rodent models of AD, which may bear relevance more to familial AD (~5% of AD) than sporadic AD (~95% of AD cases). In contrast, there is a relative paucity of studies examining the myriad of genetic and environmental factors that might confer a protective benefit that mitigates the untoward effects of alcohol consumption on the progression of cognitive dysfunction into dementia and its associated neurological diagnoses (MCI, WKS, ARBD, ADRD). To be sure, much work remains to be done.

Acknowledgments This work was supported by the National Institute on Alcohol Abuse and Alcoholism grants (P50AA017823, R01AA030469, and T32AA025606). Any opinions, findings, and conclusions or recommendations expressed in this material are those of the author(s) and do not necessarily reflect the views of the above-stated funding agencies. The authors have no conflicts of interest to declare.

References

Abdou E, Hazell AS (2015) Thiamine deficiency: an update of pathophysiologic mechanisms and future therapeutic considerations. Neurochem Res 40(2):353–361. https://doi.org/10.1007/s11064-014-1430-z

Akers KG, Martinez-Canabal A, Restivo L, Yiu AP, De Cristofaro A, Hsiang HL, Wheeler AL, Guskjolen A, Niibori Y, Shoji H, Ohira K, Richards BA, Miyakawa T, Josselyn SA, Frankland PW (2014) Hippocampal neurogenesis regulates forgetting during adulthood and infancy. Science 344(6184):598–602. https://doi.org/10.1126/science.1248903

Alberto GE, Klorig DC, Goldstein AT, Godwin DW (2023) Alcohol withdrawal produces changes in excitability, population discharge probability, and seizure threshold. Alcohol Clin Exp Res (Hoboken) 47(2):211–218. https://doi.org/10.1111/acer.15004

Alfonso-Loeches S, Pascual-Lucas M, Blanco AM, Sanchez-Vera I, Guerri C (2010) Pivotal role of TLR4 receptors in alcohol-induced neuroinflammation and brain damage. J Neurosci 30(24):8285–8295. https://doi.org/10.1523/jneurosci.0976-10.2010

Allen LH (2012) B vitamins in breast milk: relative importance of maternal status and intake, and effects on infant status and function. Adv Nutr 3(3):362–369. https://doi.org/10.3945/an.111.001172

Alosco ML, Stern RA (eds) (2019) The Oxford handbook of adult cognitive disorders. Oxford University Press, Oxford. https://doi.org/10.1093/oxfordhb/9780190664121.001.0001

Anttila T, Helkala EL, Viitanen M, Kåreholt I, Fratiglioni L, Winblad B, Soininen H, Tuomilehto J, Nissinen A, Kivipelto M (2004) Alcohol drinking in middle age and subsequent risk of mild cognitive impairment and dementia in old age: a prospective population based study. BMJ 329(7465):539. https://doi.org/10.1136/bmj.38181.418958.BE

Anzalone S, Vetreno RP, Ramos RL, Savage LM (2010) Cortical cholinergic abnormalities contribute to the amnesic state induced by pyrithiamine-induced thiamine deficiency in the rat. Eur J Neurosci 32(5):847–858. https://doi.org/10.1111/j.1460-9568.2010.07358.x

Aradóttir S, Moller K, Alling C (2004) *Phosphatidylethanol* formation and degradation in human and rat blood. Alcohol Alcohol 39(1):8–13. https://doi.org/10.1093/alcalc/agh003

Araujo I, Henriksen A, Gamsby J, Gulick D (2021) Impact of alcohol abuse on susceptibility to rare neurodegenerative diseases. Front Mol Biosci 8:643273. https://doi.org/10.3389/fmolb.2021.643273

Ariwodola OJ, Weiner JL (2004) Ethanol potentiation of GABAergic synaptic transmission may be self-limiting: role of presynaptic GABA(B) receptors. J Neurosci 24(47):10679–10686. https://doi.org/10.1523/jneurosci.1768-04.2004

Arts NJ, Walvoort SJ, Kessels RP (2017) Korsakoff's syndrome: a critical review. Neuropsychiatr Dis Treat 13:2875–2890. https://doi.org/10.2147/ndt.S130078

Attias J, Raveh E, Aizer-Dannon A, Bloch-Mimouni A, Fattal-Valevski A (2012) Auditory system dysfunction due to infantile thiamine deficiency: long-term auditory sequelae. Audiol Neurootol 17(5):309–320. https://doi.org/10.1159/000339356

Bâ A (2009) Alcohol and B1 vitamin deficiency-related stillbirths. J Matern Fetal Neonatal Med 22(5):452–457. https://doi.org/10.1080/14767050802609775

Bâ A (2011) Comparative effects of alcohol and thiamine deficiency on the developing central nervous system. Behav Brain Res 225(1):235–242. https://doi.org/10.1016/j.bbr.2011.07.015

Bâ A (2017) Alcohol and thiamine deficiency trigger differential mitochondrial transition pore opening mediating cellular death. Apoptosis 22(6):741–752. https://doi.org/10.1007/s10495-017-1372-4

Bake S, Rouzer SK, Mavuri S, Miranda RC, Mahnke AH (2023) The interaction of genetic sex and prenatal alcohol exposure on health across the lifespan. Front Neuroendocrinol 71:101103. https://doi.org/10.1016/j.yfrne.2023.101103

Barnett A, David E, Rohlman A, Nikolova VD, Moy SS, Vetreno RP, Coleman LG Jr (2022) Adolescent binge alcohol enhances early Alzheimer's disease pathology in adulthood through proinflammatory neuroimmune activation. Front Pharmacol 13:884170. https://doi.org/10.3389/fphar.2022.884170

Barrientos RM, Higgins EA, Biedenkapp JC, Sprunger DB, Wright-Hardesty KJ, Watkins LR, Rudy JW, Maier SF (2006) Peripheral infection and aging interact to impair hippocampal memory consolidation. Neurobiol Aging 27(5):723–732. https://doi.org/10.1016/j.neurobiolaging.2005.03.010

Baskerville KA, Kent C, Nicolle MM, Gallagher M, McKinney M (2006) Aging causes partial loss of basal forebrain but no loss of pontine reticular cholinergic neurons. Neuroreport 17(17):1819–1823. https://doi.org/10.1097/WNR.0b013e32800fef5a

Beardmore R, Hou R, Darekar A, Holmes C, Boche D (2021) The locus coeruleus in aging and Alzheimer's disease: a postmortem and brain imaging review. J Alzheimers Dis 83(1):5–22. https://doi.org/10.3233/jad-210191

Beresford TP, Arciniegas DB, Alfers J, Clapp L, Martin B, Du Y, Liu D, Shen D, Davatzikos C (2006) Hippocampus volume loss due to chronic heavy drinking. Alcohol Clin Exp Res 30(11):1866–1870. https://doi.org/10.1111/j.1530-0277.2006.00223.x

Bero AW, Yan P, Roh JH, Cirrito JR, Stewart FR, Raichle ME, Lee JM, Holtzman DM (2011) Neuronal activity regulates the regional vulnerability to amyloid-β deposition. Nat Neurosci 14(6):750–756. https://doi.org/10.1038/nn.2801

Bizon JL, Lee HJ, Gallagher M (2004) Neurogenesis in a rat model of age-related cognitive decline. Aging Cell 3(4):227–234. https://doi.org/10.1111/j.1474-9728.2004.00099.x

Bizon JL, LaSarge CL, Montgomery KS, McDermott AN, Setlow B, Griffith WH (2009) Spatial reference and working memory across the lifespan of male Fischer 344 rats. Neurobiol Aging 30(4):646–655. https://doi.org/10.1016/j.neurobiolaging.2007.08.004

Blazer DG, Wu LT (2009) The epidemiology of at-risk and binge drinking among middle-aged and elderly community adults: National Survey on Drug Use and Health. Am J Psychiatry 166(10):1162–1169. https://doi.org/10.1176/appi.ajp.2009.09010016

Blennow K, Mattsson N, Schöll M, Hansson O, Zetterberg H (2015) Amyloid biomarkers in Alzheimer's disease. Trends Pharmacol Sci 36(5):297–309. https://doi.org/10.1016/j.tips.2015.03.002

Bonnet U, Pohlmann L, McAnally H, Claus BB (2023) Further evidence of relationship between thiamine blood level and cognition in chronic alcohol-dependent adults: Prospective Pilot Study of an inpatient detoxification with oral supplementation protocol. Alcohol 110:23–31. https://doi.org/10.1016/j.alcohol.2023.03.001

Boots A, Wiegersma AM, Vali Y, van den Hof M, Langendam MW, Limpens J, Backhouse EV, Shenkin SD, Wardlaw JM, Roseboom TJ, de Rooij SR (2023) Shaping the risk for late-life neurodegenerative disease: A systematic review on prenatal risk factors for Alzheimer's disease-related volumetric brain biomarkers. Neurosci Biobehav Rev 146:105019. https://doi.org/10.1016/j.neubiorev.2022.105019

Braak H, Braak E (1991) Neuropathological stageing of Alzheimer-related changes. Acta Neuropathol 82(4):239–259. https://doi.org/10.1007/bf00308809

Breslow RA, Castle IP, Chen CM, Graubard BI (2017) Trends in alcohol consumption among older Americans: National Health Interview Surveys, 1997 to 2014. Alcohol Clin Exp Res 41(5):976–986. https://doi.org/10.1111/acer.13365

Broadwater MA, Liu W, Crews FT, Spear LP (2014) Persistent loss of hippocampal neurogenesis and increased cell death following adolescent, but not adult, chronic ethanol exposure. Dev Neurosci 36(3–4):297–305. https://doi.org/10.1159/000362874

Butterworth RF (1993) Maternal thiamine deficiency. A factor in intrauterine growth retardation. Ann N Y Acad Sci 678:325–329. https://doi.org/10.1111/j.1749-6632.1993.tb26135.x
Butterworth RF (2003) Thiamin deficiency and brain disorders. Nutr Res Rev 16(2):277–284. https://doi.org/10.1079/nrr200367
Butterworth RF, Kril JJ, Harper CG (1993) Thiamine-dependent enzyme changes in the brains of alcoholics: relationship to the Wernicke-Korsakoff syndrome. Alcohol Clin Exp Res 17(5):1084–1088. https://doi.org/10.1111/j.1530-0277.1993.tb05668.x
Calingasan NY, Gandy SE, Baker H, Sheu KF, Smith JD, Lamb BT, Gearhart JD, Buxbaum JD, Harper C, Selkoe DJ, Price DL, Sisodia SS, Gibson GE (1996) Novel neuritic clusters with accumulations of amyloid precursor protein and amyloid precursor-like protein 2 immunoreactivity in brain regions damaged by thiamine deficiency. Am J Pathol 149(3):1063–1071
Carlson ER, Guerin SP, Nixon K, Fonken LK (2023) The neuroimmune system—where aging and excess alcohol intersect. Alcohol 107:153–167. https://doi.org/10.1016/j.alcohol.2022.08.009
Castellani RJ, Perry G (2019) Tau biology, tauopathy, traumatic brain injury, and diagnostic challenges. J Alzheimers Dis 67(2):447–467. https://doi.org/10.3233/jad-180721
Cederbaum AI (2012) Alcohol metabolism. Clin Liver Dis 16(4):667–685. https://doi.org/10.1016/j.cld.2012.08.002
Chandrakumar A, Bhardwaj A, t Jong GW (2018) Review of thiamine deficiency disorders: Wernicke encephalopathy and Korsakoff psychosis. J Basic Clin Physiol Pharmacol 30(2):153–162. https://doi.org/10.1515/jbcpp-2018-0075
Chang Q, Gold PE (2008) Age-related changes in memory and in acetylcholine functions in the hippocampus in the Ts65Dn mouse, a model of Down syndrome. Neurobiol Learn Mem 89(2):167–177. https://doi.org/10.1016/j.nlm.2007.05.007
Cheaha D, Sawangjaroen K, Kumarnsit E (2014) Characterization of fluoxetine effects on ethanol withdrawal-induced cortical hyperexcitability by EEG spectral power in rats. Neuropharmacology 77:49–56. https://doi.org/10.1016/j.neuropharm.2013.09.020
Church MW, Abel EL, Kaltenbach JA, Overbeck GW (1996) Effects of prenatal alcohol exposure and aging on auditory function in the rat: preliminary results. Alcohol Clin Exp Res 20(1):172–179. https://doi.org/10.1111/j.1530-0277.1996.tb01061.x
Cirrito JR, Yamada KA, Finn MB, Sloviter RS, Bales KR, May PC, Schoepp DD, Paul SM, Mennerick S, Holtzman DM (2005) Synaptic activity regulates interstitial fluid amyloid-beta levels in vivo. Neuron 48(6):913–922. https://doi.org/10.1016/j.neuron.2005.10.028
Cirrito JR, Kang JE, Lee J, Stewart FR, Verges DK, Silverio LM, Bu G, Mennerick S, Holtzman DM (2008) Endocytosis is required for synaptic activity-dependent release of amyloid-beta in vivo. Neuron 58(1):42–51. https://doi.org/10.1016/j.neuron.2008.02.003
Citron M, Oltersdorf T, Haass C, McConlogue L, Hung AY, Seubert P, Vigo-Pelfrey C, Lieberburg I, Selkoe DJ (1992) Mutation of the beta-amyloid precursor protein in familial Alzheimer's disease increases beta-protein production. Nature 360(6405):672–674. https://doi.org/10.1038/360672a0
Clarke LE, Liddelow SA, Chakraborty C, Münch AE, Heiman M, Barres BA (2018) Normal aging induces A1-like astrocyte reactivity. Proc Natl Acad Sci USA 115(8):E1896–e1905. https://doi.org/10.1073/pnas.1800165115
Cnops V, Iyer VR, Parathy N, Wong P, Dawe GS (2022) Test, rinse, repeat: A review of carryover effects in rodent behavioral assays. Neurosci Biobehav Rev 135:104560. https://doi.org/10.1016/j.neubiorev.2022.104560
Cohen RM, Rezai-Zadeh K, Weitz TM, Rentsendorj A, Gate D, Spivak I, Bholat Y, Vasilevko V, Glabe CG, Breunig JJ, Rakic P, Davtyan H, Agadjanyan MG, Kepe V, Barrio JR, Bannykh S, Szekely CA, Pechnick RN, Town T (2013) A transgenic Alzheimer rat with plaques, tau pathology, behavioral impairment, oligomeric aβ, and frank neuronal loss. J Neurosci 33(15):6245–6256. https://doi.org/10.1523/jneurosci.3672-12.2013
Coleman LG Jr, Zou J, Crews FT (2020) Microglial depletion and repopulation in brain slice culture normalizes sensitized proinflammatory signaling. J Neuroinflammation 17(1):27. https://doi.org/10.1186/s12974-019-1678-y
Crary JF, Trojanowski JQ, Schneider JA, Abisambra JF, Abner EL, Alafuzoff I, Arnold SE, Attems J, Beach TG, Bigio EH, Cairns NJ, Dickson DW, Gearing M, Grinberg LT, Hof PR, Hyman BT, Jellinger K, Jicha GA, Kovacs GG, Knopman DS, Kofler J, Kukull WA, Mackenzie IR, Masliah E, McKee A, Montine TJ, Murray ME, Neltner JH, Santa-Maria I, Seeley WW, Serrano-Pozo A, Shelanski ML, Stein T, Takao M, Thal DR, Toledo JB, Troncoso JC, Vonsattel JP, White CL 3rd, Wisniewski T, Woltjer RL, Yamada M, Nelson PT (2014) Primary age-related tauopathy (PART): a common pathology associated with human aging. Acta Neuropathol 128(6):755–766. https://doi.org/10.1007/s00401-014-1349-0
Crews FT, Nixon K (2009) Mechanisms of neurodegeneration and regeneration in alcoholism. Alcohol Alcohol 44(2):115–127. https://doi.org/10.1093/alcalc/agn079
Crews FT, Qin L, Sheedy D, Vetreno RP, Zou J (2013) High mobility group box 1/Toll-like receptor danger signaling increases brain neuroimmune activation in alcohol dependence. Biol Psychiatry 73(7):602–612. https://doi.org/10.1016/j.biopsych.2012.09.030
Crews FT, Robinson DL, Chandler LJ, Ehlers CL, Mulholland PJ, Pandey SC, Rodd ZA, Spear LP, Swartzwelder HS, Vetreno RP (2019) Mechanisms of persistent neurobiological changes following adolescent alcohol exposure: NADIA Consortium findings. Alcohol Clin Exp Res 43(9):1806–1822. https://doi.org/10.1111/acer.14154

Cruz JV, Maba IK, Correia D, Kaziuk FD, Cadena S, Zampronio AR (2020) Intermittent binge-like ethanol exposure during adolescence attenuates the febrile response by reducing brown adipose tissue thermogenesis in rats. Drug Alcohol Depend 209:107904. https://doi.org/10.1016/j.drugalcdep.2020.107904
Dalçik H, Yardimoglu M, Filiz S, Gonca S, Dalçik C, Erden BF (2009) Chronic ethanol-induced glial fibrillary acidic protein (GFAP) immunoreactivity: an immunocytochemical observation in various regions of adult rat brain. Int J Neurosci 119(9):1303–1318
David J-P, Ghozali F, Fallet-Bianco C, Wattez A, Delaine S, Boniface B, Di Menza C, Delacourte A (1997) Glial reaction in the hippocampal formation is highly correlated with aging in human brain. Neurosci Lett 235(1–2):53–56
Day SM, Gironda SC, Clarke CW, Snipes JA, Nicol NI, Kamran H, Vaughan W, Weiner JL, Macauley SL (2023) Ethanol exposure alters Alzheimer's-related pathology, behavior, and metabolism in APP/PS1 mice. Neurobiol Dis 177:105967. https://doi.org/10.1016/j.nbd.2022.105967
Deak T, Savage LM (2019) Preface: Setting the stage for understanding alcohol effects in late aging: A special issue including both human and rodent studies. Int Rev Neurobiol 148:xiii–xxv. https://doi.org/10.1016/s0074-7742(19)30116-3
Deak T, Kelliher KT, Wojcik HJ, Gano A (2022) Prenatal and adolescent alcohol exposure programs immunity across the lifespan: CNS-mediated regulation. Pharmacol Biochem Behav 216:173390. https://doi.org/10.1016/j.pbb.2022.173390
de Freitas-Silva DM, Resende Lde S, Pereira SR, Franco GC, Ribeiro AM (2010) Maternal thiamine restriction during lactation induces cognitive impairments and changes in glutamate and GABA concentrations in brain of rat offspring. Behav Brain Res 211(1):33–40. https://doi.org/10.1016/j.bbr.2010.03.002
Dhir S, Tarasenko M, Napoli E, Giulivi C (2019) Neurological, psychiatric, and biochemical aspects of thiamine deficiency in children and adults. Front Psychiatry 10:207. https://doi.org/10.3389/fpsyt.2019.00207
Dickerson BC, Bakkour A, Salat DH, Feczko E, Pacheco J, Greve DN, Grodstein F, Wright CI, Blacker D, Rosas HD, Sperling RA, Atri A, Growdon JH, Hyman BT, Morris JC, Fischl B, Buckner RL (2009) The cortical signature of Alzheimer's disease: regionally specific cortical thinning relates to symptom severity in very mild to mild AD dementia and is detectable in asymptomatic amyloid-positive individuals. Cereb Cortex 19(3):497–510. https://doi.org/10.1093/cercor/bhn113
Dong H, Yuede CM, Coughlan C, Lewis B, Csernansky JG (2008) Effects of memantine on neuronal structure and conditioned fear in the Tg2576 mouse model of Alzheimer's disease. Neuropsychopharmacology 33(13):3226–3236. https://doi.org/10.1038/npp.2008.53
Donnino MW, Vega J, Miller J, Walsh M (2007) Myths and misconceptions of Wernicke's encephalopathy: what every emergency physician should know. Ann Emerg Med 50(6):715–721. https://doi.org/10.1016/j.annemergmed.2007.02.007
Doremus-Fitzwater TL, Deak T (2022) Adolescent neuroimmune function and its interaction with alcohol. Int Rev Neurobiol 161:167–208. https://doi.org/10.1016/bs.irn.2021.08.006
Doremus-Fitzwater TL, Gano A, Paniccia JE, Deak T (2015) Male adolescent rats display blunted cytokine responses in the CNS after acute ethanol or lipopolysaccharide exposure. Physiol Behav 148:131–144. https://doi.org/10.1016/j.physbeh.2015.02.032
Downs AM, Catavero CM, Kasten MR, McElligott ZA (2023) Tauopathy and alcohol consumption interact to alter locus coeruleus excitatory transmission and excitability in male and female mice. Alcohol 107:97–107. https://doi.org/10.1016/j.alcohol.2022.08.008
Eggen BJ, Raj D, Hanisch UK, Boddeke HW (2013) Microglial phenotype and adaptation. J NeuroImmune Pharmacol 8(4):807–823. https://doi.org/10.1007/s11481-013-9490-4
Evrard SG, Duhalde-Vega M, Tagliaferro P, Mirochnic S, Caltana LR, Brusco A (2006) A low chronic ethanol exposure induces morphological changes in the adolescent rat brain that are not fully recovered even after a long abstinence: an immunohistochemical study. Exp Neurol 200(2):438–459. https://doi.org/10.1016/j.expneurol.2006.03.001
Fadda F, Rossetti ZL (1998) Chronic ethanol consumption: from neuroadaptation to neurodegeneration. Prog Neurobiol 56(4):385–431. https://doi.org/10.1016/s0301-0082(98)00032-x
Fattal I, Friedmann N, Fattal-Valevski A (2011) The crucial role of thiamine in the development of syntax and lexical retrieval: a study of infantile thiamine deficiency. Brain 134(Pt 6):1720–1739. https://doi.org/10.1093/brain/awr068
Fattal-Valevski A, Kesler A, Sela BA, Nitzan-Kaluski D, Rotstein M, Mesterman R, Toledano-Alhadef H, Stolovitch C, Hoffmann C, Globus O, Eshel G (2005) Outbreak of life-threatening thiamine deficiency in infants in Israel caused by a defective soy-based formula. Pediatrics 115(2):e233–e238. https://doi.org/10.1542/peds.2004-1255
Fattal-Valevski A, Azouri-Fattal I, Greenstein YJ, Guindy M, Blau A, Zelnik N (2009) Delayed language development due to infantile thiamine deficiency. Dev Med Child Neurol 51(8):629–634. https://doi.org/10.1111/j.1469-8749.2008.03161.x
Febo M, Rani A, Yegla B, Barter J, Kumar A, Wolff CA, Esser K, Foster TC (2020) Longitudinal characterization and biomarkers of age and sex differences in the decline of spatial memory. Front Aging Neurosci 12:34. https://doi.org/10.3389/fnagi.2020.00034
Fenn AM, Henry CJ, Huang Y, Dugan A, Godbout JP (2012) Lipopolysaccharide-induced interleukin (IL)-4 receptor-α expression and corresponding sensitivity to the M2 promoting effects of IL-4 are

impaired in microglia of aged mice. Brain Behav Immun 26(5):766–777. https://doi.org/10.1016/j.bbi.2011.10.003
Fenn AM, Hall JC, Gensel JC, Popovich PG, Godbout JP (2014) IL-4 signaling drives a unique arginase+/IL-1β+ microglia phenotype and recruits macrophages to the inflammatory CNS: consequences of age-related deficits in IL-4Rα after traumatic spinal cord injury. J Neurosci 34(26):8904–8917
Fernandez GM, Savage LM (2017) Adolescent binge ethanol exposure alters specific forebrain cholinergic cell populations and leads to selective functional deficits in the prefrontal cortex. Neuroscience 361:129–143. https://doi.org/10.1016/j.neuroscience.2017.08.013
Fernández-Cabello S, Kronbichler M, Van Dijk KRA, Goodman JA, Spreng RN, Schmitz TW (2020) Basal forebrain volume reliably predicts the cortical spread of Alzheimer's degeneration. Brain 143(3):993–1009. https://doi.org/10.1093/brain/awaa012
Fischer W, Gage FH, Björklund A (1989) Degenerative changes in forebrain cholinergic nuclei correlate with cognitive impairments in aged rats. Eur J Neurosci 1(1):34–45. https://doi.org/10.1111/j.1460-9568.1989.tb00772.x
Flanigan MR, Royse SK, Cenkner DP, Kozinski KM, Stoughton CJ, Himes ML, Minhas DS, Lopresti B, Butters MA, Narendran R (2021) Imaging beta-amyloid (Aβ) burden in the brains of middle-aged individuals with alcohol-use disorders: a [(11)C]PIB PET study. Transl Psychiatry 11(1):257. https://doi.org/10.1038/s41398-021-01374-y
Fonken LK, Frank MG, Kitt MM, D'Angelo HM, Norden DM, Weber MD, Barrientos RM, Godbout JP, Watkins LR, Maier SF (2016) The alarmin HMGB1 mediates age-induced neuroinflammatory priming. J Neurosci 36(30):7946–7956. https://doi.org/10.1523/jneurosci.1161-16.2016
Fournier H, Butterworth RF (1990) Effects of thiamine deficiency on thiamine-dependent enzymes in regions of the brain of pregnant rats and their offspring. Metab Brain Dis 5(2):77–84. https://doi.org/10.1007/bf01001048
Frank MG, Barrientos RM, Biedenkapp JC, Rudy JW, Watkins LR, Maier SF (2006) mRNA up-regulation of MHC II and pivotal pro-inflammatory genes in normal brain aging. Neurobiol Aging 27(5):717–722
Frank MG, Barrientos RM, Watkins LR, Maier SF (2010) Aging sensitizes rapidly isolated hippocampal microglia to LPS ex vivo. J Neuroimmunol 226(1):181–184. https://doi.org/10.1016/j.jneuroim.2010.05.022
Frank MG, Weber MD, Fonken LK, Hershman SA, Watkins LR, Maier SF (2016) The redox state of the alarmin HMGB1 is a pivotal factor in neuroinflammatory and microglial priming: A role for the NLRP3 inflammasome. Brain Behav Immun 55:215–224. https://doi.org/10.1016/j.bbi.2015.10.009
Franke H (1995) Influence of chronic alcohol treatment on the GFAP-immunoreactivity in astrocytes of the hippocampus in rats. Acta Histochem 97(3):263–271. https://doi.org/10.1016/s0065-1281(11)80187-x
Gabriel KI, Johnston S, Weinberg J (2002) Prenatal ethanol exposure and spatial navigation: effects of postnatal handling and aging. Dev Psychobiol 40(4):345–357. https://doi.org/10.1002/dev.10023
Gage FH, Batchelor P, Chen KS, Chin D, Higgins GA, Koh S, Deputy S, Rosenberg MB, Fischer W, Bjorklund A (1989) NGF receptor reexpression and NGF-mediated cholinergic neuronal hypertrophy in the damaged adult neostriatum. Neuron 2(2):1177–1184. https://doi.org/10.1016/0896-6273(89)90184-0
Gano A, Doremus-Fitzwater TL, Deak T (2016) Sustained alterations in neuroimmune gene expression after daily, but not intermittent, alcohol exposure. Brain Res 1646:62–72. https://doi.org/10.1016/j.brainres.2016.05.027
Gano A, Doremus-Fitzwater TL, Deak T (2017) A cross-sectional comparison of ethanol-related cytokine expression in the hippocampus of young and aged Fischer 344 rats. Neurobiol Aging 54:40–53. https://doi.org/10.1016/j.neurobiolaging.2017.01.025
Gano A, Vore AS, Sammakia MN, Deak T (2019) Assessment of extracellular cytokines in the hippocampus of the awake behaving rat using large-molecule microdialysis combined with multiplex arrays after acute and chronic ethanol exposure. Alcohol Clin Exp Res 43(4):640–654. https://doi.org/10.1111/acer.13963
Gano A, Prestia L, Middleton FA, Youngentob SL, Ignacio C, Deak T (2020) Gene expression profiling reveals a lingering effect of prenatal alcohol exposure on inflammatory-related genes during adolescence and adulthood. Cytokine 133:155126. https://doi.org/10.1016/j.cyto.2020.155126
Gano A, Wojcik H, Danseglio N, Kelliher K, Varlinskaya EI, Deak T (2024) Adolescent intermittent ethanol (AIE) sensitized fever in male Sprague Dawley rats exposed to poly I:C in adulthood. Brain Behav Immun 120:82–97
Germann M, Brederoo SG, Sommer IEC (2021) Abnormal synaptic pruning during adolescence underlying the development of psychotic disorders. Curr Opin Psychiatry 34(3):222–227. https://doi.org/10.1097/yco.0000000000000696
Giannakopoulos P, Herrmann FR, Bussière T, Bouras C, Kövari E, Perl DP, Morrison JH, Gold G, Hof PR (2003) Tangle and neuron numbers, but not amyloid load, predict cognitive status in Alzheimer's disease. Neurology 60(9):1495–1500. https://doi.org/10.1212/01.wnl.0000063311.58879.01
Gibson GE, Hirsch JA, Fonzetti P, Jordan BD, Cirio RT, Elder J (2016) Vitamin B1 (thiamine) and dementia. Ann N Y Acad Sci 1367(1):21–30. https://doi.org/10.1111/nyas.13031
Gibson GE, Feldman HH, Zhang S, Flowers SA, Luchsinger JA (2022) Pharmacological thiamine levels as a therapeutic approach in Alzheimer's disease. Front Med (Lausanne) 9:1033272. https://doi.org/10.3389/fmed.2022.1033272
Ginsberg SD, Malek-Ahmadi MH, Alldred MJ, Che S, Elarova I, Chen Y, Jeanneteau F, Kranz TM, Chao MV,

Counts SE, Mufson EJ (2019) Selective decline of neurotrophin and neurotrophin receptor genes within CA1 pyramidal neurons and hippocampus proper: Correlation with cognitive performance and neuropathology in mild cognitive impairment and Alzheimer's disease. Hippocampus 29(5):422–439. https://doi.org/10.1002/hipo.22802

Godbout JP, Chen J, Abraham J, Richwine AF, Berg BM, Kelley KW, Johnson RW (2005) Exaggerated neuroinflammation and sickness behavior in aged mice following activation of the peripheral innate immune system. FASEB J 19(10):1329–1331. https://doi.org/10.1096/fj.05-3776fje

Goldstein G, Shelly C (1980) Neuropsychological investigation of brain lesion localization in alcoholism. Adv Exp Med Biol 126:731–743. https://doi.org/10.1007/978-1-4684-3632-7_54

Golomb J, de Leon MJ, Kluger A, George AE, Tarshish C, Ferris SH (1993) Hippocampal atrophy in normal aging. An association with recent memory impairment. Arch Neurol 50(9):967–973. https://doi.org/10.1001/archneur.1993.00540090066012

Grant BF, Chou SP, Saha TD, Pickering RP, Kerridge BT, Ruan WJ, Huang B, Jung J, Zhang H, Fan A, Hasin DS (2017) Prevalence of 12-month alcohol use, high-risk drinking, and DSM-IV alcohol use disorder in the United States, 2001-2002 to 2012-2013: results from the National Epidemiologic Survey on Alcohol and Related Conditions. JAMA Psychiatry 74(9):911–923. https://doi.org/10.1001/jamapsychiatry.2017.2161

Gray DT, Barnes CA (2015) Distinguishing adaptive plasticity from vulnerability in the aging hippocampus. Neuroscience 309:17–28. https://doi.org/10.1016/j.neuroscience.2015.08.001

Grothe M, Heinsen H, Teipel S (2013) Longitudinal measures of cholinergic forebrain atrophy in the transition from healthy aging to Alzheimer's disease. Neurobiol Aging 34(4):1210–1220. https://doi.org/10.1016/j.neurobiolaging.2012.10.018

Guergues J, Wohlfahrt J, Zhang P, Liu B, Stevens SM Jr (2020) Deep proteome profiling reveals novel pathways associated with pro-inflammatory and alcohol-induced microglial activation phenotypes. J Proteome 220:103753. https://doi.org/10.1016/j.jprot.2020.103753

Guerrini I, Thomson AD, Gurling HD (2007) The importance of alcohol misuse, malnutrition and genetic susceptibility on brain growth and plasticity. Neurosci Biobehav Rev 31(2):212–220. https://doi.org/10.1016/j.neubiorev.2006.06.022

Guidi M, Kumar A, Rani A, Foster TC (2014) Assessing the emergence and reliability of cognitive decline over the life span in Fisher 344 rats using the spatial water maze. Front Aging Neurosci 6:2. https://doi.org/10.3389/fnagi.2014.00002

Guo JL, Lee VM (2011) Seeding of normal Tau by pathological Tau conformers drives pathogenesis of Alzheimer-like tangles. J Biol Chem 286(17):15317–15331. https://doi.org/10.1074/jbc.M110.209296

Guo T, Noble W, Hanger DP (2017) Roles of tau protein in health and disease. Acta Neuropathol 133(5):665–704. https://doi.org/10.1007/s00401-017-1707-9

Haass C, Kaether C, Thinakaran G, Sisodia S (2012) Trafficking and proteolytic processing of APP. Cold Spring Harb Perspect Med 2(5):a006270. https://doi.org/10.1101/cshperspect.a006270

Hall JM, Savage LM (2016) Exercise leads to the re-emergence of the cholinergic/nestin neuronal phenotype within the medial septum/diagonal band and subsequent rescue of both hippocampal ACh efflux and spatial behavior. Exp Neurol 278:62–75. https://doi.org/10.1016/j.expneurol.2016.01.018

Hansson O, Zetterberg H, Buchhave P, Andreasson U, Londos E, Minthon L, Blennow K (2007) Prediction of Alzheimer's disease using the CSF Abeta42/Abeta40 ratio in patients with mild cognitive impairment. Dement Geriatr Cogn Disord 23(5):316–320. https://doi.org/10.1159/000100926

Harada CN, Natelson Love MC, Triebel KL (2013) Normal cognitive aging. Clin Geriatr Med 29(4):737–752. https://doi.org/10.1016/j.cger.2013.07.002

Harding A, Halliday G, Caine D, Kril J (2000) Degeneration of anterior thalamic nuclei differentiates alcoholics with amnesia. Brain 123(Pt 1):141–154. https://doi.org/10.1093/brain/123.1.141

Harper C (1998) The neuropathology of alcohol-specific brain damage, or does alcohol damage the brain? J Neuropathol Exp Neurol 57(2):101–110. https://doi.org/10.1097/00005072-199802000-00001

Harper C (2006) Thiamine (vitamin B1) deficiency and associated brain damage is still common throughout the world and prevention is simple and safe! Eur J Neurol 13(10):1078–1082. https://doi.org/10.1111/j.1468-1331.2006.01530.x

Harper C, Kril J (1989) Patterns of neuronal loss in the cerebral cortex in chronic alcoholic patients. J Neurol Sci 92(1):81–89

Harper CG, Giles M, Finlay-Jones R (1986) Clinical signs in the Wernicke-Korsakoff complex: a retrospective analysis of 131 cases diagnosed at necropsy. J Neurol Neurosurg Psychiatry 49(4):341–345. https://doi.org/10.1136/jnnp.49.4.341

Harris JC, Leggio L, Farokhnia M (2021) Blood biomarkers of alcohol use: a scoping review. Curr Addict Rep 8(4):500–508. https://doi.org/10.1007/s40429-021-00402-7

Harwood DG, Kalechstein A, Barker WW, Strauman S, St George-Hyslop P, Iglesias C, Loewenstein D, Duara R (2010) The effect of alcohol and tobacco consumption, and apolipoprotein E genotype, on the age of onset in Alzheimer's disease. Int J Geriatr Psychiatry 25(5):511–518. https://doi.org/10.1002/gps.2372

Hazell AS, Rao KV, Danbolt NC, Pow DV, Butterworth RF (2001) Selective down-regulation of the astrocyte glutamate transporters GLT-1 and GLAST within the medial thalamus in experimental Wernicke's encephalopathy. J Neurochem 78(3):560–568. https://doi.org/10.1046/j.1471-4159.2001.00436.x

Heinze T, Weber W (1990) Determination of thiamine (vitamin B1) in maternal blood during normal pregnancies and pregnancies with intrauterine growth retardation. Z Ernahrungswiss 29(1):39–46. https://doi.org/10.1007/bf02019533

Henry CJ, Huang Y, Wynne AM, Godbout JP (2009) Peripheral lipopolysaccharide (LPS) challenge promotes microglial hyperactivity in aged mice that is associated with exaggerated induction of both pro-inflammatory IL-1beta and anti-inflammatory IL-10 cytokines. Brain Behav Immun 23(3):309–317. https://doi.org/10.1016/j.bbi.2008.09.002

Hettinger JC, Lee H, Bu G, Holtzman DM, Cirrito JR (2018) AMPA-ergic regulation of amyloid-β levels in an Alzheimer's disease mouse model. Mol Neurodegener 13(1):22. https://doi.org/10.1186/s13024-018-0256-6

Hillbom M, Pyhtinen J, Pylvänen V, Sotaniemi K (1999) Pregnant, vomiting, and coma. Lancet 353(9164):1584. https://doi.org/10.1016/s0140-6736(99)01410-5

Hoffman JL, Faccidomo S, Kim M, Taylor SM, Agoglia AE, May AM, Smith EN, Wong LC, Hodge CW (2019) Alcohol drinking exacerbates neural and behavioral pathology in the 3xTg-AD mouse model of Alzheimer's disease. Int Rev Neurobiol 148:169–230. https://doi.org/10.1016/bs.irn.2019.10.017

Holtzman DM, Morris JC, Goate AM (2011) Alzheimer's disease: the challenge of the second century. Sci Transl Med 3(77):77sr71. https://doi.org/10.1126/scitranslmed.3002369

Hsiao K, Chapman P, Nilsen S, Eckman C, Harigaya Y, Younkin S, Yang F, Cole G (1996) Correlative memory deficits, Abeta elevation, and amyloid plaques in transgenic mice. Science 274(5284):99–102. https://doi.org/10.1126/science.274.5284.99

Hu P, Wang D, Zhang Y, Cai Z, Ye T, Tong L, Xu X, Lu J, Liu F, Lu X, Huang C (2020) Apoptosis-triggered decline in hippocampal microglia mediates adolescent intermittent alcohol exposure-induced depression-like behaviors in mice. Neuropharmacology 170:108054. https://doi.org/10.1016/j.neuropharm.2020.108054

Huang D, Yu M, Yang S, Lou D, Zhou W, Zheng L, Wang Z, Cai F, Zhou W, Li T, Song W (2018) Ethanol alters APP processing and aggravates Alzheimer-associated phenotypes. Mol Neurobiol 55(6):5006–5018. https://doi.org/10.1007/s12035-017-0703-3

Iliff JJ, Wang M, Liao Y, Plogg BA, Peng W, Gundersen GA, Benveniste H, Vates GE, Deane R, Goldman SA, Nagelhus EA, Nedergaard M (2012) A paravascular pathway facilitates CSF flow through the brain parenchyma and the clearance of interstitial solutes, including amyloid β. Sci Transl Med 4(147):147ra111. https://doi.org/10.1126/scitranslmed.3003748

Ilomaki J, Jokanovic N, Tan EC, Lonnroos E (2015) Alcohol consumption, dementia and cognitive decline: an overview of systematic reviews. Curr Clin Pharmacol 10(3):204–212. https://doi.org/10.2174/1574884711003150820145539

Isenberg-Grzeda E, Kutner HE, Nicolson SE (2012) Wernicke-Korsakoff-syndrome: under-recognized and under-treated. Psychosomatics 53(6):507–516. https://doi.org/10.1016/j.psym.2012.04.008

Ismail SK, Kenny L (2007) Review on hyperemesis gravidarum. Best Pract Res Clin Gastroenterol 21(5):755–769. https://doi.org/10.1016/j.bpg.2007.05.008

Jack CR Jr, Petersen RC, Xu Y, O'Brien PC, Smith GE, Ivnik RJ, Boeve BF, Tangalos EG, Kokmen E (2000) Rates of hippocampal atrophy correlate with change in clinical status in aging and AD. Neurology 55(4):484–489. https://doi.org/10.1212/wnl.55.4.484

Jack CR Jr, Knopman DS, Jagust WJ, Shaw LM, Aisen PS, Weiner MW, Petersen RC, Trojanowski JQ (2010) Hypothetical model of dynamic biomarkers of the Alzheimer's pathological cascade. Lancet Neurol 9(1):119–128. https://doi.org/10.1016/s1474-4422(09)70299-6

Jankowsky JL, Fadale DJ, Anderson J, Xu GM, Gonzales V, Jenkins NA, Copeland NG, Lee MK, Younkin LH, Wagner SL, Younkin SG, Borchelt DR (2004) Mutant presenilins specifically elevate the levels of the 42 residue beta-amyloid peptide in vivo: evidence for augmentation of a 42-specific gamma secretase. Hum Mol Genet 13(2):159–170. https://doi.org/10.1093/hmg/ddh019

Jessen KR, Mirsky R, Lloyd AC (2015) Schwann cells: development and role in nerve repair. Cold Spring Harb Perspect Biol 7(7):a020487. https://doi.org/10.1101/cshperspect.a020487

Jyothi HJ, Vidyadhara DJ, Mahadevan A, Philip M, Parmar SK, Manohari SG, Shankar SK, Raju TR, Alladi PA (2015) Aging causes morphological alterations in astrocytes and microglia in human substantia nigra pars compacta. Neurobiol Aging 36(12):3321–3333. https://doi.org/10.1016/j.neurobiolaging.2015.08.024

Kalant H (1998) Research on tolerance: what can we learn from history? Alcohol Clin Exp Res 22(1):67–76. https://doi.org/10.1111/j.1530-0277.1998.tb03618.x

Kalinin S, González-Prieto M, Scheiblich H, Lisi L, Kusumo H, Heneka MT, Madrigal JLM, Pandey SC, Feinstein DL (2018) Transcriptome analysis of alcohol-treated microglia reveals downregulation of beta amyloid phagocytosis. J Neuroinflammation 15(1):141. https://doi.org/10.1186/s12974-018-1184-7

Kareem O, Nisar S, Tanvir M, Muzaffer U, Bader GN (2023) Thiamine deficiency in pregnancy and lactation: implications and present perspectives. Front Nutr 10:1080611. https://doi.org/10.3389/fnut.2023.1080611

Karkhanis AN, Rose JH, Weiner JL, Jones SR (2016) Early-life social isolation stress increases kappa opioid receptor responsiveness and downregulates the dopamine system. Neuropsychopharmacology 41(9):2263–2274. https://doi.org/10.1038/npp.2016.21

Karuppagounder SS, Xu H, Shi Q, Chen LH, Pedrini S, Pechman D, Baker H, Beal MF, Gandy SE, Gibson GE (2009) Thiamine deficiency induces oxidative stress and exacerbates the plaque pathology in Alzheimer's mouse model. Neurobiol Aging 30(10):1587–1600. https://doi.org/10.1016/j.neurobiolaging.2007.12.013

Kauko A, Engler D, Niiranen T, Ortega-Alonso A, Schnabel RB (2024) Increased risk of dementia differs across cardiovascular diseases and types of dementia—data from a nationwide study. J Intern Med 295(2):196–205. https://doi.org/10.1111/joim.13733

Ke ZJ, Gibson GE (2004) Selective response of various brain cell types during neurodegeneration induced by mild impairment of oxidative metabolism. Neurochem Int 45(2–3):361–369. https://doi.org/10.1016/j.neuint.2003.09.008

Kerbler GM, Fripp J, Rowe CC, Villemagne VL, Salvado O, Rose S, Coulson EJ (2015) Basal forebrain atrophy correlates with amyloid β burden in Alzheimer's disease. Neuroimage Clin 7:105–113. https://doi.org/10.1016/j.nicl.2014.11.015

Khan KM, Bierlein-De La Rosa G, Biggerstaff N, Pushpavathi Selvakumar G, Wang R, Mason S, Dailey ME, Marcinkiewcz CA (2023) Adolescent ethanol drinking promotes hyperalgesia, neuroinflammation and serotonergic deficits in mice that persist into adulthood. Brain Behav Immun 107:419–431. https://doi.org/10.1016/j.bbi.2022.07.160

Kipp BT, Nunes PT, Galaj E, Hitchcock B, Nasra T, Poynor KR, Heide SK, Reitz NL, Savage LM (2021a) Adolescent ethanol exposure alters cholinergic function and apical dendritic branching within the orbital frontal cortex. Neuroscience 473:52–65. https://doi.org/10.1016/j.neuroscience.2021.08.014

Kipp BT, Nunes PT, Savage LM (2021b) Sex differences in cholinergic circuits and behavioral disruptions following chronic ethanol exposure with and without thiamine deficiency. Alcohol Clin Exp Res 45(5):1013–1027. https://doi.org/10.1111/acer.14594

Kirshenbaum GS, Chang CY, Bompolaki M, Bradford VR, Bell J, Kosmidis S, Shansky RM, Orlandi J, Savage LM, Harris AZ, David Leonardo E, Dranovsky A (2023) Adult-born neurons maintain hippocampal cholinergic inputs and support working memory during aging. Mol Psychiatry. https://doi.org/10.1038/s41380-023-02167-z

Kita Y, Ago Y, Higashino K, Asada K, Takano E, Takuma K, Matsuda T (2014) Galantamine promotes adult hippocampal neurogenesis via M_1 muscarinic and α7 nicotinic receptors in mice. Int J Neuropsychopharmacol 17(12):1957–1968. https://doi.org/10.1017/s1461145714000613

Kivipelto M, Rovio S, Ngandu T, Kåreholt I, Eskelinen M, Winblad B, Hachinski V, Cedazo-Minguez A, Soininen H, Tuomilehto J, Nissinen A (2008) Apolipoprotein E epsilon4 magnifies lifestyle risks for dementia: a population-based study. J Cell Mol Med 12(6b):2762–2771. https://doi.org/10.1111/j.1582-4934.2008.00296.x

Kloss O, Eskin NAM, Suh M (2018) Thiamin deficiency on fetal brain development with and without prenatal alcohol exposure. Biochem Cell Biol 96(2):169–177. https://doi.org/10.1139/bcb-2017-0082

Koch M, Fitzpatrick AL, Rapp SR, Nahin RL, Williamson JD, Lopez OL, DeKosky ST, Kuller LH, Mackey RH, Mukamal KJ, Jensen MK, Sink KM (2019) Alcohol consumption and risk of dementia and cognitive decline among older adults with or without mild cognitive impairment. JAMA Netw Open 2(9):e1910319. https://doi.org/10.1001/jamanetworkopen.2019.10319

Kopec AM, Smith CJ, Bilbo SD (2019) Neuro-immune mechanisms regulating social behavior: dopamine as mediator? Trends Neurosci 42(5):337–348. https://doi.org/10.1016/j.tins.2019.02.005

Kopelman MD (1991) Frontal dysfunction and memory deficits in the alcoholic Korsakoff syndrome and Alzheimer-type dementia. Brain 114(Pt 1A):117–137

Kopelman MD (1995) The Korsakoff syndrome. Br J Psychiatry 166(2):154–173. https://doi.org/10.1192/bjp.166.2.154

Kopelman MD, Stanhope N, Kingsley D (1999) Retrograde amnesia in patients with diencephalic, temporal lobe or frontal lesions. Neuropsychologia 37(8):939–958. https://doi.org/10.1016/s0028-3932(98)00143-2

Kopelman MD, Thomson AD, Guerrini I, Marshall EJ (2009) The Korsakoff syndrome: clinical aspects, psychology and treatment. Alcohol Alcohol 44(2):148–154. https://doi.org/10.1093/alcalc/agn118

Kotani S, Yamauchi T, Teramoto T, Ogura H (2006) Pharmacological evidence of cholinergic involvement in adult hippocampal neurogenesis in rats. Neuroscience 142(2):505–514. https://doi.org/10.1016/j.neuroscience.2006.06.035

Kress BT, Iliff JJ, Xia M, Wang M, Wei HS, Zeppenfeld D, Xie L, Kang H, Xu Q, Liew JA, Plog BA, Ding F, Deane R, Nedergaard M (2014) Impairment of paravascular clearance pathways in the aging brain. Ann Neurol 76(6):845–861. https://doi.org/10.1002/ana.24271

Kuerbis A, Sacco P, Blazer DG, Moore AA (2014) Substance abuse among older adults. Clin Geriatr Med 30(3):629–654. https://doi.org/10.1016/j.cger.2014.04.008

Kuhn HG, Dickinson-Anson H, Gage FH (1996) Neurogenesis in the dentate gyrus of the adult rat: age-related decrease of neuronal progenitor proliferation. J Neurosci 16(6):2027–2033. https://doi.org/10.1523/jneurosci.16-06-02027.1996

Lagercrantz H, Nilsson E, Redham I, Hjemdahl P (1986) Plasma catecholamines following nursing procedures in a neonatal ward. Early Hum Dev 14(1):61–65. https://doi.org/10.1016/0378-3782(86)90170-2

Langlais PJ, Savage LM (1995) Thiamine deficiency in rats produces cognitive and memory deficits on spatial tasks that correlate with tissue loss in diencephalon, cortex and white matter. Behav Brain Res 68(1):75–89. https://doi.org/10.1016/0166-4328(94)00162-9

Lao Y, Hou L, Li J, Hui X, Yan P, Yang K (2021) Association between alcohol intake, mild cognitive impairment and progression to dementia: a dose-response meta-analysis. Aging Clin Exp Res 33(5):1175–1185. https://doi.org/10.1007/s40520-020-01605-0

Latt N, Dore G (2014) Thiamine in the treatment of Wernicke encephalopathy in patients with alcohol use disorders. Intern Med J 44(9):911–915. https://doi.org/10.1111/imj.12522

Ledesma JC, Rodríguez-Arias M, Gavito AL, Sánchez-Pérez AM, Viña J, Medina Vera D, Rodríguez de Fonseca F, Miñarro J (2021) Adolescent binge-ethanol accelerates cognitive impairment and β-amyloid production and dysregulates endocannabinoid signaling in the hippocampus of APP/PSE mice. Addict Biol 26(1):e12883. https://doi.org/10.1111/adb.12883

Li SF, Jacob J, Feng J, Kulkarni M (2008) Vitamin deficiencies in acutely intoxicated patients in the ED. Am J Emerg Med 26(7):792–795. https://doi.org/10.1016/j.ajem.2007.10.003

Li H, Jia X, Qi Z, Fan X, Ma T, Ni H, Li CR, Li K (2017) Altered functional connectivity of the basal nucleus of Meynert in mild cognitive impairment: a resting-state fMRI study. Front Aging Neurosci 9:127. https://doi.org/10.3389/fnagi.2017.00127

Lippai D, Bala S, Csak T, Kurt-Jones EA, Szabo G (2013) Chronic alcohol-induced microRNA-155 contributes to neuroinflammation in a TLR4-dependent manner in mice. PLoS One 8(8):e70945. https://doi.org/10.1371/journal.pone.0070945

Listabarth S, Vyssoki B, Marculescu R, Gleiss A, Groemer M, Trojer A, Harrer C, Weber S, König D (2023) Can thiamine substitution restore cognitive function in alcohol use disorder? Alcohol Alcohol 58(3):315–323. https://doi.org/10.1093/alcalc/agad017

Liu D, Ke Z, Luo J (2017) Thiamine deficiency and neurodegeneration: the interplay among oxidative stress, endoplasmic reticulum stress, and autophagy. Mol Neurobiol 54(7):5440–5448. https://doi.org/10.1007/s12035-016-0079-9

Livingston G, Huntley J, Sommerlad A, Ames D, Ballard C, Banerjee S, Brayne C, Burns A, Cohen-Mansfield J, Cooper C, Costafreda SG, Dias A, Fox N, Gitlin LN, Howard R, Kales HC, Kivimäki M, Larson EB, Ogunniyi A, Orgeta V, Ritchie K, Rockwood K, Sampson EL, Samus Q, Schneider LS, Selbæk G, Teri L, Mukadam N (2020) Dementia prevention, intervention, and care: 2020 report of the Lancet Commission. Lancet 396(10248):413–446. https://doi.org/10.1016/s0140-6736(20)30367-6

Loane DJ, Byrnes KR (2010) Role of microglia in neurotrauma. Neurotherapeutics 7(4):366–377. https://doi.org/10.1016/j.nurt.2010.07.002

Lowe PP, Morel C, Ambade A, Iracheta-Vellve A, Kwiatkowski E, Satishchandran A, Furi I, Cho Y, Gyongyosi B, Catalano D, Lefebvre E, Fischer L, Seyedkazemi S, Schafer DP, Szabo G (2020) Chronic alcohol-induced neuroinflammation involves CCR2/5-dependent peripheral macrophage infiltration and microglia alterations. J Neuroinflammation 17(1):296. https://doi.org/10.1186/s12974-020-01972-5

Luchsinger JA, Tang MX, Siddiqui M, Shea S, Mayeux R (2004) Alcohol intake and risk of dementia. J Am Geriatr Soc 52(4):540–546. https://doi.org/10.1111/j.1532-5415.2004.52159.x

Lukoyanov NV, Andrade JP, Dulce Madeira M, Paula-Barbosa MM (1999) Effects of age and sex on the water maze performance and hippocampal cholinergic fibers in rats. Neurosci Lett 269(3):141–144. https://doi.org/10.1016/s0304-3940(99)00442-5

Lynch AM, Murphy KJ, Deighan BF, O'Reilly JA, Gun'ko YK, Cowley TR, Gonzalez-Reyes RE, Lynch MA (2010) The impact of glial activation in the aging brain. Aging Dis 1(3):262–278

Macht V, Vetreno R, Elchert N, Crews F (2021) Galantamine prevents and reverses neuroimmune induction and loss of adult hippocampal neurogenesis following adolescent alcohol exposure. J Neuroinflammation 18(1):212. https://doi.org/10.1186/s12974-021-02243-7

Macht V, Vetreno R, Elchert N, Fisher R, Crews F (2023) Indomethacin restores loss of hippocampal neurogenesis and cholinergic innervation and reduces innate immune expression and reversal learning deficits in adult male and female rats following adolescent ethanol exposure. Alcohol Clin Exp Res (Hoboken) 47(3):470–485. https://doi.org/10.1111/acer.15019

Maillard A, Laniepce A, Cabé N, Boudehent C, Chételat G, Urso L, Eustache F, Vabret F, Segobin S, Pitel AL (2021) Temporal cognitive and brain changes in Korsakoff syndrome. Neurology 96(15):e1987–e1998. https://doi.org/10.1212/wnl.0000000000011749

Maisto SA, Connors GJ, Allen JP (1995) Contrasting self-report screens for alcohol problems: a review. Alcohol Clin Exp Res 19(6):1510–1516. https://doi.org/10.1111/j.1530-0277.1995.tb01015.x

Marlatt MW, Potter MC, Lucassen PJ, van Praag H (2012) Running throughout middle-age improves memory function, hippocampal neurogenesis, and BDNF levels in female C57BL/6J mice. Dev Neurobiol 72(6):943–952. https://doi.org/10.1002/dneu.22009

Marrs C, Lonsdale D (2021) Hiding in plain sight: modern thiamine deficiency. Cells 10(10). https://doi.org/10.3390/cells10102595

Marschallinger J, Iram T, Zardeneta M, Lee SE, Lehallier B, Haney MS, Pluvinage JV, Mathur V, Hahn O, Morgens DW, Kim J, Tevini J, Felder TK, Wolinski H, Bertozzi CR, Bassik MC, Aigner L, Wyss-Coray T (2020) Lipid-droplet-accumulating microglia represent a dysfunctional and proinflammatory state in the aging brain. Nat Neurosci 23(2):194–208. https://doi.org/10.1038/s41593-019-0566-1

Marshall SA, Geil CR, Nixon K (2016) Prior binge ethanol exposure potentiates the microglial response in a model of alcohol-induced neurodegeneration. Brain Sci 6(2). https://doi.org/10.3390/brainsci6020016

Marshall SA, McClain JA, Wooden JI, Nixon K (2020) Microglia dystrophy following binge-like alcohol exposure in adolescent and adult male rats. Front Neuroanat 14:52. https://doi.org/10.3389/fnana.2020.00052

Marsland P, Parrella A, Orlofsky M, Lovelock DF, Vore AS, Varlinskaya EI, Deak T (2022) Neuroendocrine and neuroimmune responses in male and female rats: evidence for functional immaturity of the neuroimmune system during early adolescence. Eur J Neurosci 55(9–10):2311–2325. https://doi.org/10.1111/ejn.15118

Marsland P, Vore AS, DaPrano E, Paluch JM, Blackwell AA, Varlinskaya EI, Deak T (2023) Sex-specific effects of ethanol consumption in older Fischer 344 rats on microglial dynamics and Aβ((1-42)) accumulation. Alcohol 107:108–118. https://doi.org/10.1016/j.alcohol.2022.08.013

Martínez-Serrano A, Björklund A (1998) Ex vivo nerve growth factor gene transfer to the basal forebrain in presymptomatic middle-aged rats prevents the development of cholinergic neuron atrophy and cognitive impairment during aging. Proc Natl Acad Sci USA 95(4):1858–1863. https://doi.org/10.1073/pnas.95.4.1858

Matthews DB, Novier A, Diaz-Granados JL, Van Skike CE, Ornelas L, Mittleman G (2017) Impact of adolescent alcohol use across the lifespan: Long-lasting tolerance to high-dose alcohol coupled with potentiated spatial memory impairments to moderate-dose alcohol. Alcohol 61:33–42. https://doi.org/10.1016/j.alcohol.2017.01.012

Matthews DB, Scaletty S, Trapp S, Kastner A, Schneider AM, Schreiber A, Rossmann G (2022) Chronic intermittent ethanol administration during adolescence produces sex dependent impairments in behavioral flexibility and survivability. Brain Sci 12(5). https://doi.org/10.3390/brainsci12050606

Matthews DB, Scaletty S, Trapp S, Schreiber A, Rossmann G, Imhoff B, Petersilka Q, Kastner A, Pauly J, Nixon K (2023) Chronic intermittent ethanol exposure during adolescence produces sex- and age-dependent changes in anxiety and cognition without changes in microglia reactivity late in life. Front Behav Neurosci 17:1223883. https://doi.org/10.3389/fnbeh.2023.1223883

Mattsson N, Tosun D, Insel PS, Simonson A, Jack CR Jr, Beckett LA, Donohue M, Jagust W, Schuff N, Weiner MW (2014) Association of brain amyloid-β with cerebral perfusion and structure in Alzheimer's disease and mild cognitive impairment. Brain 137(Pt 5):1550–1561. https://doi.org/10.1093/brain/awu043

May PA, Chambers CD, Kalberg WO, Zellner J, Feldman H, Buckley D, Kopald D, Hasken JM, Xu R, Honerkamp-Smith G, Taras H, Manning MA, Robinson LK, Adam MP, Abdul-Rahman O, Vaux K, Jewett T, Elliott AJ, Kable JA, Akshoomoff N, Falk D, Arroyo JA, Hereld D, Riley EP, Charness ME, Coles CD, Warren KR, Jones KL, Hoyme HE (2018) Prevalence of fetal alcohol spectrum disorders in 4 US communities. JAMA 319(5):474–482. https://doi.org/10.1001/jama.2017.21896

Mayes AR, Meudell PR, Mann D, Pickering A (1988) Location of lesions in Korsakoff's syndrome: neuropsychological and neuropathological data on two patients. Cortex 24(3):367–388. https://doi.org/10.1016/s0010-9452(88)80001-7

McClain JA, Morris SA, Deeny MA, Marshall SA, Hayes DM, Kiser ZM, Nixon K (2011) Adolescent binge alcohol exposure induces long-lasting partial activation of microglia. Brain Behav Immun 25(suppl 1):S120–S128. https://doi.org/10.1016/j.bbi.2011.01.006

McCool BA, Chappell AM (2009) Early social isolation in male Long-Evans rats alters both appetitive and consummatory behaviors expressed during operant ethanol self-administration. Alcohol Clin Exp Res 33(2):273–282. https://doi.org/10.1111/j.1530-0277.2008.00830.x

McGinnis MM, Parrish BC, McCool BA (2020) Withdrawal from chronic ethanol exposure increases postsynaptic glutamate function of insular cortex projections to the rat basolateral amygdala. Neuropharmacology 172:108129. https://doi.org/10.1016/j.neuropharm.2020.108129

McQuail JA, Dunn AR, Stern Y, Barnes CA, Kempermann G, Rapp PR, Kaczorowski CC, Foster TC (2020) Cognitive reserve in model systems for mechanistic discovery: the importance of longitudinal studies. Front Aging Neurosci 12:607685. https://doi.org/10.3389/fnagi.2020.607685

Meier P, Seitz HK (2008) Age, alcohol metabolism and liver disease. Curr Opin Clin Nutr Metab Care 11(1):21–26. https://doi.org/10.1097/MCO.0b013e3282f30564

Mewton L, Visontay R, Hoy N, Lipnicki DM, Sunderland M, Lipton RB, Guerchet M, Ritchie K, Najar J, Scarmeas N, Kim KW, Riedel Heller S, van Boxtel M, Jacobsen E, Brodaty H, Anstey KJ, Haan M, Scazufca M, Lobo E, Sachdev PS (2023) The relationship between alcohol use and dementia in adults aged more than 60 years: a combined analysis of prospective, individual-participant data from 15 international studies. Addiction 118(3):412–424. https://doi.org/10.1111/add.16035

Mieling M, Meier H, Bunzeck N (2023) Structural degeneration of the nucleus basalis of Meynert in mild cognitive impairment and Alzheimer's disease—evidence from an MRI-based meta-analysis. Neurosci Biobehav Rev 154:105393. https://doi.org/10.1016/j.neubiorev.2023.105393

Mimouni-Bloch A, Goldberg-Stern H, Strausberg R, Brezner A, Heyman E, Inbar D, Kivity S, Zvulunov A, Sztarkier I, Fogelman R, Fattal-Valevski A (2014) Thiamine deficiency in infancy: long-term follow-up. Pediatr Neurol 51(3):311–316. https://doi.org/10.1016/j.pediatrneurol.2014.05.010

Moca EN, Lecca D, Hope KT, Etienne F, Schaler AW, Espinoza K, Chappell MS, Gray DT, Tweedie D, Sidhu S, Masukawa L, Sitoy H, Mathew R, Saban DR, Greig NH, Biase LMD (2022) Microglia drive pockets of neuroinflammation in middle age. J Neurosci 42(19):3896–3918. https://doi.org/10.1523/jneurosci.1922-21.2022

Molofsky AV, Deneen B (2015) Astrocyte development: A Guide for the Perplexed. Glia 63(8):1320–1329. https://doi.org/10.1002/glia.22836

Monleón S, Duque A, Vinader-Caerols C (2020) Emotional memory impairment produced by binge drinking in mice is counteracted by the anti-inflammatory indomethacin. Behav Brain Res 381:112457. https://doi.org/10.1016/j.bbr.2019.112457

Monleón S, Gómez J, Duque A, Vinader-Caerols C (2022) Effects of binge drinking and the anti-inflammatory

drug indomethacin on spatial memory and cognitive flexibility in mice. Behav Brain Res 417:113619. https://doi.org/10.1016/j.bbr.2021.113619
Moreno-Jiménez EP, Flor-García M, Terreros-Roncal J, Rábano A, Cafini F, Pallas-Bazarra N, Ávila J, Llorens-Martín M (2019) Adult hippocampal neurogenesis is abundant in neurologically healthy subjects and drops sharply in patients with Alzheimer's disease. Nat Med 25(4):554–560. https://doi.org/10.1038/s41591-019-0375-9
Morgan MY (1982) Alcohol and nutrition. Br Med Bull 38(1):21–29. https://doi.org/10.1093/oxfordjournals.bmb.a071727
Morrison C, Dadar M, Shafiee N, Collins DL (2023) Hippocampal grading provides higher classification accuracy for those in the AD trajectory than hippocampal volume. Hum Brain Mapp 44(12):4623–4633. https://doi.org/10.1002/hbm.26407
Moya M, López-Valencia L, García-Bueno B, Orio L (2022) Disinhibition-like behavior correlates with frontal cortex damage in an animal model of chronic alcohol consumption and thiamine deficiency. Biomedicines 10(2). https://doi.org/10.3390/biomedicines10020260
Musiek ES, Holtzman DM (2015) Three dimensions of the amyloid hypothesis: time, space and 'wingmen'. Nat Neurosci 18(6):800–806. https://doi.org/10.1038/nn.4018
Nahum L, Pignat JM, Bouzerda-Wahlen A, Gabriel D, Liverani MC, Lazeyras F, Ptak R, Richiardi J, Haller S, Thorens G, Zullino DF, Guggisberg AG, Schnider A (2015) Neural correlate of anterograde amnesia in Wernicke-Korsakoff syndrome. Brain Topogr 28(5):760–770. https://doi.org/10.1007/s10548-014-0391-5
Nair J, Klaassen AL, Arato J, Vyssotski AL, Harvey M, Rainer G (2018) Basal forebrain contributes to default mode network regulation. Proc Natl Acad Sci USA 115(6):1352–1357. https://doi.org/10.1073/pnas.1712431115
Nardone R, Höller Y, Storti M, Christova M, Tezzon F, Golaszewski S, Trinka E, Brigo F (2013) Thiamine deficiency induced neurochemical, neuroanatomical, and neuropsychological alterations: a reappraisal. ScientificWorldJournal 2013:309143. https://doi.org/10.1155/2013/309143
Nelson PT, Alafuzoff I, Bigio EH, Bouras C, Braak H, Cairns NJ, Castellani RJ, Crain BJ, Davies P, Del Tredici K, Duyckaerts C, Frosch MP, Haroutunian V, Hof PR, Hulette CM, Hyman BT, Iwatsubo T, Jellinger KA, Jicha GA, Kövari E, Kukull WA, Leverenz JB, Love S, Mackenzie IR, Mann DM, Masliah E, McKee AC, Montine TJ, Morris JC, Schneider JA, Sonnen JA, Thal DR, Trojanowski JQ, Troncoso JC, Wisniewski T, Woltjer RL, Beach TG (2012) Correlation of Alzheimer disease neuropathologic changes with cognitive status: a review of the literature. J Neuropathol Exp Neurol 71(5):362–381. https://doi.org/10.1097/NEN.0b013e31825018f7
Neumann H, Kotter MR, Franklin RJ (2009) Debris clearance by microglia: an essential link between degeneration and regeneration. Brain 132(Pt 2):288–295. https://doi.org/10.1093/brain/awn109
Nichols NR, Day JR, Laping NJ, Johnson SA, Finch CE (1993) GFAP mRNA increases with age in rat and human brain. Neurobiol Aging 14(5):421–429
Nicolas B, Alessandra D, Daniela P, Osman R, Sara T, Valentina G, AsDN I (2020) Basal forebrain metabolism in Alzheimer's disease continuum: relationship with education. Neurobiol Aging 87:70–77
Norden DM, Godbout JP (2013) Review: microglia of the aged brain: primed to be activated and resistant to regulation. Neuropathol Appl Neurobiol 39(1):19–34. https://doi.org/10.1111/j.1365-2990.2012.01306.x
Norden DM, Fenn AM, Dugan A, Godbout JP (2014) TGFβ produced by IL-10 redirected astrocytes attenuates microglial activation. Glia 62(6):881–895
Norden DM, Muccigrosso MM, Godbout JP (2015) Microglial priming and enhanced reactivity to secondary insult in aging, and traumatic CNS injury, and neurodegenerative disease. Neuropharmacology 96(Pt A):29–41. https://doi.org/10.1016/j.neuropharm.2014.10.028
Norden DM, Trojanowski PJ, Walker FR, Godbout JP (2016) Insensitivity of astrocytes to interleukin 10 signaling following peripheral immune challenge results in prolonged microglial activation in the aged brain. Neurobiol Aging 44:22–41. https://doi.org/10.1016/j.neurobiolaging.2016.04.014
Novier A, Van Skike CE, Diaz-Granados JL, Mittleman G, Matthews DB (2013) Acute alcohol produces ataxia and cognitive impairments in aged animals: a comparison between young adult and aged rats. Alcohol Clin Exp Res 37(8):1317–1324. https://doi.org/10.1111/acer.12110
Nunes PT, Gómez-Mendoza DP, Rezende CP, Figueiredo HCP, Ribeiro AM (2018) Thalamic Proteome Changes and Behavioral Impairments in Thiamine-deficient Rats. Neuroscience 385:181–197. https://doi.org/10.1016/j.neuroscience.2018.06.003
Nunes PT, Kipp BT, Reitz NL, Savage LM (2019) Aging with alcohol-related brain damage: Critical brain circuits associated with cognitive dysfunction. Int Rev Neurobiol 148:101–168. https://doi.org/10.1016/bs.irn.2019.09.002
Oddo S, Caccamo A, Shepherd JD, Murphy MP, Golde TE, Kayed R, Metherate R, Mattson MP, Akbari Y, LaFerla FM (2003) Triple-transgenic model of Alzheimer's disease with plaques and tangles: intracellular Abeta and synaptic dysfunction. Neuron 39(3):409–421. https://doi.org/10.1016/s0896-6273(03)00434-3
Ogershok PR, Rahman A, Nestor S, Brick J (2002) Wernicke encephalopathy in nonalcoholic patients. Am J Med Sci 323(2):107–111. https://doi.org/10.1097/00000441-200202000-00010
Ogunshola OO, Stewart WB, Mihalcik V, Solli T, Madri JA, Ment LR (2000) Neuronal VEGF expression correlates with angiogenesis in postnatal developing rat

brain. Brain Res Dev Brain Res 119(1):139–153. https://doi.org/10.1016/s0165-3806(99)00125-x

O'Neil SM, Hans EE, Jiang S, Wangler LM, Godbout JP (2022) Astrocyte immunosenescence and deficits in interleukin 10 signaling in the aged brain disrupt the regulation of microglia following innate immune activation. Glia 70(5):913–934. https://doi.org/10.1002/glia.24147

Ornelas LC, Novier A, Van Skike CE, Diaz-Granados JL, Matthews DB (2015) The effects of acute alcohol on motor impairments in adolescent, adult, and aged rats. Alcohol 49(2):121–126. https://doi.org/10.1016/j.alcohol.2014.12.002

Otto SL, Yakel JL (2019) The α7 nicotinic acetylcholine receptors regulate hippocampal adult-neurogenesis in a sexually dimorphic fashion. Brain Struct Funct 224(2):829–846. https://doi.org/10.1007/s00429-018-1799-6

Oudman E, Wijnia JW, Oey M, van Dam M, Painter RC, Postma A (2019) Wernicke's encephalopathy in hyperemesis gravidarum: A systematic review. Eur J Obstet Gynecol Reprod Biol 236:84–93. https://doi.org/10.1016/j.ejogrb.2019.03.006

Ownby RL (2010) Neuroinflammation and Cognitive Aging. Current Psychiatry Reports 12(1):39–45. https://doi.org/10.1007/s11920-009-0082-1

Palm A, Vataja R, Talaslahti T, Ginters M, Kautiainen H, Elonheimo H, Suvisaari J, Lindberg N, Koponen H (2022) Incidence and mortality of alcohol-related dementia and Wernicke-Korsakoff syndrome: A nationwide register study. Int J Geriatr Psychiatry 37(8). https://doi.org/10.1002/gps.5775

Parikh V, Howe WM, Welchko RM, Naughton SX, D'Amore DE, Han DH, Deo M, Turner DL, Sarter M (2013) Diminished trkA receptor signaling reveals cholinergic-attentional vulnerability of aging. Eur J Neurosci 37(2):278–293. https://doi.org/10.1111/ejn.12090

Pascual M, Blanco AM, Cauli O, Miñarro J, Guerri C (2007) Intermittent ethanol exposure induces inflammatory brain damage and causes long-term behavioural alterations in adolescent rats. Eur J Neurosci 25(2):541–550. https://doi.org/10.1111/j.1460-9568.2006.05298.x

Peng H, Geil Nickell CR, Chen KY, McClain JA, Nixon K (2017) Increased expression of M1 and M2 phenotypic markers in isolated microglia after four-day binge alcohol exposure in male rats. Alcohol 62:29–40. https://doi.org/10.1016/j.alcohol.2017.02.175

Perkins AE, Doremus-Fitzwater TL, Spencer RL, Varlinskaya EI, Conti MM, Bishop C, Deak T (2016) A working model for the assessment of disruptions in social behavior among aged rats: The role of sex differences, social recognition, and sensorimotor processes. Exp Gerontol 76:46–57. https://doi.org/10.1016/j.exger.2016.01.012

Perkins AE, Vore AS, Lovelock D, Varlinskaya E, Deak T (2018) Late aging alters behavioral sensitivity to ethanol in a sex-specific manner in Fischer 344 rats. Pharmacol Biochem Behav 175:1–9. https://doi.org/10.1016/j.pbb.2018.07.012

Perkins AE, Varlinskaya EI, Deak T (2019) From adolescence to late aging: A comprehensive review of social behavior, alcohol, and neuroinflammation across the lifespan. Int Rev Neurobiol 148:231–303. https://doi.org/10.1016/bs.irn.2019.08.001

Petersen RC, Caracciolo B, Brayne C, Gauthier S, Jelic V, Fratiglioni L (2014) Mild cognitive impairment: a concept in evolution. J Intern Med 275(3):214–228. https://doi.org/10.1111/joim.12190

Pfefferbaum A, Sullivan EV, Mathalon DH, Shear PK, Rosenbloom MJ, Lim KO (1995) Longitudinal changes in magnetic resonance imaging brain volumes in abstinent and relapsed alcoholics. Alcohol Clin Exp Res 19(5):1177–1191. https://doi.org/10.1111/j.1530-0277.1995.tb01598.x

Pinson MR, Chung DD, Adams AM, Scopice C, Payne EA, Sivakumar M, Miranda RC (2021) Extracellular vesicles in premature aging and diseases in adulthood due to developmental exposures. Aging Dis 12(6):1516–1535. https://doi.org/10.14336/ad.2021.0322

Pitel AL, Zahr NM, Jackson K, Sassoon SA, Rosenbloom MJ, Pfefferbaum A, Sullivan EV (2011) Signs of preclinical Wernicke's encephalopathy and thiamine levels as predictors of neuropsychological deficits in alcoholism without Korsakoff's syndrome. Neuropsychopharmacology 36(3):580–588. https://doi.org/10.1038/npp.2010.189

Pitel AL, Chételat G, Le Berre AP, Desgranges B, Eustache F, Beaunieux H (2012) Macrostructural abnormalities in Korsakoff syndrome compared with uncomplicated alcoholism. Neurology 78(17):1330–1333. https://doi.org/10.1212/WNL.0b013e318251834e

Pitel AL, Segobin SH, Ritz L, Eustache F, Beaunieux H (2015) Thalamic abnormalities are a cardinal feature of alcohol-related brain dysfunction. Neurosci Biobehav Rev 54:38–45. https://doi.org/10.1016/j.neubiorev.2014.07.023

Pitkin SR, Savage LM (2001) Aging potentiates the acute and chronic neurological symptoms of pyrithiamine-induced thiamine deficiency in the rodent. Behav Brain Res 119(2):167–177. https://doi.org/10.1016/s0166-4328(00)00350-8

Pitkin SR, Savage LM (2004) Age-related vulnerability to diencephalic amnesia produced by thiamine deficiency: the role of time of insult. Behav Brain Res 148(1–2):93–105. https://doi.org/10.1016/s0166-4328(03)00208-0

Praharaj SK, Munoli RN, Shenoy S, Udupa ST, Thomas LS (2021) High-dose thiamine strategy in Wernicke-Korsakoff syndrome and related thiamine deficiency conditions associated with alcohol use disorder. Indian J Psychiatry 63(2):121–126. https://doi.org/10.4103/psychiatry.IndianJPsychiatry_440_20

Pruckner N, Baumgartner J, Hinterbuchinger B, Glahn A, Vyssoki S, Vyssoki B (2019) Thiamine substitution in alcohol use disorder: a narrative review of medical guidelines. Eur Addict Res 25(3):103–110. https://doi.org/10.1159/000499039

Qi J, Li BZ, Zhang Y, Pan B, Gao YH, Zhan H, Liu Y, Shao YC, Zhang X (2021) Altered functional connectivity between the nucleus basalis of Meynert and anterior cingulate cortex is associated with declined attentional performance after total sleep deprivation. Behav Brain Res 409:113321. https://doi.org/10.1016/j.bbr.2021.113321

Qin L, He J, Hanes RN, Pluzarev O, Hong JS, Crews FT (2008) Increased systemic and brain cytokine production and neuroinflammation by endotoxin following ethanol treatment. J Neuroinflammation 5:10. https://doi.org/10.1186/1742-2094-5-10

Rane MA, Boorugu HK, Ravishankar U, Tarakeswari S, Vadlamani H, Anand H (2022) Wernicke's encephalopathy in women with hyperemesis gravidarum—case series and literature review. Trop Dr 52(1):98–100. https://doi.org/10.1177/00494755211014396

Ravenel JR, Perkins AE, Tomczik A, Defendini A, Strnad HK, Varlinskaya E, Deak T, Spencer RL (2024) Age-related decline in social interaction is associated with decreased c-Fos induction in select brain regions independent of oxytocin receptor expression profiles. Aging Brain 5:100107. https://doi.org/10.1016/j.nbas.2024.100107

Rehm J, Hasan OSM, Black SE, Shield KD, Schwarzinger M (2019) Alcohol use and dementia: a systematic scoping review. Alzheimers Res Ther 11(1):1. https://doi.org/10.1186/s13195-018-0453-0

Reid N, Dawe S, Shelton D, Harnett P, Warner J, Armstrong E, LeGros K, O'Callaghan F (2015) Systematic review of fetal alcohol spectrum disorder interventions across the life span. Alcohol Clin Exp Res 39(12):2283–2295. https://doi.org/10.1111/acer.12903

Reitz NL, Nunes PT, Savage LM (2021) Adolescent binge-type ethanol exposure in rats mirrors age-related cognitive decline by suppressing cholinergic tone and hippocampal neurogenesis. Front Behav Neurosci 15:772857. https://doi.org/10.3389/fnbeh.2021.772857

Reitz NL, Nunes PT, Savage LM (2024) Exercise leads to sex-specific recovery of behavior and pathological AD markers following adolescent ethanol exposure in the TgF344-AD model. Front Behav Neurosci 18:1448691. https://doi.org/10.3389/fnbeh.2024.1448691

Ren P, Ding W, Li S, Liu G, Luo M, Zhou W, Cheng R, Li Y, Wang P, Li Z, Yao L, Jiang Q, Liang X (2023) Regional transcriptional vulnerability to basal forebrain functional dysconnectivity in mild cognitive impairment patients. Neurobiol Dis 177:105983. https://doi.org/10.1016/j.nbd.2022.105983

Rice D, Barone S Jr (2000) Critical periods of vulnerability for the developing nervous system: evidence from humans and animal models. Environ Health Perspect 108(Suppl 3):511–533. https://doi.org/10.1289/ehp.00108s3511

Richter N, David LS, Grothe MJ, Teipel S, Dietlein M, Tittgemeyer M, Neumaier B, Fink GR, Onur OA, Kukolja J (2022) Age and anterior basal forebrain volume predict the cholinergic deficit in patients with mild cognitive impairment due to Alzheimer's disease. J Alzheimers Dis 86(1):425–440. https://doi.org/10.3233/jad-210261

Ridley NJ, Draper B, Withall A (2013) Alcohol-related dementia: an update of the evidence. Alzheimers Res Ther 5(1):3. https://doi.org/10.1186/alzrt157

Rintala J, Jaatinen P, Kiianmaa K, Riikonen J, Kemppainen O, Sarviharju M, Hervonen A (2001) Dose-dependent decrease in glial fibrillary acidic protein-immunoreactivity in rat cerebellum after lifelong ethanol consumption. Alcohol 23(1):1–8. https://doi.org/10.1016/S0741-8329(00)00116-6

Roberto M, Schweitzer P, Madamba SG, Stouffer DG, Parsons LH, Siggins GR (2004) Acute and chronic ethanol alter glutamatergic transmission in rat central amygdala: an in vitro and in vivo analysis. J Neurosci 24(7):1594–1603. https://doi.org/10.1523/jneurosci.5077-03.2004

Robillard KN, Lee KM, Chiu KB, MacLean AG (2016) Glial cell morphological and density changes through the lifespan of rhesus macaques. Brain Behav Immun 55:60–69. https://doi.org/10.1016/j.bbi.2016.01.006

Rock RB, Gekker G, Hu S, Sheng WS, Cheeran M, Lokensgard JR, Peterson PK (2004) Role of microglia in central nervous system infections. Clin Microbiol Rev 17(4):942–964. https://doi.org/10.1128/cmr.17.4.942-964.2004. Table of contents

Roecklein B, Levin SW, Comly M, Mukherjee AB (1985) Intrauterine growth retardation induced by thiamine deficiency and pyrithiamine during pregnancy in the rat. Am J Obstet Gynecol 151(4):455–460. https://doi.org/10.1016/0002-9378(85)90269-8

Rowan RA, Maxwell DS (1981) Patterns of vascular sprouting in the postnatal development of the cerebral cortex of the rat. Am J Anat 160(3):247–255. https://doi.org/10.1002/aja.1001600303

Rozovsky I, Finch C, Morgan T (1998) Age-related activation of microglia and astrocytes: in vitro studies show persistent phenotypes of aging, increased proliferation, and resistance to down-regulation. Neurobiol Aging 19(1):97–103

Sabia S, Fayosse A, Dumurgier J, Dugravot A, Akbaraly T, Britton A, Kivimäki M, Singh-Manoux A (2018) Alcohol consumption and risk of dementia: 23 year follow-up of Whitehall II cohort study. BMJ 362:k2927. https://doi.org/10.1136/bmj.k2927

Safaiyan S, Kannaiyan N, Snaidero N, Brioschi S, Biber K, Yona S, Edinger AL, Jung S, Rossner MJ, Simons M (2016) Age-related myelin degradation burdens the clearance function of microglia during aging. Nat Neurosci 19(8):995–998. https://doi.org/10.1038/nn.4325

Salta E, Lazarov O, Fitzsimons C, Tanzi R, Lucasses P, Choi SH (2023) Adult hippocampal neurogenesis in Alzheimer's disease: A roadmap to clinical relevance. Cell Stem Cell 30(2):120–136. https://doi.org/10.1016/j.stem.2023.01.002

Salthouse TA (2019) Trajectories of normal cognitive aging. Psychol Aging 34(1):17–24. https://doi.org/10.1037/pag0000288

Sambon M, Wins P, Bettendorff L (2021) Neuroprotective effects of thiamine and precursors with higher bioavailability: focus on benfotiamine and dibenzoylthiamine. Int J Mol Sci 22(11). https://doi.org/10.3390/ijms22115418

Savage LM, Hall JM, Resende LS (2012) Translational rodent models of Korsakoff syndrome reveal the critical neuroanatomical substrates of memory dysfunction and recovery. Neuropsychol Rev 22(2):195–209. https://doi.org/10.1007/s11065-012-9194-1

Savage LM, Nunes PT, Gursky ZH, Milbocker KA, Klintsova AY (2021) Midline thalamic damage associated with alcohol-use disorders: disruption of distinct thalamocortical pathways and function. Neuropsychol Rev 31(3):447–471. https://doi.org/10.1007/s11065-020-09450-8

Schmitz TW, Nathan Spreng R (2016) Basal forebrain degeneration precedes and predicts the cortical spread of Alzheimer's pathology. Nat Commun 7:13249. https://doi.org/10.1038/ncomms13249

Schmitz TW, Mur M, Aghourian M, Bedard MA, Spreng RN (2018) Longitudinal Alzheimer's degeneration reflects the spatial topography of cholinergic basal forebrain projections. Cell Rep 24(1):38–46. https://doi.org/10.1016/j.celrep.2018.06.001

Schwarzinger M, Pollock BG, Hasan OSM, Dufouil C, Rehm J (2018) Contribution of alcohol use disorders to the burden of dementia in France 2008-13: a nationwide retrospective cohort study. Lancet Public Health 3(3):e124–e132. https://doi.org/10.1016/s2468-2667(18)30022-7

Scott GA, Terstege DJ, Roebuck AJ, Gorzo KA, Vu AP, Howland JG, Epp JR (2021) Adult neurogenesis mediates forgetting of multiple types of memory in the rat. Mol Brain 14(1):97. https://doi.org/10.1186/s13041-021-00808-4

Segobin S, Pitel AL (2021) The specificity of thalamic alterations in Korsakoff's syndrome: Implications for the study of amnesia. Neurosci Biobehav Rev 130:292–300. https://doi.org/10.1016/j.neubiorev.2021.07.037

Segobin SH, Chételat G, Le Berre AP, Lannuzel C, Boudehent C, Vabret F, Eustache F, Beaunieux H, Pitel AL (2014) Relationship between brain volumetric changes and interim drinking at six months in alcohol-dependent patients. Alcohol Clin Exp Res 38(3):739–748. https://doi.org/10.1111/acer.12300

Segobin S, Laniepce A, Ritz L, Lannuzel C, Boudehent C, Cabé N, Urso L, Vabret F, Eustache F, Beaunieux H, Pitel AL (2019) Dissociating thalamic alterations in alcohol use disorder defines specificity of Korsakoff's syndrome. Brain 142(5):1458–1470. https://doi.org/10.1093/brain/awz056

Segobin S, Ambler M, Laniepce A, Platel H, Chételat G, Groussard M, Pitel AL (2023) Korsakoff's syndrome and Alzheimer's disease-commonalities and specificities of volumetric brain alterations within Papez circuit. J Clin Med 12(9). https://doi.org/10.3390/jcm12093147

Shao X, Kong WX, Li YT (2019) MiR-133 inhibits kidney injury in rats with diabetic nephropathy via MAPK/ERK pathway. Eur Rev Med Pharmacol Sci 23(24):10957–10963. https://doi.org/10.26355/eurrev_201912_19799

Shillinglaw JE, Morrisett RA, Mangieri RA (2018) Ethanol modulates glutamatergic transmission and NMDAR-mediated synaptic plasticity in the agranular insular cortex. Front Pharmacol 9:1458. https://doi.org/10.3389/fphar.2018.01458

Sierra A, Gottfried-Blackmore A, McEwen BS, Bulloch K (2007) Microglia derived from aging mice exhibit an altered inflammatory profile. Glia 55

Sierra A, Encinas JM, Deudero JJ, Chancey JH, Enikolopov G, Overstreet-Wadiche LS, Tsirka SE, Maletic-Savatic M (2010) Microglia shape adult hippocampal neurogenesis through apoptosis-coupled phagocytosis. Cell Stem Cell 7(4):483–495. https://doi.org/10.1016/j.stem.2010.08.014

Slawecki CJ, Roth J, Gilder A (2006) Neurobehavioral profiles during the acute phase of ethanol withdrawal in adolescent and adult Sprague-Dawley rats. Behav Brain Res 170(1):41–51. https://doi.org/10.1016/j.bbr.2006.01.023

Smith DM, Atkinson RM (1995) Alcoholism and dementia. Int J Addict 30(13-14):1843–1869. https://doi.org/10.3109/10826089509071058

Smith TJ, Johnson CR, Koshy R, Hess SY, Qureshi UA, Mynak ML, Fischer PR (2021) Thiamine deficiency disorders: a clinical perspective. Ann N Y Acad Sci 1498(1):9–28. https://doi.org/10.1111/nyas.14536

Smith SM, Pjetri E, Friday WB, Presswood BH, Ricketts DK, Walter KR, Mooney SM (2022) Aging-related behavioral, adiposity, and glucose impairments and their association following prenatal alcohol exposure in the C57BL/6J mouse. Nutrients 14(7). https://doi.org/10.3390/nu14071438

Sofroniew MV, Vinters HV (2010) Astrocytes: biology and pathology. Acta Neuropathol 119(1):7–35. https://doi.org/10.1007/s00401-009-0619-8

Soreq L, Rose J, Soreq E, Hardy J, Trabzuni D, Cookson MR, Smith C, Ryten M, Patani R, Ule J (2017) Major shifts in glial regional identity are a transcriptional hallmark of human brain aging. Cell Rep 18(2):557–570. https://doi.org/10.1016/j.celrep.2016.12.011

Spear LP (2018) Effects of adolescent alcohol consumption on the brain and behaviour. Nat Rev Neurosci 19(4):197–214. https://doi.org/10.1038/nrn.2018.10

Stanley EM, Fadel J (2012) Aging-related deficits in orexin/hypocretin modulation of the septohippocampal cholinergic system. Synapse 66(5):445–452. https://doi.org/10.1002/syn.21533

Stazi M, Wirths O (2021) Chronic memantine treatment ameliorates behavioral deficits, neuron loss, and impaired neurogenesis in a model of Alzheimer's disease. Mol Neurobiol 58(1):204–216. https://doi.org/10.1007/s12035-020-02120-z

Stepanichev M, Aniol V, Lazareva N, Gulyaeva N (2023) Decreased hippocampal neurogenesis in aged male wistar rats is not associated with memory acquisition in a water maze. Int J Mol Sci 24(17). https://doi.org/10.3390/ijms241713276

Streit WJ, Sammons NW, Kuhns AJ, Sparks DL (2004) Dystrophic microglia in the aging human brain. Glia 45(2):208–212

Sugaya K, Greene R, Personett D, Robbins M, Kent C, Bryan D, Skiba E, Gallagher M, McKinney M (1998) Septo-hippocampal cholinergic and neurotrophin markers in age-induced cognitive decline. Neurobiol Aging 19(4):351–361. https://doi.org/10.1016/s0197-4580(98)00072-4

Sullivan EV, Pfefferbaum A (2009) Neuroimaging of the Wernicke-Korsakoff syndrome. Alcohol Alcohol 44(2):155–165. https://doi.org/10.1093/alcalc/agn103

Sullivan EV, Zahr NM, Sassoon SA, Thompson WK, Kwon D, Pohl KM, Pfefferbaum A (2018) The role of aging, drug dependence, and hepatitis c comorbidity in alcoholism cortical compromise. JAMA Psychiatry 75(5):474–483. https://doi.org/10.1001/jamapsychiatry.2018.0021

Swartzwelder HS, Acheson SK, Miller KM, Sexton HG, Liu W, Crews FT, Risher ML (2015) Adolescent intermittent alcohol exposure: deficits in object recognition memory and forebrain cholinergic markers. PLoS One 10(11):e0140042. https://doi.org/10.1371/journal.pone.0140042

Telles TM, de Oliveira BM, Lomba LA, Leite-Avalca MG, Correia D, Zampronio AR (2017) Effects of binge-like ethanol exposure during adolescence on the febrile response in rats. Alcohol Clin Exp Res 41(3):507–515. https://doi.org/10.1111/acer.13333

Tevik K, Bergh S, Selbæk G, Johannessen A, Helvik AS (2021) A systematic review of self-report measures used in epidemiological studies to assess alcohol consumption among older adults. PLoS One 16(12):e0261292. https://doi.org/10.1371/journal.pone.0261292

Thomas AL, Lehn MA, Janssen EM, Hildeman DA, Chougnet CA (2022) Naturally-aged microglia exhibit phagocytic dysfunction accompanied by gene expression changes reflective of underlying neurologic disease. Sci Rep 12(1):19471. https://doi.org/10.1038/s41598-022-21920-y

Tichauer JE, Flores B, Soler B, Eugenín-von Bernhardi L, Ramírez G, von Bernhardi R (2014) Age-dependent changes on TGFβ1 Smad3 pathway modify the pattern of microglial cell activation. Brain Behav Immun 37:187–196

Tiernan CT, Ginsberg SD, He B, Ward SM, Guillozet-Bongaarts AL, Kanaan NM, Mufson EJ, Counts SE (2018) Pretangle pathology within cholinergic nucleus basalis neurons coincides with neurotrophic and neurotransmitter receptor gene dysregulation during the progression of Alzheimer's disease. Neurobiol Dis 117:125–136. https://doi.org/10.1016/j.nbd.2018.05.021

Tobin MK, Musaraca K, Disouky A, Shetti A, Bheri A, Honer WG, Kim N, Dawe RJ, Bennett DA, Arfanakis K, Lazarov O (2019) Human hippocampal neurogenesis persists in aged adults and Alzheimer's disease patients. Cell Stem Cell 24(6):974–982.e973. https://doi.org/10.1016/j.stem.2019.05.003

Toledo Nunes P, Vedder LC, Deak T, Savage LM (2019) A pivotal role for thiamine deficiency in the expression of neuroinflammation markers in models of alcohol-related brain damage. Alcohol Clin Exp Res 43(3):425–438. https://doi.org/10.1111/acer.13946

Tousley AR, Yeh PWL, Yeh HH (2023) Precocious emergence of cognitive and synaptic dysfunction in 3xTg-AD mice exposed prenatally to ethanol. Alcohol 107:56–72. https://doi.org/10.1016/j.alcohol.2022.08.003

Trostler N, Guggenheim K, Havivi E, Sklan D (1977) Effect of thiamine deficiency in pregnant and lactating rats on the brain of their offspring. Nutr Metab 21(5):294–304. https://doi.org/10.1159/000176075

Tucker AE, Alicea Pauneto CDM, Barnett AM, Coleman LG Jr (2022) Chronic ethanol causes persistent increases in Alzheimer's tau pathology in female 3xTg-AD mice: a potential role for lysosomal impairment. Front Behav Neurosci 16:886634. https://doi.org/10.3389/fnbeh.2022.886634

Vana L, Kanaan NM, Ugwu IC, Wuu J, Mufson EJ, Binder LI (2011) Progression of tau pathology in cholinergic Basal forebrain neurons in mild cognitive impairment and Alzheimer's disease. Am J Pathol 179(5):2533–2550. https://doi.org/10.1016/j.ajpath.2011.07.044

Vedder LC, Hall JM, Jabrouin KR, Savage LM (2015) Interactions between chronic ethanol consumption and thiamine deficiency on neural plasticity, spatial memory, and cognitive flexibility. Alcohol Clin Exp Res 39(11):2143–2153. https://doi.org/10.1111/acer.12859

Vemuganti R, Kalluri H, Yi JH, Bowen KK, Hazell AS (2006) Gene expression changes in thalamus and inferior colliculus associated with inflammation, cellular stress, metabolism and structural damage in thiamine deficiency. Eur J Neurosci 23(5):1172–1188. https://doi.org/10.1111/j.1460-9568.2006.04651.x

Verges DK, Restivo JL, Goebel WD, Holtzman DM, Cirrito JR (2011) Opposing synaptic regulation of amyloid-β metabolism by NMDA receptors in vivo. J Neurosci 31(31):11328–11337. https://doi.org/10.1523/jneurosci.0607-11.2011

Verkhratsky A, Semyanov A (2023) Astrocytes in ageing. Subcell Biochem 103:253–277. https://doi.org/10.1007/978-3-031-26576-1_11

Vestal RE, McGuire EA, Tobin JD, Andres R, Norris AH, Mezey E (1977) Aging and ethanol metabolism. Clin Pharmacol Ther 21(3):343–354. https://doi.org/10.1002/cpt1977213343

Vetreno RP, Crews FT (2012) Adolescent binge drinking increases expression of the danger signal receptor agonist HMGB1 and Toll-like receptors in the adult prefrontal cortex. Neuroscience 226:475–488. https://doi.org/10.1016/j.neuroscience.2012.08.046

Vetreno RP, Crews FT (2015) Binge ethanol exposure during adolescence leads to a persistent loss of neurogenesis in the dorsal and ventral hippocampus that is associated with impaired adult cognitive functioning. Front Neurosci 9:35. https://doi.org/10.3389/fnins.2015.00035

Vetreno RP, Crews FT (2018) Adolescent binge ethanol-induced loss of basal forebrain cholinergic neurons and neuroimmune activation are prevented by exercise and indomethacin. PLoS One 13(10):e0204500. https://doi.org/10.1371/journal.pone.0204500

Vetreno RP, Hall JM, Savage LM (2011) Alcohol-related amnesia and dementia: animal models have revealed the contributions of different etiological factors on neuropathology, neurochemical dysfunction and cognitive impairment. Neurobiol Learn Mem 96(4):596–608. https://doi.org/10.1016/j.nlm.2011.01.003

Vetreno RP, Ramos RL, Anzalone S, Savage LM (2012) Brain and behavioral pathology in an animal model of Wernicke's encephalopathy and Wernicke-Korsakoff Syndrome. Brain Res 1436:178–192. https://doi.org/10.1016/j.brainres.2011.11.038

Vetreno RP, Broadwater M, Liu W, Spear LP, Crews FT (2014) Adolescent, but not adult, binge ethanol exposure leads to persistent global reductions of choline acetyltransferase expressing neurons in brain. PLoS One 9(11):e113421. https://doi.org/10.1371/journal.pone.0113421

Vetreno RP, Lawrimore CJ, Rowsey PJ, Crews FT (2018) Persistent adult neuroimmune activation and loss of hippocampal neurogenesis following adolescent ethanol exposure: blockade by exercise and the anti-inflammatory drug indomethacin. Front Neurosci 12:200. https://doi.org/10.3389/fnins.2018.00200

Vetreno RP, Bohnsack JP, Kusumo H, Liu W, Pandey SC, Crews FT (2020) Neuroimmune and epigenetic involvement in adolescent binge ethanol-induced loss of basal forebrain cholinergic neurons: Restoration with voluntary exercise. Addict Biol 25(2):e12731. https://doi.org/10.1111/adb.12731

Victor M, Adams RD, Collins GH (1971) The Wernicke-Korsakoff syndrome. A clinical and pathological study of 245 patients, 82 with post-mortem examinations. Contemp Neurol Ser 7:1–206

Vongvatcharanon U, Mukem S, Udomuksorn W, Kumarsit E, Vongvatcharanon S (2010) Alcohol administration during adulthood induces alterations of parvalbumin and glial fibrillary acidic protein immunoreactivity in rat hippocampus and cingulate cortex. Acta Histochem 112(4):392–401. https://doi.org/10.1016/j.acthis.2009.04.001

Vore AS, Barney TM, Gano A, Varlinskaya EI, Deak T (2021) Adolescent intermittent ethanol (AIE) produces sex specific alterations in adult neuroimmune gene expression and ethanol sensitivity that are independent of ethanol metabolism. Neuropharmacology 195:108635. https://doi.org/10.1016/j.neuropharm.2021.108635

Walter KR, Ricketts DK, Presswood BH, Smith SM, Mooney SM (2023) Prenatal alcohol exposure causes persistent microglial activation and age- and sex-specific effects on cognition and metabolic outcomes in an Alzheimer's disease mouse model. Am J Drug Alcohol Abuse 49(3):302–320. https://doi.org/10.1080/00952990.2022.2119571

Wang Y, Mandelkow E (2016) Tau in physiology and pathology. Nat Rev Neurosci 17(1):5–21. https://doi.org/10.1038/nrn.2015.1

White AM, Orosz A, Powell PA, Koob GF (2023) Alcohol and aging—an area of increasing concern. Alcohol 107:19–27. https://doi.org/10.1016/j.alcohol.2022.07.005

Wills TA, Klug JR, Silberman Y, Baucum AJ, Weitlauf C, Colbran RJ, Delpire E, Winder DG (2012) GluN2B subunit deletion reveals key role in acute and chronic ethanol sensitivity of glutamate synapses in bed nucleus of the stria terminalis. Proc Natl Acad Sci USA 109(5):E278–E287. https://doi.org/10.1073/pnas.1113820109

Wilson S, Bair JL, Thomas KM, Iacono WG (2017) Problematic alcohol use and reduced hippocampal volume: a meta-analytic review. Psychol Med 47(13):2288–2301. https://doi.org/10.1017/s0033291717000721

Wiltfang J, Esselmann H, Bibl M, Hüll M, Hampel H, Kessler H, Frölich L, Schröder J, Peters O, Jessen F, Luckhaus C, Perneczky R, Jahn H, Fiszer M, Maler JM, Zimmermann R, Bruckmoser R, Kornhuber J, Lewczuk P (2007) Amyloid beta peptide ratio 42/40 but not A beta 42 correlates with phospho-Tau in patients with low- and high-CSF A beta 40 load. J Neurochem 101(4):1053–1059. https://doi.org/10.1111/j.1471-4159.2006.04404.x

Winblad B, Palmer K, Kivipelto M, Jelic V, Fratiglioni L, Wahlund LO, Nordberg A, Bäckman L, Albert M, Almkvist O, Arai H, Basun H, Blennow K, de Leon M, DeCarli C, Erkinjuntti T, Giacobini E, Graff C, Hardy J, Jack C, Jorm A, Ritchie K, van Duijn C, Visser P, Petersen RC (2004) Mild cognitive impairment—beyond controversies, towards a consensus: report of the International Working Group on Mild Cognitive Impairment. J Intern Med 256(3):240–246. https://doi.org/10.1111/j.1365-2796.2004.01380.x

Woods AJ, Porges EC, Bryant VE, Seider T, Gongvatana A, Kahler CW, de la Monte S, Monti PM, Cohen RA (2016) Current heavy alcohol consumption is associated with greater cognitive impairment in older adults. Alcohol Clin Exp Res 40(11):2435–2444. https://doi.org/10.1111/acer.13211

World Health Organization (2019) Global status report on alcohol and health 2018. World Health Organization

Wu Y, Bottes S, Fisch R, Zehnder C, Cole JD, Pilz GA, Helmchen F, Simons BD, Jessberger S (2023) Chronic in vivo imaging defines age-dependent alterations of neurogenesis in the mouse hippocampus. Nat Aging 3(4):380–390. https://doi.org/10.1038/s43587-023-00370-9

Xia Y, Eeles E, Fripp J, Pinsker D, Thomas P, Latter M, Doré V, Fazlollahi A, Bourgeat P, Villemagne VL, Coulson EJ, Rose S (2022) Reduced cortical cholinergic innervation measured using [(18)F]-FEOBV PET imaging correlates with cognitive decline in mild cognitive impairment. Neuroimage Clin 34:102992. https://doi.org/10.1016/j.nicl.2022.102992

Xu G, Liu X, Yin Q, Zhu W, Zhang R, Fan X (2009) Alcohol consumption and transition of mild cognitive impairment to dementia. Psychiatry Clin Neurosci 63(1):43–49. https://doi.org/10.1111/j.1440-1819.2008.01904.x

Yamada K, Holth JK, Liao F, Stewart FR, Mahan TE, Jiang H, Cirrito JR, Patel TK, Hochgräfe K, Mandelkow EM, Holtzman DM (2014) Neuronal activity regulates extracellular tau in vivo. J Exp Med 211(3):387–393. https://doi.org/10.1084/jem.20131685

Yamamoto K, Tanei ZI, Hashimoto T, Wakabayashi T, Okuno H, Naka Y, Yizhar O, Fenno LE, Fukayama M, Bito H, Cirrito JR, Holtzman DM, Deisseroth K, Iwatsubo T (2015) Chronic optogenetic activation augments aβ pathology in a mouse model of Alzheimer disease. Cell Rep 11(6):859–865. https://doi.org/10.1016/j.celrep.2015.04.017

Yan P, Bero AW, Cirrito JR, Xiao Q, Hu X, Wang Y, Gonzales E, Holtzman DM, Lee JM (2009) Characterizing the appearance and growth of amyloid plaques in APP/PS1 mice. J Neurosci 29(34):10706–10714. https://doi.org/10.1523/jneurosci.2637-09.2009

Yirmiya R, Rimmerman N, Reshef R (2015) Depression as a microglial disease. Trends Neurosci 38(10):637–658. https://doi.org/10.1016/j.tins.2015.08.001

Yoshiyama Y, Higuchi M, Zhang B, Huang SM, Iwata N, Saido TC, Maeda J, Suhara T, Trojanowski JQ, Lee VM (2007) Synapse loss and microglial activation precede tangles in a P301S tauopathy mouse model. Neuron 53(3):337–351. https://doi.org/10.1016/j.neuron.2007.01.010

Zahr NM, Pohl KM, Saranathan M, Sullivan EV, Pfefferbaum A (2019) Hippocampal subfield CA2+3 exhibits accelerated aging in Alcohol Use Disorder: A preliminary study. Neuroimage Clin 22:101764. https://doi.org/10.1016/j.nicl.2019.101764

Zhao S, Widman L, Hagström H, Shang Y (2023) Disentangling the contributions of alcohol use disorder and alcohol-related liver disease towards dementia: A population-based cohort study. Addiction. https://doi.org/10.1111/add.16395

Zheng B, Udeh-Momoh C, Watermeyer T, de Jager Loots CA, Ford JK, Robb CE, Giannakopoulou P, Ahmadi-Abhari S, Baker S, Novak GP, Price G, Middleton LT (2022) Practice effect of repeated cognitive tests among older adults: associations with brain amyloid pathology and other influencing factors. Front Aging Neurosci 14:909614. https://doi.org/10.3389/fnagi.2022.909614

Zhornitsky S, Chaudhary S, Le TM, Chen Y, Zhang S, Potvin S, Chao HH, van Dyck CH, Li CR (2021) Cognitive dysfunction and cerebral volumetric deficits in individuals with Alzheimer's disease, alcohol use disorder, and dual diagnosis. Psychiatry Res Neuroimaging 317:111380. https://doi.org/10.1016/j.pscychresns.2021.111380

Zillich L, Cetin M, Hummel EM, Poisel E, Fries GR, Frank J, Streit F, Foo JC, Sirignano L, Friske MM, Lenz B, Hoffmann S, Adorjan K, Kiefer F, Bakalkin G, Hansson AC, Lohoff FW, Kärkkäinen O, Kok E, Karhunen PJ, Sutherland GT, Walss-Bass C, Spanagel R, Rietschel M, Moser DA, Witt SH (2024) Biological aging markers in blood and brain tissue indicate age acceleration in alcohol use disorder. Alcohol Clin Exp Res (Hoboken). https://doi.org/10.1111/acer.15241

Changes in the Properties of Ethanol-Sensitive Molecular Targets During Maturation and Aging

13

Luis G. Aguayo, Lorena Armijo-Weingart, Loreto San Martin, Alejandra Guzmán Castillo, Macarena Konar-Nie, and Scarlet Gallegos

Abstract

At present, there is a good understanding of the negative neurobiological impacts that ethanol has on adolescent and adult brains; the effects of this drug on the aging brain, both normal and pathological, are only now starting to emerge. Biomedical research involving the effects of alcohol on aging is limited; however, studies in human subjects show that older adults perform worse in tests assessing working memory, attention, and cognition as compared to younger adults. The neurobiological basis for these effects in the elderly is largely unknown. In the last 30 years, important molecular targets for ethanol actions in the adolescent and adult brain have been identified. Yet, we know very little about whether these targets are still affected by ethanol in the older brain. The brain structure changes during aging, and the targets and their functional characteristics may also change. Thus, one can expect that ethanol will have distinct effects on the brain of an aged organism.

This chapter discusses the available data showing how aging influences critical proteins that affect neuronal excitability, nerve conduction, and synaptic transmission and how aging modifies the sensitivity of these proteins to ethanol. The data show limited information on ethanol's effects in the aged brains of mice and rats.

Keywords

Voltage-activated ion channel · Ligand-gated ion channel · Synapse · Ethanol · Aging

L. G. Aguayo (✉) · L. Armijo-Weingart
L. San Martin · A. Guzmán Castillo · M. Konar-Nie
S. Gallegos
Laboratory of Neurophysiology, Department of Physiology, Universidad de Concepcion, Concepcion, Chile

Programa en Neurociencia, Psiquatria y Salud Mental, Universidad de Concepcion, Concepcion, Chile
e-mail: laguayo@udec.cl

13.1 Overview of Ethanol Effects in the Brain and Body

Ethanol (C_2H_6O) is a simple aliphatic molecule that depresses some brain functions and has general anesthetic properties. However, it is not clinically used as an anesthetic agent because of its low therapeutic ratio. Other effects that might be of pharmacological and toxicological interest are that it produces a dose-dependent increase in pain threshold and a successive central nervous depression (Urban 2008; Egervari et al. 2021).

Because of its myriad of actions, including anxiolytic and rewarding effects on the central nervous system (CNS), either alone or co-abused with other drugs, this alcohol is currently the most widely used and abused legal drug (Burnette

C. F. Valenzuela et al. (eds.), *Effects of Alcohol on the Brain across the Lifespan*, Advances in Experimental Medicine and Biology 1473, https://doi.org/10.1007/978-3-031-81908-7_13

et al. 2022). More than half of the world population aged 15 years and over have reported that they consumed alcohol in the previous year, producing 3,000,000 deaths worldwide (5.3% of all global deaths) and 132.6 million disability-adjusted life years. For instance, in the United States, excessive consumption is responsible for 13% and 20% of total deaths among adults 20–64 and 20–49 years old, respectively (Esser et al. 2022a, b).

Upon oral consumption, ethanol quickly enters the blood. It reaches the central nervous system, interacting with several membrane proteins primarily involved in neuronal excitability and fast-acting neurotransmitters (e.g., GABA, glycine, glutamate). The many effects of ethanol on brain functions occur across a narrow range of concentrations, from low millimolar (5 mM) up to 100 mM, indicating a complex myriad of actions likely produced by interacting with several targets. The acute and depressive effects of ethanol on the brain grow progressively greater as its concentration increases in the blood, causing first anxiolytic and then euphoric effects (H12 mM). With higher intake, blood concentration reaches levels that cause legal intoxication (H18 mM), characterized by slower reaction times, motor incoordination, and cognitive impairment. At concentrations up to 50 mM, locomotor disruption, cognitive impairment, and sedation levels are more prominent causing falls and increasing the incidence of accidents (Fig. 13.1). Above this level, strong sedation and respiratory depression can lead to coma or death (Alifimoff et al. 1989; Abrahao et al. 2017).

It has been reported that the abuse of ethanol intake is increasing among the elderly (Han et al. 2019; Grant et al. 2017), and it is widely accepted that older brains are more sensitive to sedation and alterations in motor coordination and cognitive functions produced by ethanol (Watson et al. 2020). Evidence has emerged during the last 30 years that acute ethanol administration causes the modification of several critical molecular targets responsible for the effects in the CNS at the relevant dose range (no more than 100 mM). The accompanying long-term, chronic effects of ethanol likely result from changes originating from its sustained presence and the neuroplastic modifications produced by changes in excitability and synaptic transmission (Crews et al. 1996). This chapter will not deal with the chronic effects of ethanol, but it will focus on the main targets affected by its acute exposure, its molecular signatures, and the evidence supporting its effects on aging.

Because the conformation of molecular targets changes with age, both normal and abnormal, one can expect that their sensitivity to ethanol will also change. This concept is not unexpected because studies show that immature brains are more sensitive to the effects of ethanol (Quoilin et al. 2014; Spear and Swartzwelder 2014; Kulkarny et al. 2011). However, as will become apparent in this chapter, few studies have examined the effects of ethanol on older brains, and most experimental models to study ethanol effects have used animals 3 months old or less. Thus, little is known about ethanol on its CNS-relevant targets during aging. Interestingly, as a result of a large survey of mice ages used in pre-

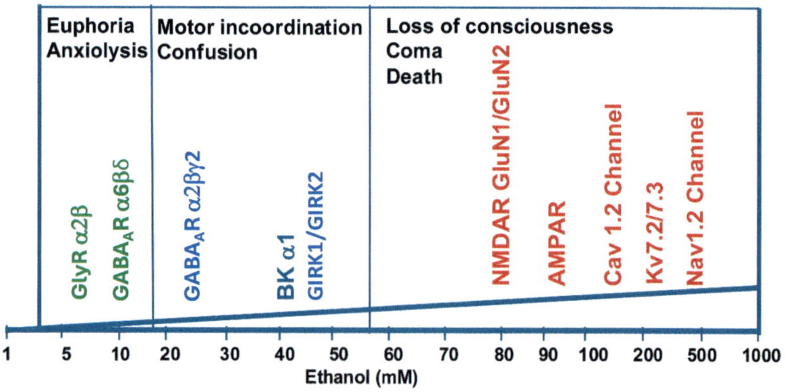

Fig. 13.1 The scheme shows the sensitivity to ethanol of critical ligand-gated and voltage-activated ion channels. The ethanol (mM) concentration indicates the associated behavioral effects it produces at the stated level

clinical research and the notion that important characteristics of brain development are still in progress at 5–6 months, it was suggested that 3–4-month-old mice should be the minimum age for an adulthood model (Jackson et al. 2017). Therefore, the most updated consensus on the stages of aging in the C57BL/6J mice considers an adolescent animal at <3 months of age, with a young adult corresponding to 3–6 months, a 12-month-old corresponding to a middle-aged adult, and an aged (old) mouse corresponding to 22–28 months (Fox 2007). These ages can be compared to mature (20–30 years), middle age (38–47 years), and old (>56 years) in humans (Konar-Nie et al. 2023; Jackson et al. 2017). This chapter will first examine the effects of ethanol on key protein targets (effectors) involved in neuronal excitability and fast-acting neurotransmitters (e.g., GABA, glycine, glutamate) and then discuss the limited data available on aged targets. The information presented will show that despite an increasingly aging population, studying what makes us sensitive to ethanol is still of limited interest to investigators and research foundations.

The molecular targets in the nervous systems where ethanol interacts and affects diverse behaviors have been defined by (1) existence of a molecular entity (protein) affected by a low pharmacologically relevant concentration of ethanol; (2) mutations of amino acids important for ethanol affecting the ethanol sensitivity of the protein; (3) structural biological evidence indicating that ethanol inhabits a putative binding site; and (4) genetic alteration of the target protein (KO or KI) that affects the ethanol phenotype (Harris et al. 2008). However, the fulfillment of all these criteria for molecular targets is not so straightforward due to the following: (1) genetic manipulations can cause significant alterations in the normal function of the protein, (2) lack of structural evidence on ethanol occupation in the mutated binding site, and (3) the criteria do not consider allosteric regulatory mechanisms on the target protein (Abrahao et al. 2017; Yevenes et al. 2008; Kuntamallappanavar and Dopico 2017).

13.2 Effects of Ethanol on Behavior with Aging

Currently, there is little information on the effects of ethanol in aged individuals. As expected, some differential neuropsychological effects have been reported at different levels of ethanol consumption. Studies in humans, for instance, showed that older individuals have worse performances in psychological tests assessing working memory, attention, and cognition when compared to younger counterparts (Garcia et al. 2020; Han and Jia 2021; Price et al. 2018). Using animal models under well-defined laboratory conditions, the main effects of ethanol that have been evaluated are cognition, ataxia, hypothermia, and sedation. For ataxia, hypothermia, and sedation, an age-dependent level of impairment was reported when comparing adolescent, adult, and aged rats (Watson et al. 2020; Matthews et al. 2019; Perkins et al. 2018). No differences were observed between adolescent and adult animals following an acute dose of ethanol on spatial memory, suggesting that aging does not affect hippocampal-dependent behaviors (Chin et al. 2011; Matthews et al. 2020). However, acute ethanol administration to the 29–33-month-old group showed significantly greater cognitive impairment in the Morris water maze (Matthews et al. 2020). Thus, ethanol targets playing roles in ataxia, hypothermia, sedation, and motor control seem to be more affected by aging and by the drug (Blednov et al. 2023; Beleslin et al. 1997; Aguayo et al. 2014; Zamudio et al. 2020). These different effects can be due to several changes occurring during brain aging, such as loss of synaptic reserves, reduction in synaptic contacts and proteins, and alterations in their molecular conformations. Here, we aim to discuss the main targets reported and changes in functions with acute ethanol. After this, and where data on ethanol actions on the aged central nervous system are available, we will review whether a particular target is more or less sensitive to ethanol and might contribute to behavioral changes after its administration.

13.3 Effects of Ethanol on Voltage-Activated Sodium Channels

Sodium channels are among the most important ion channels in neuronal excitability and nerve conductivity. Indeed, voltage-gated sodium channels mediate the rising phase of nerve depolarization during the propagation of action potentials. These proteins have been proposed as possible anesthetic targets because of their central role in neuronal communication and integration. The action potential elicited in the squid giant axon, resulting from the successive activation of inward sodium and outward potassium channels, was found to be quite resistant to high molar concentrations of ethanol (Rosenberg and Bartels 1967), suggesting that they might not serve as significant mediators of clinical anesthesia. Similarly, a high concentration (>100 mM) of ethanol did not affect the membrane conductance for sodium and potassium membrane responses (Moore et al. 1964). Studies in rats showed that ethanol acts as a general anesthetic at concentrations varying from 97 mM in adults to 235 mM in neonates (Fang et al. 1997). At this range of concentrations, it is apparent that ethanol inhibits neurotransmission in isolated rat spinal cords, indicating that ethanol can inhibit neurotransmission without changes in nerve firing and intrinsic excitability (Wong et al. 1997).

A direct analysis of the sensitivity of the predominantly adult brain-expressed sodium channel IIA (NaV1.2a) showed that they were inhibited only by high ethanol concentrations, e.g., IC_{50} of 690 mM at –120 mV and 292 mM at –60 mV (Rehberg et al. 1996). In another study where the Nav1.2 was co-expressed with the β1 subunit in *Xenopus* oocytes, it was reported that 100 mM ethanol caused less than 20% of current inhibition (Shiraishi and Harris 2004). Additional studies showed that ethanol (190 mM) inhibited several sodium channel subtypes Nav1.2, Nav1.4, Nav1.6, and Nav1.8 tested in oocytes (Horishita and Harris 2008). The data suggest that ethanol-induced anesthesia might not be associated with inhibiting these types of voltage-dependent targets in the CNS (Fig. 13.1).

Some studies have examined the effects of aging on Nav1.2 channels. For instance, wild-type dopaminergic neurons of the *substantia nigra* from 8- to 40-week-old-wild-type mice exhibited stable cell capacitance, input resistance, and spike width. Still, these membrane properties were significantly altered in a MitoPark mice model for accelerated brain aging (Branch et al. 2016).

In addition, pacemaker firing was disrupted together with an upregulation of Nav1.2, NavB3, Cav1.2, Cav1.3, and HCN1 expression in older MitoPark mice. Interestingly, a genome-wide association study of the C allele of the SCN1A variant rs10930201 was associated with poor short-term memory performance in healthy volunteers between 21 and 82 years of age. The variant rs10930201 was further determined to be associated with differences in neural activity during a working memory task and on brain structure, indicating reduced gray matter densities in the frontal and insular regions in the C allele carriers, especially during aging (Meier et al. 2012).

Regarding the excitatory role of sodium channel function on pathological brain aging, a recent study showed that Navβ2, a sodium channel accessory subunit, regulated the toxicity of Aβ1-42 hippocampal and cerebellar neurons (Li et al. 2022). For example, using a Navβ2 expression interference lentivirus, the investigators showed that neuronal cell viability was restored, together with an increased level of brain-derived growth factor. In addition, the treatment decreased the expression of the amyloid precursor protein. Another study showed that reduction of Nav1.6 following the injection of an adeno-associated virus with short hairpin RNA shRNA into the hippocampus of 5-month-old APP/PS1 transgenic mice rescued cognitive impairments and long-term potentiation (Yuan et al. 2022). In addition to Alzheimer's disease (AD), sodium channels are associated with other types of aging-related neurodegenerative diseases. For example, several subunits have been implicated in Parkinson's disease (Vaidya et al. 2024). In conclusion, sodium channels appear to influence normal and abnormal aging. However, there are no available studies indicating whether their sensitivity to ethanol is affected by age.

13.4 Effects of Ethanol on Voltage-Activated Calcium Channels

Voltage-operated calcium channels play a fundamental role in the control of neuronal excitability. Early studies showed that ethanol inhibits depolarization-induced Ca^{2+} influx through voltage-gated Ca^{2+} channels in synaptosomes and presynaptic nerve terminals (Leslie et al. 1983) (Skattebol and Rabin 1987). These studies also showed that ethanol reduced voltage-dependent Ca^{2+} influx in cultured neuronal and PC12 cells (Messing et al. 1986). Twombly et al. (1990) reported that both T-type (transient) and L-type (sustained) voltage-gated Ca^{2+} channels in N1E-115 neuroblastoma and NG108-15 neuroblastoma x glioma hybrid cells were inhibited by ethanol (>30 mM), with L-type voltage-gated Ca^{2+} channels showing significantly more inhibition at the same ethanol concentration. Concentration–response curves showed that 100 and 300 mM ethanol blocked the T- and L-current by approximately 15 and 40%, respectively. The voltage dependence of type L channel inactivation and activation was not altered by ethanol concentrations as high as 300 mM. The results did not find evidence of a use-dependent blocking action of the calcium-permeable pore by ethanol. In another early study, using whole-cell patch-clamp techniques in PC cells, it was reported that acute exposure to 25 mM ethanol inhibited macroscopic L-type Ca^{2+} (possibly Cav1.2) currents in undifferentiated PC-12 cells. In addition, intracellular application of guanosine-5′-O-(2-thio)diphosphate or pretreatment with pertussis toxin reduced the effect of ethanol inhibition in undifferentiated cells, suggesting the involvement of a G protein in ethanol inhibition of Ca^{2+} channel cells (Mullikin-Kilpatrick et al. 1995). The specific molecular targets of the inhibitory effects of ethanol reported studying whole-cell calcium current are currently unknown, but more recent studies examining mRNA and protein levels for CaV1.2 and CaV1.3 α1 subunits after ethanol administration indicated that they are potential targets (N'Gouemo et al. 2015). Interestingly, these subtypes (with the respective gene loci CACNA1C and CACNA1D) are primarily expressed in the central nervous system (Hell et al. 1993), and both have been implicated in alcohol use disorder (AUD), specifically in alcohol-dependent rats during protracted abstinence (Uhrig et al. 2017). However, the reported upregulated expression of these ion channels might represent neuroplastic changes resulting from reduced excitability during the chronic administration of ethanol (Ratkai et al. 2021) and not actual sensitive molecular targets for the drug.

Studies in aged rats (19–23 months) showed that aged hippocampal CA1 neurons had a higher somatic expression of Cav1.2 subunits and exhibited larger afterhyperpolarization (AHP) and lower excitability compared with adult (6–9 months) neurons. On the other hand, pyriform cortex neurons from aged rats showed higher excitability without differences in AHP. However, adult and aged pyriform cortex showed similar expression of Cav1.2 (Maziar et al. 2023). Additional studies in pyriform cortex neurons showed a significant increase in synaptic Cav1.2 as the rats aged from 2 to 25 months. This change was suggested to play a role in the observed shift in synaptic plasticity from an NMDA-dependent to a low threshold calcium channel-dependent long-term depression (LTD) (Rajani et al. 2021).

Using specific antibodies against Cav1.2 and Cav1.3 subunits, another study compared subunit expression in the hippocampus from young (3–4-months-old) and aged (30–32-months-old) rats. Western blot analysis of the total expression levels revealed significant reductions in both Cav 1.2 and Cav 1.3 subunits, together with a higher expression of the subunits in the plasma membrane in the CA3 region from aged rats (Nunez-Santana et al. 2014). Regulation of these channels by ethanol during aging is important because calcium serves as a second messenger, and it is believed that altered neuronal calcium homeostasis participates in brain remodeling and excitoxicity during aging in the hippocampus (Khachaturian

1987; Power et al. 2002). Additionally, reduction in the expression of CaV 1.2 prevented object recognition deficit and expression of the channels in the hippocampus of aged mice (4–18 months) (Zanos et al. 2015). More recent studies have provided further experimental confirmation that calcium dysregulation can affect brain aging. Using a transgenic mouse line that displays an enhanced expression of the L-type voltage-gated calcium, CaV1.3, it was found that pyramidal neurons in the CA1 region of the hippocampus showed a larger postburst afterhyperpolarization, together with a deficit in spatial learning and memory when compared to wild-type littermates (Moore et al. 2023). Calcium channels are also seen to play a role in neurodegeneration. For instance, silencing of striatal CACNA1D transcription with rAAV-CaV1.3-shRNA in male and female parkinsonian rats of advanced age conferred protection against dyskinesia scalation (Caulfield et al. 2023). Similar to sodium channels, no reports are available on the ethanol sensitivity of these critical ion channels in the aged brain. This information is relevant because of their key role in the calcium dysregulation hypothesis of brain accelerated aging.

13.5 Effects of Ethanol on Potassium Channels

The family of potassium channels can be allocated into four classes: (1) outward rectifiers voltage-gated K channels (Kv) responsible for repolarizing the membrane during the action potential, (2) inwardly rectifying K channels (Kir) acting to maintain a negative resting potential, (3) two-pore potassium channels (K2P) maintaining leak K currents, and (4) Ca^{2+}-activated potassium channels (KCa) with roles in synaptic transmission and microvasculature vasodilation (Taura et al. 2021).

Historically, one of the first studies that suggested that ethanol interfered with a type of potassium channel was performed in locus coeruleus neurons. These neurons show a classic type of inward rectification, which is highly sensitive to blockade by Ba^{2+}, and the application of ethanol (40–200 mM) increased the amplitude of the inward rectification in the neurons. Suggestive of a molecular interaction, the effect of ethanol was concentration-dependent and fully reversible when the tissue was exposed to an ethanol-free solution (Osmanovic and Shefner 1994). The effects of ethanol (20 mM) were also studied on hippocampal dentate granule neurons in brain slices from young-mature (6–8 months) and old (25–29 months) Fischer-344 rats. Supporting actions on potassium channels, ethanol caused a hyperpolarizing effect on the resting membrane potential and a prolongation of spike afterhyperpolarization (AHP) in young neurons. On the contrary, it caused a depolarizing action and decreased AHPs in older neurons (Niesen et al. 1988). There are no more updated indications, using more definitive molecular targets, on the differential effects of ethanol on aging neurons. Data studying neuronal excitability in the presence of ethanol indicated that it did not affect the main properties of the action potential, i.e., amplitude, threshold, or half-time, thus discounting Na^{+}, Ca^{2+}, and outward rectifying potassium channels from the list of potential ethanol-sensitive targets (Rosenberg and Bartels 1967; Moore et al. 1964).

Distinct types of G-protein-gated inwardly rectifying potassium (GIRK) subunits (GIRK1, GIRK2, and GIRK3) are expressed in the brain (Zhao et al. 2021). These ion channels can form either homotetramers of GIRK2 or heterotetramers (e.g., GIRK1/GIRK2) in distinct brain regions (Luscher and Slesinger 2010; Lujan and Aguado 2015). The opening of these potassium channels require G protein-coupled receptor (GPCR) activation leading to the dissociation of Gβγ subunits from the heterotrimeric G protein complex and the activation of GIRK channels via Gβγ binding to the channel. Studies had indicated that these potassium channels are sensitive to ethanol. For example, an electrophysiological study in oocytes showed that ethanol rapidly and reversibly enhanced the inward current associated with GIRK1/GIRK2 or GIRK1/GIRK4 subunits (50–100 mM). Interestingly, the effect of ethanol was blocked by Ba^{2+}, which was independent of the activation of a G protein-coupled receptor (GPCR) (Kobayashi et al. 1999). In contrast, the IRK3 channel was not affected by 200 mM ethanol, suggesting molecular specificity. The current induced in the GIRK1/GIRK2 complex was examined in the presence of n-alcohols, showing that from methanol to 1-butanol, an inward potassium current was induced and that the potencies of these n-alcohols increased as the carbon chain length increased, except that 1-pentanol inhibited basal GIRK1/2 currents. Finally, examination of mice with mutations in GIRK2 implicated these potassium channels in pain processing, but not in the sedative effects of ethanol. Studies using GIRK2 KO and GIRK3 KO mice showed different effects of ethanol on two behaviors. The weaver mutant mice with a missense mutation in the GIRK2 channel showed a loss of ethanol-induced analgesia (Kobayashi et al. 1999). In contrast, the GIRK3 KO mice showed increased ethanol binge-like drinking behavior (Herman et al. 2015). Not much is known about the effect of aging on the expression of these ion channels in the brain. However, a recent study in the hippocampus of wild-type and APP/PS1 Alzheimer's disease mice model reported that age reduced the expression of GIRK2, together with an increase in GIRK3 between 6 and 12 months (Temprano-Carazo et al. 2022). In addition, when the animals were challenged to perform spatial memory training, the deficit in GIRK2 was restored.

In another study, electrophysiological recordings in hippocampal slices showed that ethanol at low concentrations (22–44 mM) reduced the amplitude of the M-current amplitude at depolarized membrane potentials. In addition, 1 μM atropine (and TTX) did not alter the ethanol effect on the M-current, suggesting that the site of ethanol action is far from the muscarinic receptor (Moore et al. 1990). The nature of this slow-activating, not inactivating current is the KV7 voltage-gated subunit that is sensitive to muscarinic receptor activation. Type KV7 channels play an important role in regulating the action potential threshold in dentate gyrus granule neurons by activating at subthreshold potentials (Mateos-Aparicio et al. 2014; Cannady et al. 2018). The effect of ethanol (>100 mM) was assessed in Kv7.2/7.3 channels expressed in superior cervical neurons and HEK 293 cells (Kim et al. 2019). The study also showed that ethanol inhibited the M-type and increased neuronal excitability by resetting the resting membrane potential. Examination of a series of n-alcohols showed that ethanol depressed, whereas long-chain n-alcohols increased the amplitude of the current response (Jeong et al. 2023). The authors suggest that the effect of ethanol was coupled with a selective EtOH-binding site. A recent study has revealed an association between Kv7 channels with alcohol-related quantitative trait loci (QTLs) and chronic alcohol consumption and withdrawal (Mcguier et al. 2018).

Finally, ethanol increased the decay time of a transient outward potassium current at concentrations above 200 mM (Treistman and Wilson 1987; Weight 1992), and examination of several potassium channels in oocytes showed that subunits from the Shaker and Shab subtypes were insensitive to 200 mM (Anantharam et al. 1992).

Not many studies on the aging of potassium channels are available. For instance, the expression of potassium channels KCNQ4 (KV7.4) in outer hair and KCNQ1 (KV7.1) in marginal cells of the stria vascularis were studied at 6 weeks (juvenile), 12 weeks (young adult), and 24 weeks (adult), revealing that aging reduced their expression. How these changes in the expression of potassium channel subunits modify neuronal excitability and perhaps some ethanol behaviors is largely unknown (Peixoto Pinheiro et al. 2021).

The KCa (Ca^{2+}-activated) channels, with large conductance (BK, slo1 channels), are affected by exposure to pharmacologically relevant alcohol concentrations. For example, the current amplitude induced by BK channels was enhanced in response to rapid application to 10–100 mM ethanol in rat neurohypophysial nerve endings (Dopico et al. 1996). More recent studies using recombinant slo1 α subunit that form BK channels (Bukiya et al. 2014; Dopico et al. 2014) have provided definitive evidence for a selective interaction between ethanol and the protein. Interestingly, the current that results from the activation of native BK channels can be differentially affected by ethanol, causing an increase, decrease, or no effects on the function when these channels are exposed to ethanol, thus suggesting complex interactions with the regulatory sites in the ion channel (Dopico et al. 2016). Nevertheless, the actions of ethanol on this target appear to be selective because it occurs without altering potassium outward rectifiers (Dopico et al. 1996) or slo2 and slo3 channels (Liu et al. 2013).

Potassium channels also play a role in normal/abnormal aging. For example, in an oxidative stress-based cellular model of aging, it was reported that the activation of catalase and superoxide dismutase was accompanied by a significant reduction in the current density of Kv3.1/KCNC1 (Spinelli et al. 2024). In addition, it was reported that the inhibition of Kv3.1 was able to increase lifespan in *Caenorhabditis elegans*, affecting the production of free radicals (Admasu et al. 2022). The large-conductance BK channel also appears to regulate auditory processing in the aging brain. An age-related reduction in BK expression was reported using a selective potassium channel modulator, where equivalent doses produce a smaller effect with age (Brecht et al. 2022). Studying rat middle cerebral artery, it was found that an age-dependent increase in BK β1 protein levels paralleled the progressive increase in the susceptibility of cerebral arteries to ethanol (Bukiya et al. 2016). The results support the conclusion that this BK component regulates age-dependent ethanol-induced vasoconstriction. The function of the inward rectifying potassium channel Kir2.1 was recently studied in a model of oxidative stress-related aging neuroglia exposed to D-galactose. The results suggest that inhibition of the channel might contribute to neuronal hyperexcitability during aging (Remigante et al. 2024). The APPSw,Ind J9 transgenic AD mice (6–18 months) showed a progressive amyloid-β (Aβ) accumulation and decreased GIRK2 expression in the hippocampus (Temprano-Carazo et al. 2022). A similar reduction was also found in aged WT mice, suggesting that aging is associated with reduced GIRK2 expression. A recent report explored the link between glucose metabolism and AD by examining KATP channel subunits Kir6.2/KCNJ11 and SUR1/ABCC8 expressed in excitatory and inhibitory neurons in the human brain (Grizzanti et al. 2023), finding that their expression changed with AD pathology in human and mice models. The data in excitatory neurons showed that KCNJ11 expression was increased in early and late AD, while ABCC8 expression was decreased. The role of ATP-sensitive channels was also reported in Parkinson's disease-associated neurodegeneration. Using 3-, 6-, and 9-month-old A53T α-synuclein transgenic (α-SynA53T+/+) mice, it was reported that only the expression of the KATP subunit sulfonylurea receptor 1 (SUR1) subunit was upregulated, accompanied by neuronal damage in nigral dopaminergic neurons (Liu et al. 2022). Despite the availability of a body of information indicating that physiological and pathological aging affects the structure and function of potassium channels, to our knowledge, there are no studies on changes in ethanol sensitivity with aging.

13.6 Effects of Ethanol on *N*-Methyl-D-Aspartate (NMDA) and α-Amino-3-Hydroxy-5-Methyl-4-Isoxazolepropionic Acid Receptor (AMPA) Ion Channels

Previous electrophysiological studies showed that ethanol (100 mM) reduced the NMDA current activated in brain slices and cultured neurons by 20–35% (Xu and Woodward 2006; Lovinger et al. 1990). These results suggested that inhibition of NMDA receptors by ethanol played an important role in ethanol depressive actions (Zamudio et al. 2021; Ron and Wang 2009). A series of mutagenesis studies performed in several laboratories identified domains in GluN1 and GluN2 subunits that influenced ethanol inhibition of this receptor complex (Zamudio et al. 2020; Honse et al. 2004; Ronald et al. 2001). These data were valuable in creating KI mice to test the behavioral changes produced by the mutation. For example, GluN2A(A825W) knock-in (KI) male mice showed reduced sedation with ethanol (Zamudio et al. 2020) and decreased or delayed drinking escalation after intermittent ethanol vapor exposure (Zamudio et al. 2021). Although less is known about ethanol in AMPARs, studies in cultured neurons showed that they are inhibited at a similar concentration as NMDARs (Wirkner et al. 2000; Lovinger 1993). Nevertheless, the molecular mechanisms underlying ethanol's inhibition of AMPARs and NMDARs are largely unknown. Because some mutations that make NMDARs resistant are in transmembrane regions critical for gating and permeation, their pharmacological significance is currently discussed (Zamudio et al. 2020).

NMDARs and AMPARs are key regulators of synaptic plasticity (Jurado 2017; Henley and Wilkinson 2013) and are significantly less functional with aging, even causing a reduction in dopamine release in the nucleus accumbens (Pandey et al. 2015; Kumar and Foster 2019; Segovia and Mora 2005). Also, their reduced function likely accounts for decreased cognition with aging (Kumar and Foster 2019). For instance, impairment in memory and learning in aged rodents was associated with reduced NMDARs function and expression in the hippocampus and cortex (Guidi et al. 2015; Kumar and Foster 2019). For comprehensive recent reviews of the role of glutamatergic transmission, calcium regulation overload, and therapeutic strategies to retard normal and pathological aging, the following should be considered (Yu et al. 2023; Castillo-Vazquez et al. 2024).

Sadly, the study of ethanol actions on these important neurotransmitters signaling during aging is quite underdeveloped, thus limiting our understanding of potential neurobiological targets and mechanisms for drinking (Konar-Nie et al. 2023).

13.7 Effects of Ethanol on $GABA_A$ Cl^- Channels

Analysis of several protein targets associated with critical brain neurotransmitters has shown that ethanol can affect their functions. An early indication that ethanol directly affected the function of brain gamma amino butyric acid ($GABA_A$) receptors was reported more than 40 years ago from recordings in cat cortical neurons. This report is relevant since it was the first demonstration that ethanol selectively affected an ion channel gated by a neurotransmitter at concentrations that did not affect excitability. In addition, it initiated a long, still not completely unresolved, pathway of research that showed the complexity of defining some ethanol-sensitive targets. Using electrophysiological techniques, the study in cat cortical neurons showed that local ethanol applications enhanced GABA-mediated responses (Nestoros 1980). Supporting this finding, follow-up biochemical experiments in several laboratories showed that ethanol (>10 mM) potentiated $GABA_A$-activated $^{36}Cl^-$ flux in rat brain synaptosomes and cultured spinal neurons (Allan and Harris 1987; Mehta and Ticku 1988; Suzdak et al. 1986; Engblom and Akerman 1991). However, these studies did not provide evidence supporting a mechanism, such as pre- or postsynaptic localization for ethanol action. Electrophysiological studies performed in rat brain slices and dorsal root neurons could not confirm the existence of ethanol effects on postsynaptic $GABA_AR$ (Siggins et al. 1987; White et al. 1990), supporting the notion that this heterogeneity was due to differential expression of receptor subunits (Aguayo et al. 2002; Pritchett et al. 1989; Levitan et al. 1988). Additionally, around this time, it became evident that intracellular mechanisms, such as intracellular Ca2+ and phosphorylation states, highly regulated the function of $GABA_A$ receptors (Stelzer et al. 1988; Inoue et al. 1986).

Regarding the differential sensitivities for ethanol in the distinct $GABA_A$ subunits expressed in the brain, an early study on recombinant receptors reported that the long γ2L, but not the short γ2S splice variant of $GABA_AR$, affected the ethanol potentiation (Wafford et al. 1991, 1993). Interestingly, the phosphorylation S343 site mutation in the γ2L variant and the addition of kinase inhibitors abolished ethanol potentiation (Wafford and Whiting 1992). In another recent study supporting subunit specificity for ethanol actions, $GABA_AR$-containing δ subunits exhibited a potentiation of up to 50% with ethanol concentrations of 10 mM in oocytes (Wallner et al. 2003). However, other studies did not report potentiation using the same $GABA_A$ subunit, suggesting the role of intracellular regulations (Borghese et al. 2006; Yamashita et al. 2006). Thus, after 40 years of work in this area, we still cannot conclude if some subunit conformation of $GABA_AR$ plays a role in the effect of ethanol on this critical inhibitory neurotransmitter system.

The other alternative to explain the differences in ethanol sensitivity on this molecular target can be associated with changes in the neuronal intracellular state that affect the sensitivity of the target to ethanol, thereby explaining the diversity in the postsynaptic effects of ethanol reported in the literature. For example, ethanol potentiated the GABA-induced Cl^- current in hippocampal and cortical neurons (Aguayo 1990). This study showed cellular heterogeneity and found that ethanol affected about half of the cortical and hippocampal neurons examined. It was also demonstrated that the potentiating effects of ethanol on $GABA_AR$ were not mediated by sites related to diazepam or barbiturate actions; also, the drug itself could not gate the ion channel directly. The potentiating effects of ethanol in mouse hippocampal neurons occurred at lower concentrations (10 mM) than in rats (300 mM) (Aguayo et al. 1994). Other electrophysiological studies also reported the heterogeneity of ethanol effects using cultured and acutely dissociated Purkinje cells (Sapp and Yeh 1998). The positive modulation of the $GABA_AR$ in cerebellar Purkinje neurons by ethanol was made more apparent after sensitizing β-adrenergic receptor pathways, thus supporting the notion that intracellular pathway signaling, in this case, the cyclic AMP-PKA, played a role in the modulation of ethanol sensitivity (Lin et al. 1994; Yang et al. 1998). Other studies indicated that the potentiation of $GABA_AR$ by ethanol and flunitrazepam, two allosteric modulators, was significantly increased in PKCε KO mice mutants (Hodge et al. 1999), and the potentiation of $GABA_AR$ was affected by exogenous activation of PKC and G protein (Weiner et al. 1994a; Aguayo et al. 1994, 2002). The effects of PKC on the allosteric actions of ethanol on $GABA_AR$ were distinct in rat and mice-derived neurons. For example, PKC activation reduced ethanol sensitivity in cultured hippocampal mouse neurons, whereas it had the opposite effect in rat hippocampal slices (Weiner et al. 1994b; Aguayo and Pancetti 1994). Finally, the developmental state of the $GABA_AR$ affected its sensitivity to ethanol (Aguayo et al. 2002), similar to the sensitivity to other drugs such as benzodiazepines acting on $GABA_AR$ (Kapur and Macdonald 1999).

There is little information regarding the effects that physiological aging has on the $GABA_AR$ complex. In the normal human aging brain (40–90 years), a negative correlation between age and the level of the α1 subunit was reported in the female CA1 hippocampal region. In contrast, a negative correlation was found for β1, β3, and γ2 subunit expression in the male dentate region (Ethiraj et al. 2021). A positive correlation was reported in the female brain between age and γ2 subunit expression in the entorhinal cortex. A recent study in aging mice hippocampus, from young (6 months) to old (21 months) age groups, using Western blotting and immunohistochemistry, found that the GABAergic signaling is highly stable, with no significant age-related differences in α1, α2, α3, α5, β3, and γ2 subunit expression levels (Palpagama et al. 2019). Interestingly, an electrophysiological study in young (1–3 months) and aged rats (19–25 months) showed that aging increased the allosteric effect of midazolam in $GABA_AR$ (Griffith and Murchison 1995). The same study found that the amplitude of the current was larger and presented similar EC_{50} in the older neurons. Although limited in scope, the results indicate that $GABA_AR$ subunit expression in the normal brain is quite stable during aging. On the other hand, recent data indicate that pathological aging causes transcriptional downregulation of several $GABA_AR$ subunits, namely, α1, α2, α3, α5, β1, β2, β3, δ, γ2, γ3, and θ, in the middle temporal gyrus of AD patients (Govindpani et al. 2020). Because of the inhibitory action of $GABA_AR$ on neuronal activity, a reduction in its expression would likely lead to alterations in the excitatory/inhibitory balance that might cause the high incidence of epileptiform activity reported in AD (Vossel et al. 2017). Regarding the reduced synaptic plasticity and memory processes in AD, a disruption in $GABA_A$-mediated inhibition appears to alter key nuclear calcium signals, including neuroprotective genes in hippocampal neurons (Lobos et al. 2023).

No data on the effects of ethanol on the aged $GABA_AR$ are available. Because the elderly are more sensitive to ethanol's sedative actions, this target may be more affected by ethanol, similar to benzodiazepines (Griffith and Murchison 1995).

13.8 Effects of Ethanol on Glycine Receptors (GlyRs)

Unlike $GABA_AR$, postsynaptic GlyRs are predominantly found in spinal and medulla regions. Interestingly, upper regions of the CNS express a large number of postsynaptic GlyRs in the absence of a significant level of glycinergic neurotransmission (Muñoz et al. 2020). GlyRs are made of α1, α2, α3, and β subunits, thus generating a small number of potential conformations during development and adulthood. This is a significant advantage compared to $GABA_AR$ and makes it possible to explore more precisely their expression and sensitivity to ethanol. Mature spinal GlyRs are composed of $\alpha_1\beta$ subunits (Legendre 2001; Lynch 2009). Previous studies showed that ethanol potentiates $\alpha_1\beta$ GlyRs by a left shift on its activation curve (Aguayo et al. 1996), altering various brain functions that are affected by activation of these receptors (Schmid et al. 1991; Chang and Martin 2011). It is well known that the expression and function of GlyRs subunits are highly controlled during brain development. For instance, α2 subunits predominate in neonatal spinal cord and decline in adulthood when α1 and β predominate (Aguayo et al. 2004; Hruskova et al. 2012; Takahashi et al. 1992).

Examining GlyRs function and conformation in aging is important because they are highly sensitive to ethanol and play a role in sedation and drinking in young animals (Gallegos et al. 2021; Muñoz et al. 2020). For example, α1 and α2 β are highly sensitive ethanol targets and may be responsible for changes in ethanol-modified behaviors with aging. The role of α2 GlyRs was demonstrated by studying α2 KI and α2 KO mice that showed a reduced LORR (Gallegos et al. 2021; San Martin et al. 2020). Similarly, the α1 KI mice were sedated to a lesser extent (Aguayo et al. 2014). Notably, these mutations only affect ethanol potentiation since gating and general anesthetic actions were not affected, and KI mice did not display behavioral alterations such as those produced by mutations in TM domains (Findlay et al. 2003). It is believed that α1β and α2β are important for ethanol-induced behaviors (Blednov et al. 2015; Muñoz et al. 2021). Unlike α_{1-2}, the β subunit cannot assemble into functional homomeric complexes and is only a structural component (Van Zundert et al. 2004) that modulates GlyRs properties (Grudzinska et al. 2005; Muñoz et al. 2021) including ethanol action (Muñoz et al. 2021).

During aging, a reduction in GlyR α1 and α2 subunit levels and [3H] strychnine binding in the dorsal cochlear nucleus have been found in old rats (28–33 months old) (Wang et al. 2009). Nakayama et al. (1999) reported a significant decrease in strychnine binding in the vestibular nucleus complex of the rat as a function of age (3–26 months). The level of binding in the 26-month-old rat was about half that in the 3-month-old rat. In addition, a significant reduction in the expression and function of GlyRs, but not $GABA_AR$, was reported in a pathological aging model in the nucleus accumbens (Fernandez-Perez et al. 2020). Therefore, changes in the expression of these subunits in the aging brain might impact ethanol drinking and preference, and future studies should examine these possibilities.

Finally, GlyRs are found in synaptic and extrasynaptic (non-synaptic) sites, and several studies have shown that non-synaptic are the most ethanol-sensitive GlyRs (Muñoz et al. 2020, Mccracken et al. 2017, Maguire et al. 2014, Muñoz et al. 2018). It is currently unknown if these non-synaptic GlyRs are affected by aging, including their sensitivity to ethanol and capacity to modulate dopamine release in the nucleus

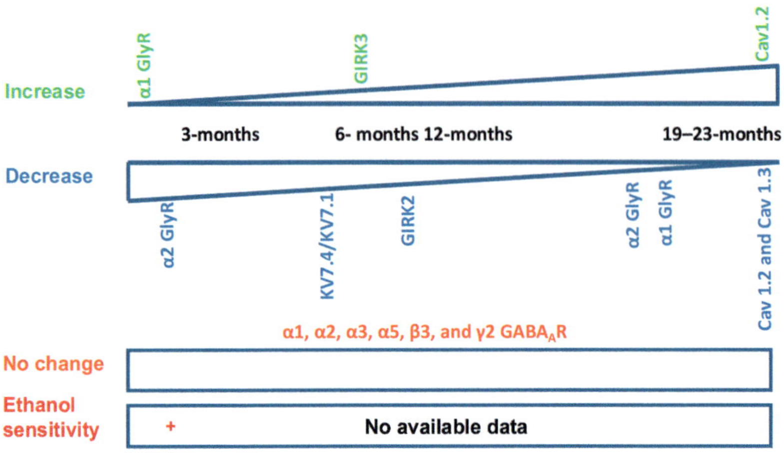

Fig. 13.2 The scheme shows how the expression of some critical ethanol-sensitive targets is affected during the life cycle (increase/decrease/no changes). Only data for changes in the ethanol sensitivity of glycine receptors with brain maturation is available. The data summarizes reported effects at 3, 6, 12, and 19–23 months in rodent models

accumbens. Because GlyRs expressed in this region are important for the release of dopamine and the intake of ethanol (Lido et al. 2011; Molander et al. 2005; Molander and Soderpalm 2005), it seems possible that changes in the conformations and density of this receptor with age might impact the pharmacological effects in the elderly (Fig. 13.2).

13.9 Conclusions

As reviewed, there is little information on the changes that occur during the complicated and little-known states of brain aging, even in a normal setting. Besides the processes of neuronal atrophy and loss of synaptic connectivity occurring with age, it is likely that the molecular targets of ethanol will undergo profound changes in their conformations. Under aged conditions, ethanol would increase its effect on the brain and further alter the functions because the native cognitive reserve is already reduced or because of changes in the sensitivity of the target protein to ethanol. The global lack of studies on the properties of these critical proteins that control excitability and connectivity with age is noteworthy. It is likely associated with the high cost of using aged animals in the laboratory and the fact that funding agencies are only now, with an increased aging population, realizing how important it is to have preclinical data on this subject. For the effects of ethanol on aging, the data are quite limited, affecting our capacity to propose new, more effective, evidence-based medical and psychological treatments for AUD in the elderly.

References

Abrahao KP, Salinas AG, Lovinger DM (2017) Alcohol and the brain: neuronal molecular targets, synapses, and circuits. Neuron 96:1223–1238

Admasu TD, Barardo D, Ng LF, Batchu KC, Cazenave-Gassiot A, Wenk MR, Gruber J (2022) A small-molecule Psora-4 acts as a caloric restriction mimetic to promote longevity in C. elegans. Geroscience 44:1029–1046

Aguayo LG (1990) Ethanol potentiates the GABAA-activated Cl- current in mouse hippocampal and cortical neurons. Eur J Pharmacol 187:127–130

Aguayo LG, Pancetti FC (1994) Ethanol modulation of the gamma-aminobutyric acidA- and glycine-activated Cl- current in cultured mouse neurons. J Pharmacol Exp Ther 270:61–69

Aguayo LG, Pancetti FC, Klein RL, Harris RA (1994) Differential effects of GABAergic ligands in mouse and rat hippocampal neurons. Brain Res 647:97–105

Aguayo LG, Tapia JC, Pancetti FC (1996) Potentiation of the glycine-activated Cl- current by ethanol in cultured mouse spinal neurons. J Pharmacol Exp Ther 279:1116–1122

Aguayo LG, Peoples RW, Yeh HH, Yevenes GE (2002) GABA(A) receptors as molecular sites of ethanol action. Direct or indirect actions? Curr Top Med Chem 2:869–885

Aguayo LG, Van Zundert B, Tapia JC, Carrasco MA, Alvarez FJ (2004) Changes on the properties of glycine receptors during neuronal development. Brain Res Brain Res Rev 47:33–45

Aguayo LG, Castro P, Mariqueo T, Muñoz B, Xiong W, Zhang L, Lovinger DM, Homanics GE (2014)

Altered sedative effects of ethanol in mice with alpha1 glycine receptor subunits that are insensitive to Gbetagamma modulation. Neuropsychopharmacology 39:2538–2548
Alifimoff JK, Firestone LL, Miller KW (1989) Anaesthetic potencies of primary alkanols: implications for the molecular dimensions of the anaesthetic site. Br J Pharmacol 96:9–16
Allan AM, Harris RA (1987) Acute and chronic ethanol treatments alter GABA receptor-operated chloride channels. Pharmacol Biochem Behav 27:665–670
Anantharam V, Bayley H, Wilson A, Treistman SN (1992) Differential effects of ethanol on electrical properties of various potassium channels expressed in oocytes. Mol Pharmacol 42:499–505
Beleslin DB, Djokanovic N, Jovanovic Micic D, Samardzic R (1997) Opposite effects of GABAA and NMDA receptor antagonists on ethanol-induced behavioral sleep in rats. Alcohol 14:167–173
Blednov YA, Benavidez JM, Black M, Leiter CR, Osterndorff-Kahanek E, Harris RA (2015) Glycine receptors containing alpha2 or alpha3 subunits regulate specific ethanol-mediated behaviors. J Pharmacol Exp Ther 353:181–191
Blednov YA, Da Costa A, Mason S, Mayfield J, Messing RO (2023) Selective PDE4B and PDE4D inhibitors produce distinct behavioral responses to ethanol and GABAergic drugs in mice. Neuropharmacology 231:109508
Borghese CM, Storustovu S, Ebert B, Herd MB, Belelli D, Lambert JJ, Marshall G, Wafford KA, Harris RA (2006) The delta subunit of gamma-aminobutyric acid type A receptors does not confer sensitivity to low concentrations of ethanol. J Pharmacol Exp Ther 316:1360–1368
Branch SY, Chen C, Sharma R, Lechleiter JD, Li S, Beckstead MJ (2016) Dopaminergic neurons exhibit an age-dependent decline in electrophysiological parameters in the MitoPark mouse model of Parkinson's disease. J Neurosci 36:4026–4037
Brecht EJ, Scott LL, Ding B, Zhu X, Walton JP (2022) A BK channel-targeted peptide induces age-dependent improvement in behavioral and neural sound representation. Neurobiol Aging 110:61–72
Bukiya AN, Kuntamallappanavar G, Edwards J, Singh AK, Shivakumar B, Dopico AM (2014) An alcohol-sensing site in the calcium- and voltage-gated, large conductance potassium (BK) channel. Proc Natl Acad Sci USA 111:9313–9318
Bukiya AN, Seleverstov O, Bisen S, Dopico AM (2016) Age-dependent susceptibility to alcohol-induced cerebral artery constriction. J Drug Alcohol Res 5
Burnette EM, Nieto SJ, Grodin EN, Meredith LR, Hurley B, Miotto K, Gillis AJ, Ray LA (2022) Novel agents for the pharmacological treatment of alcohol use disorder. Drugs 82:251–274
Cannady R, Rinker JA, Nimitvilai S, Woodward JJ, Mulholland PJ (2018) Chronic alcohol, intrinsic excitability, and potassium channels: neuroadaptations and drinking behavior. Handb Exp Pharmacol 248:311–343
Castillo-Vazquez SK, Massieu L, Rincon-Heredia R, Garcia-Delatorre P, Quiroz-Baez R, Gomez-Verjan JC, Rivero-Segura NA (2024) Glutamatergic neurotransmission in aging and neurodegenerative diseases: a potential target to improve cognitive impairment in aging. Arch Med Res 55:103039
Caulfield ME, Vander Werp MJ, Stancati JA, Collier TJ, Sortwell CE, Sandoval IM, Manfredsson FP, Steece-Collier K (2023) Downregulation of striatal Ca(V)1.3 inhibits the escalation of levodopa-induced dyskinesia in male and female Parkinsonian rats of advanced age. Neurobiol Dis 181:106111
Chang Q, Martin LJ (2011) Glycine receptor channels in spinal motoneurons are abnormal in a transgenic mouse model of amyotrophic lateral sclerosis. J Neurosci 31:2815–2827
Chin VS, Van Skike CE, Berry RB, Kirk RE, Diaz-Granados J, Matthews DB (2011) Effect of acute ethanol and acute allopregnanolone on spatial memory in adolescent and adult rats. Alcohol 45:473–483
Crews FT, Morrow AL, Criswell H, Breese G (1996) Effects of ethanol on ion channels. Int Rev Neurobiol 39:283–367
Dopico AM, Lemos JR, Treistman SN (1996) Ethanol increases the activity of large conductance, Ca(2+)-activated K+ channels in isolated neurohypophysial terminals. Mol Pharmacol 49:40–48
Dopico AM, Bukiya AN, Martin GE (2014) Ethanol modulation of mammalian BK channels in excitable tissues: molecular targets and their possible contribution to alcohol-induced altered behavior. Front Physiol 5:466
Dopico AM, Bukiya AN, Kuntamallappanavar G, Liu J (2016) Modulation of BK channels by ethanol. Int Rev Neurobiol 128:239–279
Egervari G, Siciliano CA, Whiteley EL, Ron D (2021) Alcohol and the brain: from genes to circuits. Trends Neurosci 44:1004–1015
Engblom AC, Akerman KE (1991) Effect of ethanol on gamma-aminobutyric acid and glycine receptor-coupled Cl- fluxes in rat brain synaptoneurosomes. J Neurochem 57:384–390
Esser MB, Leung G, Sherk A, Bohm MK, Liu Y, Lu H, Naimi TS (2022a) Estimated deaths attributable to excessive alcohol use among US adults aged 20 to 64 years, 2015 to 2019. JAMA Netw Open 5:e2239485
Esser MB, Sherk A, Subbaraman MS, Martinez P, Karriker-Jaffe KJ, Sacks JJ, Naimi TS (2022b) Improving estimates of alcohol-attributable deaths in the United States: impact of adjusting for the underreporting of alcohol consumption. J Stud Alcohol Drugs 83:134–144
Ethiraj J, Palpagama TH, Turner C, Van Der Werf B, Waldvogel HJ, Faull RLM, Kwakowsky A (2021) The effect of age and sex on the expression of GABA signaling components in the human hippocampus and entorhinal cortex. Sci Rep 11:21470

Fang Z, Ionescu P, Gong D, Kendig J, Harris A, Eger EI 2nd (1997) Maturation decreases ethanol minimum alveolar anesthetic concentration in mice as previously demonstrated in rats: there is no species difference. Anesth Analg 85:160–163

Fernandez-Perez EJ, Gallegos S, Armijo-Weingart L, Araya A, Riffo-Lepe NO, Cayuman F, Aguayo LG (2020) Changes in neuronal excitability and synaptic transmission in nucleus accumbens in a transgenic Alzheimer's disease mouse model. Sci Rep 10:19606

Findlay GS, Phelan R, Roberts MT, Homanics GE, Bergeson SE, Lopreato GF, Mihic SJ, Blednov YA, Harris RA (2003) Glycine receptor knock-in mice and hyperekplexia-like phenotypes: comparisons with the null mutant. J Neurosci 23:8051–8059

Fox JG (2007) The mouse in biomedical research. Academic, Amsterdam

Gallegos S, San Martin L, Araya A, Lovinger DM, Homanics GE, Aguayo LG (2021) Reduced sedation and increased ethanol consumption in knock-in mice expressing an ethanol insensitive alpha 2 subunit of the glycine receptor. Neuropsychopharmacology 46:528–536

Garcia CC, Lewis B, Boissoneault J, Nixon SJ (2020) Effects of age and acute moderate alcohol consumption on electrophysiological indices of attention. J Stud Alcohol Drugs 81:372–383

Govindpani K, Turner C, Waldvogel HJ, Faull RLM, Kwakowsky A (2020) Impaired expression of GABA signaling components in the Alzheimer's disease middle temporal gyrus. Int J Mol Sci 21

Grant BF, Chou SP, Saha TD, Pickering RP, Kerridge BT, Ruan WJ, Huang B, Jung J, Zhang H, Fan A, Hasin DS (2017) Prevalence of 12-month alcohol use, high-risk drinking, and DSM-IV alcohol use disorder in the United States, 2001-2002 to 2012-2013: results from the National Epidemiologic Survey on Alcohol and Related Conditions. JAMA Psychiatry 74:911–923

Griffith WH, Murchison DA (1995) Enhancement of GABA-activated membrane currents in aged Fischer 344 rat basal forebrain neurons. J Neurosci 15:2407–2416

Grizzanti J, Moritz WR, Pait MC, Stanley M, Kaye SD, Carroll CM, Constantino NJ, Deitelzweig LJ, Snipes JA, Kellar D, Caesar EE, Pettit-Mee RJ, Day SM, Sens JP, Nicol NI, Dhillon J, Remedi MS, Kiraly DD, Karch CM, Nichols CG, Holtzman DM, Macauley SL (2023) KATP channels are necessary for glucose-dependent increases in amyloid-beta and Alzheimer's disease-related pathology. JCI Insight 8

Grudzinska J, Schemm R, Haeger S, Nicke A, Schmalzing G, Betz H, Laube B (2005) The beta subunit determines the ligand binding properties of synaptic glycine receptors. Neuron 45:727–739

Guidi M, Kumar A, Foster TC (2015) Impaired attention and synaptic senescence of the prefrontal cortex involves redox regulation of NMDA receptors. J Neurosci 35:3966–3977

Han L, Jia J (2021) Long-term effects of alcohol consumption on cognitive function in seniors: a cohort study in China. BMC Geriatr 21:699

Han BH, Moore AA, Ferris R, Palamar JJ (2019) Binge drinking among older adults in the United States, 2015 to 2017. J Am Geriatr Soc 67:2139–2144

Harris RA, Trudell JR, Mihic SJ (2008) Ethanol's molecular targets. Sci Signal 1:re7

Hell JW, Westenbroek RE, Warner C, Ahlijanian MK, Prystay W, Gilbert MM, Snutch TP, Catterall WA (1993) Identification and differential subcellular localization of the neuronal class C and class D L-type calcium channel alpha 1 subunits. J Cell Biol 123:949–962

Henley JM, Wilkinson KA (2013) AMPA receptor trafficking and the mechanisms underlying synaptic plasticity and cognitive aging. Dialogues Clin Neurosci 15:11–27

Herman MA, Sidhu H, Stouffer DG, Kreifeldt M, Le D, Cates-Gatto C, Munoz MB, Roberts AJ, Parsons LH, Roberto M, Wickman K, Slesinger PA, Contet C (2015) GIRK3 gates activation of the mesolimbic dopaminergic pathway by ethanol. Proc Natl Acad Sci USA 112:7091–7096

Hodge CW, Mehmert KK, Kelley SP, Mcmahon T, Haywood A, Olive MF, Wang D, Sanchez-Perez AM, Messing RO (1999) Supersensitivity to allosteric GABA(A) receptor modulators and alcohol in mice lacking PKCepsilon. Nat Neurosci 2:997–1002

Honse Y, Ren H, Lipsky RH, Peoples RW (2004) Sites in the fourth membrane-associated domain regulate alcohol sensitivity of the NMDA receptor. Neuropharmacology 46:647–654

Horishita T, Harris RA (2008) n-Alcohols inhibit voltage-gated Na+ channels expressed in Xenopus oocytes. J Pharmacol Exp Ther 326:270–277

Hruskova B, Trojanova J, Kulik A, Kralikova M, Pysanenko K, Bures Z, Syka J, Trussell LO, Turecek R (2012) Differential distribution of glycine receptor subtypes at the rat calyx of Held synapse. J Neurosci 32:17012–17024

Inoue M, Oomura Y, Yakushiji T, Akaike N (1986) Intracellular calcium ions decrease the affinity of the GABA receptor. Nature 324:156–158

Jackson SJ, Andrews N, Ball D, Bellantuono I, Gray J, Hachoumi L, Holmes A, Latcham J, Petrie A, Potter P, Rice A, Ritchie A, Stewart M, Strepka C, Yeoman M, Chapman K (2017) Does age matter? The impact of rodent age on study outcomes. Lab Anim 51:160–169

Jeong DJ, Kim KW, Suh BC (2023) Dual regulation of Kv7.2/7.3 channels by long-chain n-alcohols. J Gen Physiol 155

Jurado S (2017) AMPA receptor trafficking in natural and pathological aging. Front Mol Neurosci 10:446

Kapur J, Macdonald RL (1999) Postnatal development of hippocampal dentate granule cell gamma-aminobutyric acidA receptor pharmacological properties. Mol Pharmacol 55:444–452

Khachaturian ZS (1987) Hypothesis on the regulation of cytosol calcium concentration and the aging brain. Neurobiol Aging 8:345–346

Kim KW, Kim K, Lee H, Suh BC (2019) Ethanol elevates excitability of superior cervical ganglion neurons by inhibiting Kv7 channels in a cell type-specific and PI(4,5)P(2)-dependent manner. Int J Mol Sci 20

Kobayashi T, Ikeda K, Kojima H, Niki H, Yano R, Yoshioka T, Kumanishi T (1999) Ethanol opens G-protein-activated inwardly rectifying K+ channels. Nat Neurosci 2:1091–1097

Konar-Nie M, Guzman-Castillo A, Armijo-Weingart L, Aguayo LG (2023) Aging in nucleus accumbens and its impact on alcohol use disorders. Alcohol 107:73–90

Kulkarny VV, Wiest NE, Marquez CP, Nixon SC, Valenzuela CF, Perrone-Bizzozero NI (2011) Opposite effects of acute ethanol exposure on GAP-43 and BDNF expression in the hippocampus versus the cerebellum of juvenile rats. Alcohol 45:461–471

Kumar A, Foster TC (2019) Alteration in NMDA receptor mediated glutamatergic neurotransmission in the hippocampus during senescence. Neurochem Res 44:38–48

Kuntamallappanavar G, Dopico AM (2017) BK beta1 subunit-dependent facilitation of ethanol inhibition of BK current and cerebral artery constriction is mediated by the beta1 transmembrane domain 2. Br J Pharmacol 174:4430–4448

Legendre P (2001) The glycinergic inhibitory synapse. Cell Mol Life Sci 58:760–793

Leslie SW, Barr E, Chandler J, Farrar RP (1983) Inhibition of fast- and slow-phase depolarization-dependent synaptosomal calcium uptake by ethanol. J Pharmacol Exp Ther 225:571–575

Levitan ES, Schofield PR, Burt DR, Rhee LM, Wisden W, Kohler M, Fujita N, Rodriguez HF, Stephenson A, Darlison MG et al (1988) Structural and functional basis for GABAA receptor heterogeneity. Nature 335:76–79

Li S, Yan GJ, Tan YX, Xue LL, Wang TH, Zhao HR, Lu MN, Zhang HX, Mei R, Dong XH, Liu LN, Wang D, Xiyang YB (2022) Reduced expression of voltage-gated sodium channel Beta 2 restores neuronal injury and improves cognitive dysfunction induced by Abeta1-42. Neural Plast 2022:3995227

Lido HH, Ericson M, Marston H, Soderpalm B (2011) A role for accumbal glycine receptors in modulation of dopamine release by the glycine transporter-1 inhibitor org25935. Front Psychiatry 2:8

Lin AM, Freund RK, Hoffer BJ, Palmer MR (1994) Ethanol-induced depressions of cerebellar Purkinje neurons are potentiated by beta-adrenergic mechanisms in rat brain. J Pharmacol Exp Ther 271:1175–1180

Liu J, Bukiya AN, Kuntamallappanavar G, Singh AK, Dopico AM (2013) Distinct sensitivity of slo1 channel proteins to ethanol. Mol Pharmacol 83:235–244

Liu M, Liu C, Xiao X, Han SS, Bi MX, Jiao Q, Chen X, Yan CL, Du XX, Jiang H (2022) Role of upregulation of the K(ATP) channel subunit SUR1 in dopaminergic neuron degeneration in Parkinson's disease. Aging Cell 21:e13618

Lobos P, Vega-Vasquez I, Bruna B, Gleitze S, Toledo J, Hartel S, Hidalgo C, Paula-Lima A (2023) Amyloid beta-oligomers inhibit the nuclear Ca(2+) signals and the neuroprotective gene expression induced by gabazine in hippocampal neurons. Antioxidants (Basel) 12

Lovinger DM (1993) High ethanol sensitivity of recombinant AMPA-type glutamate receptors expressed in mammalian cells. Neurosci Lett 159:83–87

Lovinger DM, White G, Weight FF (1990) NMDA receptor-mediated synaptic excitation selectively inhibited by ethanol in hippocampal slice from adult rat. J Neurosci 10:1372–1379

Lujan R, Aguado C (2015) Localization and targeting of GIRK channels in mammalian central neurons. Int Rev Neurobiol 123:161–200

Luscher C, Slesinger PA (2010) Emerging roles for G protein-gated inwardly rectifying potassium (GIRK) channels in health and disease. Nat Rev Neurosci 11:301–315

Lynch JW (2009) Native glycine receptor subtypes and their physiological roles. Neuropharmacology 56:303–309

Maguire EP, Mitchell EA, Greig SJ, Corteen N, Balfour DJ, Swinny JD, Lambert JJ, Belelli D (2014) Extrasynaptic glycine receptors of rodent dorsal raphe serotonergic neurons: a sensitive target for ethanol. Neuropsychopharmacology 39:1232–1244

Mateos-Aparicio P, Murphy R, Storm JF (2014) Complementary functions of SK and Kv7/M potassium channels in excitability control and synaptic integration in rat hippocampal dentate granule cells. J Physiol 592:669–693

Matthews DB, Schneider A, Kastner A, Scaletty S, Szenay R (2019) I can't drink what I used to: The interaction between ethanol and the aging brain. Int Rev Neurobiol 148:79–99

Matthews DB, Scaletty S, Schreiber A, Trapp S (2020) Acute ethanol administration produces larger spatial and nonspatial memory impairments in 29-33 month old rats compared to adult and 18-24 month old rats. Pharmacol Biochem Behav 199:173074

Maziar A, Critch T, Ghosh S, Rajani V, Flynn CM, Qin T, Reinhardt C, Man KNM, Lee A, Hell JW, Yuan Q (2023) Aging differentially affects LTCC function in hippocampal CA1 and piriform cortex pyramidal neurons. Cereb Cortex 33:1489–1503

Mccracken LM, Lowes DC, Salling MC, Carreau-Vollmer C, Odean NN, Blednov YA, Betz H, Harris RA, Harrison NL (2017) Glycine receptor alpha3 and alpha2 subunits mediate tonic and exogenous agonist-induced currents in forebrain. Proc Natl Acad Sci USA 114:E7179–E7186

Mcguier NS, Rinker JA, Cannady R, Fulmer DB, Jones SR, Hoffman M, Mulholland PJ (2018) Identification and validation of midbrain Kcnq4 regulation of heavy alcohol consumption in rodents. Neuropharmacology 138:10–19

Mehta AK, Ticku MK (1988) Ethanol potentiation of GABAergic transmission in cultured spinal cord neurons involves gamma-aminobutyric acidA-gated chloride channels. J Pharmacol Exp Ther 246:558–564
Meier S, Demirakca T, Brusniak W, Wolf I, Liebsch K, Tunc-Skarka N, Nieratschker V, Witt SH, Matthaus F, Ende G, Flor H, Rietschel M, Diener C, Schulze TG (2012) SCN1A affects brain structure and the neural activity of the aging brain. Biol Psychiatry 72:677–683
Messing RO, Carpenter CL, Diamond I, Greenberg DA (1986) Ethanol regulates calcium channels in clonal neural cells. Proc Natl Acad Sci USA 83:6213–6215
Molander A, Soderpalm B (2005) Accumbal strychnine-sensitive glycine receptors: an access point for ethanol to the brain reward system. Alcohol Clin Exp Res 29:27–37
Molander A, Lof E, Stomberg R, Ericson M, Soderpalm B (2005) Involvement of accumbal glycine receptors in the regulation of voluntary ethanol intake in the rat. Alcohol Clin Exp Res 29:38–45
Moore JW, Ulbricht W, Takata M (1964) Effect of ethanol on the sodium and potassium conductances of the squid axon membrane. J Gen Physiol 48:279–295
Moore SD, Madamba SG, Siggins GR (1990) Ethanol diminishes a voltage-dependent K+ current, the M-current, in CA1 hippocampal pyramidal neurons in vitro. Brain Res 516:222–228
Moore SJ, Cazares VA, Temme SJ, Murphy GG (2023) Age-related deficits in neuronal physiology and cognitive function are recapitulated in young mice overexpressing the L-type calcium channel, Ca(V) 1.3. Aging Cell 22:e13781
Mullikin-Kilpatrick D, Mehta ND, Hildebrandt JD, Treistman SN (1995) Gi is involved in ethanol inhibition of L-type calcium channels in undifferentiated but not differentiated PC-12 cells. Mol Pharmacol 47:997–1005
Muñoz B, Yevenes GE, Forstera B, Lovinger DM, Aguayo LG (2018) Presence of inhibitory glycinergic transmission in medium spiny neurons in the nucleus accumbens. Front Mol Neurosci 11:228
Muñoz B, Gallegos S, Peters C, Murath P, Lovinger DM, Homanics GE, Aguayo LG (2020) Influence of nonsynaptic alpha1 glycine receptors on ethanol consumption and place preference. Addict Biol 25:e12726
Muñoz B, Mariqueo T, Murath P, Peters C, Yevenes GE, Moraga-Cid G, Peoples RW, Aguayo LG (2021) Modulatory actions of the glycine receptor beta subunit on the positive allosteric modulation of ethanol in alpha2 containing receptors. Front Mol Neurosci 14:763868
Nakayama M, Caspary DM, Konrad HR, Milbrandt JC, Helfert RH (1999) Age-related changes in [3H]strychnine binding in the vestibular nuclei of rats. Hear Res 127:103–107
Nestoros JN (1980) Ethanol specifically potentiates GABA-mediated neurotransmission in feline cerebral cortex. Science 209:708–710
N'Gouemo P, Akinfiresoye LR, Allard JS, Lovinger DM (2015) Alcohol withdrawal-induced seizure susceptibility is associated with an upregulation of CaV1.3 channels in the rat inferior colliculus. Int J Neuropsychopharmacol 18:pyu123
Niesen CE, Baskys A, Carlen PL (1988) Reversed ethanol effects on potassium conductances in aged hippocampal dentate granule neurons. Brain Res 445:137–141
Nunez-Santana FL, Oh MM, Antion MD, Lee A, Hell JW, Disterhoft JF (2014) Surface L-type Ca2+ channel expression levels are increased in aged hippocampus. Aging Cell 13:111–120
Osmanovic SS, Shefner SA (1994) Ethanol enhances inward rectification in rat locus ceruleus neurons by increasing the extracellular potassium concentration. J Pharmacol Exp Ther 271:334–342
Palpagama TH, Sagniez M, Kim S, Waldvogel HJ, Faull RL, Kwakowsky A (2019) GABA(A) receptors are well preserved in the hippocampus of aged mice. eNeuro 6
Pandey SP, Rai R, Gaur P, Prasad S (2015) Development- and age-related alterations in the expression of AMPA receptor subunit GluR2 and its trafficking proteins in the hippocampus of male mouse brain. Biogerontology 16:317–328
Peixoto Pinheiro B, Adel Y, Knipper M, Muller M, Lowenheim H (2021) Auditory threshold variability in the SAMP8 mouse model of age-related hearing loss: functional loss and phenotypic change precede outer hair cell loss. Front Aging Neurosci 13:708190
Perkins AE, Vore AS, Lovelock D, Varlinskaya E, Deak T (2018) Late aging alters behavioral sensitivity to ethanol in a sex-specific manner in Fischer 344 rats. Pharmacol Biochem Behav 175:1–9
Power JM, Wu WW, Sametsky E, Oh MM, Disterhoft JF (2002) Age-related enhancement of the slow outward calcium-activated potassium current in hippocampal CA1 pyramidal neurons in vitro. J Neurosci 22:7234–7243
Price JL, Lewis B, Boissoneault J, Frazier IR, Nixon SJ (2018) Effects of acute alcohol and driving complexity in older and younger adults. Psychopharmacology 235:887–896
Pritchett DB, Sontheimer H, Shivers BD, Ymer S, Kettenmann H, Schofield PR, Seeburg PH (1989) Importance of a novel GABAA receptor subunit for benzodiazepine pharmacology. Nature 338:582–585
Quoilin C, Didone V, Tirelli E, Quertemont E (2014) Higher long-lasting ethanol sensitization after adolescent ethanol exposure in mice. Psychopharmacology 231:1821–1829
Rajani V, Maziar A, Man KNM, Hell JW, Yuan Q (2021) Age-dependent contributions of NMDA receptors and L-type calcium channels to long-term depression in the piriform cortex. Int J Mol Sci 22
Ratkai A, Tarnok K, Aouad HE, Micska B, Schlett K, Szucs A (2021) Homeostatic plasticity and burst activity are mediated by hyperpolarization-activated cation currents and T-type calcium channels in neuronal cultures. Sci Rep 11:3236
Rehberg B, Xiao YH, Duch DS (1996) Central nervous system sodium channels are significantly suppressed

at clinical concentrations of volatile anesthetics. Anesthesiology 84:1223–1233. Discussion 27A

Remigante A, Spinelli S, Zuccolini P, Gavazzo P, Marino A, Pusch M, Morabito R, Dossena S (2024) Melatonin protects Kir2.1 function in an oxidative stress-related model of aging neuroglia. Biofactors 50:523–541

Ron D, Wang J (2009) The NMDA receptor and alcohol addiction. In: Van Dongen AM (ed) Biology of the NMDA receptor. CRC, Boca Raton, FL

Ronald KM, Mirshahi T, Woodward JJ (2001) Ethanol inhibition of N-methyl-D-aspartate receptors is reduced by site-directed mutagenesis of a transmembrane domain phenylalanine residue. J Biol Chem 276:44729–44735

Rosenberg P, Bartels E (1967) Drug effects on the spontaneous electrical activity of the squid giant axon. J Pharmacol Exp Ther 155:532–544

San Martin L, Gallegos S, Araya A, Romero N, Morelli G, Comhair J, Harvey RJ, Rigo JM, Brone B, Aguayo LG (2020) Ethanol consumption and sedation are altered in mice lacking the glycine receptor alpha2 subunit. Br J Pharmacol 177:3941–3956

Sapp DW, Yeh HH (1998) Ethanol-GABAA receptor interactions: a comparison between cell lines and cerebellar Purkinje cells. J Pharmacol Exp Ther 284:768–776

Schmid K, Bohmer G, Gebauer K (1991) Glycine receptor-mediated fast synaptic inhibition in the brainstem respiratory system. Respir Physiol 84:351–361

Segovia G, Mora F (2005) Dopamine and GABA increases produced by activation of glutamate receptors in the nucleus accumbens are decreased during aging. Neurobiol Aging 26:91–101

Shiraishi M, Harris RA (2004) Effects of alcohols and anesthetics on recombinant voltage-gated Na+ channels. J Pharmacol Exp Ther 309:987–994

Siggins GR, Pittman QJ, French ED (1987) Effects of ethanol on CA1 and CA3 pyramidal cells in the hippocampal slice preparation: an intracellular study. Brain Res 414:22–34

Skattebol A, Rabin RA (1987) Effects of ethanol on 45Ca2+ uptake in synaptosomes and in PC12 cells. Biochem Pharmacol 36:2227–2229

Spear LP, Swartzwelder HS (2014) Adolescent alcohol exposure and persistence of adolescent-typical phenotypes into adulthood: a mini-review. Neurosci Biobehav Rev 45:1–8

Spinelli S, Remigante A, Liuni R, Mantegna G, Legname G, Marino A, Morabito R, Dossena S (2024) Oxidative stress-related cellular aging causes dysfunction of the Kv3.1/KCNC1 channel reverted by melatonin. Aging Cell 23(8):e14185

Stelzer A, Kay AR, Wong RK (1988) GABAA-receptor function in hippocampal cells is maintained by phosphorylation factors. Science 241:339–341

Suzdak PD, Schwartz RD, Skolnick P, Paul SM (1986) Ethanol stimulates gamma-aminobutyric acid receptor-mediated chloride transport in rat brain synaptoneurosomes. Proc Natl Acad Sci USA 83:4071–4075

Takahashi T, Momiyama A, Hirai K, Hishinuma F, Akagi H (1992) Functional correlation of fetal and adult forms of glycine receptors with developmental changes in inhibitory synaptic receptor channels. Neuron 9:1155–1161

Taura J, Kircher DM, Gameiro-Ros I, Slesinger PA (2021) Comparison of K(+) channel families. Handb Exp Pharmacol 267:1–49

Temprano-Carazo S, Contreras A, Saura CA, Navarro-Lopez JD, Jimenez-Diaz L (2022) Spatial memory training counteracts hippocampal GIRK channel decrease in the transgenic APP(Sw,Ind) J9 Alzheimer's disease mouse model. Int J Mol Sci 23

Treistman SN, Wilson A (1987) Effects of ethanol on early potassium currents in Aplysia: cell specificity and influence of channel state. J Neurosci 7:3207–3214

Twombly DA, Herman MD, Kye CH, Narahashi T (1990) Ethanol effects on two types of voltage-activated calcium channels. J Pharmacol Exp Ther 254:1029–1037

Uhrig S, Vandael D, Marcantoni A, Dedic N, Bilbao A, Vogt MA, Hirth N, Broccoli L, Bernardi RE, Schonig K, Gass P, Bartsch D, Spanagel R, Deussing JM, Sommer WH, Carbone E, Hansson AC (2017) Differential roles for L-type calcium channel subtypes in alcohol dependence. Neuropsychopharmacology 42:1058–1069

Urban BW (2008) The site of anesthetic action. Handb Exp Pharmacol 3–29

Vaidya B, Padhy DS, Joshi HC, Sharma SS, Singh JN (2024) Ion channels and metal ions in Parkinson's disease: historical perspective to the current scenario. Methods Mol Biol 2761:529–557

Van Zundert B, Alvarez FJ, Tapia JC, Yeh HH, Diaz E, Aguayo LG (2004) Developmental-dependent action of microtubule depolymerization on the function and structure of synaptic glycine receptor clusters in spinal neurons. J Neurophysiol 91:1036–1049

Vossel KA, Tartaglia MC, Nygaard HB, Zeman AZ, Miller BL (2017) Epileptic activity in Alzheimer's disease: causes and clinical relevance. Lancet Neurol 16:311–322

Wafford KA, Whiting PJ (1992) Ethanol potentiation of GABAA receptors requires phosphorylation of the alternatively spliced variant of the gamma 2 subunit. FEBS Lett 313:113–117

Wafford KA, Burnett DM, Leidenheimer NJ, Burt DR, Wang JB, Kofuji P, Dunwiddie TV, Harris RA, Sikela JM (1991) Ethanol sensitivity of the GABAA receptor expressed in Xenopus oocytes requires 8 amino acids contained in the gamma 2L subunit. Neuron 7:27–33

Wafford KA, Burnett D, Harris RA, Whiting PJ (1993) GABAA receptor subunit expression and sensitivity to ethanol. Alcohol Alcohol Suppl 2:327–330

Wallner M, Hanchar HJ, Olsen RW (2003) Ethanol enhances alpha 4 beta 3 delta and alpha 6 beta 3 delta gamma-aminobutyric acid type A receptors at low concentrations known to affect humans. Proc Natl Acad Sci USA 100:15218–15223

Wang H, Turner JG, Ling L, Parrish JL, Hughes LF, Caspary DM (2009) Age-related changes in glycine receptor subunit composition and binding in dorsal cochlear nucleus. Neuroscience 160:227–239

Watson MR, James K, Mittleman G, Matthews DB (2020) Impact of acute ethanol exposure on body temperatures in aged, adult and adolescent male rats. Alcohol 82:81–89

Weight FF (1992) Cellular and molecular physiology of alcohol actions in the nervous system. Int Rev Neurobiol 33:289–348

Weiner JL, Zhang L, Carlen PL (1994a) Guanosine phosphate analogs modulate ethanol potentiation of GABAA-mediated synaptic currents in hippocampal CA1 neurons. Brain Res 665:307–310

Weiner JL, Zhang L, Carlen PL (1994b) Potentiation of GABAA-mediated synaptic current by ethanol in hippocampal CA1 neurons: possible role of protein kinase C. J Pharmacol Exp Ther 268:1388–1395

White G, Lovinger DM, Weight FF (1990) Ethanol inhibits NMDA-activated current but does not alter GABA-activated current in an isolated adult mammalian neuron. Brain Res 507:332–336

Wirkner K, Eberts C, Poelchen W, Allgaier C, Illes P (2000) Mechanism of inhibition by ethanol of NMDA and AMPA receptor channel functions in cultured rat cortical neurons. Naunyn Schmiedebergs Arch Pharmacol 362:568–576

Wong SM, Fong E, Tauck DL, Kendig JJ (1997) Ethanol as a general anesthetic: actions in spinal cord. Eur J Pharmacol 329:121–127

Xu M, Woodward JJ (2006) Ethanol inhibition of NMDA receptors under conditions of altered protein kinase A activity. J Neurochem 96:1760–1767

Yamashita M, Marszalec W, Yeh JZ, Narahashi T (2006) Effects of ethanol on tonic GABA currents in cerebellar granule cells and mammalian cells recombinantly expressing GABA(A) receptors. J Pharmacol Exp Ther 319:431–438

Yang X, Knapp DJ, Criswell HE, Breese GR (1998) Action of ethanol and zolpidem on gamma-aminobutyric acid responses from cerebellar Purkinje neurons: relationship to beta-adrenergic receptor input. Alcohol Clin Exp Res 22:1655–1661

Yevenes GE, Moraga-Cid G, Peoples RW, Schmalzing G, Aguayo LG (2008) A selective G betagamma-linked intracellular mechanism for modulation of a ligand-gated ion channel by ethanol. Proc Natl Acad Sci USA 105:20523–20528

Yu SP, Jiang MQ, Shim SS, Pourkhodadad S, Wei L (2023) Extrasynaptic NMDA receptors in acute and chronic excitotoxicity: implications for preventive treatments of ischemic stroke and late-onset Alzheimer's disease. Mol Neurodegener 18:43

Yuan DJ, Yang G, Wu W, Li QF, Xu DE, Ntim M, Jiang CY, Liu JC, Zhang Y, Wang YZ, Zhu DD, Kundu S, Li AP, Xiao ZC, MA, Q. H. & Li, S. (2022) Reducing Nav1.6 expression attenuates the pathogenesis of Alzheimer's disease by suppressing BACE1 transcription. Aging Cell 21:e13593

Zamudio PA, Smothers TC, Homanics GE, Woodward JJ (2020) Knock-in mice expressing an ethanol-resistant GluN2A NMDA receptor subunit show altered responses to ethanol. Alcohol Clin Exp Res 44:479–491

Zamudio PA, Gioia DA, Lopez M, Homanics GE, Woodward JJ (2021) The escalation in ethanol consumption following chronic intermittent ethanol exposure is blunted in mice expressing ethanol-resistant GluN1 or GluN2A NMDA receptor subunits. Psychopharmacology 238:271–279

Zanos P, Bhat S, Terrillion CE, Smith RJ, Tonelli LH, Gould TD (2015) Sex-dependent modulation of age-related cognitive decline by the L-type calcium channel gene Cacna1c (Cav 1.2). Eur J Neurosci 42:2499–2507

Zhao Y, Gameiro-Ros I, Glaaser IW, Slesinger PA (2021) Advances in targeting GIRK channels in disease. Trends Pharmacol Sci 42:203–215

GPSR Compliance

The European Union's (EU) General Product Safety Regulation (GPSR) is a set of rules that requires consumer products to be safe and our obligations to ensure this.

If you have any concerns about our products, you can contact us on ProductSafety@springernature.com

In case Publisher is established outside the EU, the EU authorized representative is:

Springer Nature Customer Service Center GmbH
Europaplatz 3
69115 Heidelberg, Germany

Batch number: 10755764

Printed by Printforce, the Netherlands